BIOLOGICAL CONTROL OF TROPICAL
WEEDS USING ARTHROPODS

In the past few decades, globalization and increased trade and transportation have contributed to the rapid spread of plants, many of which have now become weeds in the introduced regions. Weeds are a major constraint to agricultural production, particularly in the developing world. Cost-efficient biological control is a self-sustaining way to reduce this problem, and produces fewer non-target effects than chemical methods, which can cause serious damage to the environment.

This book covers the origin, distribution, and ecology of 20 model invasive weed species, which occur in habitats from tropical to aquatic. Sustainable biological control of each weed using one or more arthropods is discussed. The aim is to provide ecological management models for use across the tropical world, and to assist in the assessment of potential risks to native and economic plants. This is a valuable resource for scientists and policy makers concerned with the biological control of invasive tropical plants.

RANGASWAMY MUNIAPPAN Program Director of the Integrated Pest Management Collaborative Research Program (IMP CRSP) at the Office of International Research, Education and Development at Virginia Polytechnic Institute and State University, USA, has specialized in biological control and Integrated Pest Management research in the tropics for over 35 years. He served as the Chairman of the Global Working Group on *Chromolaena* of the International Organization for Biological Control from 1992 to 2006, and is currently responsible for managing the IPM CRSP and coordinating with USAID and project partner institutions in the United States and developing countries in Asia, Africa, Eastern Europe, the Caribbean and Latin America. He has published over 200 research papers and extension articles.

GADI V. P. REDDY is an entomologist at the Agricultural Experiment Station, University of Guam, USA. His research interests include developing sex pheromones, host volatiles and other attractants for use in Integrated Pest Management, behavioral and chemical ecology of multitrophic interactions, and biological control of invasive pests and weeds. He has over 65 publications and has contributed to numerous radio shows and newspaper articles on pest management.

ANATANARAYANAN RAMAN is a Senior Lecturer of Ecological Agriculture at the Charles Sturt University & the E. H. Graham Centre for Agricultural Innovation, Australia. His research interests include arthropod–plant interactions, weed biological control, ecology of soil organisms, soil health and agroforestry. He has published over 125 research papers and authored/edited 9 reference books.

BIOLOGICAL CONTROL OF TROPICAL WEEDS USING ARTHROPODS

Edited by

RANGASWAMY MUNIAPPAN
Virginia Polytechnic Institute and State University, USA

GADI V. P. REDDY
University of Guam, USA

ANANTANARAYANAN RAMAN
*Charles Sturt University & E. H. Graham
Centre for Agricultural Innovation, Australia*

CAMBRIDGE
UNIVERSITY PRESS

CAMBRIDGE UNIVERSITY PRESS
Cambridge, New York, Melbourne, Madrid, Cape Town,
Singapore, São Paulo, Delhi, Mexico City

Cambridge University Press
The Edinburgh Building, Cambridge CB2 8RU, UK

Published in the United States of America by Cambridge University Press, New York

www.cambridge.org
Information on this title: www.cambridge.org/9781107411265

First published 2009
First paperback edition 2012

A catalogue record for this publication is available from the British Library

Library of Congress Cataloguing in Publication Data
Biological control of tropical weeds using arthropods / Rangaswamy Muniappan . . . [et al.].
p. cm.
Includes index.
ISBN 978-0-521-87791-6 (hardback)
1. Weeds – Biological control – Tropics. 2. Invasive plants – Biological control – Tropics.
3. Phytophagous insects – Tropics. 4. Insects as biological pest control agents – Tropics.
I. Muniappan, R. II. Title.
SB613.T8B56 2009
632′.5–dc22 2008052567

ISBN 978-0-521-87791-6 Hardback
ISBN 978-1-107-41126-5 Paperback

Contents

v

Contributors

Fen Beed
International Institute of Tropical Agriculture, Plot 15 Naguru East Road, PO Box 7878, Kampala, Uganda

Basavaraj S. Bhumannavar
Project Directorate of Biological Control (ICAR), Post Bag No. 2491, H. A. Farm Post, Hebbal, Bellary Road, Bangalore 560 024, Karnataka, India

Willie Cabrera–Walsh
USDA–ARS South American Biological Control Laboratory, Bolivar 1559, Hurlingham, Buenos Aires, Argentina

Ted D. Center
USDA–ARS Invasive Plant Research Laboratory, 3205 College Ave, Fort Lauderdale, FL 33314, USA

Julie A. Coetzee
Department of Zoology and Entomology, Rhodes University, P.O. Box 94, Grahamstown 6140, South Africa

Hugo A. Cordo
USDA–ARS South American Biological Control Laboratory, Bolivar 1559, Hurlingham, Buenos Aires, Argentina

Patrick Conant
Hawaii Department of Agriculture, Hilo, HI 96720, USA

Michael D. Day
Alan Fletcher Research Station, Department of Primary Industries and Fisheries, PO Box 36, Sherwood, Queensland 4075, Australia

Kunjithapatham Dhileepan
Alan Fletcher Research Station, Biosecurity Queensland, Department of Primary Industries and Fisheries, Sherwood, Queensland 4075, Australia

Thomas Dubois
IITA, Plot 15 Naguru East Road, PO Box 7878, Kampala, Uganda

Tim A. Heard
CSIRO Entomology, Long Pocket Laboratories, 120 Meiers Rd, Indooroopilly 4068, Australia

Martin P. Hill
Department of Zoology and Entomology, Rhodes University, P.O. Box 94, Grahamstown 6140, South Africa

John H. Hoffmann
Zoology Department, University of Cape Town, Rondebosch 7701, South Africa

Fiona Impson
Zoology Department, University of Cape Town, Rondebosch 7701, South Africa, and Plant Protection Research Institute, Private Bag X5017, Stellenbosch 7599, South Africa

Mic H. Julien
CSIRO Entomology European Laboratory, Campus International de Baillarguet, 34980 Montferrier sur Lez, France

Carien Kleinjan
Zoology Department, University of Cape Town, Rondebosch 7701, South Africa

Lastus S. Kuniata
Ramu Sugar Ltd, PO Box 2183, Lae, Papua New Guinea

Andrew J. McConnachie
Weed Research Division, Plant Protection Research Institute, Agricultural Research Council, Private Bag X6006, Hilton 3245, South Africa

Cliff Moran
Department of Zoology, University of Cape Town, Rondebosch 7701, South Africa

Dorette Müller-Stöver
University of Hohenheim, Institute for Plant Production and Agroecology in the Tropics and Subtropics, 70593 Stuttgart, Germany

George P. Markin
Rocky Mountain Research Station, USDA Forest Service, Bozeman, MT 59717–2780, USA

Rangaswamy Muniappan
IPM CRSP, OIRED, Virginia Tech, 526 Prices Fork Road, Blacksburg, VA 24061, USA

Peter Neuenschwander
International Institute of Tropical Agriculture, IITA–Benin, 08 BP 0932 Cotonou, Benin

Terry Olckers
School of Biological and Conservation Sciences, University of KwaZulu–Natal, Private Bag X01, Scottsville, 3209, South Africa

Warea Orapa
Plant Health Team, Land Resources Division, Secretariat of the Pacific Community, Private Mail Bag, Suva, Fiji Islands

Quentin Paynter
Landcare Research, Private Bag 92170, Auckland, New Zealand

Jebomani Rabindra
Project Directorate of Biological Control (ICAR), Post Bag No. 2491, H. A. Farm Post, Hebbal, Bellary Road, Bangalore 560 024, Karnataka, India

Anantanarayanan Raman
Charles Sturt University & E. H. Graham Centre for Agricultural innovation, PO Box 883, Orange, NSW 2800, Australia

Gadi V.P. Reddy
Western Pacific Tropical Research Center, College of Natural and Applied Sciences, University of Guam, Mangilao, Guam 96923, USA

Anthony P. Roberts
Zoology Department, University of Cape Town, Rondebosch, 7700, South Africa

Joachim Sauerborn
University of Hohenheim, Institute for Plant Production and Agroecology in the Tropics and Subtropics, 70593 Stuttgart, Germany

Shon S. Schooler
CSIRO Entomology, Long Pocket Laboratories, 120 Meiers Rd., Indooroopilly, Queensland 4068, Australia

Lorraine W. Strathie
ARC–Plant Protection Research Institute, Private Bag X6006, Hilton 3245, South Africa

Philip W. Tipping
United States Department of Agriculture, Agriculture Research Service, 3205 College Ave., Fort Lauderdale, FL 33314, USA

Rieks D. van Klinken
CSIRO Entomology, Long Pocket Laboratories, 120 Meiers Road, Indooroopilly, Queensland 4068, Australia

Costas Zachariades
Plant Protection Research Institute, Agricultural Research Council, Private Bag X6006, Hilton 3245, South Africa

Myron P. Zalucki
School of Integrative Biology, University of Queensland, St Lucia, Queensland 4072, Australia

Helmuth G. Zimmermann
Weeds Division, Plant Protection Research Institute, Private Bag X134, Queenswood 0121, South Africa

Acknowledgments

The editors greatly appreciate the contributions of the following referees and would like to thank them for reviewing different chapters in this volume.

Fen Beed, International Institute of Tropical Agriculture, Plot 15 Naguru East Road, PO Box 7878, Kampala, Uganda

Marcus J. Byrne, Animal, Plant and Environmental Sciences, University of the Witwatersrand, Johannesburg 2050, South Africa

Ted D. Center, U.S. Department of Agriculture, Agricultural Research Service, Invasive Plant Research Laboratory, 3205 College Ave, Fort Lauderdale, FL 33314, USA

Eric M. Coombs, Oregon Department of Agriculture, 635 Capitol St. NE, Salem, OR 97301–2532, USA

James P. Cuda, Biological Weed Control, University of Florida/IFAS, Entomology and Nematology Dept. Bldg. 970, Natural Area Drive, PO Box 110620, Gainesville, FL 32611–0620, USA

Thomas W. Culliney, USDA–APHIS, PPQ, Center for Plant Health Science and Technology, Plant Epidemiology and Risk Analysis Laboratory, 1730 Varsity Drive, Suite 300, Raleigh, NC 27606 USA

F. Allen Dray, Jr., USDA-ARS I Invasive Plant Research Laboratory, 3205 College Avenue, Fort Lauderdale, FL 33314, USA

Judith Hough-Goldstein, Department of Entomology and Wildlife Ecology, University of Delaware, Newark, DE 19716–2160, USA

John H. Hoffmann, Zoology Department, University of Cape Town, Rondebosch 7700, South Africa

Loke T. Kok, Department of Entomology, Virginia Tech, Blacksburg, VA 24061–0319, USA

Wilco Liebregts, Eco–Consult Pacific Co Ltd., PO Box 5406, Raiwaqa PO, Suva, Fiji Islands

Tom McAvoy, Department of Entomology, Virginia Tech, 216A Price Hall, Blacksburg, VA 24061–0311, USA

Rachel McFadyen, CRC for Australian Weed Management, Block B, 80 Meiers Road, Indooroopilly, Queensland 4068, Australia

Wondi Mersie, Associate Dean and Director of Research, Virginia State University, PO Box 9061, Petersburg, VA 23806, USA

Lytton John Musselman, Department of Biological Sciences, 110 Mills Godwin Building, Old Dominion University, 5115 Hampton Boulevard, Norfolk, VA 23529–0266, USA

James R. Nechols, Department of Entomology, Kansas State University. Waters Hall, Manhattan, KS 66506–4004, USA

Peter Neuenschwander, Plant Health Management Division, International Institute of Tropical Agriculture, B.P. 08 0932, Cotonou, Benin

Terence Olckers, University of KwaZulu–Natal (Pietermaritzburg), Faculty of Science and Agriculture, School of Biological and Conservation Sciences, Private Bag X01, Scottsville, 3209, South Africa

William A. Palmer, Alan Fletcher Research Station, 27 Magazine Street, Sherwood, Queensland 4075, Australia

Robert W. Pemberton, Invasive Plant Research Laboratory, USDA–Agricultural Research Service, 3225 College Ave, Ft. Lauderdale, FL 33312, USA

Gary L. Piper, Department of Entomology, Washington State University, Pullman, WA 99164–6382, USA

Raghu Sathyamurthy, Queensland University of Technology, GPO Box 2434, Brisbane, Queensland 4001, Australia

Robert N. Wiedenmann, Department of Entomology, University of Arkansas, Fayetteville, AR 72701, USA

Myron Zalucki, School of Integrative Biology, University of Queensland, St Lucia, Queensland 4072, Australia

1

Biological control of weeds in the tropics and sustainability

R. Muniappan, G. V. P. Reddy, and A. Raman

1.1 Introduction

Efforts to manage weeds using biological control have been gaining momentum throughout the world, especially in the recent past (Delfosse, 2004). Developed countries, which are principally distributed in the temperate regions, have been practicing classical biological control efficiently, whereas developing countries, most of which are distributed in the tropical regions and have more limited resources, have not adopted deliberate measures for biological control of invasive plants to the same extent as developed nations. The first documented case of biological control of weeds in the tropics was in 1795 and involved the invasive plant *Opuntia monacantha* (Wildenow) Haworth (Cactaceae), which was controlled serendipitously in India due to the inadvertent introduction of *Dactylopius ceylonicus* (Green) (Hemiptera: Dactylopiidae) from Brazil in mistaken identity for *Dactylopius coccus* Costa (Hemiptera: Dactylopidae) (Rabindra and Bhumannavar, this volume; Zimmerman *et al.*, this volume). Thereafter, it took more than a century for biological control of weeds to be rigorously adopted (e.g. biological control of lantana in Hawaii in 1902; biological control of cactus in Australia in 1912; Julien *et al.*, 2007) and for invasive weed species to be recognized as an international problem (Harris, 1979).

Since the early 1900s, work has been predominantly carried out on weeds of the temperate regions in countries such as Canada, New Zealand, Australia, South Africa, and the USA. The USA, Australia, and South Africa, which include tropical segments (e.g. states of Florida, Hawaii, Queensland, Northern Territory, KwaZulu-Natal, Mpuma Langa, Limpopo) have developed programs on biological control of tropical weeds. Only a few developing countries in the tropics have attempted biological control, and this has been sporadic. Moreover, most efforts have been limited to technology-transfer activities of some projects that have already been trialed and implemented in developed countries. Clearly, there is a need for developing countries to receive support from donor agencies or regional and international organizations in the form of methods for creating awareness, knowledge, technical information, training, and financial resources to implement biological control of invasive species and to maintain sustainable programs in the future.

Biological Control of Tropical Weeds using Arthropods, ed. R. Muniappan, G. V. P. Reddy, and A. Raman. Published by Cambridge University Press. © Cambridge University Press, 2009.

This book attempts to consolidate and present biological control activities that have been carried out in different parts of the tropical world on invasive weeds. It includes chapters on biological, ecological, and economic management of 20 "top-priority" weeds, of which 19 have invaded from their epicenters to other parts of the tropical world, causing serious ecological damage to the local environment and economic problems to the people. *Striga* is the singular nonexotic weed treated in this volume; it is a parasitic species and a native of the Old World tropics. Of these 19 species, 16 have been introduced from the New World into the Old World, and one each from Australia into Africa, from Asia into Australia, and from Africa into the Pacific. They include 15 terrestrial (two herbs, seven shrubs, four trees, and two vines) and five aquatic elements. Their habitats vary from arid tropical (e.g. *Parthenium hysterophorus*, Asteraceae; cacti, Cactaceae; species of *Acacia*, Mimosaceae; and species of *Striga*, Orobanchaceae) to humid tropical (e.g. *Chromolaena odorata*, Asteraceae; *Clidemia hirta*, Melastomataceae; *Coccinia grandis*, Cucurbitaceae; and *Mimosa diplotricha* and *M. pigra*, Mimosaceae). *Lantana camara* (Verbenaceae) is cosmopolitan, whereas *Passiflora mollissima* (Passifloraceae) and *Solanum mauritianum* (Solanaceae) are subtropical–tropical elements. *Ageratina adenophora* (Asteraceae) is temperate and *Azolla filiculoides* (Azollaceae), *Cabomba caroliniana* (Cabombaceae), *Eichhornia crassipes* (Pontederiaceae), *Pistia stratiotes* (Araceae), and *Salvinia molesta* (Salviniaceae) are aquatic. All of these species are weeds that have either already invaded or have the potential to invade tropical countries. Of these, 14 species are adapted to lowlands; four species lowland to mid-level altitudes (*C. odorata* and *P. mollissima*, 1000 m asl); one species (*L. camara*) lowland to higher altitudes (2000 m asl); and one species higher altitudes (*A. adenophora*) (Table 1.1). In addition to the 20 chapters on individual weeds, three chapters provide overviews of activities pertaining to biological control of tropical weeds carried out in India; by a regional organization, Secretariat of the Pacific Community (SPC); and by an international organization, International Institute of Tropical Agriculture (IITA).

Several volumes available today deal with biological control in general of insect pests and weeds (e.g. DeBach, 1974; Huffaker and Messenger, 1976; Waterhouse and Norris, 1987; Nechols *et al.*, 1995; Waterhouse, 1998; Bellows and Fisher, 1999; Gurr and Wratten, 2000; Waterhouse and Sands, 2001; Neuenschwander *et al.*, 2003; Hajek, 2004) and a few relate to biological control of weeds specifically (e.g. Waterhouse, 1994; Coombs *et al.*, 2004). In such a context, the key aim of this book is to consolidate and present the past and current research and development work which is progressing in the area of biological control of tropical weeds.

In the backdrop of brief taxonomic notes, origin, distribution, ecology, economic usefulness/uselessness and ecological criticality of the weed, biology and behaviour of the biological control agents selected, trials relating to introduction, establishment, spread, interference by local parasitoids and predators, and efficacy have been discussed at length, citing specific examples. Most importantly, benefit-cost analyses referring to the environmental and economic sustainability in the biological management of each weed have also been considered, wherever appropriate data are available.

Table 1.1. *Status of biological control of invasive weeds addressed in this volume*

Invasive weed	Origin	Invaded regions	Form	Altitude	Ecological niche where invasive	Considered invasive in:	State of the program
Acacia nilotica (Mimosaceae)	India	Australia	Tree	Lower	Arid tropics	Australia	Early stages
Ageratina adenophora (Asteraceae)	Neotropics	Australia, Asia, Africa, Hawaii, and New Zealand	Shrub	Higher	Humid temperate	Australia, New Zealand, India, China, Nepal, South Africa	In operation for 30 years
Australian *Acacias* (Mimosaceae)	Australia	South Africa	Tree	Lower	Arid tropics	France, India, Italy, New Zealand, Portugal, Reunion, South Africa, Spain, Uganda, USA	
Azolla filiculoides (Azollaceae)	Neotropics	South Africa	Aquatic herb	Lower	Tropics to subtropics	South Africa	
Cabomba caroliniana (Cabombaceae)	Neotropics	Australia and South Africa	Aquatic herb	Lower	Tropics to subtropics	Australia, Canada, China, India, Japan, Netherlands, USA	
Chromolaena odorata (Asteraceae)	Neotropics	Humid tropical Africa, Asia, Pacific, and Australia	Shrub	Lower	Humid tropics to subtropics	Western Africa, Southern Africa, South and Southeast Asia, Micronesia, Papua New Guinea, Australia	In operation for 40 years
Clidemia hirta	Neotropics	Pacific Islands	Shrub	Lower	Humid tropics	Hawaii, Fiji, Samoa, and American Samoa	In operation for 50 years
Coccinia grandis (Cucurbitaceae)	Africa	Hawaiian and Mariana Islands in the Pacific	Vine	Lower	Humid tropics	Hawaii and Mariana Islands in the Pacific	In operation for 10 years

Table 1.1. (cont.)

Invasive weed	Origin	Invaded regions	Form	Altitude	Ecological niche where invasive	Considered invasive in:	State of the program
Eichhornia crassipes (Pontederiaceae)	Neotropics	Old World, Australia and the Pacific Islands	Aquatic herb	Lower	Tropics	Australia, tropical South and North America, and Old World tropics	In operation for 60 years
Lantana camara (Verbanaceae)	Neotropics	Old World, Australia, and the Pacific Islands	Shrub	Cosmopolitan	Tropics, subtropics and temperate. Adapted to arid and humid conditions	Australia and Old World tropics	In operation over 100 years
Mimosa diplotricha (Mimosaceae)	Neotropics	Australia, Asia, and the Pacific Islands	Shrub	Lower	Humid tropics	Australia and Pacific Islands	In operation over 40 years
Mimosa pigra (Mimosaceae)	Neotropics	Australia, Southeast Asia	Tree	Lower	Humid tropics	Australia, Thailand, Philippines, and Indonesia	In operation over 30 years
Cactaceae	New World	Old World, Australia and the Pacific Islands	Shrub to tree	Lower	Arid tropics	Australia, India, Sri Lanka, South Africa, Madagascar, Hawaii, Ethiopia, Eritrea, Kenya, Yemen, Saudi Arabia, Morocco, Canary Islands, Zimbabwe, Namibia, and West Indies	In operation over 200 years

Table 1.1. (cont.)

Invasive weed	Origin	Invaded regions	Form	Altitude	Ecological niche where invasive	Considered invasive in:	State of the program
Parthenium hysterophorus (Asteraceae)	Mexico	Australia, South Asia, Eastern and Southern Africa, and some Pacific Islands	Herb	Lower	Arid tropics	Australia, India, Ethiopia, Kenya, South Africa, Botswana, Mauritius, and Madagascar	In operation for 30 years
Passiflora tripartita (Passifloraceae)	South America	Hawaii	Vine	Lower to middle	Humid tropics and subtropics	Hawaii, New Zealand	
Pistia stratiotes (Araceae)	South America	Old World, Australia, and Pacific Islands	Aquatic herb	Lower	Tropics	Old and New World tropics	
Prosopis sp. (Mimosaceae)	North America	Africa, Asia, and Australia	Tree	Lower	Arid tropics	Australia, India, South Africa, Kenya, and Ethiopia	
Salvinia molesta (Salviniaceae)	South-eastern Brazil	Old World, and Australia	Aquatic herb	Lower	Tropics	Old and New World tropics	In operation over 40 years
Solanum mauritianum (Solanaceae)	South America	Old World	Shrub	Lower to middle	Tropics to subtropics	South Africa and Australia	
Striga spp. (Orobanchaceae)	Old World	Old World	Parasitic herb	Lower	Arid tropics	Old World tropics	

Interest in ecologically sound management of invasive species is currently on the rise, among both scientists and the general public (e.g. Drake *et al.*, 1989; Devine, 1998). Among the many diverse invasive organisms, alien plants induce serious economic losses to humankind by competing for natural resources, especially in an agricultural context, which includes not only grain production but also the pasture and forestry industry, by reducing overall yield and quality through allelopathy and contamination (Dhileepan, this volume; Zachariades *et al.*, this volume). Weeds also increase the likelihood of fires, reduce property values, poison domestic and wild animals, reduce quality of milk and meat, interfere in the movement of wild animals and their breeding habits, endanger native vegetation, interfere with irrigation, navigation and recreational water bodies, inflict allergies, enhance chances for disease incidence in humans, animals, and crops by harboring disease-agent vectors, and reduce market access because of strict quarantine practices (Culliney, 2005).

1.2 Management strategies

The major weed management strategies usually applied are prevention, eradication, and control (Mack *et al.*, 2000; Monaco *et al.*, 2002; Culliney, 2005). Quarantine laws pro-mulgated with assistance from regional and international organizations regulate move-ment of weeds and products from weed-infested areas in intra- and intercountry transportation. In developed countries, clamping of strict quarantine regulations is common whenever an impending threat from an invasive species becomes obvious. Such regulations in developing countries are either rare or nonexistent. As an effort to assist smaller countries in the Pacific, such as Fiji, Kiribati, Samoa, Tonga, and Vanuatu, Pest-Alert Bulletins appear from the Secretariat of the Pacific Community (SPC) whenever a "pest" problem becomes evident in the region. This practice is done so that the member countries of SPC can promulgate their own regulations (SPC–PPS, 2003). Eradication is possible when the introduction and consequent spread of a weed species is spotted early and sustained efforts to monitor it are made. For example, when *C. odorata* and *Mikania micrantha* Kunth (Asteraceae) infestations occurred in northern Queensland in 1994 and 1998, respectively, Australia instituted immediate monitoring programs (Galway and Brooks, 2007), which paved the way for possible eradication. Early detection and determination to eradicate, backed by an adequate budget and human resources, are critical to achieve a successful program. Eradication becomes economically unviable when the weed spread is extensive (Myers and Bazely, 2003; Culliney, 2005). To prevent the spread of *P. hysterophorus* in the state of Karnataka (India), a quarantine act was passed in 1975 declaring it a noxious weed; and notices were issued to remove this weed once or twice in the 1980s by the Bangalore Municipal Corporation (Bhan *et al.*, 2007). Because of the failure to initiate the correct action in the most appropriate manner, management of *P. hysterophorus* was a failure, as it was not supported by adequate funds, human resources, nor by commitment from either people or the administration. Fur-thermore, the neighbouring states paid either little or no attention to the establishment and spread of *P. hysterophorus*. Examples of such failure to implement and follow up the correct measures (e.g. quarantine regulations) in developing countries exist plentifully.

When a weed escapes quarantine and exceeds the eradication stage, the next level of options available for management are: mechanical, chemical, cultural, and biological. Mechanical control varies from the use of hand tools to the use of heavy machinery in the removal of weeds. Slashing the weeds such as *Chromolaena odorata* in India and *Coccinia grandis* in Mariana Islands proved futile as the stubbles sprouted and the operation had little impact. Burning is one of the methods used extensively for the control of some weeds, especially in forests and rangelands; but fire is a factor that supports *C. odorata* spread. In *C. odorata*-infested areas, fire killed most of the adjoining vegetation but not the stubbles of *C. odorata*, inducing their immediate sprout soon after rains and invasion of the land occupied by native vegetation (Muniappan *et al.*, 2005). In Hawaii, *A. adenophora* was cleared from a vast spread of grazing land by farmers between 1920 and 1948 at great expense without much relief, whereas the introduction of the gall fly *Procecidochares utilis* Stone (Diptera: Tephritidae) enabled an impressive control of the weed, eliminating the need for mechanical control (Bess and Haramoto, 1958). Chemical control is widely used in croplands and rangelands, and along roadsides (Monaco *et al.*, 2002). Cultural control involves the use of mulch, cover crops, and competitive suppression. This practice is used in annual and perennial cropping systems and, to a limited extent, in vacant land areas (Mahadevappa and Ramaiah, 1988). However, these methods have only a limited effect. They are expensive and entail several repeats. Moreover, herbicides cause health problems to humans and domesticated animals, and adversely affect the environment. Most infestations of these invasive weeds are either too extensive or the land value infested by them is too marginal, thus rendering physical, cultural, and chemical control methods uneconomical and unsustainable. Benefit–cost analysis of different control options of *Salvinia molesta* in Zimbabwe showed that physical and chemical control measures were expensive and ineffective, whereas biological control was effective and inexpensive (Chikwenhere and Keswani, 1997).

1.3 Biological control

Classical biological control is the most sustainable method used in biological control of invasive, exotic weeds. This method employs the introduction of arthropod natural enemies that exist naturally in their places of origin. The method, however, involves importation, colonization, and establishment of exotic natural enemies, which include predators and parasitoids (McFadyen, 1998). This method provides long-lasting and affordable management, either alone or in combination with other methods. It is usually useful in the control of perennial weeds that infest low-productivity cropland areas, rangelands, and disturbed forests. Whereas other methods are either expensive or impractical in specific circumstances, biological control methods are affordable, safe to the environment, and economical as well.

Biological control of weeds plays a key role in the management of natural resources in Oceania (Julien *et al.*, 2007) and other parts of the world. It will usually require a long period of research and a high initial investment of capital and human resources (Culliney, 2005). A program typically requires 10–20 years to achieve satisfactory results and can easily cost US$3–8m (McFadyen, 2000). Over 350 species of natural enemies, including arthropods, pathogens, and vertebrates have been released for the biological control of weeds (Julien and Griffiths, 1998) with nearly 1000 releases made from the mid nineteenth century to the end of 1996 to control 133 weed species (Culliney, 2005). The success rate of biological control of weeds programs on the whole was 33% (Culliney, 2005). Benefit–cost analysis of weed biological control projects in Australia has been reported by Page and Lacey (2006). They quantified the overall return on investment in the form of a benefit-cost ratio of 23.1, implying that for every dollar invested in biological control of weeds there is $23.1 returned as benefits. Benefit–cost ratios available for the biological control of weeds covered in this book are given in Table 1.2.

1.4 Technology transfer

Government agencies in the USA, Canada, Australia, New Zealand, and South Africa were able to initiate new biological control projects for invasive weeds starting with identification of the native range, searching for potential natural enemies, screening for host specificity, assessing acceptable risk, establishing the agents, and evaluation of impact. Most developing countries lack the knowledge, capital, human resources, and infrastructure to carry out different steps involved in biological control. Programs on biological control of weeds carried out in developing countries involved mostly technology transfer supported by donor countries, assisted by international organizations like CAB International or through bilateral and reciprocal arrangements. Exploration, screening, introduction, and evaluation of the natural enemies introduced into these countries have already been carried out either by the developed countries or by international organizations. The advantage in such technology transfer is that only the natural enemies that have been tried elsewhere and proven effective may be selected for introduction. As noted in this volume, some spectacular successes have been achieved in controlling the invasive weeds by technology transfer, such as *Opuntia* spp. in Australia, Hawaii, Sri Lanka, and South Africa (Zimmermann *et al.*, this volume); *S. molesta* in Africa, Fiji, Papua New Guinea (PNG), Malaysia, and India (Julien *et al.*, this volume); *C. odorata* in Micronesia (Zachariades *et al.*, this volume); *M. diplotricha* in PNG, Fiji, Samoa, Solomon Islands, Pohnpei, and Yap (Kuniata, this volume); *C. hirta* in Fiji (Conant, this volume); and *E. crassipes* in PNG, and Africa (Coetzee *et al.*, this volume; Beed and Dubois, this volume). It is hoped that the information presented on the 20 tropical weeds will assist in sparking an interest to start biological control projects in developing countries through bilateral programs with the resources available within the

Table 1.2. *Benefit-cost analysis for the biological control of invasive weeds in the tropics*

Weed	Country	Benefit/cost ratio	Reference
Acacia longifolia (Andr.) Willd	South Africa	104	van Wilgen *et al.*, 2004
Acacia pycantha Benth	South Africa	665	van Wilgen *et al.*, 2004
Acacia saligna (Labill.) Wendl.)	South Africa	800	van Wilgen *et al.*, 2004
Azolla filiculoides Lamarck	South Africa	2.5	McConnachie *et al.*, 2003
		13	Hill and McConnachie, this volume
Eichhornia crassipes (Mart.) Solms-Laub.	Benin	124	De Groote *et al.*, 2003
	Australia	27.5	Page and Lacey, 2006
Lantana camara L.	South Africa	22	Van Wilgen *et al.*, 2004
	Australia	5.6	Page and Lacey, 2006
		9	AEC group, 2007
Mimosa diplotricha	Australia	17.6	Page and Lacey, 2006
Mimosa pigra	Australia	0.8	Page and Lacey, 2006
Opuntia aurantiaca Lindley	South Africa	709	van Wilgen *et al.*, 2004
Opuntia spp.	Australia	312.3	Page and Lacey, 2006
Parthenium hysterophorus L.	Australia	7.2	Page and Lacey, 2006
Pistia stratiotes	Australia	27.5	Neuenschwnader *et al.*, this volume
Prosopis spp.	Australia	0.5	Page and Lacey, 2006
Salvinia molesta D. S. Mitchell	Australia	27.5	Page and Lacey, 2006
	Sri Lanka	53	Doeleman, 1989
	Zimbabwe	10.6	Chikwenhere and Keswani, 1997

host countries; and regional and international programs with the assistance of donor agencies.

1.5 Economics of biological control of weeds

As a concept classical biological control is for the public good (Tisdell and Auld, 1990) and it is not amenable for individual profit (Culliney, 2005). Whereas chemical, cultural, and mechanical control methods benefit only the users and the geographical localities in which it is used, biological control benefits the public at large (Cullen and Whitten, 1995). Because it usually has a high initial investment and is unlikely to recover research and development expenses, biological control is unattractive as a private entrepreneurial effort (Hill and Greathead, 2000; Coombs et al., 2004). Although developing countries are unable to initiate biological control programs because of the need for a high initial investment, the technology transfer of the programs developed in other countries has benefited low-income farmers and the environment (Greathead, 1995). The economic evaluation of benefits of biological control of weeds involves considerations of esthetics, health, and natural resources (Culliney, 2005). Only in recent years has biological control of weeds been subjected to rigorous economic analysis, mostly in USA, South Africa, and Australia (van Wilgen et al., 2004; Culliney, 2005; Page and Lacey, 2006). The success rate of weed biological control programs is estimated at 17%, based on the character-ization of the small number of outstanding successes and large number of failures (Crawley, 1989), which could be due to bias in the way success has been measured traditionally. Culliney (2005) estimated the success rate of biological control to be 33% based on analysis focusing on the outcomes achieved by each program. Walton (2005) has estimated the success rate to be 80%. Success rates have varied in individual coun-tries: Hawaii – 50% (Markin et al., 1992), Australia – 51% (McFadyen, 2000), South Africa – 61% (Zimmermann et al., 2004), Mauritius – 80% (Fowler et al., 2000), and New Zealand – 81% (Fowler, 2000).

Biological control of invasive weeds is economical. In most countries, data on the agricultural and environmental impact of one or more weeds, as well as any costing done towards control, are unavailable. Whereas the benefit-cost analysis method is reliable in a broad-brush context, analyses need to be done independently for each country, because the impact of one or more weeds on agriculture and the environment is bound to vary. The cost of control will be far less and the benefit-cost ratio will be high in countries where the program had been transferred from other countries. For example, natural enemies of L. camara introduced into Guam, Micronesia, and the Solomon Islands were already host-specificity and field tested in Hawaii. Similarly, exploratory work and most of the host-specificity testing for natural enemies of Coccinia grandis introduced into Guam and Saipan were done in Hawaii. McConnachie et al. (2003) reported a benefit to cost ratio for controlling Azola filiculoides in South Africa at 2.5:1 in the year 2000, which was increased to 13:1 for 2005 by adjusting for the value of the South African R and for inflation (Hill and McConnachie, this volume). In some instances, it may not be possible to separate benefits, as examples such as Eichhornia crassipes, Salvinia molesta, and Pistia stratiotes occur in the same water bodies and locations (Julien et al., this volume). Doeleman (1989) estimated a benefit cost ratio of 53:1 in terms of money and 1671:1 in terms of labor for the complete control of the weed Salvinia molesta in Sri Lanka. These examples highlight the substantial "economic" benefits of using biological control in weed management (Julien et al., this volume). Biological control of invasive weeds is a better investment than the remaining procedures that apply to management of invasive weeds and it is needed more than ever before with the rise in the traffic of introductions of diverse plant species through extensive and rapid human movements across the continents.

1.6 Conflict of interest resolution

Conflict of interest occurs when some sectors of a community find either a use or a relevance regarding a weed or when the biological control agent begins to attack nontarget but economically important organisms (Culliney, 2005). Such a conflict usually entails sociological, political, and scientific complexities. Conflict of interest has been extensive in six out of the 20 weeds dealt with in this volume; i.e., *Chromolaena odorata* (Zachariades *et al.*, this volume) species of cacti (Zimmermann *et al.*, this volume), species of the Australian *Acacia* (Impson *et al.*, this volume), *Acacia nilotica* (Dhileepan, this volume), species of *Prosopis* (van Klinken *et al.*, this volume), and *Solanum mauritianum* (Olckers, this volume). *Opuntia elatior* has invaded Kenya and biological control was contemplated; however, commercial cultivation of *O. ficus-indica* has restricted the adoption of the currently available biological control agents. Moreover, neighboring Ethiopia decided against biological control because several rural communities depend on *O. ficus-indica* for food. Biological control of *Solanum mauritianum* has been taken up in South Africa, but most countries that either raise solanaceous crops or have other native species of *Solanum* do not consider the biological control option, because of the concern that the introduced agents may invade either the economic crops or the native species (Olckers, this volume). For biological control of the Australian *Acacia* in South Africa and *Prosopis* in Australia and South Africa, amicable resolutions have been reached. A regional project up to a value of a million dollars towards the biological control of *C. odorata* in western Africa with the prospect for funding from the Inter-African Development Bank was blocked due to a perception that this invasive plant species was considered a valuable fallow species by farmers who used shifting cultivation techniques (McFadyen, 1996). The report of feeding by *Zygogramma bicolorata* Pallister (Coleoptera: Chrysomelidae), a natural enemy introduced for the biological control of *P. hysterophorus*, on sunflower in India resulted in the cancellation of all projects for biological control of weeds for many years until it was proved that the beetle fed on sunflower leaves dusted with *P. hysterophorus* pollen from the neighborhood (Bhumannavar *et al.*, 1998).

1.7 Sustainability

In Rio de Janeiro during the Earth Summit, the gathered nations adopted Agenda 21, which made sustainable development a universal goal in June 1992 (Moldan and Billharz, 1997). The World Bank has identified economic, sociocultural and ecological dimensions as the key perspectives for an environmentally sustainable development (Moldan, 1997). The generic concept of sustainability, including the subtler aspects of sustainable growth and sustainable development, is the singular choice for the future development of any nation (Xiaomin and Li, 1997). The alteration of the population and community structure of native ecosystems by the invasion of exotic species has led to the extinction of native species and reduced local biodiversity, and has placed a heavy burden of socioeconomic investment to achieve their control and/or management (Ramakrishnan, 1991). Chemical, cultural, and mechanical control methods are limited in their application and relevance and require repeated applications and/or adoptions and therefore are not cost effective. Chemical control is amenable to the disadvantage of causing environmental and human-health problems. Classical biological control is an economically viable measure; it is environmentally safe, is long lasting, and is a sustainable method available for management of invasive weeds. It could be implemented either independently or as a component of integrated pest management (IPM) as it readily and effectively integrates several other mechanisms.

When effective, classical biological control provides the most sustained suppression of an invasive plant. The inadvertent introduction of *Dactylopius ceylonicus* into India nearly 200 years ago has effectively controlled the prickly pear (*Opuntia monacantha*), and the landscape infested by *O. monacantha* became suitable for cultivation within five to six years. The control was permanent and required no further efforts to manage the weed. Subsequent introduction of *Dactylopius ceylonicus* and similar other natural enemies into Sri Lanka, Australia, South Africa, Madagascar, and Hawaii have yielded similar sustained control of the cactus (Zimmermann *et al.*, this volume; and Rabindra and Bhumannavar, this volume). The benefit-cost ratios of 800 for control of *Acacia saligna* and 709 for control of *Opuntia aurantiaca* in South Africa (Table 1.2) are far better returns for the money invested. Biological control has offered similar results for *S. molesta* in Australia, Botswana, Congo, Fiji, Ghana, India, Kenya, Papua New Guinea, South Africa, Malaysia, Mauritania, and Namibia; for *P. stratiotes* in Australia, South Africa, Papua New Guinea, North America, Congo, Senegal, Ghana, Ivory Coast, Kenya, Benin, and Mauritania; *E. crassipes* in Papua New Guinea, Benin, Malawi, India, and Lake Victoria; *A. filiculoides* in South Africa; *C. odorata* in Micronesia; and *L. camara* in Guam and Solomon Islands. Release of the weevil *Stenopelmus rufinasus* Gyllenhal (Coleoptera: Curculionidae) for control of *Azolla filiculoides* in South Africa has completely eliminated the threat of this weed to aquatic ecosystems (Hill and McConnachie, this volume). Farmers in Maui, Hawaiian Islands, stopped controlling the Crofton weed, *Ageratina adenophora* by mechanical and chemical means about three years after establishment of the gall fly *Procecidochares utilis* Stone.

1.8 Conclusion

Classical biological control is the best among the viable options available for sustainable management of invasive weeds, especially where other technologies such as chemical and mechanical control are unacceptable due to cost and adverse impact on the environment. Available benefit-cost ratios for biological control trials made for some of the most serious weeds indicate that classical biological control has given the highest returns for the monies spent (Page and Lacey, 2006). Such ratios could be improved further in technology-transfer programs wherein the initial expenses for research and development have already been incurred elsewhere.

The techniques described for biological control of weeds in this volume could be safely and efficiently transferred to developing countries with minimal expense for the initial institutional and human-capacity building. There is a greater opportunity for individual countries and regional and international organizations to play a constructive role in implementing biological control of invasive weeds. Some of the developing countries in Asia and Africa are implementing independently or with assistance from donor agencies. The United States Agency for International Development (USAID) is assisting Ethiopia and South Africa with the biological control of *Parthenium hysterophorus*. Australian Centre for International Agricultural Research (ACIAR) is assisting in East Timor and Papua New Guinea in biological control of *Chromolaena odorata*. The World Bank is supporting biological control of *Eichhornia crassipes* in Lake Victoria. Commonweath Agricultural Bureaux International is assisting India, China, Taiwan, and Secretariat of the Pacific Community (SPC) in biological control of *Mikania micrantha*. Regional and international organizations such as SPC and International Institute of Tropical Agriculture (IITA) are also actively assisting their clientele in this activity. The role of classical biological control in sustainable integrated invasive weed management is well recognized. While there is a need for increased funding for research and development of biological control of newly emerging invasive species in the tropical world, attention should also be given to technology transfer of successful biological control programs achieved elsewhere to solve the problems of invasive species in the developing world.

References

AEC group (2007). *Economic Impact of Lantana on the Australian Grazing Industry.* Final Report for Queensland Department of Natural Resources and Water. Brisbane, Australia: AEC Group Limited, 39 pp.

Bellows, T. S. and Fisher, T. W. (1999). *Handbook of Biological Control*, Academic Press, 1046 pp.

Bess, H. A. and Haramoto, F. H. (1958). Biological control of pamakani, *Eupatorium adenophorum*, in Hawaii by a tephritid gall fly, *Procecidochares utilis*. 1. The life history of the fly and its effectiveness in the control of the weed. In *Proceedings of the Tenth International Congress of Entomology*, held in Montreal, August 17–25, 1956, Vol. 4, ed. E. C. Becker. Ottawa: Mortimer, pp. 543–548.

Bhan, V. M., Sushilkumar and Raghuwanshi, M. S. (2007). *Future strategies for Effective Parthenium Management* (www.iprng.org/IPRNG-parthenium_a&w19.htm).

Bhumannavar, B. S., Balasubramanian, C. and Ramani, S. (1998). Life table of the Mexican beetle *Zygogramma bicolorata* on parthenium and sunflower. *Journal of Biological Control*, **12**, 101–106.

Chikwenhere, G. P. and Keswani, C. L. (1997). Economics of biological control of kariba weed (*Salvinia molesta* Mitchell) at Tengwe in north-western Zimbabwe – a case study. *International Journal of Pest Management*, **43**, 109–112.

Coombs, E. M., Clark, H. K., Piper, G. L. and Cofrancesco, Jr., A. F., eds. (2004). *Biological Control of Invasive Plants in the United States*. Corvallis OR: Oregon State University Press, 467 pp.

Crawley, M. J. (1989). The successes and failures of weed biocontrol using insects. *Biocontrol News and Information*, **10**, 213–223.

Cullen, J. M. and Whitten, M. J. (1995). Economics of classical biological control: a research prespective. In *Biological Control: Benefits and Risks*, ed. H. M. T. Hokkanen and J. M. Lynch. Cambridge, UK: Cambridge University Press, pp. 270–276.

Culliney, T. W. (2005). Benefits of classical biological control for managing invasive plants, *Critical Reviews in Plant Science*, **24**, 131–150.

DeBach, P. H. (1974). *Biological Control by Natural Enemies*. New York: Cambridge University Press.

De Groote, H., Ajuonu, O., Attignon, S., Djessou, R. and Neuenschwander, P. (2003). Economic impacts of biological control of water hyacinth in Southern Benin. *Ecological Economics*, **45**, 105–117

Delfosse, E. S. (2004). Introduction. In *Biological Control of Invasive Plants in the United States*, ed. E. M. Coombs, J. K. Clark, G. L. Piper and A. F. Cofrancesco, Jr. Corrallis, OR: Oregon State University Press, pp. 1–11.

Devine, R. S. (1998). *Alien Invasion: America's Battle with Non-Native Animals and Plants*. Washington, DC: National Geographic Society, 280 pp.

Doeleman, J. A. (1989). *Biological Control of* Salvinia molesta *in Sri Lanka: An Assessment of Costs and Benefits*. Economic Assessment Series 1. Canberra, Australia: Australian Centre for International Agriculture Research.

Drake, J. A., Mooney, H. A., di Castri, F., *et al.*, eds. (1989). *Biological Invasions: A Global Perspective*. SCOPE 37. New York: John Wiley, 525 pp.

Fowler, S. V. (2000). Trivial and political reasons for the failure of classical biological control of weeds: a personal view. In *Proceedings of the X International Symposium on Biological Control of Weeds*, ed. N. R. Spencer. Great Falls, MT: Advanced Litho Printing, pp. 169–172.

Fowler, S. V., Syrett, P. and Hill, R. L. (2000). Success and safety in the biological control of environmental weeds in New Zealand. *Austral Ecology*, **25**, 553–562.

Galway, K. E. and Brooks, S. J. (2007). Control recommendations for mikania vine (*Mikania micrantha*) 106 and Siam weed (*Chromolaena odorata*) in Australia. In *Proceedings of the Seventh International Workshop on Biological Control and Management of* Chromolaena odorata *and* Mikania micrantha, ed. P-Y. Lai, G. V. P. Reddy and R. Muniappan. Taiwan: National Pingtung University of Science and Technology, pp. 106–115.

Greathead, D. J. (1995). Benefits and risks of classical biological control. In *Biological Control: Benefits and Risks*, ed. H. M. T. Hokkanen and J. M. Lynch. Cambridge, UK: Cambridge University Press, pp. 53–63.

Gurr, G. and Wratten, S., eds. (2000). *Biological Control: Measures of Success*. Boston, MD: Kluwer Academic Publishers, 429 pp.

Hajek, A., ed. (2004). *Natural Enemies: An Introduction to Biological Control*. Cambridge, UK: Cambridge University Press, 378 pp.

Harris, P. (1979). Cost of biological control of weeds by insects in Canada. *Weed Science*, **27**, 242–250.

Hill, G. and Greathead, D. (2000). Economic evaluation in classical biological control. In *The Economics of Biological Invasions*, ed. C. Perrings, M. Williamson and S. Dalmazzone. Cheltenham, UK: Edward Elgar Publishing, pp. 208–223.

Huffaker, C. B. and Messenger, P. S., eds. (1976). *Theory and Practice of Biological Control*. New York: Academic Press, 788 pp.

Julien, M. H. and Griffiths, M. W. (1998). *Biological Control of Weeds: A World Catalogue of Agents and Their Target Weeds*. Wallingford, UK: CABI Publishing 223 pp.

Julien, M. H., Scott, J. K., Orapa, W. and Paynter, Q. (2007). History, opportunities and challenges for biological control in Australia, New Zealand and the Pacific Islands. *Crop Protection*, **26**, 255–265.

Mack, R. N., Simberloff, D., Lonsdale, W. M., *et al.* (2000). Biotic invasions: causes, epidemiology, global consequences, and control. *Ecological Applications*, **10**, 689–710.

Mahadevappa, M. and Ramaiah, H. (1988). Pattern of replacement of *Parthenium hysterophorus* plants by *Cassia serecia* in waste lands. *Indian Journal of Weed Science*, **20**, 83–85.

Markin, G. P., Lai, P. Y. and Funasaki, G. Y. (1992). Status of biological control of weeds in Hawaii and implications for managing native ecosystems. In *Alien Plant Invasions in Native Ecosystems of Hawaii: Management and Research*, ed. C. P. Stone, C. W. Smith and J. T. Tunison. Honolulu, HI: University of Hawaii Cooperative National Park Resources Studies Unit, pp. 466–482.

McConnachie, A. J., de Wit, M. P., Hill, M. P. and Byrne, M. J. (2003). Economic evaluation of the successful biological control of *Azolla filiculoides* in South Africa. *Biological Control*, **28**, 25–32.

McFadyen, R. E. (1996). Biocontrol of *Chromolaena odorata*: divided we fail. In *Proceedings of the IX International Symposium on Biological Control of Weeds*, ed. V. C. Moran and J. H. Hoffmann. Cape Town, South Africa: University of Cape Town Press, pp. 455–459.

McFadyen, R. E. C. (1998). Biological control of weeds. *Annual Review of Entomology*, **43**, 369–393.

McFadyen, R. E. C. (2000). Successes in biological control of weeds. In *Proceedings of the X International Symposium on Biological Control of Weeds*, ed. N. R. Spencer. Great Falls, MT: Advanced Litho Printing, pp. 3–14.

Moldan, B. (1997). World Bank perspectives on sustainable development. In *Sustainability Indicators. Report of the Project on Indicators of Sustainable Development*, ed. B. Moldan, S. Billharz and R. Matravers. New York: John Wiley, pp. 138–141.

Moldan, B. and Billharz, S. (1997). Introduction. In *Sustainability Indicators. Report of the Project on Indicators of Sustainable Development*, ed. B. Moldan, S. Billharz and R. Matravers. New York: John Wiley, pp. 1–5.

Monaco, T. J., Weller, S. C. and Ashton, F. M. (2002). *Weed Science: Principles and Practices*. Fourth edn. New York: John Wiley.

Muniappan, R., Reddy, G. V. P. and Lai, P. Y. (2005). Distribution and biological control of *Chromolaena odorata*. In *Invasive Plants: Ecological and Agricultural Aspects*, ed. Inderjit. Basel, Switzerland: Birkhauser Verlag, pp. 223–233.

Myers, J. H. and Bazely, D. R. (2003). *Ecology and Control of Introduced Plants*. Cambridge, UK: Cambridge University Press.

Nechols, J. R., Andres, L. A., Beardsley, J. W., Goeden, R. D. and Jackson, C. G., eds. (1995). *Biological Control in the Western United States: Accomplishments and Benefits of Regional Research Project W-84, 1964–1989*. Publication 3361. Berkeley, CA: University of California Press, Division of Agricultural and Natural Resources.

Neuenschwander, P., Borgemeister, C. and Langewald, J. (2003). *Biological Control in IPM Systems in Africa*. Wallingford, UK: CABI Publishing, 414 pp.

Page, A. R. and Lacey, K. L. (2006). *Economic Impact Assessment of Australian Weed Biological Control*. Technical Series 10. Adelaide, Australia: CRC for Australian Weed Management, 150 pp.

Ramakrishnan, P. S., ed. (1991). *Ecology of Biological Invasion in the Tropics*. New Delhi: International Scientific Publications, 195 pp.

SPC-PPS. (2003). Incursion of parthenium weed (*Parthenium hysterophorus*) in Papua New Guinea. Pest Alert 30. Suva, Fiji Islands: Secretariat of the Pacific Community, Plant Protection Service, 1 p.

Tisdell, C. A. and Auld, B. A. (1990). Evaluation of biological control projects. In *Proceedings of the VII International Symposium on Biological Control of Weeds*, ed. E. S. Delfosse. Rome, Italy: Instituto Sperimentale per la Patologia Vegetale (MAF), pp. 93–100.

van Wilgen, B. W., de Wit, M. P., Anderson, H. J., *et al.* (2004). Costs and benefits of biological control of invasive alien plants: case studies from South Africa. *South African Journal of Science*, **100**, 113–122.

Walton, C. (2005). *Reclaiming Lost Provinces: A Century of Weed Biological Control in Queensland*. Brisbane, Australia: Department of Natural Resources and Mines, 104 pp.

Waterhouse, D. F. (1994). *Biological Control of Weeds: Southeast Asian Prospects*. Monograph 26. Canberra, Australia: ACIAR, 302 pp.

Waterhouse, D. F. (1998). *Biological Control of Insect Pests: Southeast Asian Prospects*. Monograph 51. Canberra, Australia: ACIAR, 548 pp.

Waterhouse, D. F. and Norris, K. R. (1987). *Biological Control Pacific Prospects*. Melbourne, Australia: Inkata Press. 454 pp.

Waterhouse, D. F. and Sands, D. P. A. (2001). *Classical Biological Control of Arthropods in Australia*. Monograph 77. Canberra, Australia: ACIAR, 560 pp.

Xiaomin, G. and Li, G. (1997). Application of indicators of sustainable development in China. In *Sustainability Indicators. Report of the Project on Indicators of Sustainable Development*, ed. B. Moldan, S. Billharz and R. Matravers. New York: John Wiley, pp. 334–339.

Zimmermann, H. G., Moran, V. C. and Hoffmann, J. H. (2004). Biological control in the management of invasive alien plants in South Africa, and the role of the working for water programme. *South African Journal of Science*, **100**, 34–40.

2

Acacia nilotica ssp. *indica* (L.) Willd. ex Del. (Mimosaceae)

K. Dhileepan

2.1 Introduction

Acacia nilotica (L.) Willd. ex Del. (Mimosaceae), known as prickly acacia in Australia, is native to the tropics and subtropics of Africa, the Middle East, and the Indian sub-continent. Under the new classification system, the subgenus *Acacia* will be *Vachellia* Wright & Arnold (Seigler and Ebinger, 2006). *Acacia nilotica* was introduced into Queensland from India in the late 1890s as an ornamental tree (Bolton, 1989). Since the mid 1920s, *A. nilotica* was widely used as a shade tree and fodder for sheep in western Queensland. A change from sheep to cattle as the predominant grazing species and a series of wet years resulted in its spread. *Acacia nilotica* was declared a noxious weed in Queensland in 1957, and is now a weed of national significance in Australia. In the Mitchell grass downs of western Queensland, which cover around 22m ha of natural grassland, *A. nilotica* has infested more than 6m ha (Fig. 2.1) and 2000 km of bore drains (Mackey, 1997; Spies and March, 2004). *Acacia nilotica* is also present in the coastal regions of Queensland, the Northern Territory, and Western Australia (Spies and March, 2004), and has the potential to infest vast areas of Australia's native grassland ecosystems (Fig. 2.2) (Kriticos *et al.*, 2003a).

Acacia nilotica populations in Australia consist of thorny large shrubs or small trees, growing 4–5 m high, occasionally reaching 10 m. Seedling recruitment is linked to rainfall pattern, and under favorable conditions, young plants attain maturity in 2–5 years (Fig. 2.3). *Acacia nilotica*, when mature, forms dense thorny thickets (900 plants/ha), and mature plants live for *c.* 40 years (Fig. 2.3). The golden-yellow flower-bearing inflorescence is ball-shaped and grows in groups of two to six on one shoot. The plants have distinct flat sickle-shaped pods, each 10–15 cm long, bearing 8–15 seeds (Spies and March, 2004). A mature tree will produce up to 30,000 seeds per year, and seeds, when buried in soil, can remain viable up to seven years (Fig. 2.3).

Biological Control of Tropical Weeds using Arthropods, ed. R. Muniappan, G. V. P. Reddy, and A. Raman. Published by Cambridge University Press. © Cambridge University Press, 2009.

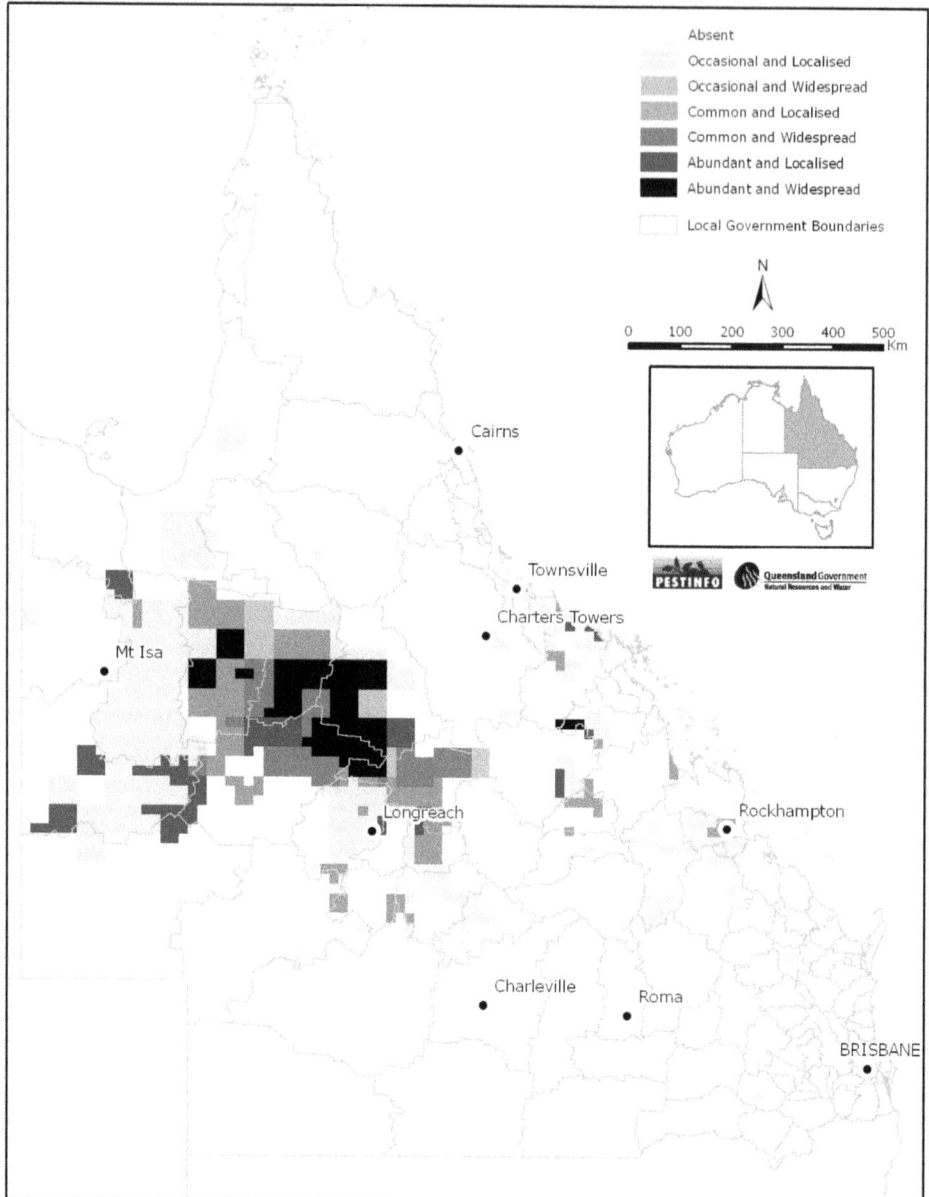

Fig. 2.1 Distribution of *Acacia nilotica* ssp. *indica* in Queensland, Australia.

2.2 Detrimental effects

Acacia nilotica is a major environmental problem in the grazing areas of western Queensland and costs primary producers AU$9m per year by decreasing pasture

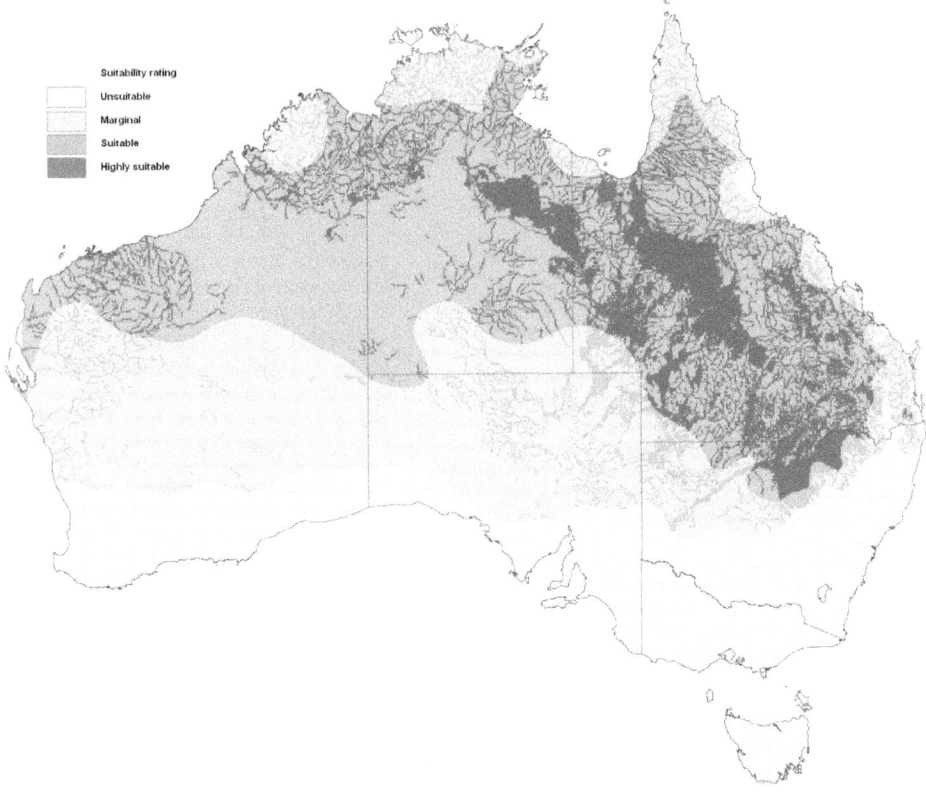

Fig. 2.2 Potential distribution range of *Acacia nilotica* ssp. *indica* in Australia.

production and hindering the mustering of livestock. In such areas, *A. nilotica* forms impenetrable thorny thickets, competes with native pasture species, facilitates the replacement of native grasses (e.g. *Astrebla* spp.) with less stable, short-lived plants, prevents the growth of native plants beneath the canopy, and restricts stock access to watercourses. In the Mitchell grass downs, *A. nilotica* poses a threat to nearly 25 rare and threatened animal species, including the endangered carnivorous marsupial Julia Creek dunnart (*Sminthopsis douglasi* Archer), and two endangered plant communities, by displacing grasslands (Spies and March, 2004).

2.3 Beneficial effects

Acacia nilotica is a multipurpose tree widely distributed, as well as cultivated, in the Indian subcontinent, the Middle East, and Africa (Dwivedi, 1993; Table 2.1). In its native ranges, *A. nilotica* is widely used in agroforestry, social forestry, reclamation of wastelands, and rehabilitation of degraded forests. In traditional agroforestry systems,

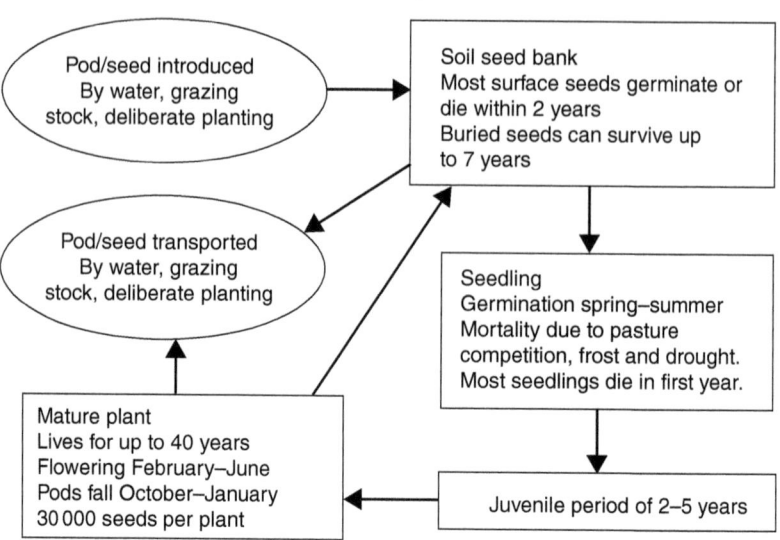

Fig. 2.3 Life cycle of *Acacia nilotica* ssp. *indica* in Australia.

A. nilotica provides fuel, fodder, gum, tannin, and timber (e.g. Puri *et al.*, 1994; Pandey *et al.*, 1999; Pandey and Sharma, 2003). Root nodulations in this species help in biological nitrogen fixation (Dreyfus and Dommergues, 1981) and enhance soil fertility (Pandey and Sharma, 2003).

Acacia nilotica was introduced to various countries including Australia, Indonesia, Iran, Iraq, Nepal, New Caledonia, Vietnam, and the West Indies as a multipurpose tree (Table 2.1). In Queensland, Australia, at low densities, *A. nilotica* provides shade, and leaves and pods are used as fodder for sheep during dry periods, thus increasing stock productivity (Carter and Cowan, 1988). *Acacia nilotica* also provides a nitrogen supplement to Mitchell grasslands.

2.4 Why biological control?

Mechanical and herbicide treatments are available to manage this weed (Jeffrey 1995; Spies and March, 2004), especially in areas with low-density infestations, but their use is not always economical (Mooy *et al.*, 1992). *Acacia nilotica* is not susceptible to fire, but fire has the potential to reduce seed banks and to promote seed germination (Radford *et al.*, 2001a), thereby allowing follow-up chemical control. Though livestock has limited impact on *A. nilotica* populations (Radford *et al.*, 2002), camels (Spies and March, 2004) and goats (Tiver *et al.*, 2001), in conjunction with traditional methods, have been shown to reduce the cost of control. Several species of insects feed on *A. nilotica* in Australia, but none of them has any major impact (Palmer *et al.*, 2005). Hence, classical biological control, a low cost and permanent alternative, is considered as a viable option for the long-term, sustainable management of *A. nilotica*.

Table 2.1 *Native and introduced ranges of subspecies of Acacia nilotica*

Subspecies	Native range	Introduced countries
A. nilotica ssp. *indica* (Benth.) Brenan	Asia (India, Pakistan, Yemen, Oman, and Myanmar)	Angola, Australia, Ethiopia, Indonesia, Iran, Iraq, Nepal, New Caledonia, Somalia, Tanzania, and Vietnam
A. nilotica ssp. *cupressiformis* (J. Stewart) Ali & Faruqi	Indian subcontinent (India and Pakistan)	Nil
A. nilotica ssp. *adstringens* (Schumach. & Thonn.) Roberty	Africa (Algeria, Cameroon, Chad, Egypt, Gambia, Libya, Mali, Nigeria, Senegal, Sudan, and Somalia).	Iran, Libya, and West Indies India and Pakistan (Hybrids??)
A. nilotica ssp. *subalata* (Vatke) Brenan	East Africa (Sudan, Ethiopia, Uganda, Kenya, and Tanzania)	India, Pakistan, and Sri Lanka??
A. nilotica ssp. *leiocarpa* Brenan	East Africa (Kenya, Somalia, Ethiopia, and Tanzania)	Nil
A. nilotica ssp. *kraussiana* (Benth.) Brenan	Africa (Angola, Botswana, Malawi, Mozambique, Namibia, South Africa-Natal & Transvaal, Zambia, Zimbabwe, Swaziland, and Tanzania)	Ethiopia, Yemen, and Oman
A. nilotica ssp. *tomentosa* (Benth.) Brenan	Africa (Senegal, Mali, Ivory Coast, Ghana, Niger, Nigeria, Sudan, and Ethiopia)	India
A. nilotica ssp. *nilotica* (L.) Willd. ex Del.	Africa (Cameroon, Chad, Egypt, Ethiopia, Sudan, Mali, Nigeria, Niger, Senegal, and Sudan). Asia (Iran, Iraq, Oman, Saudi Arabia, and Yemen)	Tanzania and Zanzibar
A. nilotica ssp. *hemispherica* Ali & Faruqi	Pakistan	Nil

2.5 Native-range studies

Acacia nilotica has a broad native range including much of Africa, the Middle East, and the Indian subcontinent (Dwivedi, 1993). It is a polytypic species exhibiting significant morphological (Brenan, 1983; Dwivedi, 1993), phenological (Marohasy, 1992) and genetic diversity (Nongonierma, 1976; Nagarajan, 2001; Nagarajan *et al.*, 2001; Wardill *et al.*, 2005a, 2005b). Nine recognized subspecies exist in its native range (Table 2.1), with each subspecies having a distinct geographic range (Brenan, 1983). All the subspecies of *A. nilotica* are

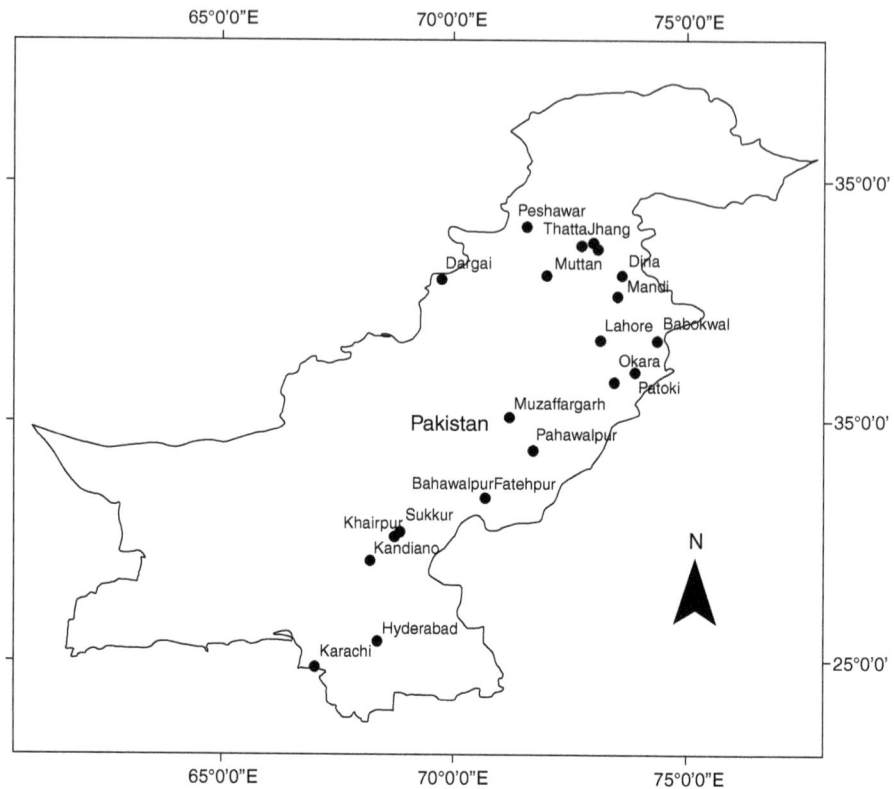

Fig. 2.4 Map of Pakistan showing sites surveyed for biological control insects.

tetraploids ($2n = 52$); except *A. nilotica* ssp. *nilotica* ($2n = 104$) and *A. nilotica* ssp. *tomentosa* ($2n = 208$) (Nongonierma, 1976). Hybridization between different subspecies is known to occur in areas where they are sympatric (Ali and Qaiser, 1980, 1992; Khatoon and Ali, 2006).

The subspecies of *A. nilotica* are genetically distinct except for *A. nilotica* ssp. *cupressiformis*, which is genetically similar to *A. nilotica* ssp. *indica*, even though they appear distinct morphologically (Wardill *et al.*, 2005a). Genetic (Wardill *et al.*, 2005a), morphological (Brenan, 1983) and biochemical (Hannan-Jones, 1999) studies indicate that the invasive *A. nilotica* population in Australia is *A. nilotica* ssp. *indica*, which is native to India and Pakistan (Wardill *et al.*, 2005a). Critical genetic differences between *Acacia nilotica* ssp. *indica* populations from different regions within its native range (Ginwal and Gera, 2000; Ginwal and Mandal, 2004) exist.

Biological control of *A. nilotica* was initiated in the early 1980s, with surveys conducted on *A. nilotica* ssp. *indica* in Pakistan (Fig. 2.4, Mohyuddin, 1981, 1986), on *A. nilotica* ssp. *subalata* and *A. nilotica* ssp. *leiocarpa* in Kenya (Fig. 2.5; Marohasy, 1992, 1995) and *A. nilotica* ssp. *kraussiana* in South Africa (Fig. 2.6; Stals, 1997). Surveys in Pakistan covering different climatic zones (Fig. 2.4), including subtropical hot

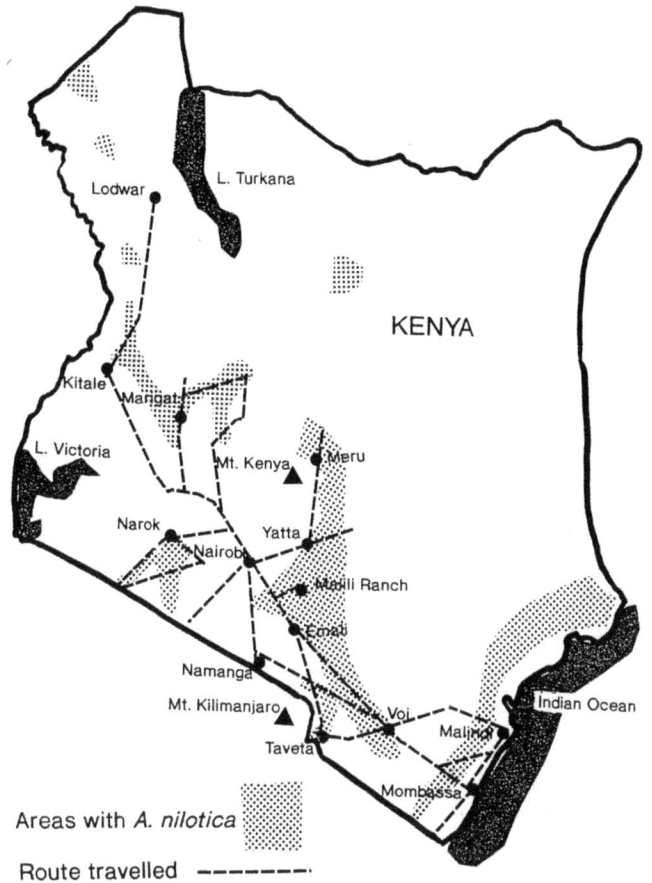

Fig. 2.5 Map of Kenya showing sites surveyed for biological control insects.

arid regions, brought to light 71 phytophagous insect species associated with *A. nilotica* ssp. *indica* in 1980–1985 (Mohyuddin, 1981, 1986). Two of the species, *Bruchidius sahlbergi* Schilsky (Coleoptera: Chrysomelidae) and *Cuphodes profluens* (Meyrick) (Lepidoptera: Gracillariidae), were introduced into Australia (Mohyuddin, 1981, 1986), but only *B. sahlbergi* has become widely established (Wilson, 1985). Surveys in Kenya in 1989–1992 (Fig. 2.5) brought to light 86 species of insects feeding on *A. nilotica* ssp. *subalata* and *A. nilotica* ssp. *leiocarpa* (Robertson, 1987; Marohasy, 1992, 1995), of which six species were tested for their host specificity (Table 2.2). No overlap of the insect fauna feeding on *A. nilotica* ssp. *indica* in Pakistan with that feeding on *A. nilotica* ssp. *subalata* and *A. nilotica* ssp. *leiocarpa* in Kenya existed, and only three of the insects found in Pakistan also occurred in Kenya (Marohasy, 1995). In South Africa (Fig. 2.6), although more than 400 species of phytophagous insects associated with *A. nilotica* ssp. *kraussiana* were recorded in 1996–1997 (Stals, 1997; Witt *et al.*, 2005, 2006), only

Table 2.2 *Biological control agents host tested for* Acacia nilotica *ssp. indica*

Insect species	Source country	Year released	Establishment status
Coleoptera: Chrysomelidae			
Bruchidius sahlbergi Schilsky	Pakistan	1982	Widespread
Homichloda barkeri(Jacoby)	Kenya	1996	No establishment to date
Lepidoptera: Gracillariidae			
Cuphodes profluens (Meyrick)	Pakistan	1983	Currently not established
Lepidoptera: Geometridae			
Chiasmia assimilis (Warren)	Kenya and South Africa	1999	Established in coastal areas
Chiasmia inconspicua (Warren)	Kenya	1998	Not established
Isturgia deeraria(Walker)	Kenya	Not released	
Isturgia disputaria(Guenée)	India	Not released	
Lepidoptera: Noctuidae			
Cometaster pyrula(Hopffer)	Kenya	2004	Too early to know
Homoptera: Psyllidae			
Acizzia melanocephala (Burckhardt & Mifsud)	Kenya	Not released	

Chiasmia assimilis (Warren) (Lepidoptera: Geometridae) was introduced in Australia, an agent that was obtained from Kenya earlier (Table 2.2).

In India, information on insects and plant pathogens associated with *A. nilotica* (e.g. Bhasin and Roonwall, 1954; Beeson, 1961; Pillai and Gopi, 1990; Dwivedi, 1993; Pillai *et al.*, 1995; Kapoor *et al.*, 2004) has been gathered from the perspective of itemizing forestry and nursery pests. India is the only country within the native range nation yet to be surveyed systematically for potential biological control agents (Dhileepan *et al.*, 2006).

2.6 Biological control agents

Thus far, six species of insects have been released in Australia (Table 2.2), but only two of them, *Bruchidius sahlbergi* from Pakistan (Wilson, 1985; Palmer, 1996) and *Chiasmia assimilis* from Kenya and South Africa (Lockett and Palmer, 2004), are established.

2.6.1 **Bruchidius sahlbergi** *Schilsky (Coleoptera: Chrysomelidae)*

In Pakistan, *B. sahlbergi* is widespread across diverse climatic conditions, and occurs in all areas wherever *A. nilotica* grows (Mohyuddin, 1981, 1986). After host-specificity

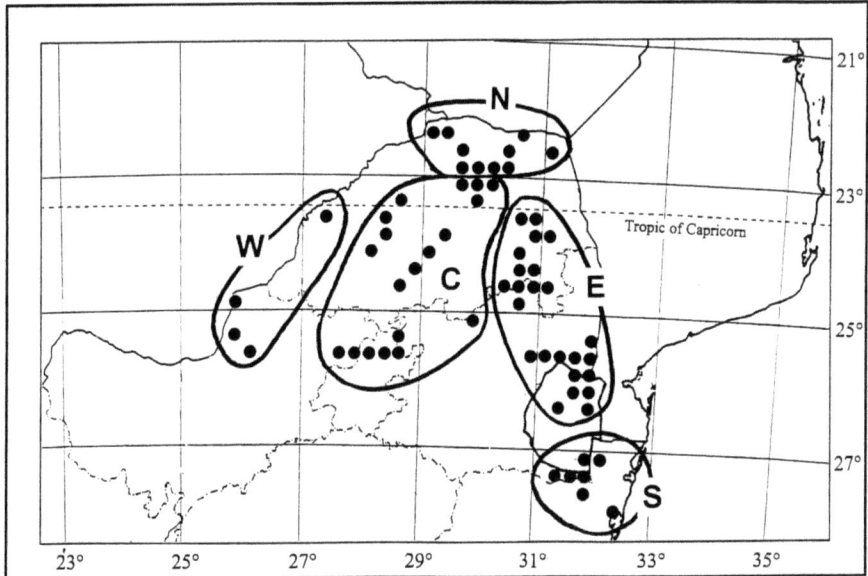

Fig. 2.6 Map of northern parts of South Africa showing sampling sites in northern (N), western (W), central (C), eastern (E) and southern (S) regions.

tests, it was released in Australia in 1982 (Mohyuddin, 1986) and became established in 80% of the sites within four years (Wilson, 1985; Marohasy, 1995). Now this agent occurs in all areas where *A. nilotica* ssp. *indica* occurs, including western Queensland where large populations of the weed occur.

2.6.2 Cuphodes profluens *(Meyrick) (Lepidoptera: Gracillariidae)*

This tip-boring moth, *Cuphodes profluens*, is native to Pakistan, where it occurs in all areas with *A. nilotica* (Mohyuddin, 1986). The moth is specific to *A. nilotica*, and causes severe damage to juveniles and adult shoots (Mohyuddin, 1986). Introduced in 1983, the moth was believed to have become established in a release site in coastal Queensland, but not in western Queensland. However, attempts to eradicate *A. nilotica* ssp. *indica* in coastal regions where this agent was believed to have become established resulted in the extermination of *C. profluens* in these areas (Mackey, 1997).

2.6.3 Homichloda barkeri *(Jacoby) (Coleoptera: Chrysomelidae)*

In Kenya, this beetle was observed defoliating *A. nilotica* ssp. *indica*, *A. nilotica* ssp. *leiocarpa* and *A. nilotica* ssp. *subalata* (Marohasy, 1994, 1995; Cox, 1997). After appropriate host-specificity tests, *Homichloda barkeri* was released in Australia in 1994 (Marohasy, 1994; Cox, 1997). Several factors, such as egg diapause (Marohasy, 1994; Nahrung and Marohasy, 1997), frequency of maternal matings (Nahrung and Merritt, 1999a), specific moisture regime requirements (Nahrung and Merritt, 1999b) and a decline in the number of adults in the laboratory colony due to inbreeding (Lockett and Palmer, 2003), hampered the large-scale release of this agent. After introducing add-itional insects from Kenya and standardizing the laboratory rearing procedures, adequate numbers of insects were released in both coastal and inland regions from 1996 to 1999 (Lockett and Palmer, 2003). However, no evidence of field establishment of *H. barkeri* is presently available.

2.6.4 Chiasmia inconspicua *(Warren) (Lepidoptera: Geometridae)*

Chiasmia inconspicua is prevalent on both *A. nilotica* ssp. *leiocarpa* and *A. nilotica* ssp. *subalata* in Kenya and Tanzania (Marohasy, 1992, 1995). Host-specificity tests conducted using insects collected from Kenya indicated that this moth has a limited host range and hence was approved for field release (Palmer *et al.*, 2007). Mass rearing of *C. inconspicua* commenced in November 1998, and it was released throughout north-western and coastal Queensland. No field establishment has resulted to date.

2.6.5 Chiasmia assimilis *(Warren) (Lepidoptera: Geometridae)*

Chiasmia assimilis, widespread on diverse subspecies of *A. nilotica* in Africa, was obtained from *A. nilotica* ssp. *leiocarpa* and *A. nilotica* ssp. *subalata* populations in Kenya in 1997 for host-specificity tests. In the laboratory, *C. assimilis* was found to feed on *A. nilotica* ssp. *indica*, and complete larval development on other species of *Acacia* (Palmer *et al.*, 2007). However, it was considered "safe" and was approved for field release. Mass rearing of *C. assimilis* commenced in 1999, and field releases were made throughout northwestern Queensland (Palmer *et al.*, 2007). Additional insects of this species were sourced from the Republic of South Africa (Lockett and Palmer, 2004), and a progeny of these insects was released in both coastal and western Queensland. The leaf-feeding moth became established in a few of the coastal sites in northern Queensland (Lockett and Palmer, 2004; Palmer *et al.*, 2007), but not in the Mitchell grass downs. Moreover, it is also not known whether progeny of the Kenyan or the South African population have become established.

2.6.6 Cometaster pyrula *(Hopffer) (Lepidoptera: Noctuidae)*

Cometaster pyrula, endemic to Africa (Mozambique, Zimbabwe, Botswana, Zambia, Malawi, Tanzania, and South Africa), was collected from *A. nilotica* ssp. *kraussiana* in South Africa (Stals, 1997; Palmer and Senaratne, 2007). Host-specificity tests confirmed that it is specific to *A. nilotica* ssp. *kraussiana* and *A. nilotica* ssp. *indica*, but larval survival on subspecies *indica* (45 ± 7.1%) was less than on subspecies *kraussiana* (70 ± 4.9%). The release of this agent in Australia commenced in 2004 (Palmer and Senaratne, 2007), but the implications of the reduced larval survival for its establishment and effectiveness are unknown presently. In view of this insect's specific climate requirements, as predicted from climate-matching models (Senaratne *et al.*, 2006), release efforts are currently focused at the more appropriate coastal regions (Palmer and Senaratne, 2007), and to a lesser extent in the inland region where *A. nilotica* ssp. *indica* is a more serious problem.

2.6.7 Acizzia melanocephala *(Burckhardt and Mifsud) (Hemiptera: Psyllidae)*

This psyllid was collected from Kenya in 1997, from a mixed stand of *A. nilotica* ssp. *leiocarpa* and *A. nilotica* ssp. *subalata* (Palmer and Witt, 2006). However, in host-specificity tests, insects collected from both Kenya and South Africa showed a high degree of specificity towards the *A. nilotica* subspecies native to Africa, and could not complete nymphal development on *A. nilotica* ssp. *indica* (Palmer and Witt, 2006). Hence, this insect was not considered as a suitable biological control agent for *A. nilotica* ssp. *indica* populations.

2.6.8 Isturgia deeraria *(Walker) (Lepidoptera: Geometridae)*

This insect was found on both *A. nilotica* ssp. *subalata* and *A. nilotica* ssp. *leiocarpa* in Kenya (Marohasy, 1995). In host-specificity tests, *Isturgia deeraria* completed larval development on 13 species of *Acacia* native to Australia (Palmer and McLennan, 2006), confirming earlier reports that this agent has a wide host range spanning on several leguminous species (Platt, 1921; Taylor, 1953; Kruger, 2001). Hence, this agent was not considered for release.

2.6.9 Isturgia disputaria *(Guenée) (Lepidoptera: Geometridae)*

This geometrid moth occurs on various subspecies of *A. nilotica* both in Africa and the Indian subcontinent. In Africa, it has also been collected on *Acacia* species native to Australia. Host-specificity tests on *Isturgia disputaria* collected from a mixed stand of *A. nilotica* ssp. *indica* and *A. nilotica* ssp. *subalata* in India confirmed the ability of this moth to complete larval development on some of the Australian *Acacia* species, and hence, no progress was made to release this agent in Australia (Palmer, 2004).

2.7 Impact of biological control

The biological control program, costing so far about AU$5.3m (Page and Lacey, 2006), has resulted in the establishment of only two agents in Australia, of which only the seed beetle occurs in all areas with *A. nilotica* ssp. *indica*, including western Queensland. However, the impact of the seed beetle on seed banks and consequently on *A. nilotica* ssp. *indica* populations has been insignificant (Marohasy, 1995; Mackey, 1997; Radford *et al.*, 2001a). As a result, the *A. nilotica* biological control program has not provided any economic benefit to date (Page and Lacey, 2006). The need for effective biological control agents continues to be a priority in the Mitchell grass downs, where the introduced agents have neither established nor been effective.

2.8 Factors influencing biological control

2.8.1 Establishment and abundance

Among the six insect species introduced so far, only the seed predator *B. sahlbergi* from Pakistan occurs in all areas where *A. nilotica* occurs in Australia, including western Queensland where the largest populations of the tree occur. However, the population densities of *B. sahlbergi* in the field remain very low (Marohasy, 1993; Mackey, 1997) and not sufficient to cause large-scale reduction in seed banks (Kriticos *et al.*, 1999a; Radford *et al.*, 2001b, 2001c) and prevent the spread of the *A. nilotica* ssp. *indica* populations (Carter and Cowan, 1988; Brown and Carter, 1998). Among the insects introduced from Africa, only one species became established and caused intense defoliation, but only along the coast, and not inland. Several of these insects are more suitable to coastal regions and they are less likely to establish in the hotter and drier climatic conditions that usually prevail in the Mitchell grass downs of western Queensland (Lockett and Palmer, 2003; Senaratne *et al.*, 2006; Palmer *et al.*, 2007).

2.8.2 Genetic constraints

Performance of insects and plant pathogens is likely to differ across plant genotypes (Dhileepan *et al.*, 2006). The nonestablishment or reduced population densities observed are possibly due to the insect"s specificity or preference for diverse subspecies of *A. nilotica* in their native ranges. For example, all insects prioritized in South Africa were collected from the subspecies *kraussiana*, whereas the insects from Kenya were collected from the mixed stand of subspecies *leiocarpa* and *subalata*. Among them, the psyllid *A. melanocephala* has an extremely narrow host range, confined to a few African subspecies of *A. nilotica*. The target subspecies *A. nilotica* ssp. *indica* in Australia thus therefore turned out to be an unsuitable host (Palmer and Witt, 2006). Similarly, *Cometaster pyrula*, collected from subspecies *kraussiana* in South Africa, showed lower survival rates and poorer development on subspecies *indica* than on subspecies *kraussiana* (Palmer and Senaratne, 2007). The seed beetle *Bruchidius grandemaculatus* (Pic.) inflicted greater damage (73% of seeds) in ssp. *indica* than in ssp. *subalata* (7% of seeds) (Marohasy, 1992). This finding supports the need to search for coevolved biological control agents in India and Pakistan (Anonymous, 1995; Wardill *et al.*, 2005a, 2005b) specific to subspecies *indica*.

2.8.3 Plant response to herbivory

Weaknesses of the target weed can be exploited to focus the search for effective agents, thereby enhancing the success rate of biological control efforts (Kriticos *et al.*, 1999a; Raghu *et al.*, 2006). Since *A. nilotica* populations are not seed limited, flower and seed feeding agents are believed not to have a major impact as weed biological control agents (Marohasy, 1992, 1995; Kriticos *et al.*, 1999a). This conclusion is supported by field studies in Kenya where flower-feeding insects caused insignificant damage, resulting in no negative impact on seed pod development (Marohasy, 1992). In Queensland, where *A. nilotica* ssp. *indica* is a serious problem in cattle grazing areas, a result of the ineffectiveness of mature-seed-feeding insects is that cattle consume large volumes of mature pods/seeds, which pass through the animals and remain viable in their dung, but escape from insect predation (Marohasy, 1992; Kriticos *et al.*, 1999a). Instead, seedlings and juveniles appear to be the most susceptible life stages for control (Kriticos *et al.*, 1999a). Hence, need to prioritize the selection of biological control agents that have the potential to target seedlings and juveniles appear imperative. In coastal areas, outbreak of *Chiasmia assimilis* resulted in 70–90% defoliation in seedlings, but resulted in reduced plant vigour only in those occurring below the canopy and not in those in open areas (C. J. Lockett and K. Dhileepan, unpublished data, 2006).

2.8.4 Abiotic factors

In classical weed biological control, success or failure of agents is often determined by climatic factors (e.g. McEvoy and Coombs, 1999). Southern Africa (Namibia, Botswana, southern Zimbabwe, northern South Africa, and Mozambique) appears to be the region climatically highly similar to northern Queensland (Marohasy, 1992; Senaratne *et al.*, 2006). However, biological control agents from South Africa and Kenya have not established to date in the arid Mitchell grass downs due to unsuitable climatic conditions (Lockett and Palmer, 2003; Senaratne *et al.*, 2006; Palmer *et al.*, 2007) and also possibly due to mismatched plant genotypes. As referred to earlier, the only established insect which is widespread throughout the *A. nilotica* infestations, including those in western Queensland, is the seed beetle from *A. nilotica* ssp. *indica* in Pakistan.

2.9 Future research

2.9.1 Plant genotypes

The probability of finding additional agents from Pakistan is low given the extensive nature of previous surveys. In India, *A. nilotica* ssp. *indica* occurs in greater habitat diversity, along with other subspecies (*A. nilotica* ssp. *subalata, A. nilotica* ssp. *cupressiformis* and *A. nilotica* ssp. *adstringens*) (Dwivedi, 1993). Therefore, future research should focus on surveys in India. Among the various subspecies of *A. nilotica* in India, *A. nilotica* ssp. *indica* is the most prevalent, occurring throughout India. It co-occurs with *A. nilotica* ssp. *cupressiformis* in Rajasthan and Karnataka States, and with *A. nilotica* ssp. *subalata* in Tamil Nadu State (Dhileepan *et al.*, 2006). Co-occurrence of various subspecies makes India an appropriate region for identifying insects and plant pathogens that are likely to be specialists on the subspecies in Australia.

2.9.2 Climatic suitability

Agents from areas with climatic conditions similar to those of the Mitchell grass downs, and the same subspecies (*A. nilotica* ssp. *indica*) are more likely to succeed as effective biological control agents in Australia (McFadyen, 1991; Marohasy, 1995). A range of climate modeling approaches has been applied to *A. nilotica* ssp. *indica*, and many suggest that regions within Pakistan and India are likely to yield potential agents that are climatically adapted to the Mitchell grasslands in Australia. Matching the climatic conditions of localities in western Queensland where *A. nilotica* ssp. *indica* is invasive (i.e. Winton, Barcaldine, and Hughenden) with regions in India indicates that most of the areas in India are suitable, with Rajasthan State being most suitable climatically for exploration (Dhileepan *et al.*, 2006). Climate modeling based on a hypothetical insect suitable to inland Queensland (e.g. Mitchell grass downs), but not to coastal Queensland, gives a more conservative prediction that India's northwest region is climatically the suitable for exploration (Senaratne *et al.*, 2006).

2.9.3 Plant response to herbivory

So far, plant responses to herbivory, or identifying the weak links in plant population dynamics, has not been used to guide insect selection, despite detailed ecological studies (Brown and Carter, 1998; Radford *et al.*, 2001b, 2001c, 2002) and the availability of a detailed demographic model of this species in Australia (Kriticos *et al.*, 1999b, 1999c, 2003a, 2003b). Future research should focus on the use of such plant-based approaches as "predictive" filters for insect prioritization (Fig. 2.7). Prerelease evaluation of the efficacy of potential biological control agents can also be used to prioritize agents in the native range (e.g. Witt *et al.*, 2005, 2006). This information, along with the results from ongoing simulated herbivory and field-exclusion trials in Australia, will be useful in assisting in agent prioritization.

2.9.4 Field host range

Surveys of field host range of potential agents in the native geographic range should be included in all future explorations to rule out potential agents prior to more expensive

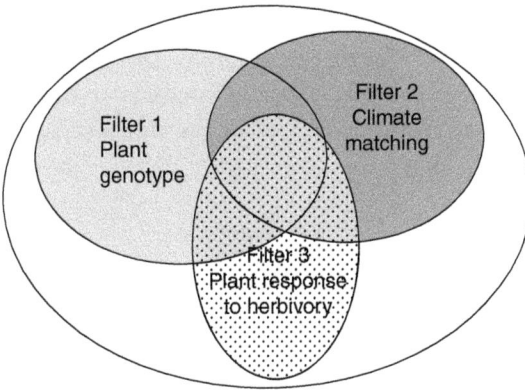

Fig. 2.7 Filters in biological control agent prioritization process.

host-specificity testing as well as to understand the ecological host range of the insect in its native range. Co-occurrence of several native *A. nilotica* subspecies and other native and nonnative *Acacia* species (including the species native to Australia) highlights the advantage of conducting surveys in India where the field host-specificity of potential agents could be determined. In Rajasthan, which has climatic conditions (hot and dry) similar to western Queensland, two subspecies of *A. nilotica* (*indica* and *cuprissiformis*) co-occur with other *Acacia* species (*Acacia senegal* (L.) Willd., *Acacia leucophloea* (Roxb.) Willd. and *Acacia tortilis* (Forsk.) Hayne) (Dhileepan *et al.*, 2006). In localities within the state of Tamil Nadu with hot and dry climate conditions, *A. nilotica* subspecies (*indica* and *subalata*) co-occur with other *Acacia* species (Dhileepan *et al.*, 2006). Testing host-use pattern by specialist insects across such climatic gradients will allow us to determine which agents are likely to be effective in the different climatic regions of Australia.

2.10 Conclusion

Adopting a more systematic approach to native-range surveys – one that incorporates plant genotype matching and climate similarities as filters, in conjunction with agent prioritization that incorporates knowledge from plant population ecology and plant–herbivore interactions – will make agent selection decisions explicit. This, in turn, will allow more rigorous evaluations of agent performance and better understanding of the reasons for success or failure of agents in weed biological control.

Acknowledgments

I thank Calvert Moya, and Nathan March for the distribution maps; Wilmot Senaratne for the CLIMEX modeling; Syed Ahmed (Arid Forest Research Institute, Jodhpur), A. Balu (Institute of Forest Genetics and Tree Breeding, Coimbatore), Sunil Puri, S. S. Shaw

(Indira Gandhi Agricultural University, Raipur), Shilesh Jadhav (Pandit Ravishankar University, Raipur) and B. S. Bhumannavar (Project Directorate of Biological Control, Bangalore) for their help in surveys in India. Comments on the earlier versions of the manuscript by Bill Palmer and Dane Panetta are gratefully acknowledged.

References

Ali, S. I. and Qaiser, M. (1980). Hybridisation in *Acacia nilotica* complex. *Botanical Journal of Linnean Society*, **80**, 69–77.

Ali, S. I. and Qaiser, M. (1992). Hybridisation between *Acacia nilotica* subsp. *indica* and subsp. *cupressiformis*. *Pakistan Journal of Botany*, **24**, 88–94.

Anonymous (1995). *Prickly Acacia Biocontrol Workshop*. Sherwood, Brisbane, Australia: Alan Fletcher Research Station, Queensland Department of Natural Resources, 13 pp.

Beeson, C. F. C. (1961). *The Ecology and Control of the Forest Insects of India and Neighbouring Countries*, 2nd edn. Dehra Dun, India: Government of India, 767 pp.

Bhasin, G. H. and Roonwall, M. L. (1954). A list of insect pests of forest plants in India and the adjacent countries. 2. List of insect pests of plant genera 'A' (*Alberia* to *Azima*). *Indian Forest Bulletin, Entomology*, **171**, 5–93.

Bolton, M. P. (1989). *Biology and ecology of prickly acacia*. Unpublished report. Richmond Prickly Acacia Field Day, 19 September 1989. Brisbane Australia: Queensland Department of Lands, pp. 21–25.

Brenan, J. P. M. (1983). Present taxonomy of four species of *Acacia* (*A. albida*, *A. senegal*, *A. nilotica* and *A. tortilis*). In *Manual on Taxonomy of Acacia Species*. Rome, Italy: Food and Agricultural Organisation of the United Nations, pp. 20–35.

Brown, J. R. and Carter, J. (1998). Spatial and temporal patterns of exotic shrub invasion in an Australian tropical grassland. *Landscape Ecology*, **13**, 93–102.

Carter, J. O. and Cowan, D. C. (1988). Phenology of *Acacia nilotica* ssp. *indica* (Benth.) Brenan. In *Proceedings of the Fifth Biennial Conference*, ed. P. E. Novelly. Longreach, Queensland, Australia: Australian Rangelands Society, pp. 9–12.

Cox, M. L. (1997). *Homichloda barkeri* (Jacoby) (Coleoptera: Chrysomelidae: Alticinae), a candidate agent for the biocontrol of prickly acacia, *Acacia nilotica* (Mimosaceae) in Australia. *Journal of Natural History*, **31**, 935–964.

Dhileepan, K., Wilmot, K. A. D. W. and Raghu, S. (2006). A systematic approach to biological control agent exploration and selection for prickly acacia (*Acacia nilotica* ssp. *indica*). *Australian Journal of Entomology*, **45**, 302–306.

Dreyfus, B. L. and Dommergues, Y. R. (1981). Nodulation of *Acacia* species by fast- and slow-growing tropical strains of *Rhizobium*. *Applied and Environmental Microbiology*, **41**, 97–99.

Dwivedi, A. P. (1993). *Babul* (Acacia nilotica): *A Multipurpose Tree of Dry Areas*. Jodhpur, India: Arid Forest Research Institute, Indian Council of Forestry Research and Education, 226 pp.

Ginwal, H. S. and Gera, M. (2000). Genetic variation in seed germination and growth performance of 12 *Acacia nilotica* provenances in India. *Journal of Tropical Forest Science*, **12**, 286–297.

Ginwal, H. S. and Mandal, A. K. (2004). Variation in growth performance of *Acacia nilotica* Willd. ex Del. provenances of wide geographical origin: six year results. *Silvae Genetica*, **53**, 264–269.

Hannan-Jones, M. A. (1999). Multivariate variation in leaf phenolics in the subspecies of *Acacia nilotica* (L.) Willd. ex Del. (Mimosaceae). In *Proceedings of the 12th Australian Weeds Conference*, ed. A. C Bishop, M. Boersma and C. D. Barnes, Hobart, Tasmania: Tasmanian Weed Society, pp. 601–604.

Jeffrey, P. L. (1995). Prickly acacia. In *Exotic Woody Weeds and Their Control in North West Queensland*, ed. N. March, Brisbare Australia: Queensland Department of Lands, pp. 3–9.

Kapoor, S., Harsh, N. S. K. and Sharma, S. K. (2004). A new wilt disease of *Acacia nilotica* caused by *Fusarium oxysporum*. *Journal of Tropical Forest Science*, **16**, 453–462.

Khatoon, S. and Ali, S. I. (2006). Hybridization in *Acacia nilotica* complex in Indo-Pakistan subcontinent: cytological evidence. *Pakistan Journal of Botany*, **38**, 63–66.

Kriticos, D., Brown, J., Maywald, G. F., *et al.* (2003b). SPAnDX: a process-based population dynamics model to explore management and climatic change impacts on an invasive alien plant, *Acacia nilotica*. *Ecological Modelling*, **163**, 187–208.

Kriticos, D., Brown, J., Radford, I. and Nicholas, M. (1999a). Plant population ecology and biological control: *Acacia nilotica* as a case study. *Biological Control*, **16**, 230–239.

Kriticos, D., Brown, J., Radford, I. and Nicholas, M. (1999b). A population model of *Acacia nilotica*: thresholds for management, In *Proceedings of the VI International Rangeland Congress*, ed. D. Eldridge and D. Freudenberger. Sydney, Australia: International Rangeland Congress, pp. 599–600.

Kriticos, D., Brown, J., Radford, I. and Nicholas, M. (1999c). A population model of *Acacia nilotica*: a tool for exploring weed management and the effects of climate change. In *Proceedings of the VI International Rangeland Congress*, ed. D. Eldridge and D. Freudenberger. Sydney, Australia: International Rangeland Congress, pp. 870–872.

Kriticos, D., Sutherst, R. W., Brown, J. R., Adkins, S. W. and Maywald, G. F. (2003a). Climatic change and the potential distribution of an invasive alien plant: *Acacia nilotica* ssp. indica in Australia. *Journal of Applied Ecology*, **40**, 111–124.

Kruger, M. (2001). A revision of the tribe Macariini (Lepidoptera: Geometridae: Ennominae) of Africa, Madagascar and Arabia. *Bulletin of the Natural History Museum (Entomology)*, **70**, 1–502.

Lockett, C. J. and Palmer, W. A. (2003). Rearing and release of *Homichloda barkeri* (Jacoby) (Coleoptera: Chrysomelidae: Alticinae) for the biological control of prickly acacia, *Acacia nilotica* ssp. *indica* (Mimosaceae) in Australia. *Australian Journal of Entomology*, **42**, 287–293.

Lockett, C. J. and Palmer, W. A. (2004). Biological control of prickly acacia (*Acacia nilotica* ssp. *indica* (Benth.) Brenan): early signs of establishment of an introduced agent. In *Proceedings of the Fourteenth Australian Weeds Conference*, ed. B. M. Sindel and S. B. Johnson. Sydney, Australia: Weed Society of New South Wales, p. 379.

Mackey, A. P. (1997). The biology of Australian weeds. 29. *Acacia nilotica* ssp. *indica* (Benth.) Brenan. *Plant Protection Quarterly*, **12**, 7–17.

Marohasy, J. (1992). *Biocontrol of* Acacia nilotica *Using Insects From Kenya*. Final Report to Australian Wool Corporation. Sherwood, Brisbane: Alan Fletcher Research Station, Queensland Department of Lands, 75 pp.

Marohasy, J. (1993). Constraints to the introduction of biological control agents for prickly acacia. In *Proceedings of the 10th Australian Weeds Conference and 14th Asian–Pacific Weed Science Society Conference*, ed. J. T. Swarbrick, C. W. L. Henderson, R. J. Jettner, L. Streit and S. R. Walker. Brisbane, Australia: Weed Society of Queensland, pp. 33–38.

Marohasy, J. (1994). Biology and host specificity of *Weiseana barkeri* (Col.: Chrysomelidae): a biological control agent for *Acacia nilotica* (Mimosaceae). *Entomophaga*, **39**, 335–340.

Marohasy, J. (1995). Prospects for the biological control of prickly acacia, *Acacia nilotica* (L.) Willd. ex Del. (Mimosaceae) in Australia. *Plant Protection Quarterly*, **10**, 24–31.

McEvoy, P. B. and Coombs, E. M. (1999). Biological control of plant invaders: regional patterns, field experiments and structured population models. *Ecological Applications*, **9**, 387–40.

McFadyen, R. E. (1991). Climate modelling and the biological control of weeds: one view. *Plant Protection Quarterly*, **6**, 14–15.

Mohyuddin, A. I. (1981). Phytophages associated with *Acacia nilotica* in Pakistan and possibilities of their introduction into Australia, In *Fifth International Symposium on Biological Control of Weeds*, ed. E. S. Delfosse. Melbourne: CSIRO, pp. 161–166.

Mohyuddin, A. I. (1986). *Investigations on the Natural Enemies of* Acacia nilotica *in Pakistan*. Final Report. Rawalpindi, Pakistan: Commonwealth Institute of Biological Control, 116 pp.

Mooy, L. M., Scanlan, J. C., Bolton, M. P. and Dorney, W. J. (1992). Management guidelines derived from ecological studies of prickly acacia (*Acacia nilotica*). In *Proceedings of the First International Weed Control Congress*, Vol. 2, ed. R. G. Richardson. Melbourne, Australia: International Weed Science Society and Weed Science Society of Victoria, pp. 347–349.

Nagarajan, B. (2001). Genetic variation study in *Acacia nilotica* (L.) Willd. subspecies complex using randomly amplified polymorphic DNA markers. In *Proceedings of the International Symposium on Tropical Forestry Research: Challenges in the New Millennium*, ed. R. V. Varma, K. V. Bhat, E. M. Muralidharan and J. K. Sharma. Peechi, India: Kerala Forest Research Institute, pp. 99–105.

Nagarajan, B., Varghese, M., Nicodemus, A. and Mandal, A. K. (2001). Genetic resources of *Acacia nilotica*: patterns of diversity and its implications. In *Forest Genetic Resources: Status, Threats and Conservation Strategies*, ed. U. Shaanker, K. N. Ganeshaiah and K. S. Bawa. New Delhi, India: Oxford and IBH Publishing, pp. 85–97.

Nahrung, H. F. and Marohasy, J. (1997). Maternal frass is necessary for embryonic development in *Weiseana barkeri* Jacoby (Coleoptera: Chrysomelidae). *Australian Journal of Entomology*, **36**, 95–96.

Nahrung, H. F. and Merritt, D. (1999a). Effects of mate availability on female longevity, fecundity and egg development of *Homichloda barkeri* (Jacoby) (Coleoptera: Chrysomelidae). *Coleopterists Bulletin*, **53**: 329–332.

Nahrung, H. F. and Merritt, D. (1999b). Moisture is required for the termination of egg diapause in the chrysomelid beetle, *Homichloda barkeri*. *Entomologia Experimentalis et Applicata*, **93**, 201–207.

Nongonierma, A. (1976). Contribution à l'étude du genre *Acacia* Miller en Afrique occidentale. H. Caractères des inflorescences, et des fleurs. *Bulletin de l'IFAN Serie A*, **38**, 487–657.

Page, A. R. and Lacey, K. L. (2006). *Economic Impact Assessment of Australian Weed Biological Control*. Technical Series 10. Adelaide, Australia: CRC for Australian Weed Management, 150 pp.

Palmer, W. A. (1996). Biological control of prickly acacia in Australia. In *Proceedings of the Eleventh Australian Weeds Conference*, ed. R. C. H. Shepherd. Melbourne: Weed Society of Victoria, pp. 239–242.

Palmer, W. A. (2004). Risk analysis of recent cases of non-target attack by potential biological control agents in Queensland. In *Proceedings of the XI International Symposium on Biological Control of Weeds*, ed. J. M. Cullen, D. T. Briese, D. J. Kriticos, *et al.* Canberra, Australia: CSIRO Entomology, pp. 305–309.

Palmer, W. A. and McLennan, A. (2006). The host range of *Isturgia deeraria*, an insect considered for the biological control of *Acacia nilotica* in Australia. *African Entomology*, **14**, 141–145.

Palmer, W. A. and Senaratne, K. A. D. W. (2007). The host range and biology of *Cometarser pyrula*, a biocontrol agent for *Acacia nilotica* subsp. *indica* in Australia. *BioControl*, **52**, 129–143.

Palmer, W. A. and Witt, A. B. R. (2006). On the host range and biology of *Acizzia melanocephala* (Hemiptera: Psyllidae) an insect rejected for the biological control of *Acacia nilotica* subsp. *indica* (Mimosaceae) in Australia. *African Entomology*, **14**, 387–390.

Palmer, W. A., Lockett, C. J., Senaratne, K. A. D. W. and McLennan, A. (2007). The introduction and release of *Chiasmia inconspicua* and *C. assimilis* (Lepidoptera: Geometridae) for the biological control of *Acacia nilotica* in Australia. *Biological Control*, **41**, 368–378.

Palmer, W. A., Vitelli, M. P. and Donnelly, G. P. (2005). The phytophagous insect fauna associated with *Acacia nilotica* subsp. *indica* (Mimosaceae) in Australia. *Australian Entomologist*, **32**, 173–180.

Pandey, C. B. and Sharma, D. K. (2003). Residual effect of nitrogen on rice productivity following removal of *Acacia nilotica* in traditional agroforestry system in central India. *Agriculture, Ecosystems and Environment*, **96**, 133–139.

Pandey, C. B., Pandya, K. S., Pandey, D. and Sharma, D. K. (1999). Growth and productivity of rice (*Oryza sativa*) as affected by *Acacia nilotica* in a traditional agroforestry system. *Tropical Ecology*, **40**, 109–117.

Pillai, S. R. M. and Gopi, K. C. (1990). Further records of insect pests on *Acacia nilotica* (Linn.) Willd. ex Del. in nurseries and young plantations and the need for control measures. *Indian Journal of Forestry*, **13**, 8–13.

Pillai, S. R. M., Balu, A., Singh, R., *et al.* (1995). *Pest Problems of Babul* (Acacia nilotica *ssp.* indica) and Their Management. Technical Bulletin 1. Coimbatore, India: Institute of Forest Genetics and Tree Breeding, Indian Council of Forestry Research and Education, 16 pp.

Platt, E. E. (1921). List of food plants of some South African lepidopterous larvae. *South African Journal of Natural History*, **3**, 65–138.

Puri, S., Singh, S. and Kumar, A. (1994). Growth and productivity of crops in association with an *Acacia nilotica* tree belt. *Journal of Arid Environments*, **27**, 37–48.

Radford, I. J., Nicholas, D. M. and Brown, J. R. (2001a). Impact of prescribed burning on *Acacia nilotica* seed banks and seedlings in the *Astrebla* grasslands of northern Australia. *Journal of Arid Environments*, **49**, 795–807.

Radford, I. J., Nicholas, D. M. and Brown, J. R. (2001b). Assessment of biological control impact of seed predators on the invasive shrub *Acacia nilotica* (prickly acacia) in Australia. *Biological Control*, **20**, 261–268.

Radford, I. J., Nicholas, D. M., Brown, J. R. and Kriticos, D. J. (2001c). Paddock-scale patterns of seed production and dispersal in the invasive shrub *Acacia nilotica* (Mimosaceae) in northern Australian rangelands. *Austral Ecology*, **26**, 338–348.

Radford, I. J., Nicholas, D. M., Tiver, F., Brown, J. R. and Kriticos, D. J. (2002). Seedling establishment, mortality, tree growth rates and vigour of *Acacia nilotica* in different *Astrebla* grassland habitats: implications for invasion. *Austral Ecology*, **27**, 258–268.

Raghu, S., Wilson, J. R. and Dhileepan, K. (2006). Refining the process of agent selection through understanding plant demography and plant response to herbivory. *Australian Journal of Entomology*, **45**, 307–315.

Robertson, I. A. D. (1987). *A Preliminary Survey of the Natural Enemies of* Acacia nilotica *in Kenya*. Nairobi, Kenya: CAB International Institute of Biological Control, Kenyan Station, 15 pp.

Seigler, D. S. and Ebinger, E. J. (2006). New combinations in the genus *Vachellia* (Fabaceae: Mimosoideae) from the new world. *Phytologia*, **87**, 139–178.

Senaratne, K. A. D. W., Palmer, W. A. and Sutherst, R. W. (2006). Case study: use of CLIMEX modelling to identify prospective areas for exploration to find new biological control agents for prickly acacia. *Australian Journal of Entomology*, **45**, 298–302.

Spies, P. and March, N. (2004). *Prickly Acacia: National Case Studies Manual*. Brisbane, Australia: Natural Heritage Trust and Queensland Department of Natural Resources and Mines, 97 pp.

Stals, R. (1997). *A Survey of Phytophagous Organisms Associated with* Acacia nilotica *in South Africa*. Final Report to the Queensland Department of Natural Resources. Pretoria, South Africa: ARC-Plant Protection Research Institute, 113 pp.

Taylor, J. S. (1953). Notes on Lepidoptera in the Eastern Cape Province, Part III. *Journal of the Entomological Society of South Africa*, **16**, 143–167.

Tiver, F., Nicholas, M., Kriticos, D. and Brown, J. R. (2001). Low density of prickly acacia under sheep grazing in Queensland. *Journal of Range Management*, **54**, 382–389.

Wardill, T. J., Graham, G. C., Manners, A., *et al.* (2005a). The importance of species identity in the biocontrol process: identifying the subspecies of *Acacia nilotica* (Leguminosae: Mimosoideae) by genetic distance and the implications for biological control. *Journal of Biogeography*, **32**, 2145–2159.

Wardill, T. J., Graham, G. C., Playford, J., *et al.* (2005b). The importance of species identity in the biocontrol process. In *Proceedings of the Fourteenth Australian Weeds Conference*, ed. B. M. Sindel and S. B. Johnson. Sydney, Australia: Weed Society of New South Wales, pp. 364–367.

Wilson, B. W. (1985). The biological control of *Acacia nilotica* in Australia. In *Proceedings of the VI International Symposium for Biological Control of Weeds*, ed. E. S. Delfosse. Vancouver, Canada: Agriculture Canada, pp. 849–853.

Witt, A. B. R., Docherty, S. and Palmer, W. A. (2005). Distribution and aspects of the biology of *Chlamisus malvernensis* Bryant (Coleoptera: Chrysomelidae) collected on *Acacia nilotica* (L.) Wild. ex Del. ssp. *kraussiana* (Benth.) Brenan (Mimosaceae) in South Africa. *African Entomology*, **13**, 376–379.

Witt, A. B. R., Steenkamp, H. E. and Palmer, W. A. (2006). Initial screening of *Pseudomalegia* cf. *lefevrei* (Coleoptera: Chrysomelidae), a potential biological control agent for *Acacia nilotica* (L.) Willd. ex Del. ssp. *indica* (Benth.) Brenan (Mimosaceae) in Australia. *African Entomology*, **14**, 384–386.

3

Australian *Acacia* species (Mimosaceae) in South Africa

Fiona Impson, John Hoffmann, and Carien Kleinjan

3.1 Introduction

Acacia, sensu lato, is a large and cosmopolitan genus containing some 1350 described species (Ebinger *et al.*, 2000; Orchard and Wilson, 2001; Maslin *et al.*, 2003; Orchard and Maslin, 2003; www.worldwidewattle.com). This genus is associated in particular with savannas and open woodlands in the tropical, subtropical and warm temperate parts of the world. The highest number of species occurs in Australia, followed by the Americas and Africa, and to a lesser extent in Asia (Ross, 1981; Maslin, 2001). They are also found in parts of Indonesia and islands in the Pacific and Indian oceans, and although possibly present in New Zealand millions of years ago, only naturalized Australian species now remain there.

The acacias all belong to the family Mimosaceae. The taxonomy of the genus has been revised several times over the last 30 years, most notably by Vassal (1972) and Pedley (1978, 1986), and more recently by Seigler and Ebinger (2005) and Seigler *et al.* (2006). The genus was formally described in 1754, being based on the African species *Acacia nilotica* (Linn.). The word "acacia" was derived from the Greek word *akis* meaning "sharp point," referring to the spiny stipules which are characteristic of *A. nilotica*. Historically their classification was based on morphological characteristics of the inflorescences and foliage; however, more recent descriptions consider chemical, pollen, seed, and germination traits (Evans *et al.*, 1977; New, 1984).

The genus currently comprises three subgenera, *Acacia, Aculeiferum*, and *Phyllodineae*, each differing substantially in their geographic distributions and biological characteristics. The majority of the Australian species belong to the *Phyllodineae*, having flattened leaf rachises, or phyllodes, rather than bipinnate leaves, and are taxonomically distinct from their African and American congeners which belong to the subgenera *Acacia* and *Aculeiferum* respectively (Pedley, 1978; Ross, 1979; New, 1984; Maslin *et al.*, 2003). This taxonomic distinction shows that the two groups have dichotomized and acquired distinct features, which are likely to prevent specific herbivorous insects from utilizing species across the two groups (Janzen, 1969, 1971, 1980; Kergoat *et al.*,

Biological Control of Tropical Weeds using Arthropods, ed. R. Muniappan, G. V. P. Reddy, and A. Raman. Published by Cambridge University Press. © Cambridge University Press, 2009.

2007). Despite uncertainty regarding the evolution of the genus, it has been suggested that the early plants which gave rise to the acacias probably originated in the tropical areas of what was the African and South American part of Gondwanaland (Pedley, 1986).

In their natural habitats, *Acacia* species have been widely used by indigenous peoples, probably from the start of human history. As early as Neolithic times they were utilized in Africa and India in construction activities and furniture and tannin production; other uses include the production of fuel, fibers, gum arabic, medicines, dyes, and soaps (New, 1984). Over the years, the useful attributes of many of these plants have led to their deliberate introductions into countries where they would not normally occur, and these introductions have in turn often had negative ecological and environmental impacts.

Almost without exception, the Australian *Acacia* species which have become invasive are fast growing, reaching reproductive maturity within a few years following germination. High levels of seed production, coupled with persistence of seed in the soil, have contributed to their success as invasive species in introduced areas (Milton, 1980; Milton and Hall, 1981; Clark, 1998; Caswell *et al.*, 2003).

The Australian acacias, commonly known as wattles, include approximately 900 species (many not formally described) (New, 1984), and since the early eighteenth century, a number of these species have been introduced into countries in Africa, Europe, North and South America, and Asia (Tutin *et al.*, 1968; Anonymous, 1977; Smith, 1979; Milton, 1980) (Table 3.1). There have been various reasons for these introductions, including use as either agroforestry or ornamental species, stabilization of drift sands, and as human food and animal fodder crops. The intentions for these introductions were well meant, but several of the species are now considered to be invasive and problematic in a number of regions, where they have negative social impacts, affect conservation and agriculture efforts, and deplete water resources. Despite their proven negative impacts, Australian *Acacia* species continue to be planted around the world and now occur in over 80 countries, predominantly as a result of their success as agroforestry species. They have become prominent in the subtropical and tropical regions of Asia, Africa, and central and South America.

Their suitability for marginal land and their vigorous growth, together with a capacity for prolific seed production, has led to their ready use in commercial plantations. There are believed to be over 2m ha of commercial plantings, as well as amenity and "landcare" plantings. Unfortunately, in many instances, the characteristics of species selected for agroforestry favour the potentially invasive species, and potential for weediness should always be included in any risk analysis prior to large-scale planting. Moreover, differences in perceptions of a desirable versus an undesirable species exist in poor and rich nations. In countries where deforestation of natural forests is high, local demand for wood in combination with good market prices frequently drive utilization of invasive species. In developing nations, unless alternatives for invasive species are proposed, the invasive species, which regenerate readily and hence are cost effective, will always be viewed as being useful, and will be promoted enthusiastically.

Table 3.1 *Worldwide distribution records[a] for the invasive Australian* Acacia *species against which biological control projects have been undertaken in South Africa*

Acacia species	Considered invasive in:	Present in:
A. mearnsii	France, Hawaii, India, Israel, Italy, New Zealand, Portugal, Reunion, South Africa, Spain, Uganda, USA. (California)	Argentina, Bolivia, Brazil, Chile, China, Corsica, Cook Islands, Ecuador, Ethiopia, Jamaica, Japan, Kenya, Macaronesia, Madagascar, Madeira, Malawi, Malesia, Morocco, Nepal, Pakistan, Rwanda, Ryukyu Island, Seychelles, Sri Lanka, Sudan, Swaziland, Taiwan, Tanzania, USA (Florida, Texas), Zambia, and Zimbabwe
A. dealbata	Portugal, South Africa, Spain, USA (California)	Argentina, Azerbaijan, Azores, Bolivia, Canary Islands, Chile, China, Costa Rica, Ethiopia, France, Georgia, Guatemala, Gruzia, India, Israel, Italy, Jordan, Lesotho, Macaronesia, Madagascar, Mauritius, Mozambique, Nepal, New Zealand, Reunion, Romania, Sardinia, Sri Lanka, Swaziland, Tanzania, Uganda, USA (Florida, Oregon, Texas), Yugoslavia, and Zimbabwe
A. decurrens	South Africa, USA (California)	Brazil, Ethiopia, China, Columbia, Ecuador, Haiti, India, Indonesia, Java, New Zealand, Portugal, Pakistan, Spain, Sri Lanka, Tanzania, USA (Florida, Texas), and Venezuela
A. melanoxylon	Hawaii, South Africa, USA (California)	Algeria, Argentina, Azores, Belgium, Bhutan, Bolivia, Brazil, Chile, China, Columbia, Ecuador, Ethiopia, France, Gruzia, Israel, Jordan, India, Italy, Kenya, Lesotho, Luxembourg, Mauritius, Macaronesia, Moldova, Nepal, New Caledonia, New Zealand, Pakistan, Peru, Portugal, Reunion, St Helena Island, Spain, Sri Lanka, Swaziland, Tanzania, and USA (Florida, Texas), Uruguay, Venezuela
A. longifolia	New Zealand, Portugal, South Africa, Spain, USA (California, Florida)	Argentina, Brazil, Columbia, Dominican Republic, Ecuador, France, India, Indonesia, Israel, Italy, Java, Jordan, Kenya, Mauritius, Myanmar, Reunion, Sri Lanka, and Uruguay
A. saligna	Israel, Malta, Portugal, Sardinia, South Africa, Spain, USA (California)	Albania, Algeria, Argentina, Bolivia, Canary Islands, Chile, Corsica, Cyprus, Crete, Egypt, Ethiopia, France, Gibraltar, Greece, India, Iraq, Italy, Jordan, Kenya, Libya, Macaronesia, Mauritius, Mozambique, Namibia, New Zealand,

Table 3.1 (cont.)

Acacia species	Considered invasive in:	Present in:
A. cyclops	South Africa, USA (California)	Pakistan, Saudi Arabia, Sicily, and Tanzania, Tunisia, Turkey, Uganda, USA (Florida), Yemen, and Zambia Azores, Canary Islands, Cyprus, Ethiopia, Gibraltar, Macaronesia, Morocco, Namibia, Portugal, and Spain
A. pycnantha	South Africa	Cyprus, India, Indonesia, Italy, Java, New Zealand, Portugal, Sardinia, Spain, Tanzania, and USA (California)
A. podalyriifolia	Hawaii, South Africa	Argentina, Brazil, Ethiopia, India, Indonesia, Java, Kenya, Malawi, Malaysia, Mauritius, New Zealand, Reunion, Sri Lanka, Tanzania, Uganda, USA (California), Zimbabwe, and widely planted in the Pacific
A. baileyana	South Africa	Argentina, Columbia, Costa Rica, India, Indonesia, Java, Kenya, New Zealand, Swaziland, and Zimbabwe

[a] Records were primarily extracted from the International Legume Database, http://www.ildis.org, and the Global Invasive Species Database, http://www.issg.org/database

Despite campaigns in some countries to remove and control invasive acacias using mechanical and chemical control methods, such methods are expensive and temporary because constant regeneration of seedlings from the large seed-banks takes place, requiring frequent follow-up clearing operations. An integrated approach using biological control and other methods is, therefore, the only long-term and potentially sustainable method of managing such plants (Dennill *et al.*, 1999; Impson, 2005). To date, however, biological control of Australian *Acacia* species has been implemented only in South Africa. This chapter provides an overview of the invasive Australian *Acacia* species in South Africa and reviews the history and effectiveness of the biological control program against them.

3.2 Biology of the invasive Australian *Acacia* species

Of all the Australian *Acacia* species that are considered either problematic or invasive worldwide, ten, in addition to the closely related *Paraserianthes lophantha* (Willd.) Nielsen, are present in South Africa (Table 3.1). All of these are unarmed, evergreen trees. They thrive in a wide range of habitats, growing in disturbed areas, rangelands and/ or grasslands, riparian zones, urban areas, coastal and mountain habitats, natural and cultivated forests, and mesic habitats, and are also well suited to a variety of climates from dry warm temperate to tropical, and can even tolerate mild frost. Several of the *Acacia* species are relatively short lived; however, some (e.g. *Acacia melanoxylon* R. Br.) live for more than 100 years, particularly in the more arid areas.

The trees are fast growing and can produce seeds within three years. Most species flower profusely during spring. Flowers are bisexual and *Acacia* species are able to self-pollinate, although insects also carry pollen between flowers. In most species, relatively few pods develop from the large numbers of flowers. Pod development and ripening can range from a few months to a year, depending on the species, and upon ripening the pods split longitudinally, to release between one and 15 seeds.

All acacia seeds have a hard, water-impermeable seed coat, and most can survive in a dormant condition for many years (Rolston, 1978). Mechanisms that control seed dormancy within the *Acacia* species are varied, but heat and/or scarification are the main factors required to break dormancy (Clemens *et al.*, 1977; Auld, 1986; Jeffery *et al.*, 1988; Morrison *et al.*, 1992). In the absence of such factors, seeds frequently accumulate in the soil for several years before either germinating or rotting, and soil-stored seed banks of up to 45 800 seeds m^{-2} have been recorded for some species in South Africa (Holmes *et al.*, 1987).

Dispersal mechanisms in seeds of *Acacia* species are varied. Besides being passively dispersed, many of the species have seed with attachments (arils or eliasomes), which are attractive to birds and rodents (Middlemiss, 1963; Turcek, 1963; Winterbottom, 1970; Carr, 1976; Glyphis *et al.*, 1981; Impson, 2005), or ants (Bond and Slingsby, 1983; Holmes, 1990). Seeds also spread rapidly via watercourses (de Wit *et al.*, 2001), or by sticking on animals (Milton *et al.*, 2003), and also through humans collecting twigs and logs for firewood (Sankaran, 2002).

Several of the invasive *Acacia* species have the ability to alter nutrient cycling in the nutrient-poor environments they invade. Symbiotic associations with soil-borne bacteria (e.g. *Rhizobium*) in plant roots allow the fixation of atmospheric nitrogen, and landscapes previously devoid of trees change dramatically, usually to the detriment of the indigenous vegetation (van Wilgen *et al.*, 2002).

3.3 Conflicts of interest: the South African situation

There will undoubtedly be conflicts of interest if biological control is to be implemented in countries that experience problems with Australian acacias. The South African example therefore serves as a model as to how the process can be facilitated and unnecessary delays can be avoided.

Australian *Acacia* species were first introduced into South Africa in the early nineteenth Century (Roux, 1961; Stirton, 1978; Shaughnessy, 1980). In the following years, planting of these species was encouraged for timber requirements in the construction of wagons and for use as fence posts and firewood, and also for production of tannin from the bark (Shaughnessy, 1980). In addition to natural spread, sale and the redistribution of acacia seeds played a key role in dispersal and widespread establishment of several species (Dennill and Donnelly, 1991; Morris, 1991). Climatic and edaphic factors in

South Africa were also highly suitable for the Australian acacias, which in time became major invaders in practically every biome in the country (Macdonald and Jarman, 1984).

Exploration for natural enemies of the invasive Australian *Acacia* species in South Africa began in the early 1970s (Neser and Annecke, 1973; van den Berg 1973, 1977, 1980a, b, c, d, 1982a, b, c). Those surveys resulted in long lists of natural enemies (van den Berg 1980a, b, c, 1982a, b, c). However, the economic importance of several of the Australian *Acacia* species (particularly *Acacia mearnsii* De Wild. and *A. melanoxylon*) in South Africa defined the criteria for the selection of agents (Impson and Moran, 2004). Consequently, the emphasis for commercially important species was on insects that could reduce the reproductive capabilities of the plants without otherwise damaging them (Dennill and Donnelly, 1991).

In the mid 1970s, discussions held between representatives of the Plant Protection Research Institute and Wattle Research Institute resulted in the agreement that seed destruction up to 80% would be acceptable to the wattle industry (van den Berg, 1973). This agreement was subsequently revoked because it was felt that seed-reducing agents could impact negatively on the reestablishment of wattle plantations after felling. At this time the surveys for potential natural enemies were well underway in Australia and substantial efforts were made by the wattle industry to halt the progress of the program. Stubbings (1977) argued that the problems caused by wattle invasions were exaggerated because of the "ill-defined and largely unsubstantiated status of black wattle as a weed," especially when considered against the substantial value of the wattle industry, which at the time earned an annual income of about US$30 million. In addition, it was mooted that arthropod seed consumers would not be effective biological control agents anyway, based on the premise that a good agent should have the ability to destroy existing stands through either direct or indirect action (Huffaker, 1964).

Although Annecke (1975) gave assurance that all interested parties would be consulted prior to the release of any biological control agents in South Africa, groups who had commercial interests in wattle were sceptical and negotiations had to be resumed before the conflicts of interest were seemingly resolved (Anonymous, 1978). By 1985, three insect species (*Trichilogaster acaieaelongifoliae* Froggatt, *Melanterius ventralis* Lea and *Rayieria* sp.) had been brought into quarantine in South Africa and been tested for host specificity. Two of these (*T. acaciaelongifoliae* and *M. ventralis*) had obtained approval for release.

In 1987, concerns about the safety of biological control were once again raised by wattle growers, because results of host-specificity tests in quarantine had shown that a seed-feeding weevil, *Melanterius servulus* (Pascoe), being tested for *P. lophantha*, a close relative of Australian *Acacia* species, was able to survive and destroy the seeds of *A. mearnsii* in caged conditions (Donnelly, 1992). This finding should not have been a problem because an agreement had been reached in 1977 that seed-feeding insects which fed on *A. mearnsii* would be permitted in South Africa (Anonymous, 1978). However, new concerns emerged that the weevils would attack seeds in the relatively small area (40 ha) of black wattle orchards which supplied stock to the wattle industry. The South African Wattle Growers Union strongly opposed the release of the seed weevil, resulting in a temporary suspension of the biological control program (Anonymous, 1987). Following further negotiations a compromise was reached when it was demonstrated that insecticides could be used to exclude weevils from seed orchards (Donnelly *et al.*, 1992).

In the following years several species of biological control agents were released on various Australian *Acacia* species in South Africa, and in 1993 the first release of a weevil was made onto black wattle (*A. mearnsii*), the central subject of the conflict for almost 20 years. Initial releases were restricted to the Western Cape Province (about 1000 km away from commercial wattle plantations) until it could be shown that under local conditions the insects would have no detrimental effect on *A. mearnsii*. Release of the weevils has been extended to cover much of South Africa in recent years.

Several lessons were learned from the conflicts in the acacia biological control program in South Africa, notably: (i) ecological studies on the invasive species prior to introduction are necessary to confirm the existence of a problem and that biological control is either the best or the only management option, (ii) perceptions that possible agents (in this case seed feeders) will be ineffective need to be substantiated before rejection of the nominated agents, (iii) all interested and affected parties should communicate openly and continuously to exchange information and to prevent, or at least minimize, controversy, and (iv) agreements about what can and cannot be done should be drawn up prior to the implementation of the biological control program.

3.4 Biological Control

3.4.1 Acquired natural enemies

Wide-ranging records of both vertebrate and invertebrate natural enemies of Australian acacias in their introduced range in South Africa are available. An unidentified tortricid moth (Tortricidae: Olethreutinae) has been recorded from green seed pods of *Acacia cyclops* A. Cunn. ex G. Don (Donnelly and Stewart, 1990), and adults and nymphs of an unnamed pentatomid have been observed feeding on the green ripening seed pods of the same plant (Impson, 2005). Several species from at least four genera of Alydidae (Hemiptera) occur on exotic acacias (Schaefer, 1980; Schaffner, 1987). An Australian psyllid species (*Acizzia uncatoides* (Ferris and Kylver)) is frequently abundant and found feeding on new vegetative growth of several of the introduced acacias (unpublished data). This species was accidentally introduced at some unknown time, but first records of it date from the 1980s.

Natural pests of black wattle are also well documented by the South African wattle industry, and *Chaliopsis junodi* (Heylaerts) (Lepidoptera: Psychidae), *Lygidolon laevigatum* Reuter (Hemiptera: Miridae), and fungal diseases such as wattle wilt and gummosis, are a few of the key herbivores and diseases of *A. mearnsii* in South Africa. A variety of vertebrate granivores, including rodents (David, 1980) and birds (Winterbottom, 1970; Glyphis *et al.*, 1981) feed and destroy a substantial proportion of the seeds of some of the Australian *Acacia* species. More recently an indigenous root fungus (*Pseudolagarobasidium acaciicola* s. nov) (Cyphellaceae) has been isolated and described from *A. cyclops*, with field trials showing up to 100% mortality of inoculated trees, making the fungus a potential bioherbicide (Wood and Ginns, 2006). None of these locally acquired natural enemies has had a substantial impact on any of the invasive Australian acacias, and therefore, the focus since the outset has been on classical biological control with introductions of herbivorous species from Australia into South Africa.

3.5 Classical biological control

South Africa is currently the only country actively engaged in biological control of Australian *Acacia* species. New Zealand is the only other country which benefits from the presence of Australian natural enemies which have become naturalized on their own on some of the introduced *Acacia* species (A. J. Gordon, PPRI, personal communication; Hill *et al.*, 2000). For reasons outlined above (i.e. conflict resolution), the South African biological control campaign has been limited, with one exception, to the use of insects that damage the reproductive structures of their host plants rather than the vegetative parts of the plants.

Initial surveys in Australia during the 1970s identified a large number of natural enemies associated with buds, flowers, and seeds of the Australian *Acacia* species that are the most important invaders in South Africa (van den Berg, 1977, 1980a, b, c, 1982a, b, c). A few other potential agents (arthropods and pathogens) have been found during subsequent collecting trips. Several of these potential agents have been introduced into quarantine facilities in South Africa for host-specificity tests. This process has been complicated because of the need to have plants which are bearing flowers and seeds, which is not readily achieved on potted plants. As a result, much of the host-specificity testing has had to be done by confining the insects in mesh sleeves on branches of trees growing under natural conditions. To minimize the consequences of inadvertent escapes, only insects that have been investigated in their native habitats and that have shown substantive evidence of being host specific are considered for sleeve tests. The primary purpose of these tests is to satisfy procedural protocols required by regulatory agencies.

The natural arthropod enemies imported into South Africa for assessment as biological control agents against various *Acacia* species are described below. Although not included in this chapter, a rust fungus, *Uromycladium tepperianum* (Sacc.) McAlp. (Pileolariaceae), has been used as an effective classical biological control agent against *Acacia saligna* (Labill.) H. Wendl. in South Africa (Morris, 1991, 1999).

3.5.1 *The bud-galling wasps,* Trichilogaster *spp. (Hymenoptera: Pteromalidae)*

Trichilogaster spp. are gall wasps endemic to Australia. The immature stages develop primarily within the flower buds of *Acacia* species. (Noble, 1940; Dennill and Donnelly, 1991; Dorchin *et al.*, 2006). Female wasps lay one to several eggs in the immature flower buds, generally during spring (September–November) and summer (December–February); however, when wasps are abundant and flower buds are in short supply, vegetative buds will also be used. Adults are short-lived (2–3 days), and females lay approximately 400 eggs during this time. The eggs remain dormant until the following spring when they hatch. Larval feeding causes the buds to become distorted and induces them to develop into galls. The galls have 1–18 discrete chambers, each containing a single wasp larva. The *Trichilogaster* species are mostly univoltine. Some species reproduce largely parthenogenetically, with males making up only a small proportion of numbers.

Trichilogaster acaciaelongifoliae *Froggatt*

The bud-galling wasp *T. acaciaelongifoliae* was released as a biological control agent for *Acacia longifolia* (Andr.) Willd. in 1982. It was the first agent to be released onto any of the Australian acacias in South Africa. Specificity tests were conducted, in quarantine, on a range of potted plants. Gall development was only confirmed on *A. longifolia*, despite some oviposition probing on *A. melanoxylon* (van den Berg, 1980d). Following its release, the wasp established readily throughout the range of the weed through natural dispersal and manual distribution of gall material, and populations increased exponentially (Dennill, 1987, 1988). Negligible gall development by *T. acaciaelongifoliae* has been recorded on *A. melanoxylon* and *P. lophantha* (which was not one of the test plants), when these plants grow in close proximity to heavily galled *A. longifolia* trees. However, neither of these plant species acts as a suitable and permanent host for *T. acaciaelongifoliae* (Dennill et al., 1993).

Trichilogaster acaciaelongifoliae reduces seed production of *A. longifolia* by >95% (Dennill, 1987, 1988). Gall induction also suppresses vegetative growth by inducing phyllode abscission and shoot dieback, in severe cases resulting in tree mortality when environmental conditions are stressful (Dennill, 1985, 1988). Although widely established, *T. acaciaelongifoliae* is not consistently effective throughout its range, and appears to do less well in the hot inland valleys and the elevated cooler mist belt regions, a factor which has been attributed to poor climatic matching (Dennill and Gordon, 1990). Plants seem to have the ability to compensate for high gall loads when growing in areas where there is no water stress, and under these conditions, pods may still develop despite high numbers of galls. Since its introduction into South Africa the wasp has also acquired several indigenous parasitoids. Initial levels of parasitism were measured at 1.6% (Dennill, 1987), although subsequent studies measure levels of parasitism up to 21.3% in the Western Cape (Manongi and Hoffmann, 1995) and 80% in the Eastern Cape (Baars, 1994). Parasitism is mostly limited to larvae developing late in the season and those in small galls or in chambers close to the surface of larger galls. Consequently the effectiveness of *T. acaciaelongifoliae* as a biological control agent has not been negatively affected.

Trichilogaster acaciaelongifoliae has been imported from South Africa into a quarantine facility in Portugal, where it is currently undergoing host-specificity testing (Marchante et al., 2006).

Trichilogaster signiventris *(Girault)*

The success of *T. acaciaelongifoliae* on *A. longifolia* paved the way for the introduction of a related species, *Trichilogaster signiventris*, for use against *Acacia pycnantha* Benth. (Dennill and Gordon, 1991). Host-specificity tests were undertaken and in 1987 a small number of adult wasps were placed in mesh sleeves on trees growing under natural conditions. Surveys in the following summer indicated that the wasp had failed to induce galls and *A. pycnantha* in South Africa was thought to be an unsuitable host for the strain of *T. signiventris*. In 1992 a further consignment of wasps was received and once again females were placed in sleeves on branches, but apparently without success. It was assumed that the wasp had not survived. However, a small number of galls were unexpectedly discovered on *A. pycnantha* plants in the vicinity of the original release sites in 1995. Over the next two years levels of gall induction increased dramatically and by 1997 manual distribution of mature galls began. The insects have now become established throughout the range of *A. pycnantha* in South Africa (Dennill *et al.*, 1999).

Studies have shown that *T. signiventris* has significantly reduced the reproductive output of trees (Hoffmann *et al.*, 2002). When particularly abundant the weight of galls causes collapse of branches and trees. Unlike its counterpart on *A. longifolia*, no parasitoids have become associated with *T. signiventris* in South Africa (N. Dorchin, personal communication).

3.5.2 The leaf-spotting bug, **Rayieria** sp. (Hemiptera: Miridae)

Rayieria sp. is native to eastern, central and western Australia. Adults, which can fly, live for 4–8 weeks and mate repeatedly, and females have an approximate 7-day pre-oviposition period. Females lay *c.* 80 eggs, which are embedded in groups of up to seven within the stems of young host plants. Oviposition sites are covered with a "waxy" substance, through which respiratory tubes protrude characteristically (Donnelly, 1986). In cold climates the eggs overwinter, although no obligatory diapause phase occurs, and under laboratory conditions hatching occurs within approximately 21 days. Five nymphal stages follow, each lasting three days.

Adults and nymphs of *Rayieria* sp. feed on the phyllodes. Clear watery lesions (2–4 mm diameter) form around the insertion point of the stylets within about an hour of commencement of feeding. After removal of the stylets the lesions dry, forming characteristic brown feeding "damage" points. Feeding damage leads to browning of the phyllodes, and defoliation in severe instances.

Rayieria sp. was one of the first three insect biological control agents to be introduced into quarantine in South Africa. Host-specificity tests confirmed that feeding was restricted to Australian *Acacia* species, but this was not limited to the phyllodinous acacias, and feeding was recorded on *A. mearnsii*, *A. cyclops*, *Acacia implexa* Benth., and

A. melanoxylon (in addition to records from *A. longifolia* and *A. saligna* from Australia). As a result of potential damage to commercial wattle (*A. mearnsii*), *Rayieria* sp. was rejected as being unsuitable for release (Donnelly, 1986).

3.5.3 *The seed-feeding weevils,* Melanterius *species (Coleoptera: Curculionidae)*

Melanterius is a genus of 88 species which, with the exception of two species from New Caledonia and Papua New Guinea, occur only in Australia. All of the known species are seed feeders and univoltine, with breeding commencing in early spring (September), coinciding with the onset of pod maturation of host plants. Adults feed on and destroy developing seeds. Mating and oviposition follow. Female weevils chew small holes through the walls of the green pods, into which they insert a single egg. The eggs are deposited either on or close to the seed surface, after which the oviposition hole is usually sealed with regurgitated substances. Neonate larvae burrow into the seeds where they feed and complete several larval instars in 4–6 weeks.

Usually one larva develops per seed, during which time it consumes the entire contents of the seed, leaving only the hard, outer seed coat. In some species, notably those on hosts with small seeds, the larvae feed on two or more adjacent seeds. Once fully developed, the larvae chew their way out of the seed and surrounding fruit tissue, and drop to the soil, where they tunnel 5–10 cm, and pupate within fragile "chambers" of compacted soil. Although some larvae may remain in the soil until the following spring, fully developed adult weevils usually emerge from the soil 6–8 weeks later. These reproductively immature adults remain largely inactive from the end of summer to late winter (January– August), sheltering either on host plants or in cracks and crevices under bark of nonhosts. The adults occasionally emerge to feed on reproductive buds or flowers, or young vegetative growth of their host plant, and also feed on the extrafloral nectaries. However, adults are never evident on the plants in any great numbers until spring months (September–November).

Although some *Melanterius* species are very host specific and are only ever associated with a particular species of *Acacia*, others can feed and develop on several different acacia hosts (New, 1979, 1983; Auld, 1983; Donnelly, 1992; Dennill *et al.*, 1999). *Melanterius servulus* and *Melanterius maculatus* Lea, which have been imported into South Africa, have been recorded on several different Australian acacia hosts (Dennill *et al.*, 1999; Impson and Moran, 2004). Trials involving the transfer of beetles between host plants suggest that there are host-specific entities within these species, at least in terms of larval feeding, and there may be several strains or biotypes involved. However, taxonomic and molecular studies (Oberprieler and Zimmerman, 2001; Clarke, 2002) could find no intraspecific differences between the groups and their nomenclature stands. Unfortunately host-plant and phylogenetic relationships between the Australian acacias and *Melanterius* weevils are poorly understood (Impson and Moran, 2004). To date, some patterns of host association are clearly evident. For example, *M. ventralis*, which is morphologically and phylogenetically distant from other *Melanterius* species used for biological control in South Africa (Clarke, 2002), is specific to *A. longifolia*, the only target species in the country belonging to the section Juliflorae. *Melanterius maculatus*, which appears to be less specific, has host plants (*A. mearnsii*, *Acacia dealbata* Link, *Acacia decurrens* (J. C. Wendl.), *Acacia baileyana* F. Muell.) all belonging to the Botrycephalae (Oberprieler and Zimmerman, 2001). However, associations between *Melanterius* species and their host plants can really only be accurately determined from a comprehensive study of specimens reared from seeds of acacia host plants.

Melanterius ventralis was the first of the seed-feeding weevils to be imported into South Africa, and was released against *A. longifolia* in 1985 (Dennill and Donnelly, 1991). Although populations were slow to increase, the weevils established readily at all release sites, and after three years the levels of seed destruction varied from 15% to 80% (Dennill and Donnelly, 1991). More importantly, the weevils were particularly useful at destroying the seeds on *A. longifolia* growing close to rivers, where despite extensive gall induction by *T. acaciaelongifoliae*, the trees were able to produce many more pods than those trees growing in dry regions.

Following the ready establishment and success of *M. ventralis*, the next 20 years of the biological control program against Australian acacias were dominated by research on several other species within the *Melanterius* genus. *Melanterius acaciae* Lea was released against *A. melanoxylon* in 1986. This was followed by introductions of *M. servulus* against *P. lophantha* in 1989, and the same species against *A. cyclops* in 1991. *Melanterius maculatus* was released onto *A. mearnsii* in 1993, with further releases of the same species carried out onto *A. dealbata* and *A. decurrens* in 2001, *A. pycnantha* in 2005 and *A. baileyana* in 2006. *Melanterius compactus* Lea was released on *A. saligna* in 2001.

Much of the recently published work on the impact of *Melanterius* species has focused on *M. servulus* on *A. cyclops* (Impson *et al.*, 2004; Impson, 2005), although studies have also been done on *A. longifolia* (Dennill, 1985, 1988; Dennill and Donnelly, 1991; Donnelly and Hoffmann, 2004), *A. melanoxylon* (Donnelly, 1995a), *P. lophantha* (Schmidt *et al.*, 1999), and unpublished data exist for ongoing evaluation on a number of the species (F. Impson). The data accumulated so far indicate that the various *Melanterius* species have similar impacts on their different acacia hosts in South Africa, so it is possible to generalize about what could be expected from biological control using these species.

Establishment of the weevils is generally followed by slowly increasing levels of seed damage over several years, and then gradual dispersal of the weevils away from the release sites (Dennill and Donnelly, 1991; Donnelly, 1995b; Impson, 2005). Levels of damage tend to fluctuate annually in different regions, but in areas where the beetles have been established for several years, levels of damage regularly reach 85%, and occasionally 100%. The levels of damage are influenced by extraneous disturbances (e.g. fires, land-clearing operations, and dust from roads), and habitat characteristics such as soil type and climate may also play a key role in the overall success of any of the species concerned (Impson *et al.*, 2004; Impson, 2005). *Melanterius* species have not been found to be affected by any native parasitoids in their introduced ranges.

3.5.4 The flower-galling midges, Dasineura species (Diptera: Cecidomyiidae)

Most Australian *Acacia* species are hosts to Cecidomyiidae, which induce galls on their reproductive parts, phyllodes, and twigs, and a single plant species may be host to either one or a complex of cecidomyiid species (Adair *et al.*, 2000; Kolesik *et al.*, 2005). Two species show promise as biological control agents in South Africa and are discussed below.

Dasineura dielsi *Rübsaamen on* A. cyclops

Dasineura dielsi is a multivoltine species, which targets open florets of *A. cyclops*. Up to five generations occur per year (Adair, 2005), enabling it to capitalize on the extended flowering period characteristic of *A. cyclops*. The third-instar larvae exhibit a quiescent stage of variable length, which results in staggered emergence of adults originating from the same egg cohort. For a few individuals, this quiescent phase exceeds 12 months. The peak flowering time of *A. cyclops* varies between localities and occasional flowers are produced throughout the year. The combination of multivoltinism and staggered emergence ensure that at all times *D. dielsi* adults are present within a site and able to utilize any flowers occurring.

The adult midges are tiny, highly vagile and short-lived. Following mating the females lay one or more eggs within the perianth tube of mature florets. Each female produces 233 ± 13 S.E. ($n = 16$) eggs (Adair, 2005). Eggs hatch within a few days, and the newly hatched larvae initiate feeding on the surface of the ovary, inducing the ovaries to evaginate rapidly, creating an individual gall chamber within which a single larva develops. Each individual gall chamber has an opening at its apex (the ostiole) which is protected by a ring of inward-extending white hairs. The number of chambers that develop on a single floret varies from 1 to 16. Inflorescences of *A. cylops* consist of multiple florets on a stalk and since many florets are often affected by gall induction simultaneously, a convoluted "gall complex" results. The overall appearance of the gall complex is highly variable as it depends on several factors including the number of chambers per floret and the total number of galled florets. Three larval instars occur; the third-instar larva seals the ostiole with a fine silk mesh and then creates a compartment at the base of the chamber by the formation of a silken cap, partial or complete cocoon. The third-instar larva remains quiescent within the compartment until pupation and subsequent emergence of the adult.

Acacia cyclops is the primary host of *D. dielsi*. In South Australia, *D. dielsi* is occasionally associated with *Acacia sophorae* (Labill.), *Acacia papyrocarpa* Benth., *Acacia oswaldii* (F. Muell.) and *Acacia ligulata* (Cunn. ex Benth), when these species grow in close proximity to *A. cyclops* (Adair, 2005). Adair (2005) conducted no-choice host-specificity tests on a limited number of species in South Africa. Gall induction occurred, but was rare, on *Acacia elata* A. Cunn. ex Benth. and *A. implexa*. No galling was induced on the seven African *Acacia* species tested. Subsequent to the release of *D. dielsi* in South Africa in 2001, gall induction was recorded on *A. melanoxylon*, *A. longifolia*, *A. saligna*, and *A. implexa*, with successful development and emergence of adults ensuing (unpublished data). Gall induction has also been recorded on *Acacia pendula* A. Cunn ex G. Don, but successful development of adults has yet to be confirmed for this species. Within three years of release, *D. dielsi* dispersed from a single release locality and spread throughout the South African range of *A. cyclops*. The initial buildup of *D. dielsi* populations resulted in high levels of gall induction and very low levels of seed-pod production on *A. cyclops*.

The success of cecidomyiids as biological control agents has often been compromised, as they have suffered high levels of mortality from generalist parasitoids and predators acquired after introduction into the new country (Goeden and Louda, 1976; McFadyen, 1985; Wehling and Piper, 1988; Carlson and Mundal, 1990; Harris and Shorthouse, 1996). Not surprisingly therefore, *D. dielsi* was rapidly utilized by native parasitoids subsequent to its introduction in South Africa, with four species emerging from galls collected in April 2003 at three sites in the Western Cape: Lyndoch (33° 58′ S, 18° 46′ E), Strand Beach (34° 06′ S, 18° 48′ E) and Somerset West (34° 03′ S, 18° 51′ E) (Adair, 2005). More than 20 parasitoid taxa have been reared from subsequent collections of *D. dielsi* galls and a definitive association with *D. dielsi* has been established for four of these species (unpublished results). *Dasineura dielsi* galls also provide a potential refuge or resource for numerous other types of arthropods including phytophagous, predatory, and possibly saprophagous species. The list includes Lepidoptera, Homoptera (Psyllidae – *A. uncatoides*), Heteroptera (Anthocoridae, Pentatomidae, Lygaeidae – *Macchiademus diplopterus*, Alydidae), Neuroptera (Chrysopidae), Diptera (Syrphidae), Coleoptera (Coccinellidae, Curculionidae – *M. servulus*), Thysanoptera, Acarina, and Araneae.

In spite of all these associations, *D. dielsi* remains abundant and has inflicted a substantial overall decline in seed production on *A. cyclops* in South Africa. Initially the abundance of *D. dielsi* was high and there were correspondingly low levels of pod set on trees. Subsequently the fly populations have been less stable and considerable variation in the quantity of pods produced between sites and between years (unpublished results) has occurred. Ongoing studies are investigating the impact of acquired parasitoids and predators, including "itch mites," thrips, and anthocorids, as well as edaphic factors, on the population dynamics of *D. dielsi*. Of concern is the possibility that the fluctuations in pod set will destabilize populations of *M. servulus* and render the beetles unable to exploit, and destroy, the surfeit of seeds that develop when cecids are scarce. Studies are currently underway to investigate this relationship.

Dasineura rubiformis *Kolesik on* A. mearnsii

The biology of this species was described by Adair (2004). *Dasineura rubiformis* is a univoltine species. Similar to *D. dielsi*, the adults are small and short-lived. Each female lays 90 eggs on average, depositing them among clusters of mature flowers of the host plant. Larval feeding induces gall development with up to five larval chambers per floret. Larvae develop within the ovoid gall chambers over several months, and in winter (June–July), mature third-instar larvae emerge from their chambers and drop to the soil where they pupate in the leaf litter within silken cocoons. Adults emerge in late spring and summer (September–December).

In Australia, *D. rubiformis* is widespread on *A. mearnsii*, both within its natural distribution in eastern Australia, and also in Western Australia where *A. mearnsii* has become naturalized. In eastern Australia, galls similar to those of *D. rubiformis* have been recorded from several other *Acacia* species within the Botrycephalae (Kolesik *et al.*, 2005). No-choice host testing in South Africa, using a number of native African and introduced Australian acacias, demonstrated that *D. rubiformis* was restricted to *A. mearnsii* (Adair, 2004).

The narrow host range and the small size and biomass of *D. rubiformis* galls indicate that these insects are unlikely to have any resource-loading impact on the growth of *A. mearnsii*, and are thus compatible with the commercial exploitation of the plants (Adair, 2004). Small populations of this insect became established near Stellenbosch, South Africa, in 2006, but nothing can be said yet about their impact and dispersal.

3.6 The seed-feeding wasps *Bruchophagus* species Ashmead (Hymenoptera: Eurytomidae)

Bruchophagus species are 4–5 mm-long wasps, which lay their eggs in summer (October–November) on the pods of acacia host plants, selecting pods as the seeds start to swell before ripening. Approximately 40 eggs are laid by each female. Each larva develops rapidly within the seed, eating out the entire content and creating two holes in the distal ends through which frass is ejected, forming water-resistant plugs. The seed then falls to the ground and the fully developed larva spends the remainder of summer, autumn, and winter (December–August) within the seed cases. Most larvae pupate in late spring and emerge as adults soon afterward, but a proportion of the larval population enters into diapause, which lasts until the following spring or even longer (Kluge, 1990; Hill, 1999).

As a result of taxonomic uncertainty within the genus, much of the preliminary work on *Bruchophagus* has focused on determining the identities and host ranges of *Bruchophagus* species collected from 15 different *Acacia* species in Australia, and from New Zealand, where both the host plants and the insect natural enemies have become naturalized (Neser, 2002; Neser and Prinsloo, 2002). Seven *Bruchophagus* species were collected, six of which were undescribed at the outset of the study. Two of these have received consideration for biological control in South Africa.

Bruchophagus orarius *Neser and Prinsloo on* A. longifolia

Preliminary host-specificity tests showed that *Bruchophagus orarius* from *A. longifolia* is host specific and a potentially suitable candidate for biological control (Kluge, 1990). Further trials have not been carried out and the insect has never been released, possibly due to the success of the two agents already present on *A. longifolia*.

Bruchophagus acaciae *(Cameron) on* A. dealbata

Bruchophagus acaciae was considered as a potential biocontrol agent of *A. dealbata* in the late 1990s. It was collected from *A. baileyana*, *Acacia caesiella* Maiden & Blakely, *A. dealbata*, *A. decurrens*, *Acacia fimbriata* A. Cunn. ex G. Don, *A. mearnsii*, *Acacia myrtifolia* (Smith) Willd., *Acacia rubida* A. Cunn., and *Acacia silvestris* Tind., and has subsequently also been reared from seed of *A. pycnantha* and *Acacia podalyriifolia* A. Cunn. ex G. Don (S. Neser, Plant Protection Research Institute, personal communication). This study, together with an additional study commissioned by South African researchers to be done in New Zealand (Hill, 1999) indicated that although the host range of *B. acaciae* was clearly wider than the records suggest, it is restricted to *Acacia* species of Australian origin. Examination of the field host range in New Zealand demonstrated that among 28 *Acacia* species and *P. lophantha*, *B. acaciae* was present only on six, and at extremely low levels (<0.1%) on two of these (*A. mearnsii* and *A. rubida*) (Hill, 1999).

The results so far indicate that *B. acaciae* could become an excellent biological control agent for use against Australian acacias in South Africa.

3.7 Impacts of seed-reducing agents against invasive tree species

The effectiveness of seed-reducing biological control agents has long been the subject of discussion and debate. Many biological control practitioners have not supported their use (Huffaker, 1964; Harris, 1973; Goeden, 1983), and theoretical studies have indicated that such agents need to consistently destroy a high proportion of the annual seed crop to have any critical impacts on the population density, particularly where seed banks are large and long-lived (Hoffmann and Moran, 1991; Sheppard *et al.*, 1994; Kriticos *et al.*, 1999). In some instances, moderate to high levels of seed reduction in weeds have even resulted in increases in population densities (Myers *et al.*, 1990; Kelly and McCallum, 1995; Myers and Risley, 2000). Other models have, however, indicated that a reduction in reproductive capacity could reduce density, providing disturbance rates are high and plant fecundity and seedling survival are low (Rees and Paynter, 1997; Hoffmann and Moran, 1998). The role of seed-reducing agents in reducing dispersal beyond the limits of existing infestations should also be considered, particularly for weeds such as the acacias that have long-distance dispersal of seeds via wind, water, or birds. Despite ongoing debate, biological control practitioners should probably consider each program in isolation rather than generalizing about seed-reducing agents as a whole.

The economics of weed invasions and their control are increasingly being recognized, and evaluation studies identifying the costs of such invasions to the environment (i.e. through water loss, interference with agriculture, and loss of biodiversity) as well as the costs of clearing and reducing spread, are critical aspects, although for the most part they are poorly understood. In many situations, and particularly where seed feeders are concerned, the benefits of a biological control agent are difficult to evaluate. Simulation models can be useful for predicting the potential of agents (Withers *et al.*, 2004), the rates of spread of invasions (Higgins *et al.*, 1996; Clark *et al.*, 2003; Rouget and Richardson, 2003), and also the impacts of invasions on biodiversity (Higgins *et al.*, 1997). Such tools can provide effective means of estimating economic consequences of alien trees and their control.

It is also important to clarify from the beginning of a program what the expectations are for effective management (Moran *et al.*, 2004). In the past many biological control practitioners have measured success as being a reduction in the original population density of the invasive plant as a result of the activity of one or more biological control agents. Using this measure, and looking specifically at the invasive *Acacia* species, the introduction of one, or even more, seed-reducing agents is unlikely to ever achieve the goal. Even with high levels of seed destruction, the large seed loads allow sufficient seed to escape destruction and find safe sites, and the self-thinning principle also applies (Silverton and Lovette Doust, 1993). A reasonable goal post, in the context of seed-reducing agents being used, would be a reduction in the spread, or the rates of spread, of the target weed, and also to see biological control as being an integral part in the overall management of invasive weeds, through integrated biological, chemical, and mechanical control operations (Moran *et al.*, 2004).

Bearing in mind that there were several constraints placed on the types of biological control agents that could be used against *Acacia* species in South Africa, and also that the ultimate "success" of these agents would not necessarily be measured by their ability to reduce existing populations of the weeds, the use of seed-reducing agents was not only well founded, but has been a positive step in the program. In the last ten years large-scale manual clearing operations have been undertaken as part of a government initiative to alleviate poverty in South Africa. Presence of seed-reducing agents in the system, combined with such clearing methods, could be playing a substantial role in the overall management of invasive plants. A reduction in seed numbers can effectively be translated into a corresponding reduction in costs and effort in follow-up clearing treatments when removing seedlings or young plants from previously cleared areas (Moran *et al.*, 2004).

There is a great deal of potential for future use of the agents outlined in this chapter in other countries where acacias are recognized as being problematic, particularly in countries such as South Africa, where the commercial and social benefits of such invasive plants remain important. In the past, limited attention has been given to invasive woody plants in the tropics and reports of invasive *Acacia* species remain relatively scarce. Binggeli *et al.* (2004) ascribe this to factors such as poor records and reporting, restricted circulation of publications, and also to the perceptions and financial resources that undoubtedly play a role. Although some countries will not only recognize the need but also have the resources to initiate and carry through integrated control programs against invasive acacias, the challenge will be to introduce and maintain such concepts in the poorer tropical and subtropical regions of the world.

References

Adair, R. J. (2004). Seed-reducing Cecidomyiidae as potential biological control agents for invasive Australian wattles in South Africa, particularly *A. mearnsii* and *A. cyclops*. Ph.D. thesis, University of Cape Town, South Africa, 224 pp.

Adair, R. J. (2005). The biology of *Dasineura dielsi* (Rübsaamen (Diptera: Cecidomyiidae) in relation to the biological control of *Acacia cyclops* (Mimosaceae) in South Africa. *Australian Journal of Entomology*, **44**, 446–456.

Adair, R. J., Kolesik, P. and Neser, S. (2000). Australian seed-preventing gall midges (Diptera: Cecidomyiidae) as potential biological control agents for invasive *Acacia* spp. in South Africa. In *Proceedings of the X International Symposium on Biological Control of Weeds*, held July 4–9, 1999, Bozeman, Montana, USA, ed. N. R. Spencer. Sidney, MT: USDA-ARS, pp. 605–614.

Annecke, D. P. (1975). Biological control of Australian *Acacia* species. Official communication Ref. 13/3/417/6 dated 26 November, 1975 from Director, Plant Protection Research Institute, Pretoria, to Director, Wattle Research Institute, University of Natal, Pietermaritzburg.

Anonymous (1977). Planting forage species in Region IV. *Chile Forestal*, **2**(18), 12–13.

Anonymous (1978). Biological control of Australian acacias. *Wattle Growers News*, **62**, 10–12.

Anonymous (1987). Editorial. The history of the conflict of interest regarding the biological control of alien acacias in South Africa. *South African Institute of Ecologists Bulletin* (ed. B. W. van Wilgen and B. McKenzie), **6**, 1–9.

Auld, T. D. (1983). Seed predation in native legumes of south-eastern Australia. *Australian Journal of Ecology*, **8**, 367–376.

Auld, T. D. (1986). Dormancy and viability in *Acacia suaveolens* (Sm.) Willd. *Australian Journal of Botany*, **34**, 463–472.

Baars, J. R. (1994). Parasitization of a biological control agent *Trichilogaster acaciaelongifoliae* by native parasitoids: is there cause for concern? Unpublished Honours project, Department of Zoology and Entomology, Rhodes University, Grahamstown, South Africa.

Binggeli, P., Hall, J. B., Healey, J. R. and Hamilton, A. C. (2004). Invasive woody plants in tropical Africa. Indicators and tools for restoration and sustainable management of forests in East Africa. I-TOO Working paper 15. Freiburg, Germany: Waldbau Institute.

Bond, W. J. and Slingsby, P. (1983). Seed dispersal by ants in shrublands of the Cape Province and its evolutionary implications. *South African Journal of Science*, **79**, 231.

Carlson, R. B. and Mundal, D. (1990). Introduction of insects for the biological control of leafy spurge in North Dakota. *North Dakota Farm Research*, **47**, 7–8.

Carr, J. D. (1976). *The South African Acacias*. Johannesburg: Conservation Press, 323 pp.

Caswell, H., Lensink, R. and Neubert, M. G. (2003). Demography and dispersal: life table response experiments for invasion speed. *Ecology*, **84**, 1968–1978.

Clark, J. S. (1998). Why trees migrate so fast: confronting theory with dispersal biology and the paleorecord. *The American Naturalist*, **152**, 204–224.

Clark, J. S., Lewis, M., McLachlan, J. S. and HilleRisLlambers, J. (2003). Estimating population spread: what can we forecast and how well? *Ecology*, **84**, 1979–1988.

Clarke, G. M. (2002). *Molecular Phylogenetics and Host Ranges of the* Melanterius *Weevils Used as Biological Control Agents of Australian Acacias in South Africa.* Canberra, Australia: CSIRO Entomology.

Clemens, J., Jones, P. G. and Gilbert, N. H. (1977). Effect of seed treatments on germination in *Acacia*. *Australian Journal of Botany*, **25**, 269–276.

David, J .H. M. (1980). Demography and population dynamics of the striped field mouse, *Rhabdomys pumilio*, in alien acacia vegetation on the Cape Flats, Cape Peninsula, South Africa. Ph.D. thesis, University of Cape Town, South Africa.

Dennill, G. B. (1985). The effect of the gall wasp *Trichilogaster acaciaelongifoliae* (Hymenoptera: Pteromalidae) on reproductive potential and vegetative growth of the weed *Acacia longifolia*. *Agriculture, Ecosystems and Environment*, **14**, 53–61.

Dennill, G. B. (1987). The importance of understanding host plant phenology in the biological control of *Acacia longifolia*. *Annals of Applied Biology*, **111**, 661–666.

Dennill, G. B. (1988). Why a gall former can be a good biocontrol agent – the gall wasp *Trichilogaster acaciaelongifoliae* and the weed *Acacia longifolia*. *Ecological Entomology*, **13**, 1–9.

Dennill, G. B. and Donnelly, D. (1991). Biological control of *Acacia longifolia* and related weed species (Fabaceae) in South Africa. *Agriculture, Ecosystems and Environment*, **37**, 115–135.

Dennill, G. B. and Gordon, A. J. (1990). Climate-related differences in the efficacy of the Australian gall wasp (Hymenoptera: Pteromalidae) released for the control of *Acacia longifolia* in South Africa. *Environmental Entomology*, **19**, 130–136.

Dennill, G. B. and Gordon, A. J. (1991). *Trichilogaster* sp. (Hymenoptera: Pteromalidae), a potential biocontrol agent for the weed *Acacia pycnantha* (Fabaceae). *Entomophaga*, **36**, 295–301.

Dennill, G. B., Donnelly, D. and Chown, S. L. (1993). Expansion of host-plant range of a biocontrol agent *Trichilogaster acaciaelongifoliae* (Pteromalidae) released against the weed *Acacia longifolia* in South Africa. *Agriculture, Ecosystems and Environment*, **37**, 115–135.

Dennill, G. B., Donnelly, D., Stewart, K. and Impson, F. A. C. (1999). Insect agents used for the biological control of Australian *Acacia* species and *Paraserianthes lophantha* (Willd.) Nielsen (Fabaceae) in South Africa. *African Entomology Memoir*, **1**, 45–54.

de Wit, M. P., Crookes, D. J. and van Wilgen, B. W. (2001). Conflicts of interest in environmental management: estimating the costs and benefits of a tree invasion. *Biological Invasions*, **3**, 167–178.

Donnelly, D. (1986). *Rayieria* sp. (Heteroptera: Miridae): host specificity, conflicting interests, and rejection as a biological control agent against the weed *Acacia longifolia* (Andr.) Willd. in South Africa. *Journal of the Entomological Society of Southern Africa*, **49**, 183–191.

Donnelly, D. (1992). The potential host range of three seed-feeding *Melanterius* spp. (Curculionidae), candidates for the biological control of Australian *Acacia* spp. and *Paraserianthes* (*Albizia*) *lophantha* in South Africa. *Phytophylactica*, **24**, 163–167.

Donnelly, D. (1995a). Good news for blackwood control. *Plant Protection News*, **41**, 4.

Donnelly, D. (1995b). Host searching behaviour of the seed weevil, *Melanterius ventralis*: implications for the biological control of *Acacia longifolia* in South

Africa. In *Proceedings of the VIII International Symposium on Biological Control of Weeds*, held 2–7 February 1992, Lincoln University, Canterbury, New Zealand, ed. E. S. Delfosse and R. R. Scott. Australia: Melbourne, DSIR/CSIRO, p. 577

Donnelly, D. and Hoffmann, J. H. (2004). Utilization of an unpredictable food source by *Melanterius ventralis*, a seed-feeding biological control agent of *Acacia longifolia* in South Africa. *BioControl*, **49**, 225–235.

Donnelly, D. and Stewart, K. (1990). An indigenous tortricid moth on the seeds of the alien weed *Acacia cyclops* in South Africa: a potential for biological control. *Journal of the Entomological Society of Southern Africa*, **53**, 199–202.

Donnelly, D., Calitz, F. J. and Van Aarde, I. M. R. (1992). Insecticidal control of *Melanterius servulus* (Coleoptera: Curculionidae), a potential biocontrol agent of *Paraserianthes lophantha* (Leguminosae) in commercial seed orchards of black wattle, *Acacia mearnsii* (Leguminosae). *Bulletin of Entomological Research*, **82**, 197–202.

Dorchin, N., Cramer, M. D. and Hoffmann, J. H. (2006). Photosynthesis and sink activity of wasp-induced galls in *Acacia pycnantha*. *Ecology*, **87**, 1781–1791.

Ebinger, J. E., Seigler, D. S. and Clarke, H. D. (2000). Taxonomic revision of South American species of the genus *Acacia* subgenus *Acacia* (Fabaceae: Mimosoideae). *Systematic Botany*, **25**, 588–617.

Evans, C. S., Qureshi, M. Y. and Bell, E. A. (1977). Free amino acids in the seeds of *Acacia* species. *Phytochemistry*, **16**, 565–570.

Glyphis, J. P., Milton, S. J. and Siegfried, W. R. (1981). Dispersal of *Acacia cyclops* by birds. *Oecologia*, **48**, 138–141.

Goeden, R. D. (1983). Critique and revision of Harris' scoring system for selection of insect agents in biological control of weeds. *Protection Ecology*, **5**, 287–301.

Goeden, R. D. and Louda, S. M. (1976). Biotic interference with insects imported for weed control. *Annual Review of Entomology*, **21**, 325–342.

Harris, P. (1973). The selection of effective agents for the biological control of weeds. *Canadian Entomologist*, **105**, 1495–1503.

Harris, P. and Shorthouse, J. D. (1996). Effectiveness of gall inducers in weed biological control. *Canadian Entomologist*, **128**, 1021–1055.

Higgins, S. I., Azorin, E. J., Cowling, R. M. and Morris, M. J. (1997). A dynamic ecological-economic model as a tool for conflict resolution in an invasive-alien-plant, biological control and native-plant scenario. *Ecological Economics*, **22**, 141–154.

Higgins, S. I., Richardson, D. M. and Cowling, R. M. (1996). Modelling invasive plant spread: the role of plant environment interactions and model structure. *Ecology*, **77**, 2043–2054.

Hill, R. L. (1999). The host range and biology of *Bruchophagus acaciae* (Cameron) (Hymenoptera: Eurytomidae) in New Zealand. Unpublished report to the Plant Protection Research Institute, Pretoria, South Africa, September 1999, 46 pp.

Hill, R. L., Gordon, A. J. and Neser, S. (2000). The potential role of *Bruchophagus acaciae* (Cameron) (Hymenoptera: Eurytomidae) in the integrated control of *Acacia* species in South Africa. In *Proceedings of the X International Symposium on Biological Control of Weeds*, 4–14 July 1999, Bozeman, Montana, USA, ed. N. R. Spencer, Sidney, MT: USDA-ARS, pp. 919–929.

Hoffmann, J. H. and Moran, V. C. (1991). Biocontrol of a perennial legume, *Sesbania punicea*, using a florivorous weevil, *Trichapion lativentre*: weed population dynamics with a scarcity of seeds. *Oecologia*, **88**, 574–576.

Hoffmann, J. H. and Moran, V. C. (1998). The population dynamics of an introduced tree, *Sesbania punicea*, in South Africa, in response to long-term damage caused by different combinations of three species of biological control agents. *Oecologia*, **114**, 343–348.

Hoffmann, J. H., Impson, F. A. C., Moran, V. C. and Donnelly, D. (2002). *Trichilogaster* gall wasps (Pteromalidae) and biological control of invasive golden wattle trees (*Acacia pycnantha*) in South Africa. *Biological Control*, **25**, 64–73.

Holmes, P. M. (1990). Dispersal and predation in alien acacia. *Oecologia*, **83**, 288–290.

Holmes, P. M., Dennill, G. B. and Moll, E. J. (1987). Effects of feeding by native alydid insects on the seed viability of an alien invasive weed, *Acacia cyclops*. *South African Journal of Science*, **83**, 580–581.

Huffaker, C. B. (1964). Fundamentals of biological weed control. In *Biological Control of Insect Pests and Weeds*, ed. P. De Bach. London: Chapman and Hall, pp. 631–649.

Impson, F. (2005). Biological control of *Acacia cyclops* in South Africa: the role of an introduced seed-feeding weevil, *Melanterius servulus* (Coleoptera: Curculionidae) together with indigenous seed-sucking bugs and birds. M.Sc. thesis, University of Cape Town, South Africa, 91 pp.

Impson, F. A. C. and Moran, V. C. (2004). Thirty years of exploration for and selection of a succession of *Melanterius* weevil species for biological control of invasive Australian acacias in South Africa: should we have done anything differently? In *Proceedings of the XI International Symposium on Biological Control of Weeds*, held April 27–May 2, 2003, Canberra, Australia, ed. J. M. Cullen, D. T. Briese, D. J. Kriticos, *et al.* Canberra, Australia: CSIRO Entomology, pp. 127–134.

Impson, F. A. C., Moran, V. C. and Hoffmann, J. H. (2004). Biological control of an alien tree, *Acacia cyclops*, in South Africa: impact and dispersal of a seed-feeding weevil, *Melanterius servulus*. *Biological Control*, **29**, 375–381.

Janzen, D. H. (1969). Seed-eaters versus seed size, number, toxicity and dispersal. *Evolution*, **23**, 1–27.

Janzen, D. H. (1971). Seed predation by animals. *Annual Review of Ecology and Systematics*, **2**, 465–492.

Janzen, D. H. (1980). Specificity of seed-attacking beetles in a Costa Rican deciduous forest. *Journal of Ecology*, **68**, 929–952.

Jeffery, D. J., Holmes, P. M. and Rebelo, A. G. (1988). Effects of dry heat on seed germination in selected indigenous and alien legume species in South Africa. *South African Journal of Botany*, **54**, 28–34.

Kelly, D. and McCallum, K. (1995). Evaluating the impact of *Rhinocyllus conicus* on *Carduus nutans* in New Zealand. In *Proceedings of the VIII International Symposium on Biological Control of Weeds*, held 2–7 February 1992, Lincoln University, Canterbury, New Zealand, ed. E. S. Delfosse and R. R. Scott. Melbourne, Australia: DSIR/CSIRO, pp. 205–211.

Kergoat, G. J., Silvain, J-F., Buranapanichpan, S. and Tuda, M. (2007). When insects help to resolve plant phylogeny: evidence for a paraphyletic genus *Acacia* from the sytematics and host-plant range of their seed-predators. *Zoologica Scripta*, **36**, 143–152.

Kluge, R. L. (1990). The seed-attacking wasp *Bruchophagus* sp. (Hymenoptera: Eurytomidae) and its potential for biological control of *Acacia longifolia* in South Africa. In *Proceedings of the VII International Symposium on Biological Control of Weeds*, held 6–11 March 1988, Rome, ed. E. S. Delfosse, Rome, Italy: Istituto Sperimentale per la Patologia Vegetale Ministero dell'Agricoltura e delle Foreste, pp. 349–356.

Kolesik, P., Adair, R. J. and Eick, G. (2005). Nine new species of *Dasineura* (Diptera: Cecidomyiidae) from flowers of Australian *Acacia* (Mimosaceae). *Systematic Entomology*, **30**, 454–479.

Kriticos, D., Brown, J., Radford, I. and Nicholas, M. (1999). Plant population ecology and biological control: *Acacia nilotica* as a case study. *Biological Control*, **16**, 230–239.

Macdonald, I. A. W. and Jarman, M. L., eds. (1984). Invasive alien organisms in the terrestrial ecosytems of the fynbos biome, South Africa. South African National Scientific Programmes Report 85. Pretoria, South Africa: CSIR, 72 pp.

Manongi, F. S. and Hoffmann, J. H. (1995). The incidence of parasitism in *Trichilogaster acaciaelongifoliae* (Froggatt) (Hymenoptera: Pteromalidae), a gall-forming biological control agent of *Acacia longifolia* (Andr.) Willd. (Fabaceae) in South Africa. *African Entomology*, **3**, 147–151.

Marchante, H., Marchante, E., Hoffmann J. H. and Freitas, H. (2006). Potential use of *Trichilogaster acaciaelongifoliae* as a biocontrol agent of *Acacia longifolia* in Portugal. *Biocontrol News and Information*, **27**, 31N–32N.

Maslin, B. (2001). *Acacia. Flora of Australia*, Vol. 11A, 11B. In *Mimosaceae*, Acacia. Melbourne, Australia: ABRS/CSIRO Publishing, pp. 3–13.

Maslin, B. R., Miller, J. T. and Seigler, D. S. (2003). Overview of the generic status of *Acacia* (Leguminosae: Mimosoideaea). In *Advances in Legume Systematics*. Vol. 9. *Phylogeny*, ed. P. Herendeen and A. Bruneau. London: Royal Botanic Gardens, Kew, pp. 181–200.

McFadyen, P. J. (1985). Introduction of the gall fly *Rhopalomyia californica* from the USA into Australia for the control of the weed *Baccharis halimifolia*. In *Proceedings of the VI International Symposium on Biological Control of Weeds*, held 19–25 August 1984, ed. E. S. Delfosse. Agriculture Canada: Vancouver, Canada, pp. 779–796.

Middlemiss, E. (1963). The distribution of *Acacia cyclops* in the Cape Peninsula area by birds and other animals. *South African Journal of Science*, **59**, 419–420.

Milton, S. J. (1980). Australian acacias in the South Western Cape: pre-adaption, predation and success. In *Proceedings of the Third National Weeds Conference of South Africa*, ed. S. Neser and A. L. P. Cairns. Cape Town: A. A. Balkema, pp. 69–78.

Milton, S. J. and Hall, A. V. (1981). Reproductive biology of Australian acacias in the south-western Cape Province, South Africa. *Transactions of the Royal Society of South Africa*, **44**, 465–487.

Milton, S. J., Dean, W. R. J. and Richardson, D. M. (2003). Economic incentives for restoring natural capital in southern Africa rangelands. *Frontiers in Ecology and the Environment*, **1**, 247–254.

Moran, V. C., Hoffmann, J. H. and Olckers, T. (2004). Politics and ecology in the management of alien invasive woody trees: the pivotal role of biological control agents that diminish seed-production. In *Proceedings of the XI International Symposium on Biological Control of Weeds*, held 27 April–2 May 2003, ed. J. M. Cullen, D. T. Driese, D. J. Kriticos, *et al*. Canberra, Australia: CSIRO Entomology, pp. 434–439.

Morris, M. J. (1991). The use of plant pathogens for biological weed control in South Africa. *Agriculture, Ecosystems and Environment*, **37**, 239–255.

Morris, M. J. (1999). The contribution of the gall-forming rust fungus *Uromycladium tepperianum* (Sacc.) McAlp. to the biological control of *Acacia saligna* (Labill.) Wendl. (Fabaceae) in South Africa. *African Entomology Memoir*, **1**, 125–128.

Morrison, D. A., Auld, T. D., Rish, S., Porter, C. and McClay, K. (1992). Patterns of testa-imposed seed dormancy in native Australian legumes. *Annals of Botany Company*, **70**, 157–163.

Myers, J. H., Risley, C. and Eng, R. (1990). The ability of plants to compensate for insect attack: why biological control of weeds with insects is so difficult. In *Proceedings of the VII International Symposium on Biological Control of Weeds*, held 6–11 March 1988, ed. E. S. Delfosse, Rome, Italy: Istituto Sperimentale per la Patologia Vegetale, MAF, pp. 57–73.

Myers, J. H. and Risley, C. (2000). Why reduced seed production is not necessarily translated into successful biological weed control. In *Proceedings of the X International Symposium on Biological Control of Weeds*, held 4–14 July 1999, Montana State University, Bozeman, Montana, USA, ed. N. R. Spencer, Sidney MT: USDA-ARS, pp. 569–581.

Neser, S. (2002). Studies on the identity and host range of *Bruchophagus* spp. (Hymenoptera: Eurytomidae) with a view to curbing seeding of Australian acacias. In Bi-annual report on the Biological Control of Weeds for the Working for Water Programme, Department of Water Affairs and Forestry, Pretoria. Tender WF7374. pp. 91–94.

Neser, S. and Annecke, D. P. (1973). Biological control of weeds in South Africa. *Entomology Memoir* (Department of Agricultural Technical Services, Republic of South Africa), 28, 27.

Neser, O. C. and Prinsloo, G. L. (2002). Report on a study to determine the identity and natural host ranges of *Bruchophagus* spp. (Hymenoptera: Eurytomidae) that develop in the seeds of Australian acacias with a view to facilitating host specificity studies of those from species invasive in South Africa. In Bi-annual report on the Biological Control of Weeds for the Working for Water Programme, Department of Water Affairs and Forestry, Pretoria. Tender WF7374. pp. 95–98.

New, T. R. (1979). Phenology and relative abundance of Coleoptera on some Australian acacias. *Australian Journal of Zoology*, **27**, 9–16.

New, T. R. (1983). Seed predation of some Australian acacias by weevils (Coleoptera: Curculionidae). *Australian Journal of Zoology*, **31**, 345–352.

New, T. R. (1984). *A Biology of Acacias*. Melbourne: Oxford University Press.

Noble, N. S. (1940). *Trichilogaster acaciaelongifoliae* (Froggatt) (Hymenoptera: Chalcidoidea) a wasp causing galls on the flower buds of *Acacia longifolia* Willd., *Acacia floribunda* Sieber and *Acacia sophorae* R. Br. *Transactions of the Entomological Society of London*, **90**, 13–38.

Oberprieler, R. G. and Zimmerman, E. C. (2001). Identification and host ranges of the *Melanterius* weevils used as biocontrol agents of invasive Australian acacias in South Africa, Canberra, Australia: CSIRO Entomology.

Orchard, A. E. and Wilson, A. J. G. (2001). *Flora of Australia*, Vols. IIA and IIB. *Mimosaceae*, Acacia parts 1 & 2. Canberra and Melbourne: ABRS/CSIRO Publishing, pp. 437–466.

Orchard, A. E. and Maslin, B. R. (2003). Proposal to conserve the name *Acacia* (Leguminosae: Mimosoideae) with a conserved type. *Taxon*, **52**, 362–363.

Pedley, A. (1978). Revision of *Acacia* Mill. in Queensland. *Austrobaileyana*, **1**, 77–234.

Pedley, A. (1986). Derivation and dispersal of *Acacia* (Leguminosae) with particular reference to Australia, and the recognition of *Senegalia* and *Racosperma*. *Botanical Journal of the Linnean Society*, **92**, 219–254.

Rees, M. and Paynter, Q. (1997). Biological control of Scotch broom: modelling the determinants of abundance and the potential impact of introduced insect herbivores. *Journal of Applied Ecology*, **34**, 1203–1221.

Rolston, M. P. (1978). Water impermeable seed dormancy. *The Botanical Review*, **44**, 365–396.

Ross, J. H. (1979). A conspectus of the African *Acacia* species. *Memoirs of the Botanical Survey of South Africa*, **44**, 1–155.

Ross, J. H. (1981). An analysis of the African *Acacia* species: their distribution, possible origins and relationships. *Bothalia*, **13**, 389–413.

Rouget, M. and Richardson, D. M. (2003). Inferring process from pattern in plant invasions: a semimechanistic model incorporating propagule pressure and environmental factors. *The American Naturalist*, **162**, 713–724.

Roux, E. R. (1961). History of the introduction of Australian acacias on the Cape Flats. *South African Journal of Science*, **57**, 99–102.

Sankaran, K. V. (2002). Black wattle problem emerges in Indian forests. *Biocontrol News and Information*, **23**, 5N.

Schaefer, C. W. (1980). Host plants of the Alydinae, with a note on heterotypic feeding aggregations (Hemiptera: Coreoidea: Alydidae). *Journal of the Kansas Entomological Society*, **53**, 115–122.

Schaffner, J. C. (1987). The genus *Zulubius* Bergroth (Heteroptera: Alydidae). *Journal of the Entomological Society of Southern Africa*, **50**, 313–322.

Schmidt, F., Hoffmann, J. H. and Donnelly, D. (1999). Levels of damage caused by *Melanterius servulus* Pascoe (Coleoptera: Curculionidae), a seed-feeding weevil introduced into South Africa for biological control of *Paraserianthes lophantha* (Fabaceae). *African Entomology*, **7**, 107–112.

Seigler, D. S. and Ebinger, J. E. (2005). New combinations in the genus *Vachellia* (Fabaceae: Mimosoideae) from the New World. *Phytologia*, **87**, 139–178.

Seigler, D. S., Ebinger, J. E. and Miller, J. T. (2006). The genus *Senegalia* (Fabaceae: Mimosoideae) from the New World. *Phytologia*, **88**, 38–93.

Shaughnessy, G. L. (1980). Historical ecology of alien woody plants in the vicinity of Cape Town, South Africa. Ph.D. thesis, University of Cape Town, South Africa.

Sheppard, A., Cullen, J. and Aeschlimann, J. (1994). Predispersal seed predation on *Carduus nutans* (Asteraceae) in southern Europe. *Acta Oecologica*, **15**, 529–541.

Silverton, J. and Lovette Doust, J. (1993). *Introduction to Plant Population Biology.* Oxford, UK: Blackwell Scientific Publications.

Smith, M. N. (1979). Report from the escaped exotics committee, California. *Fremontia*, **6**, 18–19.

Stirton, C. H., ed. (1978). *Plant Invaders: Beautiful But Dangerous.* Cape Town, South Africa: Department of Nature and Environmental Conservation of the Cape Provincial Administration, 175 pp.

Stubbings, J. A. (1977). A case against controlling introduced acacias. In *Proceedings of the Second National Weeds Conference of South Africa*, held in Stellenbosch, South Africa. Cape Town, South Africa: A. A. Balkema, pp. 89–107.

Turcek, F. J. (1963). Color preference in fruit- and seed-eating birds. In *Proceedings of the Thirteenth International Ornithological Congress*. Ithaca, NY: IOC 285–292.

Tutin, T. G., Heywood, V. H., Burges, N. A., *et al.* eds. (1968). *Flora Europaea*, Vol. 2. Cambridge, UK: Cambridge University Press.

van den Berg, M. A. (1973) Indringing van bosse in plantasies en inheemse flora. Report of the Agricultural Technical Services, Pretoria, South Africa, 6 pp.

van den Berg, M. A. (1977). Natural enemies of certain acacias in Australia. In *Proceedings of the Second National Weeds Conference of South Africa*, held in Stellenbosch, South Africa. Cape Town, South Africa: A. A. Balkema, pp. 75–82.

van den Berg, M. A. (1980a). Natural enemies of *Acacia cyclops* A. Cunn. ex G. Don and *Acacia saligna* (Labill.) Wendl. in Western Australia. I. Lepidoptera. *Phytophylactica*, **12**, 165–167.

van den Berg, M. A. (1980b). Natural enemies of *Acacia cyclops* A. Cunn. ex G. Don and *Acacia saligna* (Labill.) Wendl. in Western Australia. II. Coleoptera. *Phytophylactica*, **12**, 169–171.

van den Berg, M. A. (1980c). Natural enemies of *Acacia cyclops* A. Cunn. ex G. Don and *Acacia saligna* (Labill.) Wendl. in Western Australia. III. Hemiptera. *Phytophylactica*, **12**, 223–226.

van den Berg, M. A. (1980d). *Trichilogaster acaciaelongifoliae* (Frogatt) (Hymenoptera: Pteromalidae): a potential agent for the biological control of *Acacia longifolia* in South Africa. In *Proceedings of the Third National Weeds Conference of South Africa*, ed. S. Neser and A. L. P. Cairns. Cape Town, South Africa: A. A. Balkema, pp. 61–64.

van den Berg, M. A. (1982a). Lepidoptera attacking *Acacia dealbata* Link., *Acacia decurrens* Willd., *Acacia longifolia* (Andr.) Willd., *Acacia mearnsii* De Wild. and *Acacia melanoxylon* R. Br. in eastern Australia. *Phytophylactica*, **14**, 3–46.

van den Berg, M. A. (1982b). Hemiptera attacking *Acacia dealbata* Link., *Acacia decurrens* Willd., *Acacia longifolia* (Andr.) Willd., *Acacia mearnsii* De Wild. and *Acacia melanoxylon* R. Br. in eastern Australia. *Phytophylactica*, **4**, 47–50.

van den Berg, M. A. (1982c). Coleoptera attacking *Acacia dealbata* Link., *Acacia decurrens* Willd., *Acacia longifolia* (Andr.) Willd., *Acacia mearnsii* De Wild. and *Acacia melanoxylon* R. Br. in eastern Australia. *Phytophylactica*, **14**, 51–55.

van Wilgen, B. W., Richardson, D. M., Le Maitre, D. C., Marais, C. and Magadlela, D. (2002). The economic consequences of alien plant invasions: examples of impacts and approaches to sustainable management in South Africa. In *Biological Invasions: Economic and Environmental Costs of Alien Plant, Animal, and Microbe Species*, ed. D. Pimental. Boca Raton, FL: CRC Press: pp. 243–265.

Vassal, J. (1972). Apport des recherches ontogéniques et séminologiques à l'étude morphologique taxonomique et phylogénique du genre *Acacia*. *Travaux du Laboratoire Forestier de Toulouse*, **8**, 1–128.

Wehling, W. F. and Piper, G. L. (1988). Efficacy diminution of the rush skeletonweed gall midge, *Cystiphora schmidti* (Diptera: Cecidomyiidae), by an indigenous parasitoid. *Pan-Pacific Entomologist*, **64**, 83–85.

Winterbottom, J. M. (1970). The birds of the alien *Acacia* thickets of the south-western Cape. *Zoologica Africana*, **5**, 49–57.

Withers, T., Richardson, B., Kimberley, M., *et al.* (2004). Can population modelling predict potential impacts of biocontrol? A case study using *Cleopus japonicus* on *Buddleja davidii*. In *Proceedings of the XI International Symposium on Biological Control of Weeds*, held 27 April–2 May 2003, ed. J. M Cullen, D. T. Briese, D. J. Kriticos, *et al.* Canberra, Australia: CSIRO Entomology, pp. 57–62.

Wood, A. R. and Ginns, J. (2006). A new dieback disease of *Acacia cyclops* in South Africa caused by *Pseudolagarobasidium acaciicola* s. nov. *Canadian Journal of Botany*, **84**, 750–758.

4

Ageratina adenophora (Sprengel) King and Robinson (Asteraceae)

R. Muniappan, A. Raman, and G. V. P. Reddy

4.1 Introduction

Ageratina adenophora (Sprengel) King and Robinson (= *Eupatorium adenophorum, E. glandulosum, E. pasadense*) (Asterales: Asteraceae) is popularly known as the Crofton weed; other common names are eupatory, sticky snakeroot, cat weed, hemp agrimony, sticky agrimony, Mexican devil, and sticky eupatorium in different parts of the world (Hoshovsky and Lichti, 2007). In Hawaii it is known as *Maui pāmakani* and *pāmakani haole* and in Nepal as *banmara* (killer of the forests). Usually, it grows into an erect herb (occasionally into a subshrub) of one to three meters in height, with trailing purplish to chocolate-brown branches that strike roots upon contact with soil, resulting in dense thickets (Bess and Haramoto, 1958). The base of the plant is woody and densely clothed with stalked glandular hairs. Leaves are dark green, opposite, deltoid-ovate, serrate, and purple underneath, and each grows to about 10 cm in length. Flowers are white and borne terminally in compound clusters in spring and summer. The seed is an achene, varying from elliptic to oblanceolate, often gibbous, 1.5–2 mm long, 0.3–0.5 mm wide; with five prominent ribs and five to 40 pappi with slender scabrous bristles (Hickman, 1993). Dispersal occurs by wind-borne seeds and each plant produces about 100 000 seeds per season. Seeds are also spread by water, as contaminants of agricultural produce, via sand and gravel used in road preparation, via soil sticking to animals, machinery, and vehicles, and by adhering to footwear or clothing of farm workers (Parsons, 1992). Seeds are set without either pollination or fertilization, and 15–30% of seeds are usually not viable. Dense stands can contribute up to 60 000 viable seeds per square meter to the seed bank. Light is essential for seed germination, so unshaded contexts, such as vacant soil, are essential for the establishment of seedlings of *A. adenophora*. Once germinated, seedlings withstand considerable levels of shading, by increasing leaf area to compensate for reduced light intensity. The weed grows rapidly, forming dense thickets (Parsons, 1992).

Ageratina adenophora is a native of Mexico, but has naturalized in many countries. It was introduced into several parts of the world as an ornamental in the nineteenth century

Biological Control of Tropical Weeds using Arthropods, ed. R. Muniappan, G. V. P. Reddy, and A. Raman. Published by Cambridge University Press. © Cambridge University Press, 2009.

and is now an established invasive weed in many subtropical regions of Asia (India, Nepal, China, the Philippines, Thailand, and Brunei), Oceania (Hawaiian Islands, Tahiti, New Zealand, Australia, and Papua New Guinea), Africa (Nigeria, Zimbabwe, and South Africa), and Europe (France, Greece, Portugal, and Spain) (Morris, 1989; Kluge, 1991; Parsons, 1992; Waterhouse, 1993; Wagner *et al.*, 1999). *Ageratina adenophora* was brought to the island of Maui (Hawaii) as an ornamental by Captain James Makee in 1860, and subsequently became a weed occupying rangelands and roadsides. It was introduced into California around 1849 and the first field collection was made in 1878. By 1920, it had spread throughout the mountains on the northern side of the Los Angeles Basin. It has flourished in areas which receive year-round rainfall, edging out native vegetation (Fuller, 1981). Robbins (1940) has reported the spread of *A. adenophora* as a "rare escape" in the San Francisco Bay area and along the southern coast of California. It was introduced to Australia as an ornamental under the name of "*E. riparium*," probably as early as 1875 (Shephard, 1875, cited in Auld and Martin, 1975). In 1943, it was proclaimed as a noxious plant in New South Wales, while in southeastern Queensland the population exploded in 1949. It took complete possession of large tracts of pasture and horticultural land along the border of New South Wales and Queensland and the spread was so fast that in some areas farmers had to abandon their holdings (Everist, 1959; Dodd, 1961; Auld and Martin, 1975). Its occurrence in tropical countries such as India, Nepal, the Philippines, and Thailand is limited to elevations between 1000 and 2000 m in the hills (Borthakur, 1977; Sharma and Chhetri, 1977).

Ageratina adenophora is considered to be a serious weed in agriculture, especially in rangelands where it often replaces either the more-desirable vegetation or native species (Bess and Haramoto, 1958), but also in forests (Sharma and Chhetri, 1977). It is generally unpalatable to grazing animals, but goats graze on this plant infrequently (Wilson *et al.*, 1985). It is fatally toxic to horses and causes the "blowing disease" in Hawaii and "Numinbah disease" or "Tollebudgera horse disease" in Australia. The disease may take several years to become evident in horses (O'Sullivan, 1979). Symptoms such as coughing, difficulty in breathing, and violent blowing after exertion are the result of acute lung edema leading to hemorrhage (O'Sullivan, 1985). This plant reduces growth of nearby vegetation by releasing allelopathic compounds (Kaul and Bansal, 2002) and altering the soil microbial communities (Niu *et al.*, 2007). It is a problem weed in forest plantations as it infests disturbed areas and prevents self-seeding of cultivated trees, and hence it is known as *banmara* in Nepal (Sharma and Chhetri, 1977; Morris, 1989). Moreover, it reduces biodiversity by suppressing native vegetation, interfering with the movement of wildlife, depleting soil nutrients and clogging irrigation channels (Sharma and Chhetri, 1977, Wilson *et al.*, 1985).

Although *A. adenophora* replaces native vegetation such as grasses that protect soil from erosion, because of its dense canopy it protects soil from splash and rill erosion. In Nepal, the leafy stems are harvested and used as cattle bedding and the dry brittle stems are used as fuel (Wilson *et al.*, 1985). It is capable of replacing other invasive weeds such as *Imperata cylindrica* (L.) Beauv. (Poaceae) (Falvey, 1982) and *Lantana camara* L.

(Verbenaceae) (Dhyani, 1978). However, none of these benefits is sufficient to detract from the plant's status as an invasive weed.

4.2 Biological control initiatives

Mechanical control is difficult to practice as this plant grows on slopes (Hoshovsky and Lichti, 2007). Between 1920 and 1945, hundreds of acres of grazing land in Hawaii were reclaimed from *A. adenophora* by mechanical removal, which proved expensive. In Australia slashing followed by plowing is carried out (Parsons, 1992). Chemicals such as 2,4-D (Borthakur, 1977), glyphosate, 'dicamba+MCPA' and 'picloram+triclopyr' (Parsons, 1992) were recommended. These methods are expensive and temporary, and most countries where this weed is a problem have adopted classical biological control. When Crofton weed became a serious pest in agriculture and forestry in Hawaii, the gall-inducing fly, *Procecidochares utilis* Stone (Diptera: Tephritidae) was imported from Mexico in 1945 by the then Territorial Board of Agriculture and Forestry, Hawaii. The fly has since established in Maui and within three years, farmers had stopped managing this plant (Bess and Haramoto, 1958). The fly was imported into Queensland (Australia) from Hawaii in 1952 and it established easily. Populations of *P. utilis* increased substantially and suppressed the weed at sites closer to the release area (Dodd, 1961). It was later introduced into New Zealand from Queensland in 1958 and the first release was made on the Coromandel Peninsula, from where the fly spread naturally throughout the peninsula. In 1963, it was released and established over the rest of New Zealand (Hill, 1989). *Procecidochares utilis* was introduced from New Zealand into India and released in the states of Tamil Nadu, Assam, West Bengal, and Utter Pradesh. By 1971, it had spread widely in the hill ranges of Nilgris and Darjeeling and provided localized control (Rao *et al.*, 1971). The fly reached Nepal from India and established in the Ilam, Terhathum, and Dhankuta districts by 1973 (Sharma and Chhetri, 1977). It then crossed the Himalaya and reached China in 1984 (Wan and Wang, 1991). It was introduced into South Africa from Australia and released at seven sites around Pietermaritzburg and one around Muden between October 1984 and March 1986 (Bennett, 1986). After three generations, galls induced by *P. utilis* on *A. adenophora* were spotted up to one kilometer radius from the release sites and after 20 months up to six kilometers further. It also established around Stellenbosch in 1986 (Kluge, 1991). Several releases have since been undertaken in other areas that have recently become invaded by the weed in South Africa and the agent is now widespread.

4.2.1 *Biology of* Procecidochares utilis

The female flies are capable of laying up to 160 eggs (mean number of eggs = 74). Eggs are laid in batches of 2–23 in the terminal vegetative buds. The female usually inserts the ovipositor through one of the second pair of leaves from the top and lays eggs in between the first pair of young leaves. Eggs are creamy-white, elongate, 0.6 mm in length and coated with a mucoid secretion that adheres the eggs together when laid, thus offering a clumped appearance (Bess and Haramoto, 1958; Sharma and Chhetri, 1977; Bennett and Van Staden, 1986). As many as 20 eggs are laid at the tip of a plant, although the average is seven. The egg stage lasts three to four days during summer and six to eight days in winter in Honolulu, Hawaii (Bess and Haramoto, 1958), three to five days in Yunnan province in China (Zhang *et al.*, 1988) and five to eight days in Nepal (Sharma and Chhetri, 1977). Upon hatching from the eggs, the maggots migrate downward to the base of the leaves, mine into the apical meristem, feed on plant tissue and induce a gall. Occasionally galls develop on leaf petioles and/or leaf midribs. Three larval instars develop; the mature maggot (four mm long) usually excavates a tunnel from the larval chamber to the exterior of the gall before pupation, leaving the epidermis intact to form a "window." The larval stages last for about 20 days in the summer months in Hawaii (Bess and Haramoto, 1958) and 25–30 days in Yunnan (Zhang *et al.*, 1988). The puparia are blackish and are formed within the gall chamber, and the pupal stage lasts 14–21 days in Hawaii (Bess and Haramoto, 1958) and 20–25 days in Yunnan (Zhang *et al.*, 1988). Flies emerge from the galls by breaking the epidermis at the "window." In Nepal, the average time for development from egg to adult is 56 and 60 days for males and females, respectively. Dodd (1961) reported 41 days for males and 43 days for females to complete their life cycles, although Bennett and Van Staden (1986) found considerable variation under identical climatic conditions. The sex ratio of flies emerging from the galls was 1:1 irrespective of the number of larvae per gall. Mating occurs on the same day of emergence and oviposition may also commence on the same day; oviposition continues for up to three weeks, but the majority of the eggs are laid in the first week. The average number of eggs laid per female was 171. On average, adults live for two weeks, rarely extending to three weeks (Bess and Haramoto, 1958).

4.2.2 Gall development

The first sign of gall development is evident a week after oviposition when the young leaves at the oviposition site turn crinkled and chlorotic. Three or four days later the stem bends to about 45° at the point where oviposition occurred (Bennett and Van Staden, 1986). In two weeks from the time of egg laying, gall initiation becomes evident with a red and pink pigmentation. Gall size depends on (1) the number of larvae and (2) the vigor of the plant. With one larva in the gall, the gall size is usually 15×10 mm, whereas, when multiple larvae occur, the gall may reach a size of 35×17 mm. The average number of larvae found in a gall was three, even though a maximum of 11 flies emerged from a gall (Bess and Haramoto, 1958). During favorable seasons and environmental conditions, oviposition continues and individual stems are attacked repeatedly, resulting in compound galls which are caused by smaller galls (those induced later in time) coalescing with previously induced ones. Larvae remain confined to individual chambers, but in larger galls the larval chambers may coalesce.

4.2.3 Effect of gall induction on the plant

In a favorable environment such as the Ulupalakua area in Maui, Hawaiian Islands, *P. utilis* is abundant throughout the year and all shoots are attacked during plant growth. In some instances, the shoot tips died because of intense oviposition (Bess and Haramoto, 1958). Galls function as nutrient sinks and the nutritive tissue establishes when the gall becomes visible. The larva that induces the gall derives its nourishment from the specialized tissue in this gall chamber, which is composed of proliferating parenchyma cells along the walls of the chamber (Meyer and Maresquelle, 1983; Bronner, 1992). Such specialized nutritive tissue becomes a metabolic sink for energy nutrients and accumulates minerals (Ca, Cu, Fe, Mg, Mn, Ni, and Zn) from adjacent plant tissues or other parts of the plant (Abrahamson and Weis, 1987; Raman, 1994; Raman and Abrahamson, 1995; Cruz *et al.*, 2006; Raman *et al.*, 2006). Gall induction also reduces several vital metabolic and transpiration efficiencies, stomatal conductance and water potential (Florentine *et al.*, 2001, 2005). In addition, galls reduce shoot height, and production of leaves, flowers, and seeds (Raman and Abrahamson, 1995; Cruz *et al.*, 2006). The exit holes cut by the inhabiting larvae enable access by microorganisms that induce decay. High galling intensity results in plant mortality.

4.2.4 Effect of parasitism on **P. utilis**

In its native range in Mexico, *P. utilis* is attacked by the parasitoids *Eurytoma obtusiventris* Gahan (Hymenoptera: Eurytomidae), *Eupelmus cyaniceps* (Ashm.) (Hymenoptera: Eupelmidae), *E. allynii* (French) (Hymenoptera: Eupelmidae), *Torymus umbilicatus* (Gahan) (Hymenoptera: Torymidae), *Galeopsomopsis* sp. (Hymenoptera: Eulophidae), and *Zatropis* sp. (Hymenoptera: Pteromalidae). Five species of parasitoids, *Opius tryoni* Cameron (Hymenoptera: Braconidae), *Opius longicaudatus* (Ashmead) (Hymenoptera: Braconidae), *Bracon terryi* (Bridwell) (Hymenoptera: Braconidae), *Eupelmus cushmani* (Crawford) (Hymenoptera: Eupelmidae), and *Eurytoma tephritidis* Fullaway (Hymenoptera: Eurytomidae) were reared from *P. utilis* in Hawaii. The two *Opius* spp. were imported into Hawaii to control tephritid pests but became casual parasitoids of *P. utilis*. Parasitism was higher in warmer months, increasing up to 93% in some localities and averaging 50% and 60% in 1950–1957 and 1966–1971 surveys, respectively. Despite high parasitism, *P. utilis* could still eliminate *A. adenophora* over large areas in Maui; the lack of success in other areas was attributed to heavy rainfall and wet conditions and not to recruited parasitoids (Bess and Haramoto, 1959, 1972).

In the high-altitude regions of Tamil Nadu (India) (2000–2300 m asl), four hymenopteran parasitoids, *Diameromicrus kiesenwetteri* (Meyr) (Hymenoptera: Torymidae), *Syntomopus* sp. (Hymenoptera: Pteromalidae), *Bracon* sp. (Hymenoptera: Braconidae) and *Eurytoma* sp. (Hymenoptera: Eurytomidae) have been recorded on *P. utilis* (Swaminathan and Raman, 1981). Parasitism by *Bracon* sp. was as high as 80% and was considered to be the primary cause for the failure of the gall fly to control Crofton weed in India. In Nepal, the parasitoids *Eurytoma* sp. (Hymenoptera: Eurytomidae) and *Dimeromicrus vibidia* (Walker) (Hymenoptera: Torymidae) have been reported to parasitize up to 17.5% and 30%, respectively, of *P. utilis* populations (Sharma and Chhetri, 1977; Kapoor and Malla, 1979).

Nearly two years after the release and establishment of *P. utilis* in Australia, eight species of indigenous hymenopteran parasitoids were found attacking it. Of these, *Megastigmus* sp. (Hymenoptera: Torymidae), *Macrodontomerus australiensis* Gir. (Hymenoptera: Torymidae) and *Campyloneurus* sp. (Hymenoptera: Braconidae) were significant. In particular, *Megastigmus* sp. caused 90% parasitism (Dodd, 1961). Species of minor importance included one species each of *Campyloneurus* and Pteromalidae, and three species of Eupelmidae (Dodd, 1961). The reported population decline of the fly in Australia was due to parasitism by these diverse parasitoids; however, the fly still provided partial control of the weed. As a result, the rapid spread of the weed was halted and its vigor, growth and density have been reduced. The gall fly was introduced to New Zealand in 1958, and for many years no parasitoids were recorded on it. In 1964, some parasitism was recorded but was regarded as insignificant. However, in 1972, *Megastigmus* sp., the same parasitoid reported in Queensland, Australia, was found parasitizing up to 71% of the fly's population in New Zealand (Hill, 1989). In South Africa, *P. utilis* is attacked by *Dimeromicrus* sp. (Hymenoptera: Torymidae), *Eupelmus* sp. (Hymenoptera: Eupelmidae) and an unidentified species of Pteromalidae, and rate of parasitism varied from 26% to 52% (Bennett, 1986).

Other organisms that have been reported to feed on the galls include: larvae of *Heliothis* spp. (Lepidoptera: Noctuidae) in Australia (Dodd, 1961); larvae of *Spodoptera litura* (Lepidoptera: Noctuidae) and slugs in Nepal (Sharma and Chhetri, 1977); and mice in Oahu, Hawaii (Bess and Haramoto, 1958).

4.2.5 Other natural enemies recorded on A. adenophora

Oidaematophorus beneficus *Yano and Heppner (Lepidoptera: Pterophoridae)*

This insect was introduced to Hawaii from Mexico for the control of *Ageratina riparia* (Regal) R. M. King and H. Robinson (Asteraceae) (Nakao *et al.*, 1975). The larva is a leaf feeder and causes smooth edged holes. In 1991, Conant (1998) reported finding a few specimens feeding on *A. adenophora* in Hawaii. It is possibly a spillover feeding incidence than a true association (T. Olckers, personal communication).

Dihammus argentatus *Auriv. (Coleoptera: Cerambycidae)*

Dihammus argentatus is an indigenous Australian cerambycid that has been recorded on *A. adenophora* since 1950 (Dodd, 1961). Stem-boring larvae are found in the rootstock and base of the stems of larger plants, while smaller plants are not favored. During the rainy season, the infested plants are not seriously weakened but in the dry season they suffer damage and mortality in extreme cases. This insect has been observed damaging cultivated dahlias (Dodd, 1961) and thus cannot be considered for introduction into other countries as a "new association" biological control agent.

Phaeoramularia *sp. (Fungi: Ascomycota)*

A leaf-spot fungus, isolated at gall-fly release sites in Queensland, Australia, in 1954 suggested that the spores were passively transmitted by *P. utilis* adults from Hawaii. By 1957, the leaf-spot disease occurred in all Crofton weed areas in Australia. Originally this fungus was determined to be *Cercospora eupatorii* Peck (Dodd, 1961), but has since been assigned to *Phaeoramularia* and is probably a new species (Morris, 1991). This leaf-spot fungus could have originated from tropical Central America (Julien and Griffiths, 1998) and appears to have been introduced accidentally into Hawaii, Australia, New Zealand, India, and Nepal, wherever gall-fly releases took place.

For introduction into South Africa, a single-spore isolate of the pathogen was obtained from infected leaf material from Queensland in 1984 and used for host-specificity studies and subsequent releases (Morris, 1991). Several species of Asteraceae were tested and all, except *A. adenophora*, were found resistant to the fungus. It was originally released at Stellenbosch and Pietermaritzburg (South Africa) between 1987 and 1989, but has since been redistributed to other areas. The fungus has established well and has caused partial defoliation of plants at Pietermaritzburg, but not at Stellenbosch, as the fungus is not adapted to the Mediterranean climate of the southwestern Cape (Morris, 1991).

4.2.6 Other candidate agents

The recent expansion of *A. adenophora* in South Africa has raised concerns that the weed is emerging as a more serious problem and that the two established agents are having little impact. Consequently, funds have been secured for the importation of additional biocontrol agents. In 2007, a trip was undertaken to Mexico to survey for promising insect and pathogen agents. Some of the more promising agents included stem-boring and defoliating insect species as well as a range of pathogens (S. Neser, personal communication). More collecting trips have been planned and introductions of new agents are imminent.

4.3 Biological control and the status of the weed

In Hawaii, prior to the introduction of *P. utilis*, the weed had developed into dense thickets and grown up to three meters tall on commercially important grazing land; infested ranchland was reclaimed by expending labor and funds for mechanical removal of the weed. Within a few years after the introduction of the fly, the plant was effectively controlled in several thousand acres of rangeland. In some areas, the fly eliminated the weed completely. Even though parasitism of the fly was 50% or more in some areas, this did not diminish its efficacy in Hawaii. Although the fly did not prove effective in the wet, steep slopes of east Maui, it was generally considered to be an outstanding success (Bess and Haramoto, 1958, 1959, 1972). The critical returns on the introduction of *P. utilis* into Hawaii included savings due to the reduced need for mechanical equipment and labor for removal of the weed, improvement in biodiversity, reduction in animal toxicity and prevention of further spread of the weed.

In Australia, partial control of the weed was achieved due to the introduction of the fly and was aided by the inadvertently introduced leaf-spot fungus and an indigenous cerambycid stem borer, *D. argentatus* (Dodd, 1961). Since 1952, the spread of this weed has not increased (Page and Lacey, 2006). In New Zealand, the fly inflicted significant damage on the weed for five years after its introduction. The abundance and importance of the weed have declined in the last 25 years and it is no longer considered to be economically important (Hill, 1989). Introduction of the gall fly into India and its eventual movement into Nepal has resulted in some reduction in vigor, growth, and density of the plant; however, the heavy incidence of parasitism has reduced the efficiency of *P. utilis* (Sankaran, 1973). The gall fly dispersed from Nepal and has established in an area near Tibet in 1984. The flies were also collected and released in Yunnan and neighboring provinces in southern China for control of the weed (Zhang *et al.*, 1988). In South Africa, recent studies have suggested the fly is having a limited impact and has not curtailed the spread of the weed, making the introduction of new agents a priority (S. Neser and A. B. R. Witt, personal communication).

The cost of introduction of *P. utilis* to various countries has been minimal. When the fly was collected on *A. adenophora* in Mexico in 1944, an immediate decision was made to import it into Hawaii by the Territorial Board of Agriculture and Forestry, and by 1945, it was released and established. Although some expenditure was incurred for initial host-specificity studies in Hawaii (including tests for Australia), no costs have accrued to other countries except for minor transportation expenses. In many countries, *P. utilis* has suppressed the weed and contained its spread, despite heavy parasitism by local parasitoids in some countries. Besides South Africa, none of the countries affected by *A. adenophora* has considered further introductions of additional natural enemies.

References

Auld, B. S. and Martin, P. M. (1975). The autecology of *Eupatorium adenophorum* Spreng. in Australia. *Weed Research*, **15**, 27–31.

Abrahamson, W. G. and Weis, A. E. (1987). Nutritional ecology of arthropod gall makers. In *The Nutrional Ecology of Insects*, ed. F. Slansky, Jr. and J. G. Rodriguez. New York: Wiley, pp. 236–258.

Bennett, P. H. (1986). An investigation into the biological control of Crofton weed, *Ageratina adenophora* (K & R.) (syn. *Eupatorium adenophorum* Spreng.) by the gall fly *Procecidochares utilis* Stone. M.Sc. thesis, University of Natal, Pietermaritzburg, 205 pp.

Bennett, P. H. and Van Staden, J. (1986). Gall formation in Crofton weed, *Eupatorium adenophorum* Spreng. (syn. *Ageratina adenophora*), by the Eupatorium gall fly

Procecidochares utilis Stone (Diptera: Trypetidae). *Australian Journal of Botany*, **34**, 473–480.

Bess, H. A. and Haramoto, F. H. (1958). Biological control of pamakani, *Eupatorium adenophorum*, in Hawaii by a tephritid gall fly, *Procecidochares utilis*. 1. The life history of the fly and its effectiveness in the control of the weed. *Proceedings of the Tenth International Congress of Entomology*, Vol. 4, ed. E. C. Becker. Ottawa, Canada: Mortimer, 543–548.

Bess, H. A. and Haramoto, F. H. (1959). Biological control of pamakani, *Eupatorium adenophorum*, in Hawaii by a tephritid gall fly, *Procecidochares utilis*. 2. Population studies of the weed, the fly, and the parasites of the fly. *Ecology*, **40**, 244–249.

Bess, H. A. and Haramoto, F. H. (1972). Biological control of pamakani, *Eupatorium adenophorum*, in Hawaii by a tephritid gall fly, *Procecidochares utilis*. 3. Status of the weed, fly and parasites of the fly in 1966–71 versus 1950–57. *Proceedings of the Hawaiian Entomological Society*, **21**, 165–178.

Borthakur, D. N. (1977). *Mikania* and *Eupatorium*, two noxious weeds of NE region. *Indian Farming*, **26**, 48–49.

Bronner, R. (1992). The role of nutrient cells in the nutrition of cynipids and cecidomyiids. In *Biology of Insect Induced Galls*, ed. J. D. Shorthouse and O. Rohritsch. New York: Oxford University Press, pp. 118–140.

Conant, P. (1998). A new host record for *Oidaemetophorus beneficus* Yana & Heppner (Lepidoptera: Pterophoridae). *Proceedings of the Hawaiian Entomological Society*, **33**, 151–152.

Cruz, Z. T., Muniappan, R. and Reddy, G. V. P. (2006). Establishment of *Cecidochares connexa* (Diptera: Tephritidae) in Guam and its effect on the growth of *Chromolaena odorata* (Asteraceae). *Annals of the Entomological Society of America*, **99**, 845–850.

Dhyani, S. K. (1978). Allelopathic potential of *Eupatorium adenophorum* on seed germination of *Lantana camara* var. *aculeata*. *Indian Journal of Forestry*, **1**, 311.

Dodd, A. P. (1961). Biological control of *Eupatorium adenophorum* in Queensland. *Australian Journal of Science*, **23**, 356–365.

Everist, S. L. (1959). Strangers within the gates (Plants naturalized in Queensland). *Queensland Naturalist*, **16**, 49–60.

Falvey, J. L. (1982). Factors limiting cattle production in the Thai highlands. *Journal of the Australian Institute of Agricultural Science*, **48**, 51.

Florentine, S. K., Raman, A. and Dhileepan, K. (2001). Gall-inducing insects and biological control of *Parthenium hysterophorus* L. (Asteraceae). *Plant Protection Quarterly*, **16**, 63–68.

Florentine, S. K., Raman, A. and. Dhileepan, K. (2005). Effects of gall induction by *Epiblema styenuana* on gas exchange, nutrients, and energetics in *Parthenium hysterophorus*. *BioControl*, **50**, 787–801.

Fuller, T. C. (1981). Introduction and spread of *Eupatorium adenophorum* in California. *Proceedings of the Eighth Asian-Pacific Weed Science Society Conference*, held in Bangalore, November 1981, pp. 277–280.

Hickman, J. C. (1993). *The Jepson Manual: Higher Plants of California*. Berkeley, CA: University of California Press, 1400 pp.

Hill, R. L. (1989). *Ageratina adenophora* (Sprengel) R. King & H. Robinson, Mexican devil weed (Asteraceae). In *A Review of Biological Control of Invertebrate Pests and Weeds in New Zealand 1874 to 1987*, ed. P. J. Cameron, R. L. Hill, J. Bain and

V. P. Thomas, Technical Communication 10. Wallingford, UK: CAB International, pp. 317–320.

Hoshovsky, M. C. and Lichti, R. (2007). *Ageratina adenophora*. http://ucce.ucdavis.edu/datastore/detailreport.cfm?usernumber=2&surveynumber=182

Julien, M. H. and Griffiths, M. W. (1998). *Biological Control of Weeds: A World Catalogue of Agents and their Target Weeds*. Wallingford, UK: CABI Publishing.

Kapoor, V. C. and Malla, Y. K. (1979). New record of *Dimeromicrus vibidia* (Walker) (Hymenoptera: Torymidae), a parasite of the gall fly *Procecidochares utilis* (Stone) (Diptera: Tephritidae) from Nepal. *Journal of Bombay Natural History Society*, **75**, 932.

Kaul, S. and Bansal, G. L. 2002. Allelopathic effect of *Ageratina adenophora* on growth and development of *Lantana camara*. *Indian Journal of Plant Physiology*, **7**, 195–197.

Kluge, R. L. (1991). Biological control of Crofton weed, *Ageratina adenophora* (Asteraceae), in South Africa. *Agriculture, Ecosystems and Environment*, **37**, 187–191.

Meyer, J. and Maresquelle, H. J. (1983). *Anatomie des Galles*. Stuttgart, Germany: Gebrüder Borntraeger.

Morris, M. J. (1989). Host specificity studies of a leaf spot fungus, *Phaeoramularia* sp., for the biological control of Crofton weed (*Ageratina adenophora*) in South Africa. *Phytophylactica*, **21**, 281–283.

Morris, M. J. (1991). The use of plant pathogens for biological weed control in South Africa. *Agriculture, Ecosystems and Environment*, **37**, 239–255.

Nakao, H. K., Funasaki, G. Y. and Davis, C. J. (1975). Introductions for biological control in Hawaii, 1973. *Proceedings of the Hawaiian Entomological Society*, **22**, 109.

Niu, H. B., Liu, W. X., Wan, F. H. and Liu, B. (2007). An invasive aster (*Ageratina adenophora*) invades and dominates forest understories in China: altered soil microbial communities facilitate the invader and inhibit natives. *Plant and Soil*, **294**, 73–85.

O'Sullivan, B. M. (1979). Crofton weed (*Eupatorium adenophorum*) toxicity in horses. *Australian Veterinary Journal*, **55**, 19–21.

O'Sullivan, B. M. (1985). Investigation into Crofton weed (*Eupatorium adenophorum*) toxicity in horses. *Australian Veterinary Journal*, **62**, 30–32.

Page, A. R. and Lacey, K. L. (2006). *Economic Impact Assessment of Australian Weed Biological Control*. Glen Osmond, Australia: Technical Series 10. CRC for Australian Weed Management, 150 pp.

Parsons, W. T. (1992). *Noxious Weeds of Australia*. Melbourne: Inkata Press, 692 pp.

Raman, A. (1994). Adaptational integration between gall-inducing insects and their host plants. In *Functional Dynamics of Phytophagous Insects*, ed. T. N. Ananthakrishnan. New Delhi: Oxford & IBH Publishing Company, pp. 249–276.

Raman, A. and Abrahamson, W. G. (1995). Morphometric relationships and energy allocation in the apical rosette galls of *Solidago altissima* (Asteraceae) induced by *Rhopalomyia solidaginis* (Diptera: Cecidomyiidae). *Environmental Entomology*, **24**, 635–639.

Raman, A., Madhavan, S., Florentine, S. K. and Dhileepan, K. (2006). Stable-isotope ratio analyses of metabolite mobilization in the shoot galls of *Parthenium hysterophorus* (Asteraceae) induced by *Epiblema strenuana* (Lepidoptera, Tortricidae). *Entomologia Experimentalis et Applicata*, **119**, 101–107.

Rao, V. P., Ghani, M. A., Sankaran, T. and Mathur, K. C. (1971). *A Review of the Biological Control of Insects and Other Pests in Southeast Asia and the Pacific*

Region. Technical Communication. Slough, UK: Commonwealth Institute of Biological Control, 149 pp.

Robbins, W. W. (1940). Alien plants growing without cultivation in California. *Bulletin of the California Agricultural Experiment Station*, **637**, 1–128.

Sankaran, T. (1973). Biological control of weeds in India. A review of introductions and current investigations of natural enemies. In *Proceedings of the Second International Symposium on Biological Control of Weeds*, ed. P. H. Dunn. Slough, UK: Commonwealth Agricultural Bureaux, pp. 82–88.

Sharma, K. C. and Chhetri, G. K. K. (1977). Reports on studies on the biological control of *Eupatorium adenophorum*. *Nepalese Journal of Agriculture*, **12**, 135–157.

Swaminathan, S. and Raman, A. (1981). On the morphology of the stem-galls of *Eupatorium adenophorum* Spreng. (Compositae) and the natural enemies of the cecidozoan, *Procecidochares utilis* Stone (Tephritidae, Diptera). *Current Science*, **50**, 294–295.

Wagner, W. L., Herbst, D. R. and Sohmer, S. H. (1999). *Manual of the Flowering Plants of Hawai'i*. Revised edition. Honolulu, HI: University of Hawai'i Press, 254 pp.

Wan, F. and Wang, R. (1991). Achievements of biological weed control in the world and its prospects in China. *Chinese Journal of Biological Control*, **7**, 81–87.

Waterhouse, D. F. (1993). *The Major Invertebrate Pests and Weeds of Agriculture in Southeast Asia*. Canberra: The Australian Centre for International Agricultural Research, 141 pp.

Wilson, E., Walisiewicz, M., Harvey, S., Gay, H. and Shrestha, K. (1985). The report of the Oxford University expedition to Nepal. Unpublished Report. 40 pp.

Zhang, Z., Wei, Y. and He, D. (1988). Biology of a gall fly, *Procecidochares utilits* (Dip.: Tephritidae) and its impact on Crofton weed, *Eupatorium adenophorum*. *Chinese Journal of Biological Control*, **4**, 10–13.

5

Azolla filiculoides Lamarck
(Azollaceae)

M. P. Hill and A. J. McConnachie

5.1 Introduction

Azolla Lam. is an aquatic fern taxon, which grows in symbiotic association with *Anabaena azolla* Strasburger (Nostocales: Nostocaceae) within the dorsal leaf lobe cavities (Ashton and Walmsley, 1976, 1984). *Anabaena azolla* can fix atmospheric nitrogen and is able to fulfill nitrogen requirements of *Azolla*, making it able to thrive in nitrogen-deficient waters (Ashton, 1974, 1978). *Azolla* is economically important and has been used in Southeast Asia as a green manure associated with wetland rice cultivation for the last 200 years (Lumpkin and Plucknett, 1982). However, the wider utilization of *Azolla* for agricultural purposes has been constrained by various biological factors including low tolerance to high temperatures and insect damage (Van Cat *et al.*, 1989) and application of ammonia-based fertilizers (Lumpkin and Plucknett, 1982).

5.2 Taxonomy

Early classifications of *Azolla* were based mainly on vegetative characteristics, in particular, using the form and size of leaves (Svenson, 1944). This, however, has led to considerable confusion, since the phenotypes of *Azolla* are plastic, varying under environmental influences (Ashton, 1978; Wantanabe and Berja, 1983; Moretti and Gigliano, 1988). Zimmermann *et al.* (1989) reclassified *Azolla* using electrophoretic techniques, whereas Nayak and Singh (1989) used cytological techniques. Traditionally 25 fossil and seven extant species of *Azolla* are recognized (Hills and Gopal, 1967; Lumpkin and Plucknett 1980; Ashton and Walmsley, 1984), which are divided into two sections:

1. *Azolla* (*Euazolla*) Lam. that includes *A. filiculoides* Lamarck, indigenous to South America and western North America, but introduced into Europe, southern Africa, China, Japan, and southern Australia, and is most commonly used as green manure; *A. caroliniana* Willd., which is indigenous to the eastern United States, but found elsewhere in North, Central, and South America, and Europe; *A. mexicana* Presl., which is distributed from northern South America to North America extending up to British Columbia; *A. ruba* R. Br. which is usually regarded as a variety of *A. filiculoides* and is

Biological Control of Tropical Weeds using Arthropods, ed. R. Muniappan, G. V. P. Reddy, and A. Raman. Published by Cambridge University Press. © Cambridge University Press, 2009.

recorded only in Australia and New Zealand; and *A. microphylla* Kaulf., which is recorded from western and northern South America, Central America, and the West Indies.

2. *Rhizosperma* (Mey.) Mett., which includes *A. pinnata* R. Br., distributed widely in tropical Africa, Australasia, and Southeast Asia, and *A. nilotica* Decne. ex. Mett., found only in Africa from Egypt to South Africa (Stergianou and Fowler, 1990).

As to the status of *A. pinnata*, some confusion prevails. Initially it was regarded as a complex of *A. pinnata*, *A. africana* Desv. and *A. imbricata* Roxb. ex Griff., but later reduced to one species with two varieties, *A. pinnata* var. *imbricata* and *A. pinnata* var. *africana* (also called var. *pinnata*) (Sweet and Hills, 1971; Stergianaou and Fowler, 1990). Nayak and Singh (1989), using karyological and morphological results, suggested that *A. pinnata* var. *africana* was sufficiently morphologically different form the other two varieties and should once again be accorded the species status, as *A. africana*.

A revision of the section *Rhizosperma* (Saunders and Fowler, 1992) shows considerable intraspecific variation within *A. pinnata*. Three main geographically related intraspecific groups are evident: African, Asian, and Australasian. The "distinct" morphology of the variety "Asian" has long been recognized either as a variety of *A. pinnata*, that is var. *imbicata* (e.g. Sweet and Hills, 1971; Lumpkin and Plucknett, 1982; Tan *et al.*, 1986), or as a distinct species, *A. imbricata* (Lin, 1980). However, the integrity of the Asian variant of *A. pinnata* is less evident when specimens from the Indian subcontinent were considered, which displayed morphological characters resembling closely those of the African variant indicating that the Asian variant should be *A. pinnata* subsp. *asiatica* R. M. K. Saunders & K. Fowler, the African variant should be *A. pinnata* subsp. *africana* (Desv.) R. M. K. Saunders & K. Fowler, and the Australasian variant should be *A. pinnata* subsp. *pinnata* R. Brown (Saunders and Fowler, 1992).

A recent revision of species of American *Azolla* based on leaf trichomes and glochidia indicates that only two species exist in America: *A. filiculoides* (= *A. caroliniana*) and *A. cristata* (= *A. mexicana*, *A. microphylla* and *A. caroliniana*) (Evrard and van Hove, 2004). Accordingly *A. caroliniana* refers to *A. caroliniana sensu* Willdenow (= *A. filiculoides*) and *A. caroliniana sensu* Mettenius (= *A. cristata*).

5.3 Plant biology

Azolla filiculoides is an aquatic perennial heterosporous fern, rarely larger than 25 mm in overall width (O'Keeffe, 1986), native to South America and southern North America (Lumpkin and Plucknett, 1980), but now widely distributed throughout the world, and often gaining the status of a weed (Ashton, 1983). Every plant consists of a short, branched rhizome that bears small, alternate, overlapping leaves and roots hanging into the water (Ashton, 1974,1978).

Azolla filiculoides can reproduce vegetatively rapidly throughout the year through elongation and fragmentation of the small fronds, and under optimal conditions, the everyday rate of increase can exceed 15%. Under ideal conditions the doubling time of populations can be 4–5 days (Lumpkin and Plucknett, 1982). Moreover, the fern reproduces sexually via spores in spring and summer, which overwinter and that are resistant to extreme desiccation, thus enabling the fern to reestablish after any spells of drought.

5.4 Impact

This plant is most notably considered a weed in the Republic of South Africa, but is also considered a weed of minor importance in Portugal, Ireland, and the United Kingdom (Gassmann *et al.*, 2006). The increasing abundance of *A. filiculoides* in conservation, agricultural, recreational, and suburban areas since the late 1980s was a matter of concern. Among the major consequences of the dense mats (5–20 cm thick) of the weed on slow-moving and still waters are: reduction in quality of drinking water due to unpleasant odor, color and turbidity, promotion of water-related human diseases, increased siltation of rivers and dams, loss of water by evapotranspiration through the weed surfaces, reduction in water-surface area used for recreation (fishing, swimming, and water skiing) and water transport, deterioration of aquatic biodiversity, clogging of irrigation pumps, and reduction in the water flow in canals used for irrigation (McConnachie *et al.*, 2003).

5.5 Utilization

Azolla has been used as a green manure in rice paddies in China and Vietnam over the last 200-odd years (Lumpkin and Plucknett, 1980, 1982): *A. filiculoides* and *A. pinnata* var. *imbricata* are used in this practice (Lumpkin and Plucknett, 1982). The most widely used system is to raise *Azolla* in rice paddies, by floating *Azolla* in the paddy prior to rice planting; the paddy is drained after 6–8 weeks, and *Azolla* is subsequently ploughed into the soil. This improves the soil quality by increasing organic-nitrogen levels, improving water-holding and cation-exchange capacities of the soil. Under such conditions, *Azolla* contributes as much as 180 kg of organic nitrogen/hectare/year to the soil and increases rice yields by 100%. Intercropping of *Azolla* with rice by not draining the paddy is also possible, but the organic nitrogen becomes available only later in the season; however, in such circumstances, *Azolla* will compete for phosphates with rice seedlings. The techniques required to utilize *Azolla* as a green manure are complex and labour intensive; but they are viable in regions where commercially produced nitrogen-based fertilizers are expensive.

   Azolla can also be used as a fodder for swine, poultry, cattle, and fish, but cannot be used as the only protein source and should be supplemented with other feeds. The advantages are that it has a high nutrient content, it grows quickly on natural water bodies, it is available throughout the year and does not need processing. *Azolla* is compostable, used as an ornamental on ponds and fish tanks; in India, *Azolla* is eaten; it can be used in mosquito control, because a complete mat disrupts larval development (however, an incomplete mat increases mosquito problems as it affords the larvae protection from predation).

   Taking into consideration the potential for utilization of *A. filiculoides* and its severe impact on aquatic ecosystems in southern Africa, chances for resolving the conflict of interest between those wanting to utilize *A. filiculoides* and those wanting to control it are slim. Control methods, whether chemical, mechanical, or biological, can never result in the total eradication of *Azolla*; and any utilization program can never control it totally.

5.6 Management

Techniques for the management of *A. filiculoides* fall into three categories. These are mechanical control, chemical control, and biological control.

5.6.1 Mechanical

Mechanical control is a labour-intensive method, but has the advantage of being eco-logically benign. Small infestations of *Azolla* in accessible areas can be removed with rakes and fine-mesh nets, and used as either fodder or compost. The disadvantage of this method is that under optimal conditions, *Azolla* can double its population every 4–5 days (Lumpkin and Plucknett, 1982), necessitating a concerted effort. Even if total eradication could be achieved, re-establishment of the weed from spores resident in the substratum is inevitable. *Azolla* is susceptible to fragmentation because of physical disturbances and the fragments are sensitive to high-light intensity and killed by direct sunlight. Capitalizing on this behavior, Ashton (1992) proposed a mechanical agitator, which can provide enough turbulence to fragment *Azolla* stolons; however, cost of such a control method, even on a small scale, is prohibitive.

5.6.2 Chemical

Chemical control of *A. filiculoides* using the either glyphosate (Steyn *et al.*, 1979; Ashton, 1992) or paraquat and diquat (Axelsen and Julien, 1988), or kerosene mixed with a surfactant (Diatloff and Lee, 1979) is recommended. Chemical control has a few weaknesses: expensive, especially in view of the extensive follow-up program necessary to monitor and eradicate plants germinating from spores; treated plants remaining in the water cause extensive deoxygenation within the system, affecting water quality; possibility of spray drift onto non-target vegetation, water cannot be used for either irrigation or watering stock until the breakdown of the herbicide; and the need for trained personnel to administer spraying operations.

5.6.3 Biological

On the grounds of insufficient research and the level of risk involved, biological control of *A. filiculoides* was considered an inappropriate method (Ashton, 1992). However, in view of the difficulties involved with mechanical control and the expense, risk, and variable results of chemical control programs, biological control is the only sustainable method for controlling this strongly invasive weed.

Phytophagous species associated with Azolla filiculoides

As *A. filiculoides* is a problem only in South Africa, there has only been one dedicated survey in the native range of *Azolla* for potential control agents (McConnachie and Hill, 2005). However, as it is utilized in other parts of the world, inventories of insects associated with *A. filiculoides* are available. Lumpkin and Plucknett (1982) and Calilung and Lit (1986) recorded six species of Lepidoptera, five species of Diptera, two species of Orthoptera, one species of an aphid, and several snail species that warranted control when trying to utilize the plant in Southeast Asia. During a preintroductory survey in South Africa, Hill (1998a) recorded *Rhopalosiphum nymphaeae* (L.) (Hemipetra: Aphidoidea) and *Nymphula (Synclita) obliteralis* (Walker) (Lepidopetra: Pyralidae) feeding and developing on the plant. However, all of these are considered generalist herbivores that have utilized *A. filiculoides* in its regions of introduction. Four specialist species have been recorded on the plant: two species of chrysomelid beetles, *Pseudolampsis guttata* (Leconte) from Florida (Habeck, 1979) and *Pseudolampsis darwini* (Scherer) from Brazil and northern Argentina (Casari and Duckett, 1997), and two weevil species, *Stenoplemus rufinasus* Gyllenhal from Florida (Richerson and Grigarick, 1967), and *Stenopelmus brunneus* (Hustache) from northern Argentina (McConnachie and Hill, 2005), Brazil, and Peru (C. O'Brien, personal communication). Of these, *S. rufinasus* and *P. guttata* have been tested as possible biological control agents for *A. filiculoides* in South Africa.

Stenoplemus Rufinasus *Gyllenhal (Coleoptera: Curculionidae)*

The frond-feeding weevil *Stenopelmus rufinasus* was imported into South Africa from Florida, USA, in the late 1995 (Hill, 1998b). The adults are about 1.7 mm in length, gray-black, and covered with red, black, and white scales in varying patterns (Fig. 5.1). The sexes appear to be similar, but in males the first abdominal sternite is either flat or slightly concave along the midline and in females it is strongly convex. Adults live up to several months and the females produce on average 350 eggs. The female usually chews a hole in the tip of the frond into which a yellow egg is inserted and the hole is then covered with frass. The average incubation period is four days. Three larval instars occur, which feed voraciously on the fronds and the rhizome of the plant. Older larvae are capable of consuming several plants/day. The duration of each instar is 2–3 days. Pupation occurs in a black, ovoid chamber constructed within *A. filiculoides*. The pupal period is 4–6 days long. The duration of development of the immature stages, from egg to eclosion as adult, ranges over 12–23 days (Hill, 1998b; McConnachie, 2004).

The host range of *S. rufinasus* was determined through adult "no-choice" trials on 31 species of plants from 19 families (Hill, 1998b). Although some feeding and development occurred on other species of *Azolla*, they proved to be far inferior hosts in comparison with *A. filiculoides* and would not be able to sustain populations of the weevil under field conditions, thus presenting no danger to nontarget plant species. The weevil was cleared for release in South Africia in late 1997 (Hill, 1999).

Fig. 5.1 Adult *Stenopelmus rufinasus* on *Azolla filiculoides*.

Pseudolampsis guttata *(Hustache) (Coleoptera: Chrysomelidae)*

The flea beetle *Pseudoplampsis guttata* was collected on *Azolla caroliniana* (now considered to be synonymous with *A. filiculoides* (Evrard and van Hove, 2004)) in Florida, USA, and imported into South Africa in late 1997; the biology of this beetle is available in Buckingham and Buckingham (1981). The adults are 2–2.5 mm long, light brown, and covered in golden setae. Females lay eggs between lower overlapping lobes of the fronds, the eggs are yellow and incubation requires about a week. Three larval instars occur feeding on fronds and the mature third instar forms a cocoon in the fronds, within which pupation occurs. Pupation lasts 3–5 days. *Pseudoplampsis guttata* larvae inflict greater damage to *A. filiculoides* than those of *S. rufinasus*. A female beetle lays about 670 eggs in a 130-day oviposition period and the adults live up to 200 days. The total generation time is 14–24 days (Buckingham and Buckingham, 1981).

  Host-specificity screening of *P. guttata* was conducted in South Africa on 18 species in 10 families (Hill and Oberholzer, 2002). Considerable adult feeding and oviposition and larval development occurred on *Azolla nilotica*, *A. pinnata*, and *Salvinia hastata* Desv., which are indigenous to southern Africa. Although *A. filiculoides* was the most suitable host, the potential threat to the nontarget species was considered substantial and therefore *P. guttata* was rejected (Hill and Oberholzer, 2002).

Implementing biological control

Stenoplemus rufinasus was released at 112 sites (representing 208 ha) of *A. filiculoides*-infested water bodies in South Africa between 1997 and 2002 and one (7 ha) site in

Zimbabwe in 1999. The weevils were posted initially in batches of 500 adults. However, these starter cultures were later reduced to 100 (which were found to be sufficient to establish viable field populations). The starter colonies were sent to landowners through the general postal service. The adults were sent without food to prevent the spread of other aquatic aliens such as duckweeds (e.g. *Lemna* sp. and *Spirodela* sp.). A central record of weevil establishment and the impact of the weevil on the weed (changes in the area of the water body covered, time taken for the weed to disappear, reappearance of the weed and recolonization by the weevil) were maintained at the Plant Protection Research Institute in Pretoria. The effect of the weevils on *A. filiculoides* was recorded using "before" and "after" fixed-point photographs. Involvement of the public in this program contributed to its success as individual landowners took ownership of the weevils and consequent control of the program.

Ninety-one field sites (out of the 112 trialled) were completely cleared of the weed by the weevils in about 7 months (McConnachie *et al.*, 2004). This appears to be an impressive example in weed biological control considering the speed with which the weed was brought under control and the fact that at the release sites, no residual population of the weed remained (Fig. 5.2). Moreover the weevil dispersed to other sites where they had not previously been released. Between 1997 and 2007, *A. filiculoides* has returned to about 50% of the initial release sites. However, these infestations have not reached the pre-1997 levels, in a sense that thick mats of *A. filiculoides* have not reappeared and the weevils have recolonized weed at all of these sites. The biological control program against *A. filiculoides* in South Africa has been highly successful. The release of the weevil has resulted in the weed no longer posing a problem to waterways in that country.

Azolla filiculoides is a minor pest in the United Kingdom, where it invades small water bodies, especially in the south of the country in summer months (Janes, 1998). The fern is able to overwinter as spores, although the water body freezes in winter. The incidence and impact of the weed had increased since about 2000 and biological control was considered (Gassmann *et al.*, 2006). Although *Stenopelmus rufinasus* was recorded in the United Kingdom in the early 1920s (Janson, 1921), it does not appear to have had the same impact as it did in South Africa. However, an outbreak of the weevil in southern UK in 2002 led to the widespread effort to control *A. filiculoides* (Gassmann *et al.*, 2006). Currently the weevil is being reared and released by CAB-Bioscience in the UK. The weevil was also recorded in the south of France on *A. filiculoides* in 2007; its mode of introduction is unknown, but based on the weevil dispersal data from South Africa, this population could have originated from the UK population.

5.7 Economics of biological control efforts

McConnachie *et al.* (2003) undertook an economic evaluation of the biological control program on *A. filiculoides* in South Africa. Affected water users were surveyed using a questionnaire to assess the importance of the weed. The impact of the weed included loss

(a)

(b)

Fig. 5.2 Before and after photographs of the impact of *Stenopelmus rufinasus* on *Azolla filiculoides* in the field in South Africa. (a) and (b): Witmos, Eastern Cape Province – 312 days to clearance. (c) and (d): Slykspruit River, Free State Province – 271 days to clearance. (e) and (f): Sasolburg Nature Reserve Dam, Free State Province – 270 days to clearance. Note secondary infestation of *Wolffia* sp., *Spirodela* sp. and *Lemna* sp. in foreground of (f).

(c)

(d)

Fig. 5.2 (cont.)

(e)

(f)

Fig. 5.2 (cont.)

of livestock, clogging of water pumps, building alternative water supply reservoirs, a drop in tourism, and loss of biodiversity. Unfortunately no monetary value was attributed to loss of biodiversity.

All costs were adjusted to South African R and values for 2000, which took into consideration adjustments made for inflation. The cost savings (per ha per year) resulting from the biological control program included a reduction of on-site damages caused by the weed was US$589/ha/year. The average cost (per ha/year) for the biological control program for 1995–2000 was US$278 (per ha/year). For the year 2000, the "benefit to cost" ratio was calculated at 2.5:1, which increased to 13:1 by 2005, and was estimated to be 15:1 by 2010 as the costs of the biological control program have declined over time and the benefits continue to accrue.

5.8 Sustainability of the *Azolla filiculoides* control program

Ten years after the release of the weevil *Stenopelmus rufinasus*, *A. filiculoides* no longer poses a threat to the aquatic ecosystems of South Africa. This weevil has been able to establish and control the weed across the entire climatic range of the weed. McConnachie (2004) showed that the adult weevil had a lower lethal temperature of about −12 °C and an upper lethal temperature of ±40 °C. While the insect has never been exposed to this range of temperatures in field conditions in South Africa, it demonstrates that the weevil populations are limited only by the host plant densities and not by the climate. No other intervention methods are required to reduce *A. filiculoides* populations that recur from spore germination. The weed is therefore considered to be under complete biological control in South Africa (Hoffmann, 1995).

5.9 Conclusion

The biological-control program on aquatic weeds, including salvinia, water lettuce, water hyacinth, alligator weed, parrot's feather, and hydrilla have generally been successful (McFadyen, 1998) and that on red water fern is no exception. The biological control program against *A. filiculoides* in South Africa is, to date, one of the most dramatic examples of biological control in the 80-year history of science in South Africa. In fact it can be rated as a "unique" project. Rapid increases in weevil populations instigating local extinction of the weed within short periods of time make this program unique. The interaction between *S. rufinasus* and its host appears to be far more stochastic than programs on other weeds. The weevil finds dense mats of the weed, resulting in rapid population increases due to high fecundity and short generation times. The high feeding rates cause extensive damage to the mats, making them sink, thus leaving no residual red water fern populations. The insect population then undergoes massive larval and pupal mortality with the sinking of *A. filiculoides* mats, but the adults are capable of dispersing to locate other *Azolla* mats.

About 10 years after the first releases of *S. rufinasus* in *A. filiculoides*, the weed is under complete control in South Africa. However, with the decline of red water fern mats, other aquatic plant taxa (*Lemna* sp., *Wolffia* sp., *Spirodela* sp. and algae) have taken the vacated niche. The successful control of these three species will rely on a commitment to reducing eutrophication in the aquatic ecosystems in tropical countries of the world (Hill, 2003).

Acknowledgments

Water Research Commission of South Africa supported the biological control program against red water fern in South Africa.

References

Ashton, P. J. (1974). The effect to some environmental factors on the growth of *Azolla filiculoides* Lam. In *Orange River Progress Report*, ed. E. M. v. Zinderen-Bakker, Sr. Bloemfontein, South Africa: Institute for Environmental Sciences, University of the Orange Free State, pp. 123–138.

Ashton, P. J. (1978). Factors affecting the growth and development of *Azolla filiculoides* Lam. *Proceedings of the Second National Weeds Conference of South Africa*, held at Stellenbosch University, 1997. cape Town, South Africa: A. A. Balkema, pp. 249–268.

Ashton, P. J. (1983). The autecology of *Azolla filiculoides* Lamarck with special references to its occurrence in the Hendrick Verwoerd catchment area. Ph.D. thesis, Rhodes University, Grahamstown, South Africa.

Ashton, P. J. (1992). Azolla *Infestations in South Africa: History of the Introduction, Scope of the Problem and Prospects for Management*. Water Quality Information Sheet. Pretoria, South Africa: Department of Water Affairs and Forestry.

Ashton, P. J. and Walmsley, R. D. (1976). The aquatic fern *Azolla* and its *Anabaena* symbiont. *Endeavour*, **35**, 39–43.

Ashton, P. J. and Walmsley, R. D. (1984). The taxonomy and distribution of *Azolla* species in southern Africa. *Botanical Journal of the Linnean Society* **89**, 239–247.

Axelsen, S. and Julien, C. (1988). Weed control in small dams. II. Control of salvinia, azolla and water hyacinth. *Queensland Agricultural Journal*, **September–October**, 291–298.

Buckingham, G. R. and Buckingham, M. (1981). A laboratory biology of *Pseudolampsis guttata* (Leconte) (Coleoptera: Chrysomelidae) on waterfern, *Azolla caroliniana* Willd. (Pteridoptera: Azollaceae). *The Coleopterists Bulletin*, **35**, 181–188.

Calilung, V. J. and Lit, I. T. (1986). Studies on the insect fauna and other invertebrates associated with *Azolla* spp. *Philippines Agriculturalist*, **69**, 513–520.

Casari, S. A. and Duckett, C. (1997). Description of immature stages of two species of *Pseudolampsis* (Coleoptera: Chrysomelidae) and the establishment of a new combination in the genus. *Journal of the New York Entomological Society*, **105**, 50–64.

Diatloff, G. and Lee, A. (1979). A new approach for control of *Azolla filiculoides*. In *Proceedings of the 7th Asian-Pacific Weed Science Society Conference*, held in Sydney, Australia, 1979.

Evrard, C. and van Hove, C. (2004). Taxonomy of American *Azolla* species (Azollaceae): a critical review. *Systematics and Geography of Plants*, **74**, 301–318.

Gassmann, A., Cock, M. J., Shaw, R. and Evans, H. (2006). The potential for biological control of invasive alien aquatic weeds in Europe: a review. *Hydrobiologia*, **570**, 217–222.

Habeck, D. H. (1979). Host plants of *Pseudolampsis guttata* (Leconte (Coleoptera: Chrysomelidae). *The Coleopterists Bulletin*, **33**, 150.

Hill, M. P. (1998a). Herbivorous insect fauna associated with *Azolla* species in southern Africa. *African Entomology*, **6**, 370–372.

Hill, M. P. (1998b). Life history and laboratory host range of *Stenopelmus rufinasus*, a natural enemy for *Azolla filiculoides* in South Africa. *BioControl*, **43**, 215–224.

Hill, M. P. (1999). Biological control of red water fern, *Azolla filiculoides* Lamarck (Pteridophyta: Azollaceae), in South Africa. In *Biological Control of Weeds in South Africa (1990–1998)*. ed. T. Olckers and M. P. Hill: African Entomology Memoir 1. Johannesgws, South Africa: Entomological Society of Southern Africa, pp. 119–124.

Hill, M. P. (2003). The impact and control of alien aquatic vegetation in South African aquatic ecosystems. *African Journal of Aquatic Science*, **28**, 19–24.

Hill, M. P. and Oberholzer, I. G. (2002). Laboratory host range testing of the flea beetle, *Pseudolampsis guttata* (LeConte) (Coleoptera: Chrysomeliadae), a potential natural enemy for red water fern, *Azolla filiculoides* Lamarck (Pteridophyta: Azollaceae) in South Africa. *The Coleopterists Bulletin*, **56**, 79–83.

Hills, L. V. and Gopal, B. (1967). *Azolla primaeva* and its phylogenetic significance. *Canadian Journal of Botany*, **45**, 1179–1191.

Hoffmann, J. H. (1995). Biological control of weeds: the way forward, a South African perspective. In *Weeds in a Changing World*, ed. C. H. Stirton. BCPC Symposium Proceedings 64. Alton, UK: British crop protection council, pp. 77–89.

Janes, R. (1998). Growth and survival of *Azolla filiculoides* in Britain. *New Phytologist*, **138**, 377–384.

Janson, O. E. (1921). *Stenopelmus rufinasus* Gyll. an addition to the list of British Coleoptera. *Entomologists Monthly Magazine*, **57**, 225–226.

Lin, Y. X. (1980). A systemic study of the family Azollaceae with reference to the extending utilization of certain species in China. *Acta Phytotaxonomica Sinica*, **18**, 450–456.

Lumpkin, T. A. and Plucknett, D. L. (1980). *Azolla*: botany, physiology and use as green manure. *Economic Botany*, **34**, 111–153.

Lumpkin, T. A. and Plucknett, D. L. (1982). *Azolla* as a Green Manure: Use and Management in crop production. Westview Tropical Agriculture Series 5. Bolder, CO: Westview Press, 230 pp.

McConnachie, A. J. (2004). Post release evaluation of *Stenoelmus rufinasus* Gyllenhal (Coleoptera: Curculionidae) – a natural enemy released against the red water fern, *Azolla filiculoides* Lamarck (Pteridophyta: Azollaceae) in South Africa. Unpublished Ph.D. thesis, University of the Witwatersrand, Johannesburg, South Africa.

McConnachie, A. J. and Hill, M. P. (2005). *Biological Control of Red Water Fern in South Africa*. Report to the Water Research Commission of South Africa, WRC Report KV 158/05, 108 pp.

McConnachie, A. J., de Wit, M. P., Hill, M. P. and Byrne, M. J. (2003). Economic evaluation of the successful biological control of *Azolla filiculoides* in South Africa. *Biological Control*, **28**, 25–32.

McConnachie, A. J., Hill, M. P. and Byrne, M. J. (2004). Field assessment of a frond-feeding weevil, a successful biological control agent of red water fern, *Azolla filiculoides*, in southern Africa. *Biological Control*, **29**, 326–331.

McFadyen, R. E. (1998). Biological control of weeds. *Annual Review of Entomology*, **43**, 369–393.

Moretti, A. and Gigliano, G. S. (1988). Influence of light and pH on growth and nitrogenise activity of temperate grown *Azolla*. *Biology and Fertility of Soils*, **6**, 131–136.

Nayak, S. K. and Singh, P. K. (1989). Cytological studies in the genus *Azolla*. *Cytologia*, **54**, 275–286.

O"Keeffe, J. H. (1986). Ecological research on South African rivers – a preliminary synthesis. *South African National Scientific Programmes Report*, **121**, 1–121.

Richerson, P. J. and Grigarick, A. A. (1967). The life history of *Stenopelmus rufinasis* (Coleoptera: Curculionidae). *Annals of the Entomological Society of America*, **60**, 351–354.

Saunders, R. M. K. and Fowler, K. (1992). A morphological taxonomic revision of *Azolla*. *Linnaean Society*, **109**, 329–357.

Stergianou, K. K. and Fowler, K. (1990). Chromosome numbers and taxonomic implications in the fern genus *Azolla* (Azollaceae). *Plant Systematics and Evolution*, **173**, 223–239.

Steyn, D. J., Scott, W. E., Ashton, P. J. and Vivier, F. S. (1979). *Guide to the Use of Herbicides on Aquatic Plants*. Technical Report TR95. Pretoria, South Africa: Department of Water Affairs, pp. 1–29.

Svenson, H. K. (1944). The new world species of *Azolla*. *American Fern Journal*, **34**, 69–84.

Sweet, A. and Hills, L. V. (1971). A study of *Azolla pinnata* R. Brown. *American Fern Journal*, **61**, 1–13.

Tan, B. C., Payawal, P., Watanabe, I., Lacdan, N. and Ramirez, C. (1986). Modern taxonomy of *Azolla*: a review. *Philippine Agriculturalist*, **69**, 491–512.

Van Cat, D., Watanabe. I., Zimmerman, W. J., Lumpkin, T. A. and de Waha Baillonville, T. (1989). Sexual hybridization among *Azolla* species. *Canadian Journal of Botany*, **67**, 3482–3485.

Watanabe, I. and Berja, N. S. (1983). The growth of four species of *Azolla* affected by temperature. *Aquatic Botany*, **15**, 175–185.

Zimmermann, W. J., Lumpkin, T. A. and Watanabe, I. (1989). Classification of *Azolla*. *Euphytica*, **43**, 223–232.

6

Cabomba caroliniana Gray (Cabombaceae)

Shon Schooler, Willie Cabrera-Walsh, and Mic Julien

6.1 Introduction

Cabomba (*Cabomba caroliniana* Gray, Cabombaceae), or water fanwort, is a fast-growing submerged aquatic plant that has the potential to infest permanent water bodies in a range of regions – from tropical to cool temperate – throughout the world. It is considered a serious pest in the United States, Canada, the Netherlands, Japan, India, China, and Australia, and is present in Hungary, South Africa, and the United Kingdom. Cabomba grows well in slow-moving water bodies, preferring areas of permanent standing water less than 4 m deep; however, it can also grow at depths up to 6 m in Australia (Schooler and Julien, 2006). The weed is recognized by its opposing pairs of finely dissected underwater leaves that are feathery or fan-like in appearance (Fig. 6.1 and Fig. 6.2). Small white flowers bearing three petals and three sepals extend above the water surface, making infestations more visible in summer months (Fig. 6.3). Reproduction is almost entirely vegetative throughout most of the introduced localities and any fragment that includes nodes can grow into a new plant (Sanders, 1979).

Cabomba originates from South America (Orgaard, 1991). The plant's tolerance of fragmentation and delicate appearance make it a desirable aquarium plant (Hiscock, 2003) and consequently it was brought into many countries through the aquarium trade. Cabomba was subsequently introduced into lakes and streams both accidentally, through the dumping of aquarium water, and on purpose, to enable cultivation for later collection and sale. It is primarily spread across catchments by humans on watercraft, boat trailers, and eel-trapping cages.

Cabomba negatively affects the environment, recreational activities, public safety, and water quality (Mackey and Swarbrick, 1997). The weed can smother native submerged plants such as pondweeds (*Potamogeton* spp., Potamogetonaceae), stoneworts (*Chara* spp., Charophyceae), hornwort (*Ceratophyllum demersum* Linn., Hydrocharitaceae), and water nymph (*Najas tenuifolia* R. Br., Najadaceae). Cabomba infestation may also reduce germination of desirable native emergent plants. Alteration of the flora by cabomba is thought to have reduced populations of platypus (*Ornithorhynchus anatinus* Shaw,

Biological Control of Tropical Weeds using Arthropods, ed. R. Muniappan, G. V. P. Reddy, and A. Raman. Published by Cambridge University Press. © Cambridge University Press, 2009.

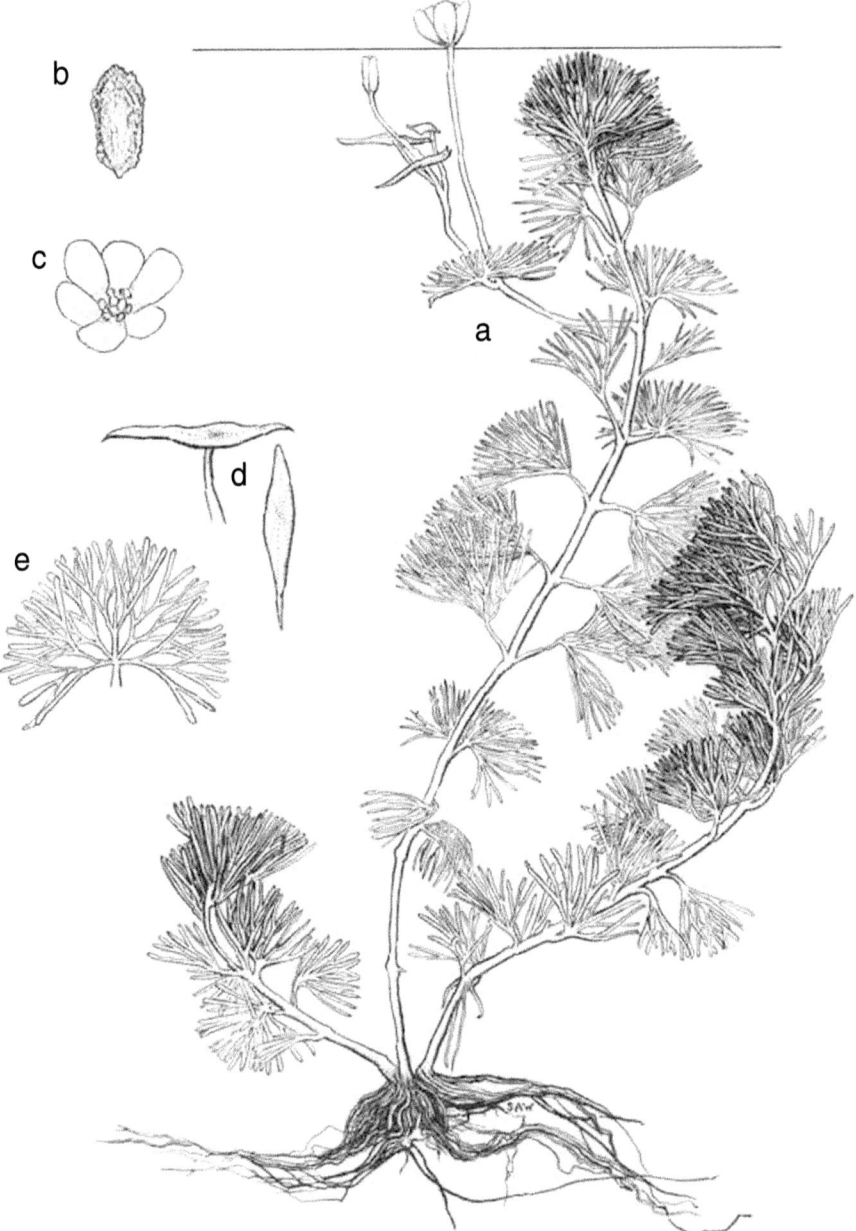

Fig. 6.1 Line drawing of *Cabomba caroliniana* illustrating (a) whole plant, (b) fruit, (c) flower, (d) entire upper leaves, and (e) dissected lower leaves.

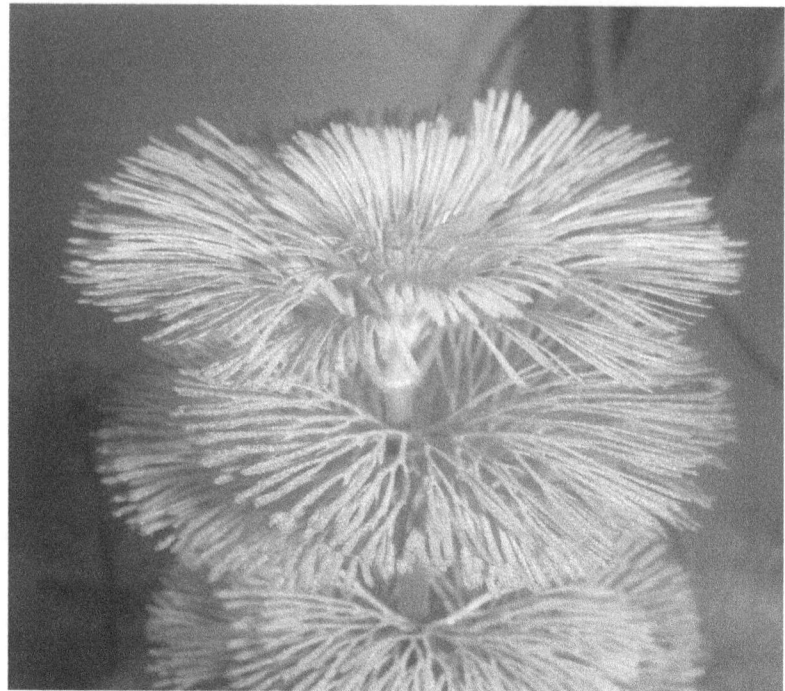

Fig. 6.2 Cabomba has highly dissected leaves that are fan-like in appearance. (Photo by S. Schooler, CSIRO Entomology.)

Ornithorhynchidae) and water rats (*Hydromys chrysogaster* Geoffroy, Muridae) in northern Queensland, Australia (Mackey and Swarbrick, 1997). In southern Queensland, cabomba appears to negatively affect populations of the endangered Mary River cod (*Maccullochella peelii mariensis* Rowland, Percichthyidae) (T. Anderson, personal communication). A study of impacts in Canada found changes in light penetration and composition of biological communities in cabomba beds when compared with native macrophyte communities (Hogsden *et al.*, 2007). The long stems of cabomba impede the movement of boats and can get tangled in propellers, paddles, and fishing lines. This makes many recreational activities less desirable in areas infested with cabomba and thereby impacts the tourism economy. In addition, cabomba is a potential danger to swimmers who could also become entangled in the long stems. It also interferes with dam machinery, such as valves, pumps, and aerators, which leads to increased costs of maintenance. It taints water and adds to the cost of treatment of potable water. In its native range, however, cabomba seldom covers large expanses or becomes a problematic plant, except in a few irrigation ditches.

It is difficult to assess the value of the damage that cabomba is causing to the environment, economy, and quality of life. However, there are data available that document the cost of managing cabomba infestations. Generally herbicides are largely ineffective

Fig. 6.3 Cabomba infestation at a small farm dam in Queensland, Australia. Only the white flowers extend above the water's surface making infestations more visible during the summer months. (Photo by S. Schooler, CSIRO Entomology.)

and herbicide use is severely regulated in or around public water supplies (Anderson and Diatloff, 1999). However, in 2003 the Northern Territory, Australia, successfully eradicated cabomba from a lagoon near Darwin (Marlow Lagoon, 12.488° S, 130.968° E) using herbicides. Drawdown and mechanical control initially cost US$340 000 without success. Subsequently, a single application of herbicide costing US$2 500 was effective at removing the cabomba with no new plants observed up to present (2007). They are currently attempting to eradicate cabomba from the Darwin River (12.521° S, 131.055° E) using herbicides. However, supplying water to residents, water quality testing, public education, and repeated applications of herbicide have cost over US$300 000 in the 2006–2007 financial year alone (Steve Wingrave, personal communication, 2007). The infestation has been reduced to less than 1% of former levels, but the presence of viable seeds in this population is making eradication difficult.

Physical control methods are also being used to manage cabomba. In Queensland, Australia, Caloundra Shire Council currently spends over US$170 000 per year to physically remove cabomba around public areas of Ewen Maddock Reservoir (26.797° S, 152.990° E) using SCUBA divers and suction pumps (R. Rainbird, personal communication, 2007). Another physical removal method uses floating mechanical "harvesters" to cut (to a depth of up to 1 m below water surface), collect, and remove cabomba. Such harvesters are expensive to purchase and operate and are generally only usable in channels of considerable depth and width. Noosa Shire Council spends in excess of US $85 000 per year (with US$180 000 initial investment) to mechanically remove cabomba from a small fraction of Lake Macdonald (26.385° S, 152.929° E) using a floating weed harvester (B. McMullen, personal communication, 2007). The only effective and sustainable method in managing cabomba appears to be biological control (Culliney, 2005).

6.2 Phylogeny

Few plants are closely related to *Cabomba caroliniana*. This is conducive to finding safe biological control agents because host specificity is linked to phylogenetic relatedness (Pemberton, 2000). Cabomba occurs within Nymphaeales, which consists of two families, Nymphaeaceae and Cabombaceae (Podoplelova and Ryzhakov, 2005). Cabombaceae consists of the two genera, *Cabomba* and *Brasenia*. The genus *Cabomba* consists of *C. caroliniana* along with four other species (*C. aquatica* Aublet, *C. palaeformis* Fassett, *C. furcata* Schultes and Schultes, and *C. haynesii* Wiersema), all of which are endemic to Central and South America (Orgaard, 1991). The genus *Brasenia* consists of one species (*B. schreberi* J. F. Ginel) which has a worldwide distribution, except South America (Tur, 1987).

6.3 Origins of *Cabomba*

Cabomba caroliniana apparently has a disjunct native range; in South America centered around northern Argentina and in the southern USA along the Atlantic and Gulf Coasts (Wain *et al.*, 1983; Orgaard, 1991). However, it is recognized as a noxious weed in Canada and in the USA north of the Carolinas and along the west coast. Three varieties of Cabomba are recognized throughout the native range; two in South America (*C. caroliniana* Gray var. *caroliniana* Gray and *C. c.* var. *flavide* Orgaad) and one in North America (*C. c.* var. *pulcherrima* Harper).

As part of a biological control project for Australia, surveys have delimited the native range in South America. Surveys covered most of northern and central Argentina, southern Paraguay, southern Brazil, and Uruguay. Waterways and water bodies along more than 21 000 km of road stretches were surveyed. Cabomba was present only in specific environments (e.g. large and shallow lakes), which occur in Argentina in the province of Corrientes (and in isolated locations in the adjacent provinces of Formosa, Chaco, and Entre Ríos), southern Paraguay, and as few isolated populations on the southeast coast of Brazil. Mostly, but not exclusively, it occurred in clear-water lakes and streams with low current velocity at 1–4 m depth.

6.4 Biology and ecology of Cabomba

6.4.1 *Biology of* Cabomba caroliniana *in its native range*

In its native range in South America there are two varieties of cabomba: *C. caroliniana* Gray var. *caroliniana* Gray, a stout plant with white flowers, and the smaller and more

Table 6.1 *List of chemical, physical and descriptive features measured in water bodies within the distribution range of* Cabomba caroliniana

Chemistry	Descriptive	Physical
Nitrate (N_2O_3 ppm)	Bed (sand, clay, rock, silt)	Conductivity (μ Siemens)
Nitrite (N_2O_2 ppm)	Sediments (clean; some; deep layer)	Clarity (Secchi disk depth in cm)
Ammonia (NH_4 ppm)	Banks (steep; gradual; beach)	Water column depth (meters)
Phosphorus (PO_4 ppm)	Water level (stable; irregular; flood plain)	Temperature (°C)
pH	Co-occurring plant species	
Dissolved oxygen (ppm)	Type of plants (submerged; floating; both; none)	
	Current (slow or none; fast)	
	Climate (humid, dry season, semiarid)	
	Surrounding topography (plains; hills)	

delicate *C. caroliniana* var. *flavida* Orgaard that has yellow flowers. These varieties have never been collected together. *C. c. flavida* occurs only in small streams, shallow ponds, and irrigation ditches, whereas *C. c. caroliniana* occurs in larger and deeper water bodies. In North America there is another variety, *C. c.* var. *pulcherrima* Harper, known only from the southeastern states (from North Carolina to Florida). In this chapter, whenever we refer to cabomba, we are referring to *C. c. caroliniana*, which is currently the most common weedy variety around the world. No known reports of *C. c. flavida* or *pulcherrima* being invasive exist. We suspect *C. c. caroliniana* is a better competitor in permanent water bodies while *C. c. flavida* is better suited to fluctuating or temporary water bodies. Propagation of cabomba seems to be mostly from cloning and rooting of fragments, although viable seed is produced throughout its native distribution.

During surveys, water quality data was taken at every cabomba location and at most other water bodies within cabomba's presumed natural area of distribution. In addition, other physical and/or topographic traits, such as sediment type, current velocity, and surrounding topography were also assessed (Table 6.1).

Pearson correlation matrices were used to examine associations between the presence of cabomba and continuous environmental variables and Spearman rank-order correlation coefficients were used to examine these relationships for categorical data (SYSTAT, 2004). The results indicate that the chemical and physical factors had little bearing on predicting cabomba presence in the native range. Some moderate association was observed with descriptive factors such as current speed, sediment depth, bank type, and water level fluctuation (Table 6.2). However, there was relatively little variation among the water bodies sampled throughout our surveys. A larger data set, covering more diverse aquatic environments, may provide a better picture of the environmental constraints of cabomba.

Table 6.2 *Variables correlated to C. caroliniana presence in its native range*

Variable	No. of sites	Correlation statistic
Sediment depth	70	0.351
Current speed	70	0.349
Bank type	70	0.327
Water level fluctuation	70	0.326

Cabomba caroliniana in South America is limited to aquatic environments of the Paraná/Paraguay and Uruguay River basins that are relatively clear and have acidic low-nutrient waters which restrict the growth of floating aquatic plants. Floating plants dominate the more turbid and nutrient-rich water bodies of these basins (Carignan and Neiff, 1992). Cabomba appears to prefer the protected inlets of lakes and rivers where gentle currents allow soil sediments to accumulate and where the relatively fragile stems of cabomba are not disturbed. The plant rarely occurs in water bodies with weather-beaten shores or with large fluctuations in water level.

In its native range in South America, cabomba grows down to a maximum of 3.5 m depth, often growing on a layer of organic silt (up to 2 m thick). The maximum depth in which cabomba can grow is negatively correlated with water turbidity. Turbid waters have high light attenuation and a shallow euphotic depth (the depth at which surface light is reduced by 99%). Therefore, even in shallow water, light penetration to the substrate may be too low for plants to become established. However, if either water clarity increases or the water level drops, plants may become established and grow upwards into the photic zone (Schooler *et al.*, 2006).

Several studies have found that dissolved carbon dioxide (CO_2) and pH are correlated with cabomba growth and abundance (Tarver, 1977; Sanders, 1979). There are two probable reasons for this. First, CO_2 diffusing from the air does not supply enough carbon for the growth of most submerged plants (Bowes, 1993; Hiscock, 2003). In natural systems atmospheric CO_2 is supplemented by microbial decomposition, which is regulated by temperature, oxygen, and available organic material. Secondly, pH affects CO_2 availability. When pH increases above 6, dissolved CO_2 concentrations decline precipitously and most carbon is in the form of bicarbonate (Maberly and Spence, 1983). Many submerged aquatic plants cannot use bicarbonate as a carbon source and cabomba is one of these soft water plants that prefers low pH (5–6) and needs supplemental CO_2 when grown in aquaria (Hiscock, 2003). Humans are decreasing the pH of many water bodies through acid rain while simultaneously increasing atmospheric CO_2 concentrations. This will likely increase available CO_2 concentrations in aquatic systems and thus increase the biomass of many submerged aquatic plants where the current limitation on plant abundance is inorganic carbon supply.

In its South American range, cabomba coexists with several other submerged plants such as *Ceratophyllum demersum* Linn. (Ceratophyllaceae), *Utricularia brevicaspa* Wright ex Grisebach (Lentibulariaceae), *U. foliosa* Linn., *U. platensis* Spegazzini,

U. gibba Linn., *U. poconensis* Fromm-Trinta, and *Chara* sp. among the unattached submerged plants, and *Egeria najas* Planchon (Hydrocharitaceae), *Potamogeton gayi* Bennett (Potamogetonaceae), *P. pusillus* Linn., *P. ferrugineus* Hagstrom, *P. illinoensis* Morong, and *P. nodosus* Poiret among the rooted submerged plants. Since our sampling began in 2003, cabomba has been the dominant submerged plant species within its range, except some sites where temporary *C. demersum* outbreaks were observed and also in some small lakes and ponds where cabomba stands appear to be undergoing replacement by *E. najas*. At the sampling sites in the province of Corrientes a different successional pattern has been observed within the submerged plant community in large lakes such as Iberá and Santa Lucía. Here, coverage of *E. najas* (measured by point-intercept sampling) tends to increase during the summer from near 0% in early spring to 20–40% in autumn. This correlates with increasing water turbidity due to high water temperatures and lower drainage rates of the lakes in summer. Increased turbidity appears to favor shade-tolerant *E. najas* plants. Consistent with this, the cabomba stands found in more open and clear waters had less *E. najas* interspersed among them.

6.4.2 Biology in the introduced range

Cabomba has a wide potential distribution. Current latitudinal distribution is from monsoonal tropical (Darwin River, Darwin, Australia, 12°) to cold temperate (Loosdrecht Lakes, Holland, 52°) environments and it can persist under ice in continental temperate climates (Kasshabog Lake, Peterborough, Canada, 45°) (Hogsden *et al.*, 2007; Wilson *et al.*, 2007). It is primarily a problem in lakes and reservoirs, but can also establish populations in streams, rivers, and irrigation canals. It is often found near bridges, either because there are pools of slow moving water and/or because it was planted for later collection and sale. Therefore, cabomba has the potential to colonize most water bodies throughout the world. Environmental factors that appear to have the greatest effect on cabomba abundance include substrate type, water movement, water turbidity, dissolved carbon dioxide, and pH.

Cabomba prefers areas with fine and soft sediments. In clayey or sandy soils the thin hair-like roots struggle to anchor the plants. In human-made lakes with hard clay bottoms cabomba occurs in topographic depressions where sediment has accumulated. Lack of thick roots also limits cabomba to areas with slow-moving water such as lakes, ponds, reservoirs, and slack water pools along streams and rivers.

Colonization occurs when plant fragments or seeds are introduced to a new habitat. A single stem node can produce a new plant (Sanders, 1979). Individual plant fragments dry quickly when exposed to air and rarely remain viable for more than 24 hours. However, bunches of fragments or fragments in mud can remain moist for weeks even under extremely hot and dry conditions (S. Schooler, unpublished data). Some populations produce viable seed and some do not. Seeds are commonly found in the southern USA with production peaking in May and October (Sanders, 1979). In Australia, viable seeds have only been found in the Northern Territory. Within the Northern Territory,

(a)   (b)

Fig. 6.4 Illustration of the clonal spread of cabomba by attached stems. (a) In the summer the buoyant stems keep the tips in a vertical position. (b) Stem tips lose buoyancy during the winter and drop to the sediment. In the spring, nodes near the tip then form roots and a new growing tip. Eventually the connecting stem disintegrates, separating the mother from the daughter plants.

populations in the Darwin River have produced viable seed while those nearby in Marlow Lagoon have not (P. Clifton, personal communication, 2006). This suggests that seed viability is not linked with environmental conditions but may be caused by hybrid sterility (infertile karyotype). A preliminary genetic analysis indicates that most Australian populations are hybrids of the varieties *C. c. caroliniana* and *C. c. pulcherrima* which supports the hypothesis that lack of viable seeds may be caused by hybrid sterility (A. Weiss, University of Connecticut). Ongoing genetic studies in the USA (A. Weiss) and karyotype analysis in Argentina (E. Greizerstein, University of Buenos Aires) may soon provide a more complete answer to this question.

Once a plant is established, the population can spread throughout the water body in three ways: viable seeds, broken fragments, or attached stems. Seeds drop to the sediment after maturing and remain viable for at least two years (Sanders, 1979). Seeds dried and stored in a greenhouse before planting in an aquarium generally showed a higher germination rate (85%) than those seeds that were kept moist before planting (25%) (Sanders, 1979). This suggests that drying stimulates germination and that seeds are a means of overcoming fluctuating water levels where conditions would not support vegetative reproduction alone. Fragments broken from the main plant float to new locations. Plants usually tend to lose buoyancy and turn brittle in the autumn and winter. This increases fragmentation in the autumn, after which the fragments settle on the sediment at the bottom of the water body in winter and grow the following spring. Attached stem tips that lose buoyancy during the autumn can also settle to the substrate and these nodes will also produce new plants

(Fig. 6.4). Clonal subsidization may occur where the mother plant subsidizes the daughter plant in deep water (delivering resources through the attached stem), which may allow the daughter plant to grow into the photic zone.

Therefore, seeds provide a means of overcoming disturbances that would destroy vegetative propagules (drought, turbidity, herbicides) and may also be involved in long distance dispersal, for example in mud on vehicles or animals and possibly in the digestive tracts of waterfowl. Buoyant stem fragments can be carried long distances across lakes or down rivers, but are less robust than seeds. Finally, attached stems allow plants to efficiently colonize local canopy gaps, much like terrestrial weeds such as lantana (*Lantana* spp.) and blackberry (*Rubus* spp.).

6.4.3 Population dynamics

Earlier observations suggested that cabomba populations decline in abundance during winter months. Because fluctuating resource availability may affect the efficacy of biological control efforts, we carried out a regular sampling program to quantify season-induced changes in biomass at different depths (Schooler and Julien, 2006). Samples were collected at 2–3 month intervals from September 2004 to February 2007 at three sites in southeast Queensland, Australia: Lake Macdonald (26.386° S, 152.929° E), Ewen Maddock Reservoir (26.797° S, 152.990° E), and Seibs Dam (26.494° S, 152.972° E).

Although populations of cabomba were variable, they exhibited no discernable pattern with respect to season at the sites studied. This is a favorable trait in the context of biological control because host-plant resources will be available to the agents throughout the year, thus promoting a stable equilibrium between host-plant and herbivore populations. In systems with high seasonal host-plant fluctuations, like salvinia (*Salvinia molesta* D. S. Mitchell) (Salviniaceae) in Kakadu National Park (Northern Territory, Australia), biological control is less effective because the agents overexploit their resource as host-plant populations decline during the dry season (Julien and Storrs, 1996). This creates an unstable situation when the populations of the control agents dramatically decrease as resources become scarce and must increase again from low numbers as the host-plant population increases during the following wet season. This induces a lag in control, which may be considerable if the host plant grows rapidly, as salvinia does (Storrs and Julien, 1996). We found no seasonal fluctuation in cabomba biomass in Queensland, although more information is needed for temperate climates where temperature fluctuates more widely.

Water depth was the main environmental variable associated with variation in cabomba abundance. Plant size increases with water depth to 4 m and then declines while plant density decreases with water depth (Schooler and Julien, 2006). Consequently, cabomba can grow to depths of 6 m in southeast Queensland but biomass is greatest at depths of 2–3 m (Fig. 6.5). In contrast, other rooted aquatic plants are rarely found at depths greater than 2 m (Sainty and Jacobs, 2003). Maximum depth of rooted cabomba plants correlated with turbidity (measured as Secchi disk depth) across the three sites, suggesting that maximum depth is limited by light penetration (Fig. 6.6).

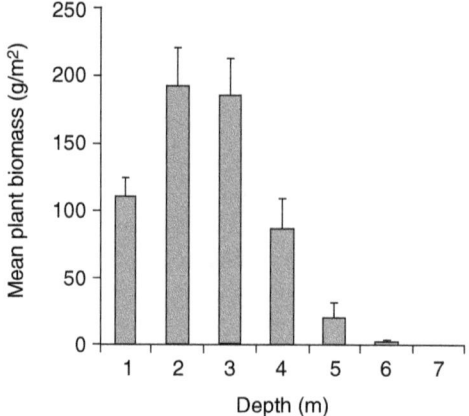

Fig. 6.5 Relationship between depth and mean plant biomass (standard error bars shown) at three sites in subtropical Australia. Biomass is dry above-substrate weight per m² ($n=20$).

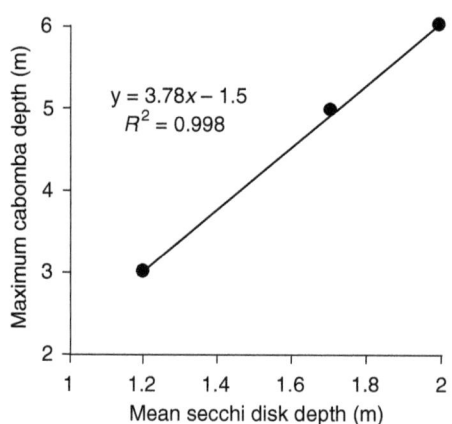

Fig. 6.6 Correlation between turbidity (Secchi disk depth) and maximum depth of rooted cabomba plants ($n = 3$).

6.4.5 Comparison between native and introduced ranges

There are four main differences between cabomba populations in its native and introduced range. First, the climatic variation in the native range is very narrow (restricted to a small area of northern Argentina), while climatic variation in the introduced range is extreme (tropical to cold continental climates). Probably this is due to the ameliorating effect of the surrounding water (even under ice the temperature does not decrease below 4 °C). However, this may make finding biological control agents that can survive these extremes more difficult. Second, cabomba is rarely found in depths of 0–2 m in the native range because floating vegetation usually occurs around the edges of lakes and ponds. This vegetation greatly reduces light penetration and consequently prevents the growth of cabomba. However, floating vegetation is rare in most water bodies in subtropical Australia and cabomba grows up to the water's edge (Fig. 6.5). Third, cabomba tends to grow in oligotrophic conditions in the native range, whereas it grows in eutrophic conditions in its introduced range. This is also related to competition with floating vegetation where floating vegetation is either absent or intensively controlled (primarily through biological control or physical removal) in the introduced range. Fourth, populations are patchy in the native range but approach monospecific stands in the introduced range. This dearth of large dense stands in the native range is presumably due to the combination of herbivory by natural enemies and competition (and disturbance) caused by floating vegetation.

Comparable biomass measurements have not been possible in the native range. The only large cabomba populations in Argentina occur in lakes at depths of 3–5 m because the lake shores are often covered with wide mats of floating vegetation. Further, this floating vegetation can be mobile (floating islands moved by wind and currents) and so, over time, some patches close over and other areas open up. This suggests that the community dynamics of submerged aquatics in that part of the world may be based on an intrinsic instability: every *Cabomba/Egeria/Potamogeton* patch is periodically disturbed by these floating islands. Presumably the fragments left behind sprout again once the floating masses pass over, while the fragments they drag grow into new patches. It appears that there is no stable climax vegetation pattern for submerged plants in such lakes. This may explain why cabomba patches are always found against an island front, and why thick, large, widespread patches of *Cabomba/Egeria/Potamogeton* are not found. The exception to this situation occurs in the rivers that drain the lakes of the Iberá wetlands. These have stable cabomba populations along a depth gradient. However, here the cabomba stands are thick near the banks where the current is gentle, and shorter and sparse in the deeper parts where the current is stronger. But this situation cannot be directly compared with the Australian reservoirs because biomass and stem length is not so much affected by depth (light availability), but by current speed (disturbance).

6.5 The Australian biological control project

In 2003, Australia's Commonwealth Scientific and Industrial Research Organisation (CSIRO) began a project to discover and test biological control agents from the native range in South America in an effort to find a long-term, sustainable solution to cabomba in its introduced range. We are completing the surveys for herbivores in the native range and are currently planning host-specificity tests on selected agents for quarantine facilities in Brisbane, Australia.

Hydrilla verticillata (L.f.) Royle (Hydrocharitaceae) is the only other submerged weed for which biological control has been attempted, with four agents released in the USA (Julien and Griffiths, 1998). So far that project has not resulted in effective control. Possibly this is because the specific life cycle requirements of the agent are not met in the introduced range (i.e. many agents require a drawdown to complete their life cycle) (Center *et al.*, 2002). The challenge is to find potential agents for cabomba that will perform well in the diverse environments encountered in the introduced range.

6.6 Ecology of potential biological control agents

Surveys for potential biological control agents are almost complete for the native range of *C. caroliniana* in South America. We have located the area with the greatest density of cabomba sites (presumably the center of origin of *C. caroliniana*) and have sampled numerous lakes in this area throughout the year for several years to ensure that we do not miss potential agents with differing life cycles. Through this process we have found three potential agents. We are currently studying their life-cycles, identifying methods to rear them under laboratory conditions, and conducting preliminary host-specificity tests at the USDA-ARS South American Biological Control Laboratory in Buenos Aires, Argentina. We have focused our surveys on arthropods because pathogens are difficult to recognize when sampling submerged aquatic plants with highly dissected leaves. In addition, plant leaves and stems are usually coated with algae.

6.7 Natural enemies found in the native range

Among many obviously generalist herbivores, such as snail and limpet species, midges (Chironomidae), soldier flies (*Hedriodiscus* sp., Stratiomyidae), aphids (*Rhopalosiphum nymphaeae* Linn., Aphidae), and shore flies (Ephydridae), three potentially specialist phytophagous insects have been found. These include an aquatic weevil (*Hydrotimetes natans* Kolbe, Curculionidae) and two moth species (*Paracles* sp. Arctiidae and *Paraponyx diminutalis* (Snellen), Pyralidae).

6.7.1 Hydrotimetes natans *(Coleoptera: Curculionidae)*

This weevil feeds on plant tips as an adult, while the larvae mine inside the plant stems. Development, from egg to adult, requires about 40 days in the laboratory. At high densities adults can cause extensive tip damage, whereas the larvae can induce stem decay. In field conditions, adults are present year-round. They survive for approximately one year in the laboratory. During the summer mating season (December–February) adults occur on flowers at the water's surface, but remain underwater the rest of the year. Larvae have been found within plant stems from October to May with populations peaking at the beginning of summer (Fig. 6.7). The same behaviour has been observed in the laboratory in 3-liter containers exposed to natural light and ambient temperatures, suggesting the weevil responds to photoperiod.

In order to determine host range of the weevil in the field, samples of submerged plants occurring near cabomba were collected and suspended in cloth Berlese funnels. Thus

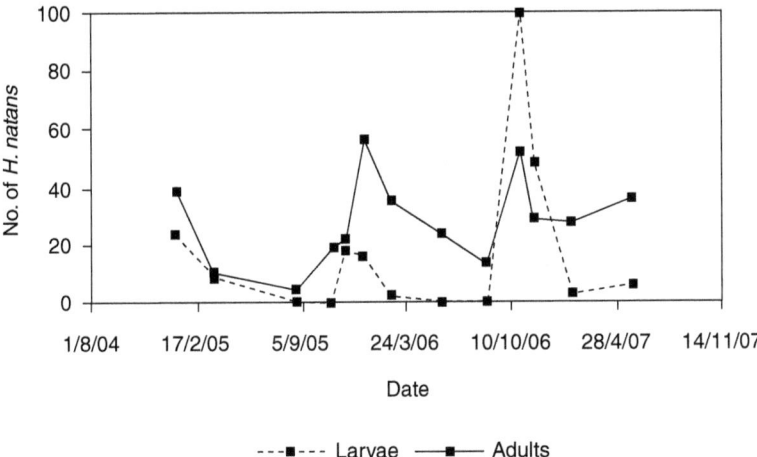

Fig. 6.7 Seasonal abundance of *Hydrotimetes natans* larvae and adults. Abundance data were obtained from 24-kg samples, fresh biomass.

far, *H. natans* larvae have only been extracted from cabomba samples. Some adult weevils (< 3%) were found on *Egeria najas* when the plant was growing intertwined with cabomba in the field. However, we have found no evidence that this weevil uses *E. najas* as a host.

We plan to conduct host range tests in the laboratory to confirm our field observations, but we have not been able to rear *H. natans* in the laboratory in containers small enough to closely observe its development and behavior. We have been able to rear colonies outdoors in 1000-liter glass tanks in Buenos Aires where stems with larvae, pupae, and adults have been observed for the past year (2006). In these tanks we also grew cabomba alongside *Egeria densa* Planchon (Hydrocharitaceae), *Potamogeton illinoensis*, and *P. pusillus*, thus exposing these plants *H. natans*. These are key test-plant species as they are the plants available in Argentina that are closely related to potential hosts in Australia. Several native *Potamogeton* species occur in Australia. *Egeria densa* is a close relative, and is similar in many morphological and ecological aspects to many of the native Australian Hydrocharitaceae (e.g. *H. verticillata*, *Maidenia rubra* Rendle, *Blyxa* spp.). During the 11-month trial period no *H. natans* of any stage and no damage attributable to *H. natans* was observed on plants other than cabomba.

6.7.2 Paracles sp. (Lepidoptera: Arctiidae)

This large aquatic caterpillar was collected in several locations of Corrientes province. It causes heavy defoliation on cabomba and seems to prefer the leaves near the apical tips. It feeds underwater, keeping air bubbles amidst the short hairs on its dorsum. Its cocoon is also aquatic, resembling a canoe, and pupating larvae sometimes form a "raft" of cocoons woven together in which only silk tufts at the ends remain dry and above water surface.

Table 6.3 *Results of host range testing of* Paracles *sp. on aquatic plants that co-occur with cabomba in its native range in Argentina*

		Instar I and II		Instars III +	
Host plant	Field feeding	Feeding	Development	Feeding	Development
Cabomba caroliniana	yes	yes	normal	yes	full
Egeria densa	no	yes	normal	yes	full
Panicum sp.	yes	no	1−2 molts[a]	yes	full
Egeria najas	no	no	no	yes	1−2 molts[a]
Eichhornia azurea (Sw.) Kunth	yes	yes	1−2 molts[a]	yes	1−2 molts[a]
Myriophyllum aquaticum (Vell.) Verde	no	no	no	yes	no[b]
Ceratophyllum demersum	no	yes	1 molt[a]	yes	1−2 molts[a]
Nymphoides indica (Linn.) Kuntze	no	no	no	no	no[b]
Potamogeton spp.	no	no	no	yes	no[b]
Hydrocleys nymphoides (Humb, & Bonpl. ex Willd) Buch.	yes	no	no	yes	no[b]
Ludwigia peploides (Kunth) Raven	no	no	no	no	no[b]

[a] Abnormal development followed by death.
[b] No molting to next instar was observed despite feeding.

The adult, a brown moth with a 25 mm wingspan, stretches it wings on the floating cocoon upon emergence. The life cycle of *Paracles* sp. lasts about 40 days. Mating occurs within 2 days after emergence and the female then lays 20–70 whitish eggs (in a mass covered with pale-orange scales) on any vertical surface over the water. The eggs hatch in approximately eight days and the larval and pupal stages last 22 and 10 days, respectively, at ambient temperature.

Laboratory no-choice and preference host range tests were performed with this moth on 11 aquatic plants (Table 6.3). Test plants were selected based on damage observed in the field and their co-occurrence with cabomba. When first-instar larvae were transferred to test plants in no-choice trials a narrow host range was observed. However, feeding of mature larvae was observed in the field on some plants that had been rejected in the laboratory (Table 6.3). Further choice tests, in the presence of cabomba, were conducted in interconnected 1000-liter tanks under cages. Adults oviposited indiscriminately on any

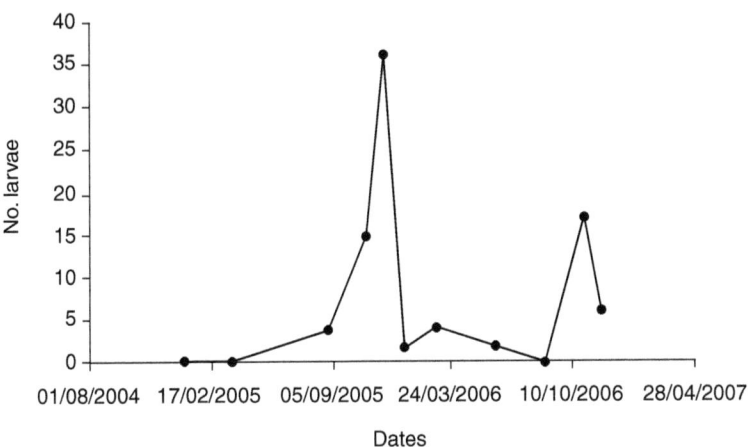

Fig. 6.8 Seasonal abundance of *Paraponyx diminutalis* larvae. Data were obtained from 24-kg samples, fresh biomass.

vertical surface and eggs were laid in every tank. Complete development from first instar was obtained both on cabomba and on *E. densa*. In addition, mature larvae could develop into fertile adults on aquatic grasses (e.g. *Panicum* spp., Poaceae) and live for a considerable length of time on three other species (Table 6.3). We still do not know, however, if *Paracles* spp. can accept these plants as permanent hosts in the field and we have never observed it on *E. densa* in the field. However, because Australia has several native aquatics belonging to Hydrocharitaceae (e.g. *Hydrilla*, *Blyxa*, *Vallisneria*, and *Maidenia*), we have reduced the priority of this potential agent in favor of the weevil.

6.7.3 Paraponyx diminutalis *(Lepidoptera: Pyralidae)*

The gilled larvae of this moth feed on the terminal shoots of cabomba stems, inducing a specific type of damage which stunts stem growth. Because of its cryptic habits the larvae are hard to detect despite being greater than 15 mm in length in later instars. At the end of its larval stage (45 days), it spins an irregular cocoon attached to the plant tips and wrapped with live leaves. In the field it is quite common during spring and early summer, but its presence is erratic (Fig. 6.8).

Although *P. diminutalis* is a well-known natural enemy of several aquatic plants in its native range of Southeast Asia (from Indonesia to Pakistan) and East Africa (Balloch and Sana-Ullah, 1974; Buckingham and Bennett, 1989) and in its accidentally introduced range in the southeastern USA and Panama (Buckingham and Bennett, 1989; Center *et al.*, 2002), this species has not been found on other plants collected together with cabomba in Argentina. This fact, added to it not being recorded in South America previously, has led us to continue working on this species in the event that unexpected genetic or taxonomic differences might prove it to be an eligible candidate. We are currently conducting a genetic analysis on populations of this moth to determine if it is indeed a single species.

6.7.4 Surveys on other cabomba species

The suite of specialist natural enemies found on cabomba in Argentina is relatively small, as expected of a plant growing in such a specialized habitat. Consequently, we have begun surveying populations of four closely related cabomba species searching for additional potential biological control agents. Since no closely related native or economically important plants occur in Australia, we may find effective agents that are host specific to the *Cabomba* genus and that constitute low risk to nontarget species. Preliminary surveys were conducted in three areas of Venezuela (Lower Orinoco River and Delta, Gran Sabana, and Lago Maracaibo) for cabomba species and associated herbivores in 2006. Six populations of cabomba were found in Venezuela, including three species (*C. furcata*, *C. aquatica*, and *C. haynseii*). We identified several herbivorous insects including two weevils and one moth. Specimens are currently housed at the CSIRO Mexican Field Station and are in the process of being identified. More substantive surveys will be conducted in the future and surveys are planned for Costa Rica, Mexico, and Puerto Rico.

6.8 Conclusions

Since its inception, this project was approached from several fronts. No single aspect of it had priority over the others and the work on the recipient side of the problem (Australia) was as intense as the work on the donor side (southern South America). The interaction between both "shores" of the project, although involving two separate institutions, enhanced each other so that the objectives and priorities of each partaking laboratory pivoted on the information provided by the partner in a dynamic process.

Biological studies in the native range of cabomba and its introduced range in Australia were also oriented to identifying the plant's critical transitional stages (Briese, 2006) and chemical and physical constraints. A sampling scheme was applied in Australia to detect those environmental factors that produce variations in plant biomass. Apart from the water depth and clarity factors mentioned above, no seasonal related factors could be identified in subtropical habitats (Schooler *et al.*, 2006). This is likely not the case in temperate climates but has yet to be assessed for Australia. A similar assessment in China determined that the plant presented maximum biomass in summer (Yu *et al.*, 2004, from an abstract) suggesting the behavior of the plant will vary according to climate throughout its adventive distribution. However, again, once established it seems to hold its ground.

In Argentina, this sampling system could not be replicated because the physical characteristics of the aquatic environments where cabomba is abundant are not conducive to the same type of study. However, we found evidence that, regardless of seasonal variations, disturbance regimes prevented the establishment of climax communities in these habitats. In addition, the chemical characteristics of the water observed between cabomba sites indicate that, within the chemical variations observed in its native range, these hold little bearing on the plant's abundance. Instead, physical environmental traits seem to be much more important, a factor observed previously for other communities of submerged macrophytes (Hudon *et al.*, 2000).

Community structure frequently changes as a result of an invading organism. This is the obvious result of new interactions developing between previously unconnected species. In the cabomba project, however, we also encountered a drastic modification in the growing conditions of plant populations brought about by an environmental factor. So whereas in its homeland the plant is found in periodically disturbed unstable patches, in Australia it appears in dense and temporally stable monospecific stands. Under this scenario, it is expected that an agent that is present year-round and host-specific should exhibit density-dependent populations that fluctuate with its host population. We predict that the result will be a reduction of cabomba abundance, particularly in areas of deep water where disturbance will exacerbate the effects of herbivore damage and in shallow areas where damage to cabomba will increase the relative competitive ability of native plant species (Schooler *et al.*, 2006).

Acknowledgments

We thank Russell Rainbird, Peter Clifton, Colin and Kay Seibs, Tom Anderson, the Noosa Shire Council, and the Lake Macdonald Catchment Care Group (LMCCG) for their support. We thank Ricardo Segura for helping to organize and conduct surveys in tropical South America and Federico Mattioli and Joaquín Sacco for assistance in the field and the laboratory. This work was supported by the Australian Department of Environment and Heritage (DEH), the Australian Department of Agriculture, Fisheries and Forestry (DAFF), the Burnett Mary Regional Group (BMRG), and a number of stakeholder groups administered by the LMCCG.

References

Anderson, T. and Diatloff, G. (1999). Cabomba management attempts in Queensland. In *Proceedings of the 10th Biennial Noxious Weeds Conference*, held in Ballina, Australia, ed. J. Quinn; Gosford, Australia: NSW Agriculture, pp. 42–45.

Balloch, G. M. and Sana-Ullah. (1974). Insects and other organisms associated with *Hydrilla verticillata* (L.f.) L.C. (Hydrocharitaceae) in Pakistan. In *Proceedings of the 3rd International Symposium on the Biological Control of Weeds, Montpellier, France*, ed. A. J. Wapshere. Miscellaneous Publication 8. Farnham Royal, UK: Commonwealth Institute of Biological Control, pp. 61–66.

Bowes, G. (1993). Facing the inevitable: plants and increasing atmospheric CO_2. *Annual Reviews in Plant Physiology and Plant Molecular Biology*, **44**, 309–332.

Buckingham, G. R. and Bennett, C. A. (1989). Laboratory host range of *Paraponyx diminutalis* (Lepidpotera: Pyralidae), an Asian aquatic moth adventive in Florida and

Panama on *Hydrilla verticillata* (Hydrocharitaceae). *Environmental Entomology*, **18**, 526–530.

Briese, D. T. (2006). Can an *a priori* strategy be developed for biological control? The case of *Onopordum* spp. thistles in Australia. *Australian Journal of Entomology*, **45**, 317–323.

Carignan, R. and Neiff, J. J. (1992). Nutrient dynamics in the floodplain ponds of the Paraná river (Argentina) dominated by the water hyacinth *Eichhornia crassipes*. *Biogeochemistry*, **17**, 85–121.

Center, T. D., Dray, F. A., Jubinsky, G. P. and Grodowitz, M. J. (2002). *Insects and Other Arthropods That Feed on Aquatic and Wetland Plants*. Technical Bulletin 1870. Springfield, VA: U.S. Department of Agriculture, Agricultural Research Service.

Culliney, T. W. (2005). Benefits of classical biological control for managing invasive plants. *Critical Reviews in Plant Sciences*, **24**, 131–150.

Hiscock, P. (2003). *Encyclopedia of Aquarium Plants*. Hauppauge, NY: Barron's Educational Series, Inc.

Hogsden, K. L., Sager, E. C. and Hutchinson, T. C. (2007) The impacts of the non-native macrophyte *Cabomba caroliniana* on littoral biota of Kasshabog Lake, Ontario. *Journal of Great Lakes Research*, **33**, 497–505.

Hudon, C., Lalonde, S. and Gagnon, P. (2000). Ranking the effects of site exposure, plant growth form, water depth, and transparency on aquatic plant biomass. *Canadian Journal of Fisheries and Aquatic Sciences*, **57**, 31–42.

Julien, M. H. and Griffiths, M. W., eds. (1998). *Biological Control of Weeds: A World Catalogue of Agents and Their Target Weeds*. Fourth ed. Wallingford, UK: CABI Publishing, 223 pp.

Julien, M. H. and Storrs, M. J. (1996). Integrating biological and herbicidal controls to manage salvinia in Kakadu National Park, northern Australia. In *Proceedings of the IX International Symposium on Biological Control of Weeds*, ed. V. C. Moran and J. H. Hoffman. Rondebosch, South Africa: University of Cape Town, pp. 445–449.

Maberly, S. C. and Spence, D. H. N. (1983). Photosynthetic inorganic carbon use by freshwater plants. *Journal of Ecology*, **71**, 705–724.

Mackey, A. P. and Swarbrick, J. T. (1997). The biology of Australian weeds. 32. *Cabomba caroliniana* Gray. *Plant Protection Quarterly*, **12**, 154–165.

Orgaard, M. (1991). The genus *Cabomba* (Cabombaceae): a taxonomic study. *Nordic Journal of Botany*, **11**, 179–203.

Pemberton, R. W. (2000). Predictable risk to native plants in weed biological control. *Oecologia*, **125**, 489–494.

Podoplelova, Y. and Ryzhakov, G. (2005). Phylogenetic analysis of the order Nymphaeales based on nucleotide sequences of the chloroplast ITS2–4 region. *Plant Science*, **169**, 606–611.

Sainty, G. R. and Jacobs, S. W. L. (2003). *Waterplants in Australia*. Potts Point, Australia: Sainty and Associates, pp. 336–353.

Sanders, D. R. (1979). The ecology of *Cabomba caroliniana*. In *Weed Control Methods for Public Health Applications*, ed. E. O. Gangstad. Boca Raton, FL: CRC Press, pp. 133–146

Schooler, S. S. and Julien, M. H. (2006). Effects of depth and season on the population dynamics of *Cabomba caroliniana* in SE Queensland. In *Proceedings of the 15th Australian Weed Conference*, ed. C. Preston, J. H. Watts and N. D. Crossman. Adelaide, Australia: Weed Management Society of South Australia, pp. 768–771.

Schooler, S. S., Julien, M. H. and Walsh, G. C. (2006). Predicting the response of *Cabomba caroliniana* to biological control agent damage. *Australian Journal of Entomology*, **45**, 326–329.

Storrs, M. J. and Julien, M. H. (1996). *Salvinia: A Handbook for the Integrated Control of* Salvinia molesta *in Kakadu National Park*. Northern Landscapes Occasional Series 1. Darwin, Australia: Australia Nature Conservation Agency.

SYSTAT Software, Inc. (2004). Systat Version 11. (www.systat.com).

Tarver, D. P. (1977). Selected life cycle features of fanwort. *Journal of Aquatic Plant Management*, **15**, 18–22.

Tur, N. (1987). Cabombaceae. In *Flora Ilustrada de Entre Ríos (Argentina)*, ed. A. Burkart, N. S. T. de Burkart and N. M. Bacigalupo. Buenos Aires, Argentina: Colección Cientifica del I.N.T.A., Vols. VI, III, pp. 303–305.

Wain, R. P., Haller, W. T. and Martin, D. F. (1983). Genetic relationship between three forms of *Cabomba*. *Journal of Aquatic Plant Management*, **21**, 96–98.

Wilson, C. E., Darbyshire, S. J., and Jones, R. (2007). The biology of invasive alien plants in Canada. 7. *Cabomba caroliniana* A. Gray. *Canadian Journal of Plant Science*, **87**, 615–638.

Yu, J., Ding, B., Yu, M., *et al.* (2004). The seasonal dynamics of the submerged plant communities invaded by *Cabomba caroliniana* Gray. *Acta Ecologica Sinica/ Shengtai*, **24**, 2149–2156.

7

Invasive cactus species (Cactaceae)

Helmuth Zimmermann, Cliff Moran, and John Hoffmann

7.1 Introduction

Approximately 1600 species are recognized in the family Cactaceae and, with one possible exception, all are native to the New World (Gibson and Nobel, 1986; Wallace and Gibson, 2002). The exception is an epiphyte, *Rhipsalis baccifera* (previously *cassutha*) (J. S. Miller) Stern, which appears to be indigenous in Africa and Madagascar (Wallace and Gibson, 2002).

Most species of the Cactaceae have leafless photosynthetic stems that bear spines on modified axillary buds called areoles. The cacti are predominantly succulent and adapted to survive in extreme xeric habitats. Because of their often bizarre structures and appearances, many species are now cultivated widely around the world mostly as curiosities and ornamentals. Some species of cacti are used as a source of fruit and fodder, and as hedge plants (Casas and Barbera, 2002). The most common and commercially important species in the Cactaceae is *Opuntia ficus-indica* (Linnaeus) Miller, usually known as either "prickly pear" or "cactus pear". This species is cultivated in many countries for its fruit and as fodder (Barbera *et al.*, 1995) but also as the main host for production of the carmine cochineal insect, *Dactylopius coccus* Costa (Homoptera: Dactylopiidae) (Casas and Barbera, 2002), and supports a flourishing dye industry in Peru, Chile, Bolivia, and the Canary Islands (Flores-Flores and Tekelenburg, 1995).

Several species of Cactaceae, introduced either deliberately or accidentally into countries outside the Americas, have become invasive. Indeed, some of the earliest records of alien-plant invasions refer to *Opuntia* species (Tryon, 1910). Their success as invasive species is attributed to their ability to: (i) survive dry periods by retaining moisture in their succulent stems and by using the CAM (Crassulacean Acid Metabolism) mode of photosynthesis; (ii) thrive and outcompete other plants under disturbed conditions; and (iii) reproduce both sexually and asexually. In addition, like many exotic species, cacti suffer very low levels of damage from phytophagous insects and pathogenic fungi, outside their native range.

Biological Control of Tropical Weeds using Arthropods, ed. R. Muniappan, G. V. P. Reddy, and A. Raman. Published by Cambridge University Press. © Cambridge University Press, 2009.

Biological control against invasive and problematic Cactaceae has had a long history and has included the use of a wide range of insect agents. Mann (1969) and Moran (1980) provide accounts on the biology of the insect species involved and their respective interactions with their cactus hosts. Moran and Zimmermann (1984) published a comprehensive review of the biological control of Cactaceae since the late eighteenth and early nineteenth centuries that includes lists of all the invasive cactus species targeted, and the organisms that have been used against them in various countries.

The objectives of this chapter are to: (i) highlight some of the "milestones" in the biological control of Cactaceae; (ii) provide an updated account of the biological control of cactus weeds; and (iii) to discuss lessons that have been learned from these programmes that are applicable to the practice of biological control of invasive plants.

7.2 The cactus species targeted for biological control

At least 49 cactus species (approximately 3% of the recognized flora, depending on taxonomic interpretations) are regarded as invasive, mainly in regions outside of their natural ranges, but several species are also considered as problematic within their natural distributions. Biological control has been used against 23 of the invasive species, with 19 species of insects and mites deployed over a period that commenced over 200 years ago. Details are provided in Tables 7.1 and 7.2. Moran and Zimmermann (1984) concluded that 11 of these biological control attempts have been "completely" successful, seven "substantially" successful and five have had "negligible" impact.

Biological control against cactus weeds has generally been easier than for most other taxa because, with one possible exception, there are no cactaceous species that are native outside of the Americas. This has allowed the safe use of phytophagous insects and mites that are less host-specific (i.e. oligophagous species) than is usual in other biological control efforts, because outside of the Americas there are no related indigenous plant species that might serve as host plants (Zimmermann and Granata, 2002).

The countries that pioneered the use of biological control for the management of cactus weeds are India, Sri Lanka, Australia, South Africa, Madagascar, and Hawaii. Other countries, including Ethiopia, Eritrea, Kenya, Yemen, Saudi Arabia, Morocco, Canary Islands, Zimbabwe, and Namibia, are experiencing increasing problems with cactus invasions, but have not yet initiated biological control programs.

Some species of Cactaceae have become problematic within their native ranges, largely the consequence of extensive disturbance of habitats through overgrazing. This has occurred on several islands in the West Indies, and on Santa Cruz Island, California, as well as some mainland areas in Argentina, Bolivia, Peru, Canada, and Texas in the United States of America (Moran and Zimmermann, 1984; Brutsch and Zimmermann, 1995). Biological control has been used, unwisely, to regulate invasions of native cactus species in the West Indies and on Santa Cruz Island (Simmonds and Bennett, 1966; Goeden *et al.*, 1967; Moran and Zimmermann, 1984).

Table 7.1 *Total number of invasive species in genera of three subfamilies in the Cactaceae that have been subjected to biological control*

Subfamily and genera	Species	Invasive species	No. species subjected to biological control
PERESKIOIDEAE			
Pereskia	15	1	1
OPUNTIOIDEAE			
Austrocylindropuntia	11	2	0
Cylindropuntia	33	8	4
Opuntia	138	25	15
Tephrocactus	6	1	0
CACTOIDEAE			
Cereus	28	1	1
Echinopsis	129	2	0
Harrisia	20	4	2
Acanthocereus	6	1	0
Hylocereus	18	1	0
Peniocereus	18	1	0
Selenicereus	29	1	0
Epiphyllum	19	1	0
Totals	**470**	**49**	**23**

7.3 Milestones in the biological control of cactus weeds

7.3.1 1796–1850. Projects that started the practice of biological control of invasive alien plants

The problem of dense infestations of the prickly pear *Opuntia monacantha* (Wildenow) Haworth, in India during the middle of the nineteenth century, resulted in the first inadvertent, but successful, biological control of a weedy plant species (Tryon, 1910; Rao *et al.*, 1971). The agent involved was a species of *Dactylopius* (Homoptera: Dactylopiidae) which comprises a small, distinctive genus of wax-covered soft scales, called the "cochineal insects." All live exclusively on cactus species and feed on the phloem of their cactus host-plants. At high population densities, cochineal insects can overwhelm and kill their hosts.

In the 1500s, Spanish conquerors discovered that cochineal insects had already been exploited for many centuries by the native peoples of South America as a source of red colorant (carminic acid). This discovery became an incentive for the shipment and cultivation of cactus plants in attempts to establish dye-production facilities in many countries around the world (Baranyovits, 1978; Greenfield, 2006). One of these attempts

Table 7.2 *Insect species and mites that have been intentionally released and become established as biological control agents against 23 species of invasive cacti, their countries of introduction, and the outcome*

Agent species and country of origin	Cactus weed species targeted	Country of introduction	Damage to weed[a]
Coleoptera: Cerambicidae			
Alcidion cereicola	*Harrisia martinii*	Australia	Considerable
(Argentina)		South Africa	Moderate
Archlagocheirus funestus	*H. bonplandii*	Australia	Moderate
(Mexico)	*H. tortuosus*	Australia	Trivial
	Cereus jamacaru	South Africa	Moderate
Moneilema ulkei	*Opuntia ficus-indica*	South Africa	Trivial
(Texas, USA)	*O. tomentosa*	Australia	Trivial
	O. streptacantha	Australia	Trivial
Moneilema variolare	*O. stricta* var. *inermis*	Australia	Trivial
(Mexico)	*O. streptacantha*	Australia	Trivial
	O. tomentosa	Australia	Trivial
	O. stricta var. *inermis*	Australia	Trivial
	O. stricta var. *stricta*	Australia	Trivial
Chrysomelidae			
Phenrica guerini	*Pereskia aculeata*	South Africa	Trivial
Curculionidae			
Eriocereophaga humeridens	*H. martinii*	Australia	Trivial
(Brazil)			
Metamasius spinolae	*O. ficus-indica*	South Africa	Considerable
(Mexico)			
Lepidoptera: Pyralidae			
Cactoblastis cactorum	Small *Opuntia* spp.	Australia	Extensive
(Argentina)		South Africa	Considerable
		New Caledonia	Considerable
		St. Helena	Considerable
	Large *Opuntia* spp.	Australia	Considerable
		South Africa	Moderate
		Hawaii	Considerable
		Mauritius	Considerable
	Cylindropuntia imbricata	South Africa	Trivial
	O. tuna	St. Helena	Considerable
	O. triacantha	Cuba[b]	Considerable
		Nevis[b]	Extensive
		St. Kitts[b]	Extensive
		Antigua[b]	Extensive
	O. dillenii	Cuba[b]	Considerable
		Nevis[b]	Considerable

Table 7.2 (cont.)

Agent species and country of origin	Cactus weed species targeted	Country of introduction	Damage to weed[a]
		St. Kitts[b]	Extensive
		Montserrat[b]	Considerable
		Antigua[b]	Considerable
		Gr. Cayman[b]	Extensive
Olycella junctolineella	*O. stricta* var. *stricta*	Australia	Trivial[c]
(USA)	*O. stricta* var. *inermis*	Australia	Trivial[c]
Tucumania tapiacola	*O. aurantiaca*	Australia	Moderate
(Argentina)			
Heteroptera: Coreidae			
Chelinidea tabulata	*O. stricta* var. *stricta*	Australia	Trivial
(USA)	*O. stricta* var. *inermis*	Australia	Trivial
	O. tomentosa	Australia	Trivial
	O. streptacantha	Australia	Trivial
Chelinidea vittiger	*O. stricta* var. *stricta*	Australia	Trivial[c]
(USA)	*O. streptacantha*	Australia	Trivial[c]
Homoptera: Dactylopiidae			
Dactylopius austrinus	*O. aurantiaca*	Australia	Considerable
(Argentina)		South Africa	Considerable
Dactylopius ceylonicus	*O. monacantha*	Australia	Extensive
(South America)		India	Extensive
		Sri Lanka	Extensive
		South Africa	Extensive
		Madagascar	Extensive
		Kenya	Moderate
		Australia	Trivial[c]
Dactylopius confusus	*O. dillenii*	India	Trivial[c]
(South America)		Sri Lanka	Extensive
		Australia	Trivial
Dactylopius opuntiae	*O. elatior*	India	Extensive
(USA)		Indonesia	Extensive
	O. engelmannii	South Africa	Trivial
	O. lindheimeri	South Africa	Moderate
	O. ficus-indica	South Africa	Considerable
		Australia	Moderate
		Hawaii	Moderate
	O. humifusa	South Africa	Trivial
	O. littoralis[b]	Santa Cruz Island	Extensive
	O. oricola[b]	Santa Cruz Island	Extensive
	O. streptacantha	Australia	Moderate
	O. stricta var. *stricta*	South Africa	Extensive

Table 7.2 (cont.)

Agent species and country of origin	Cactus weed species targeted	Country of introduction	Damage to weed[a]
		Australia	Considerable
	O. stricta var. *inermis*	Australia	Considerable
		India	Extensive
		Sri Lanka	Extensive
	O. tomentosa	Australia	Moderate
	O. tuna	Mauritius	Considerable
	O. sp.	Madagascar	Extensive
	O. streptacantha	Australia	Moderate
Dactylopius tomentosus	*Cyl. imbricata*	South Africa	Moderate
(USA)		Australia	Extensive
	Cyl. fulgida	South Africa	Trivial
	Cyl. rosea	Australia	Trivial
	Cyl. leptocaulis	South Africa	Extensive
Pseudococcidae			
Hypogeococcus festerianus	*Acanthocereus* sp.	Australia	Trivial
(Argentina)	*H. martinii*	Australia	Considerable
		South Africa	Considerable
	H. tortuosus	Australia	Considerable
	H. bonplandii	Australia	Considerable
	Cereus jamacaru	South Africa	Moderate
Diaspididae			
Diaspis echinocacti	All Cactaceae	Worldwide	Trivial
(unknown: inadvertent introductions)			
Acari: Tetranychidae			
Tetranychus opuntiae	*O. stricta* var. *stricta*	Australia	Trivial
(=*T. desertorum*)	*O. stricta* var. *inermis*	Australia	Moderate
(Texas, USA)			

[a] *Extensive*: Very high levels of damage. Few plants survive or growth is arrested or almost no seeds are produced. *Considerable*: High levels of damage. Some plants may survive but growth rates are noticeably slower or seed production is reduced by more than 50%. *Moderate*: Perceivable damage, but most plants survive. Growth may be slowed to some extent or seed production is reduced by less than 50%. *Trivial*: Some damage, but survival, growth and seed production of the plants is almost normal (Olckers and Hill, 1999).

[b] Target species in native range.

[c] Established but eventually died out.

was based on the use of *O. monacantha* that had been cultivated and became invasive in the southern Punjab region and Assam State of India (Ramakrishna Ayyar, 1931). Several introductions of various cochineal species were made to initiate dye production on a commercial scale on *O. monacantha*, although the exact identity of the species of cochineal insects that were used is still in doubt.

It seems almost certain that a consignment of cochineal insects from Rio de Janeiro that a Captain R. Neilson delivered to officials in Calcutta, in 1795, was of special significance. The cochineal species involved is now known to have been *Dactylopius ceylonicus* (Green), which thrived on *O. monacantha* in India; this species was subsequently collected in large quantities, dried and exported to England where it was known in the trade as the "Madras cochineal." Two years after its introduction, more than two tons of the Madras cochineal were exported, and in the following years exports increased to more than 18 tons (Lounsbury, 1915). In spite of this impressive start, the venture failed for two reasons: (i) Madras cochineal was not popular among buyers in England because each female yielded only a limited quantity, about a quarter of the volume, of carminic acid that could be obtained from *D. coccus*, which was known as "Grana Fina" because of its superior yields (Green, 1912; Lounsbury, 1915); and (ii) unlike *D. coccus*, which has no detrimental effect on its host plant, feeding damage caused by high numbers of *D. ceylonicus* eventually kills *O. monacantha*. Thus, within a few years of its introduction to India, *D. ceylonicus* drove the density of the cactus to very low levels and caused the demise of the short-lived commercial dye-production industry in that country. Despite all efforts, over many decades, a viable cochineal dye industry has never been established in either India or Sri Lanka (Ceylon). Many other similar attempts to produce cochineal dye also failed, for example those by the French in Madagascar (Middleton, 1999).

The destruction of cactus infestations by cochineal insects encouraged the authorities to spread the cochineal insect *D. ceylonicus*, to problematic infestations of *O. monacantha* throughout India and Sri Lanka. In some cases the repeated redistributions of these cochineal insects were misguided attempts to revive cochineal dye production. Eventually the cochineal insects "disappeared [locally] with the disappearance of the cactus" (Tryon, 1910). Ramakrishna Ayyar (1931) reported that *O. monacantha* had become extinct in southern India. Thus, the largely inadvertent manipulations of cochineal insect species, leading to the eventual control of *O. monacantha* in India between 1796 and 1809, became the first documented record of biological control of an alien invasive plant anywhere in the world.

Many subsequent attempts, in India, to introduce *D. ceylonicus* to suppress other weedy *Opuntia* species, notably *Opuntia dillenii* (Ker Gawler) Haworth, failed because this cactus species is an unsuitable host for *D. ceylonicus* (Tryon, 1910; Green, 1912; Ramakrishna Ayyar, 1931). All but one of the introductions of yet other species of cochineal from Mexico in 1821 also failed for the same reason (Tryon, 1910). The one cochineal insect that did succeed against *O. dillenii* was presumably *Dactylopius opuntiae* (Cockerell), and not *Dactylopius tomentosus* (Lamark), as De Lotto (1974) noted had previously been erroneously recorded. The introductions of various cochineal insect species into different countries, either to produce dye, or to control invading cactus species, have highlighted the host specificity of species within the genus *Dactylopius*. This phenomenon is summarized by Burkill (1911) (in Ramakrishna Ayyar, 1931) who, referring to the failure of *D. ceylonicus* to control some target species of cacti, noted that there had been a "waste of money in fruitless attempts to destroy *Opuntia* weeds with inappropriate [species of cochineal] insects."

7.3.2 1900–1960: spectacular successes using biological control against cactus weeds

The astonishing destruction of prickly pear in India and Sri Lanka by cochineal insects eventually came to the attention of entomologists in Australia, and later in South Africa. As a result, *D. ceylonicus* was imported from India and Sri Lanka into Australia and South Africa to control *O. monacantha*, in 1903 and 1913, respectively. The insects failed to have any effect in Australia (Tryon, 1910), but spread rapidly and provided excellent control of the weed in South Africa (Lounsbury, 1915). It was the South African success that prompted Australia to try the insect again, and releases made there in 1914 resulted in excellent control of the weed (Dodd, 1927). Although these were technically not the first cases of biological control of a weed they were certainly the earliest cases where the introduction of a biological control agent was deliberately planned and executed with the sole purpose of controlling an alien invasive plant species.

The early successes against *O. monocantha* inspired authorities in Australia to adopt biological control as a potential solution to the extensive problems being caused by other introduced and highly invasive *Opuntia* species, notably *O. stricta* (Haworth) Haworth (var. *stricta* and var. *dillenii*). By 1930, these species had invaded more than 24 m ha, mainly in Queensland and New South Wales (Dodd, 1940). Half of this area was so densely infested "that the land was useless from a productive viewpoint" (Dodd, 1940). As a consequence the Australian "Prickly Pear Destruction Commission" was initiated and coordinated an ambitious program which is still unsurpassed in the history of weed biological control. Extensive surveys for potential agents were undertaken by Australian scientists on *Opuntia* species in both South and North America between 1921 and 1939. In all, about 150 species of herbivorous insects and mites were collected and identified from various cactus hosts in the Opuntioideae (Wallace and Gibson, 2002). Fifty-six of these were shipped to Australia for investigation as potential biological control agents (Dodd, 1940; Mann, 1969, 1970; Julien and Griffiths, 1998). Eventually, 17 of the introduced species established in Australia (Mann, 1969; Julien and Griffiths, 1998) (see Table 7.2), but only four of these contributed to the control of the most problematic *Opuntia* species.

In Australia, the cactus moth, *Cactoblastis cactorum* (Bergroth) (Lepidoptera: Pyralidae) was undoubtedly the most damaging on the main invasive cactus species, *O. stricta* var. *stricta* and *O. stricta* var. *inermis* (now known as var. *dillenii*), followed by the cochineal insect *D. opuntiae* (Dodd, 1940). Ironically, *C. cactorum* was only introduced towards the end of the project after all the seemingly more promising agents associated with the target weeds, for example *Melitara*, *Olycella*, and *Megastes* (Lepidoptera: Pyraustidae), and *Moneilema* sp. (Coleoptera: Curculionidae), were released (Dodd, 1940; Mann, 1969). It remains a mystery as to why these other, apparently "perfect" biological control agents that established in Australia, particularly those that have similar biological characteristics and life histories to those of *C. cactorum*, did not perform better. It is possible that some of the earlier failures with some of the Lepidoptera and Coleoptera species were a consequence of the importation of maladapted biotypes (see below). Some of the agents that did establish had a negligible impact on suppression of the weeds because of the dramatic demise of their host-plant populations caused by *C. cactorum* (Mann, 1969).

Dodd (1940) reported that for some years, until 1933, the extent of the *Cactoblastis* biological control operation in Australia was "vast" and the scenery changed rapidly:

from flourishing [prickly] pear to dead [prickly] pear . . . to crops and fodder grasses . . . The celerity with which the insect multiplied and spread from many release centres is illustrated by the situation along the Moonie River . . . In August 1930, for 150 miles [240 km] along the river the pest [*O. stricta*] was in its full vigour, its continuity almost unbroken by cleared land; the pastoral properties had been overrun and mainly deserted, former large holdings having become mere names on a map . . . in August 1932, 90 per cent of the [prickly] pear had collapsed. The change in exactly two years was extraordinary.

Its [i.e. the cactus moth's] progress has been spectacular, its achievements border on the miraculous. . . . The prickly pear territory has been transformed as though by magic from a wilderness to a scene of prosperous endeavour . . . the most optimistic scientific opinions could not have foreseen the extent and completeness of the destruction. The spectacle of mile after mile of heavy [prickly] pear growth collapsing *en masse* and disappearing in the short space of a few years did not appear to fall within the bounds of possibility".

Today the Boonaga "Cactoblastis Memorial Hall" and the "Cactoblastis Cairn" in Dalby, Queensland are among the memorabilia celebrating these extraordinary events (Zimmermann *et al.*, 2001).

An unusual aspect of this project was that *C. cactorum* came from South America while the target weed was a North American cactus species, *O. stricta*. The cactus moth and its novel host were brought together for the first time in Australia, a so-called "new association" (Dennill and Moran, 1989; Hokkanen and Pimentel, 1989). The use of *C. cactorum* to control *O. stricta* was possible because the moth is an oligophagous species whose larvae can develop on several different *Opuntia* species. This oligophagy has allowed *C. cactorum* to be deployed successfully on a wide range of *Opuntia* species in several countries around the world, including South Africa, Mauritius, Hawaii, Ascension Island, and some Caribbean islands (Julien and Griffiths, 1998; Zimmermann *et al.*, 2001). The dramatic impact of *C. cactorum* has overshadowed the effects of the other biological control agents released in Australia, and led to speculation as to whether these biological control projects would have been successful in the absence of the cactus moth (Mann, 1969).

The Australian experiences encouraged and facilitated the adoption of biological control of cactus weeds in South Africa in the 1930s. The main target then was the tree-like prickly pear *O. ficus-indica*. Two of the natural enemies that had proved so successful in Australia were obtained from that country and deployed, namely *C. cactorum*, in 1933, followed by the cochineal, *D. opuntiae*, in 1938 (Pettey, 1948; Annecke and Moran, 1978). In this instance, *D. opuntiae* proved to be the better biological control agent, because *C. cactorum* suffers from high levels of predation in South Africa (Pettey, 1948; Robertson, 1988; Robertson and Hoffmann, 1989) and because its larvae are unable to cope with the tough, woody stems of mature *O. ficus-indica* plants (Zimmermann and Malan, 1981). This constraint prompted South African entomologists to search in Mexico during the 1950s for natural enemies that would feed on the woody stems of *O. ficus-indica*, which resulted in the release of a stem-boring weevil, *Metamasius spinolae* Gyllenhall (Coleoptera: Curculionidae) (Annecke and Moran, 1978). Initial reports ranked the weevils as of limited use because they remained localized and population increases were very gradual, apparently due to predation by vervet monkeys (Annecke and Moran, 1978). However, the weevils have spread, albeit slowly (at about 10 km in 40 years), and have increased in numbers, killing large plants and clearing extensive infestations of the weed around the original release sites. Today, due to the combined damage caused by these three species of insect agents, *O. ficus-indica* is considered to be under excellent control in South Africa and is only a relatively minor problem in some localities (Annecke and Moran, 1978).

Other early, but less dramatic successes, in Australia and South Africa, were the biological control campaigns against the far smaller, scrambling, thorny, *Opuntia aurantiaca* Lindley, using both *C. cactorum* and a cochineal, *Dactylopus austrinus* De Lotto (Moran and Annecke, 1979; Hosking and Deighton, 1981), and against *Cylindropuntia* (sometimes included in the genus *Opuntia*) *imbricata* (Haworth) Knuth, using *D. tomentosus* (Moran and Zimmermann, 1991).

7.3.3 More recent biological control programs against cactus weeds

Many of the recent projects against cactus weeds include species that are not in the taxon Opuntioideae. The species involved mainly originate from introduced ornamentals within the Cereinae (the columnar cactus group) and include the genera *Harrisia*, *Acanthocereus*, and *Cereus*. Surveys for cactus-feeding insects made by Australian, South African, and British entomologists between 1964 and 1985, led to the selection and identification of potential biological control candidates for use against *Harrisia martinii* (Labouret) Britton, *Pereskia aculeata* Miller, and to supplement the biological control of *O. aurantiaca* (Zimmermann *et al.*, 1974; McFadyen and Tomley, 1981a, b; Moran and Zimmermann, 1984; Moran and Zimmermann, 1991). Table 7.2 provides lists of the agent species that were introduced and eventually released and also ranks their respective contributions to the biological control of the target weed species.

Releases of the stem-borer *Alcidion cereicola* Fisher (Coleoptera: Cerambycidae), followed by *Hypogeococcus festerianus* (Lizer and Trelles) (Homoptera: Pseudococci-dae), on *Harrisia martinii* in 1979 have resulted in excellent control of the weed in most of Queensland (McFadyen and Tomley, 1981a, b; Tomley and McFadyen, 1985). In South Africa the introduction of *H. festerianus* has resulted in reasonable control of harrisia cactus, *H. martinii*, while *Hypogeococcus festerianus* and *A. cereicola* are still contributing to the control of *Cereus jamacaru* De Candolle (Moran and Zimmermann,1991; Klein, 1999).

Pereskia aculeata is the only weed species in the Pereskioideae that has recently been considered for biological control. *Phenrica guérini* Bechyné (Coleoptera: Chrysomelidae) has become established, albeit in restricted areas, on *P. aculetata* in South Africa but the level of control achieved by this agent is ranked as "trivial" (Klein, 1999). Nonetheless, research into the biological control of *P. aculeata* is ongoing.

7.4 Lessons learned from the biological control of cactus weeds

Experience gained over more than 200 years of biological control of cactus weeds has revealed several critical issues and key principles that are of relevance to the biological control of alien invasive plant species generally, and which are discussed below.

7.4.1 Taxonomy and species biotypes

Almost from the outset of biological control of invasive alien plants, practitioners in India and Sri Lanka realized that cochineal insects are highly host specific and that many early failures could be traced back to misidentifications of the agents leading to the introduction of inappropriate taxa for a particular target weed species (Tryon, 1910; Green, 1912; Ramakrishna Ayyar, 1931). The problem was exacerbated by the complicated and con-fusing taxonomy of the cactus host plants themselves, for example among the 200 species in the Opuntiodeae (Wallace and Gibson, 2002). Some resolution of these taxonomic difficulties was achieved with the completion of the study of the Dactylopiidae (De Lotto, 1974), and of the Cactaceae (Anderson, 2001). Many of the earlier failures in cactus biological control using Lepidoptera and Coleoptera were a consequence of the import-ation of agents that were not adapted to their host plants. More recently, Moran *et al.* (1976) and Annecke and Moran (1978), for example, have highlighted how the confusing taxonomy of *O. ficus-indica* and *O. aurantiaca* respectively, hindered the search for new biological control agents.

Precise alpha-taxonomy is essential, but is not sufficient. Successful biological control of cacti, and of other plants, often depends on the correct identification of the most effective biological control agent biotypes and of their primary host-plant species. For example, although *O. stricta* var. *stricta* had been recorded in South Africa for many years, it was only recognized as a problem in the 1970s, with one of the main infestations located in the country's premier conservation area, the Kruger National Park (KNP). The obvious solution was biological control and *C. cactorum* was introduced into the park in 1987 (Hoffmann *et al.*, 1998a). Although the moth has played an important role in supplementing the management of the weed (Hoffmann *et al.*, 1998b), the outcome of its release was disappointing when compared to outcomes achieved 50 years earlier on *O. stricta* in Australia. In South Africa, mainly because of predation of the cactus moth eggs by ants and baboons (Hoffmann *et al.*, 1998a), the effects of the cactus moth were far less impressive than they had been in Australia. To supplement the actions of the cactus moth, the cochineal insect, *D. opuntiae*, was introduced into the KNP in 1997. Initial results were discouraging because, after some generations, populations of the insects died out. This led to an extensive investigation, which showed that distinct biotypes of this species of cochineal insect exist, each of which is associated with a specific *Opuntia* species, and that it is essential to match the agent biotype with its host (Githure *et al.*, 1999; Hoffmann *et al.*, 1999; Volchansky *et al.*, 1999; Hoffmann *et al.*, 2002). This information enabled the release of the correct biotype of *D. opuntiae* into the KNP, resulting in a dramatic decline in the abundance of the weed (Hoffmann *et al.*, 1999; J. H. Hoffmann, unpublished results).

Further research will focus on finding new cochineal biotypes. Host-specific biotypes have been demonstrated in other cochineal species, for example *D. tomentosus*, which utilizes species of *Cylindropuntia* (Anderson, 2001) as hosts (C. Mathenge, unpublished results). This finding explains why *D. tomentosus* is associated with *C. imbricata* but does not survive on *Cylindropuntia fulgida* (Engelmann) in South Africa, although both plant species have been recorded as suitable hosts for the insect. Biotypes of *D. tomentosus* have been collected on *C. fulgida* (formerly known as *C. cholla*) and on *Cylindropuntia rosea* (De Candolle) in Mexico: both are specific to their respective hosts and cannot develop on the other *Cylindropuntia* species. These realizations have increased the chances of improved biological control of several new *Cylindropuntia* weeds in Australia and South Africa.

7.4.2 *Research and evaluation*

Sustained research on the ecology and the evaluation of the impacts of biological control agents is required to enhance the effectiveness of established biological control agents and to improve the overall management of cactus weeds.

For example, because of the apparently unsatisfactory level of biological control achieved on *O. aurantiaca* in South Africa, by the cactus moth and by the cochineal *D. austrinus*, surveys were made for new agents in South America between 1970 and 1985. Three additional agents were introduced and released between 1979 and 1983, none of which established. Two of these introduced species, *Mimorista pulchellalis* (Dyar) (Lepidoptera: Pyraustidae) and *Tucumania tapiacola* Dyar (Lepidoptera: Pyralidae), were collected from the target weed in Argentina while the third, *Nanaia* sp., from Peru, came from *Opuntia pubescens* Wendland ex Pfeiffer (formerly known as *O. pascoensis*).

Research revealed the probable causes for the failure of these insects, thereby obviating repeated and expensive efforts expended on the recollection and reintroduction of the agents (Hoffmann, 1988; Moran and Zimmermann, 1991).

Subsequently the ecology of the established biological control agents and their hosts was investigated (Zimmermann, 1979; Moran and Zimmermann, 1991) as part of a study to determine how the insects coped with herbicides, which were in use to control *O. aurantiaca*. The results of these studies showed that herbicides were hindering the efficacy of the cochineal insects, such that their populations never increased to levels that were lethal to the host. Further studies revealed that herbicides were no longer needed in some areas and that biological control could keep the weed in check (Zimmermann and Moran, 1982; Moran and Zimmermann, 1991). In some regions infested by the weed, biological control could be improved by modifying the technique of herbicide application (Zimmermann and Malan, 1980). In recent years, an increased reliance on *D. austrinus* prevails and the cochineal insects are now mass-reared and distributed to landowners to control *O. aurantiaca*. In Australia, herbicides are being phased out and a greater reliance is now placed on biological control in managing *O. aurantiaca* (Hosking and Zimmermann, 1996).

Similarly, the use of herbicides for the control of *O. ficus-indica* has decreased considerably in South Africa and the remaining problem populations of the plants are destroyed using stem injections of systemic herbicides on plants bearing 14 cladodes or more (Zimmermann, 1989). The rationale for this approach is that the smaller plants are killed by *C. cactorum* and *D. opuntiae* (Zimmermann and Malan, 1981), thus rendering herbicidal control unnecessary. The status of *O. ficus-indica* as an invader has been further diminished by research aimed at the improved utilization of the remaining prickly pear populations as a source of food for human and animal consumption, and for the production of carmine dye derived from the cochineal insect *D. coccus* (Zimmermann and Zimmermann, 1987; Zimmermann, 1989; Brutsch and Zimmermann, 1993).

7.4.3 Conflicts of interest

Some of the large tree-like *Opuntia* species, particularly the many cultivars of *O. ficus-indica*, provide sustenance for people and livestock (Casas and Barbera, 2002). This context has been a source of conflict when biological control measures have been contemplated. Emerging cactus crops in *Hylocereus*, *Selenicereus*, *Cereus*, and *Stenocereus* are also assuming greater relevance (Nerd *et al.*, 2002) and could be a point of conflict in future in some countries.

For example, at the peak of its invasion, in the early 1900s, the prickly pear, *O. ficus-indica*, was devastating to people and their livestock in the Eastern Cape and Karroo regions of South Africa (Annecke and Moran, 1978; van Sittert, 2002). However, small interest groups who used the plants as fodder originally opposed the release of biological control agents in South Africa. Eventually a parliamentary committee authorized biological control because the problems of dense prickly pear invasions to people individually, and to the affected societies, their livestock, and agriculture generally, far outweighed the benefits of these plants (van Sittert, 2002). The introduced biological control agents have provided acceptable levels of control of *O. ficus-indica* over about 90% of its formerly infested area (nearly 1 m ha), although residual populations of plants remain in sufficient abundance for people and animals to utilize them as a food source (Annecke and Moran, 1978). In plantations of cultivated prickly pears, conventional insecticidal and manual control methods are used to suppress the biological control agents, *C. cactorum* and *D. opuntiae* (Pretorius *et al.*, 1986; Pretorius and van Ark, 1992; Bloem *et al.*, 2005). The informal prickly pear cactus industry in South Africa is dependent on the residual prickly pear populations for the sale of fruit, livestock fodder, and to a far lesser extent, the use of the young cladodes as a vegetable and as a source of food for some wild animals. Thus, fortuitously, a compromise situation acceptable to most of the stakeholders has been achieved.

Some further current examples are illustrative of the complexities and considerations involved:

- Many African and Arabian countries increasingly rely on *O. ficus-indica*, and other species such as *Opuntia engelmannii* Salm Dyck, *Opuntia exaltata* Berger, and *Opuntia elatior* Miller, as multipurpose crop plants. These species have the potential to become highly invasive. Fortunately, host-adapted biotypes of the biological control agents may still be an option for management of these plants, without affecting the economic value of *O. ficus-indica*.
- Other developing countries (e.g. Ethiopia, Madagascar, Eritrea, and Yemen) are already suffering from massive and severely debilitating infestations of *O. ficus-indica*. Biological control is not considered as an option in these regions because many rural communities are dependent on the cactus (Ellenberg, 1982; Brutsch and Zimmermann, 1993, 1995; Haile *et al.*, 2002; Middleton, 2002). Instead, these countries are striving to increase utilization of the resource to such an extent that infestations will be reduced to acceptable levels. It remains a matter of conjecture as to whether or not this will be possible without biological control interventions. The South African precedent could guide these countries in finding a solution if increased utilization efforts fail (Zimmermann, 1997).
- Kenya is being invaded by *O. elatior* and biological control, including the use of *C. cactorum*, could eventually be the best, if not the only, management option, provided that no commercial cultivations of *O. ficus-indica* are contemplated for the future. The proximity of Ethiopia, which has decided not to use biological control as a management tool for *O. ficus-indica*, could prevent the release of any biological control agents in Kenya unless an agent specific to *O. elatior* can be found. This will be difficult. Again, compromises may be possible through the careful selection of agents that would reduce the aggressiveness of the weeds, but that would allow the continued utilization of the resource.
- *Hypogeococcus festerianus* and *A. cereicola* that were used for biological control of *Cereus jamacaru* in South Africa could also attack the cultivated conspecific varieties that are now commercially cultivated for fruits in countries such as Israel (Nerd *et al.*, 2002).

In summary, the biological control of cactus weeds is entering a new phase where conflicts of interest will dictate what is possible and thus present a daunting array of attendant sociological, political, and pragmatic complexities.

7.4.4 Controversial biological control efforts against cactus weeds

The use of oligophagous species, such as the cactus moth, *C. cactorum*, for the biological control of cactus weeds has never been a contentious issue in countries outside of the Americas which are devoid of native cactus species (e.g. South Africa and Australia), but recent developments in the New World have been perceived as a setback for biological control. The problem started when *C. cactorum* was introduced, by the then Commonwealth Institute for Biological Control (CIBC), into a few Caribbean islands (Nevis, St. Kitts, Antigua, Montserrat, and Grand Cayman) for the biological control of some indigenous cactus species (Simmonds and Bennett, 1966) that had become invasive because of pasture mismanagement. The decision to release *C. cactorum* to control native *Opuntia* species was not contested at the time. The deliberate spread and natural dispersal of *C. cactorum* to other islands in the Caribbean, some of which had problems with *Opuntia* species (e.g. Cuba) (Blanco and Vazquez, 2001), was overlooked even though the cactus moth had induced a decline in some nontarget native *Opuntia* species (Zimmermann *et al.*, 2005).

In retrospect, it is easy to say that the use of *C. cactorum* to control native cactus species in their natural habitats should never have happened. The basic problem caused by native cacti in the Caribbean was inappropriate agricultural practices that encouraged the invasive growth of native cacti. Trying to solve the problem by treating the symptoms and not the cause was conceptually flawed and unwise. The probability that other nontarget native cactus species would be destroyed by the cactus moths, at least in the immediate vicinity of the release areas, should have been foreseen as a virtual certainty and should have been considered to be unacceptable. However, in the mid 1950s, no reservations were expressed by any of the participating authorities or scientists about the wisdom of these projects and the programs were rated a resounding success. The very recent discovery of *C. cactorum* in Jamaica, the Dominican Republic, and Guadeloupe, despite its presence there for probably two decades or more, is proof of the ignorance regarding the importance and value of indigenous *Opuntia* species in this region (Zimmermann *et al.*, 2005).

The originally positive perceptions of the utility and effectiveness of biological control against cactus weeds in the West Indies changed completely with the later discovery of *C. cactorum* in Florida in 1989 (Bennett and Habeck, 1995; Simberloff and Stiling, 1996; Zimmermann *et al.*, 2001; Moran *et al.*, 2005). The mechanism of the movement and establishment of *C. cactorum* onto the North American mainland remains uncertain, but contaminated horticultural cactus plants imported from the Caribbean by the nursery industry are the most likely source (Pemberton, 1995). All six of the native *Opuntia* species in Florida are now serving as hosts for *C. cactorum* as it spreads through the southern United States of America, destroying native North American cactus species. A further major concern is that *C. cactorum* will eventually reach Mexico with its highly prized and extensively utilized cactus flora (Vigueras and Portillo, 2001; Zimmermann *et al.*, 2004). The discovery of *C. cactorum* on Isla Mujeres, 8 km from the Mexican mainland, at Cancun, in August 2006, has considerably exacerbated this concern, but concerted and urgent interventions are underway.

Although the introduction of *C. cactorum* into the Caribbean was a completely aberrant program, the practice of biological control had nothing to do with the arrival of the cactus moth in Florida. The insect could have moved as easily with nursery stock directly from South America. Nevertheless, these events have been used by antagonists to fuel the debate about the risks of nontarget effects in biological control (Zimmermann *et al.*, 2001). Indeed, the negative publicity may eventually result in unrealistic constraints being imposed on biological control in the future (McEvoy and Coombs, 1999).

7.4.5 Economic assessments of the benefits of biological control against cactus weeds

Although the classical and extensive biological control campaigns against cactus weeds in Australia and South Africa were relatively expensive, the cost-benefit ratios are still overwhelmingly in favour of biological control. In the case of *O. aurantiaca* in South Africa, for example, which was not rated as a particularly successful project (Moran and Annecke, 1979), the cost-benefit ratio, up to 2000, was calculated at an extraordinary 1:709 (Van Wilgen *et al.*, 2004). A somewhat lower, but still exceptional, cost-benefit ratio of 1:312, using an 8% discount rate, was calculated for the biological control programs against the four most important cactus species in Australia, but even this high figure is considered an underestimate of the total benefits (Page and Lacey, 2006). The cost-benefit ratio of the more recent biological control program against *H. martinii* in Australia is given as 1:23.5, also calculated at an 8% discount rate. Unquestionably, the biological control of cactus weeds has been a good investment.

7.5 Conclusions

In the case of most biological control projects against cactus weeds, the decline in populations of the target plants has been rapid (e.g. the decline of *O. stricta* due to *C. cactorum*, in Australia). In contrast, the full impacts of biological control have sometimes taken decades to manifest themselves, for example in the case of the stem-boring weevil *Metamasius spinolae*, on *O. ficus-indica* in South Africa. The course and the outcomes of biological control are almost impossible to predict. This raises an important point: in some countries, for example New Zealand (Sheppard *et al.*, 2003), it is a prerequisite for the release of a new biological control agent that, in addition to rigorous testing for the safety of an agent, motivations for release must include predictions of the damage levels that might be achieved if the agents were to be released. This requirement seems entirely reasonable, in principle, but a century of precedents in biological control of cactus weeds suggest that accurate and thus useful predictions will be far easier in theory than in practice.

In any event, it is extremely encouraging that the dramatic effects of the agents used in cactus control (usually the cactus moth, in combination with appropriate species of cochineal insects) have been sustained for many years: for nearly two centuries in India and Sri Lanka, and for nearly 100 years in South Africa and in Australia. Ironically the now classical and outstanding early successes in Australia, particularly, and in South Africa, have often imposed unrealistically high expectations on the practice of biological control in general and encouraged the mistaken notion that biological control is an easy and inexpensive panacea for tackling weed problems (Hoffmann, 1995). This mis-perception has led to many of the more-recent programs against other weeds (i.e. not including cacti) having been rated, unjustifiably, as disappointing or as outright failures. Recent detailed studies of the ecology of the interactions between biological control agents and their cactus host plants, and of the impacts of biological control in cactus management have provided a different perspective. Weed populations do not have to collapse completely over wide areas for biological control to be rated a success. Often the effects of biological control are far more subtle (and include, for example, reductions in seeding and thus invasiveness, or reductions in the frequency, rate of application, and costs of herbicides) but, nonetheless, are of enormous benefit in containing the problem of invasive alien plant species and in enhancing integrated management practices (Hoffmann and Moran, 2008).

Careful, long-term evaluation studies that reveal the full benefits of biological control in suppressing problem plants are the challenge for the future.

References

Anderson, E. F. (2001). *The Cactus Family*. Portland, OR: Timber Press, 776 pp.

Annecke, D. P. and Moran, V. C. (1978). Critical review of biological pest control in South Africa. 2. The prickly pear, *Opuntia ficus-indica* (L.) Miller. *Journal of the Entomological Society of Southern Africa*, **41**, 161–188.

Baranyovits, F. L. C. (1978). Cochineal carmine: an ancient dye with a modern role. *Endeavour*, **2**, 85–92.

Barbera, G., Inglese, P., Pimienta-Barrios, E. and Arias-Jimenez, E. de J., eds. (1995). *Agro-ecology, Cultivation and Uses of Cactus Pear*. FAO Plant Production and Protection Paper 132. Rome, Italy: FAO, 216 pp.

Bennett, F. D. and Habeck, D. H. (1995). *Cactoblastis cactorum*: a successful weed control agent in the Caribbean, now a pest in Florida? In *Proceedings of the VII International Symposium on Biological Control of Weeds*, ed. E. S. Delfosse and R. R. Scott, 1992. Melbourne, Australia: CSIRO, pp. 21–26.

Blanco, E. and Vazquez, L. L. (2001). Analisis de los riesgos fitosanitarios asociados al uso de *Cactoblastis cactorum* (Berg) (Lepidoptera: Pyralidae: Phycitinae) como agente de control biológico de *Opuntia dillenii* (Cactaceae) en Cuba. *Fitosanidad*, **5**, 63–73.

Bloem, S., Mizell, R. F., Bloem, S. K. A., Hight, D. and Carpenter, J. E. (2005). New insecticides for the control of the invasive cactus moth, *Cactoblastis cactorum* (Lepidoptera: Pyralidae) in Florida. *Florida Entomologist*, **88**, 400–407.

Brutsch, M. O. and Zimmermann, H. G. (1993). The prickly pear (*Opuntia ficus-indica*) (Cactaceae), in South Africa: utilization of the naturalized weed and of the cultivated plant. *Economic Botany*, **47**, 154–156.

Brutsch, M. O. and Zimmermann, H. G. (1995). Control and utilization of wild opuntias. In *Agro-ecology, Cultivation and Uses of Cactus Pear*, ed. G. Barbera, P. Inglese, E. Pimienta-Barrios and E. de J. Arias-Jimenez. FAO Plant Production and Protection Paper 132. Rome, Italy: FAO, pp. 155–166.

Casas, A. and Barbera, G. (2002). Mesoamerican domestication and diffusion. In *Cacti: Biology and Uses*, ed. P. S. Nobel. Berkeley, CA: University of California Press, pp. 143–162.

De Lotto, G. (1974). On the status and identity of the cochineal insects (Homoptera: Dactylopiidae). *Journal of the Entomological Society of Southern Africa*, **37**, 167–193.

Dennill, G. B. and Moran, V. C. (1989). On insect–plant associations in agriculture and the selection of agents for weed biocontrol. *Annals of Applied Biology*, **114**, 157–166.

Dodd, A. P. (1927). The biological control of prickly pear in Australia. *Australian Council of Scientific and Industrial Research Bulletin*, **34**, 44 pp.

Dodd, A. P. (1940). *The Biological Campaign Against Prickly Pear*. Commonwealth Prickly-pear Board Bulletin. Brisbane, Australia: Government Printer, 177 pp.

Ellenberg, H. (1982). *Opuntien-probleme und Wege zu deren Lösung*. Deutsche Gesellschaft für Technische Zusammenarbeit (GTZ) 73.2109.4. Eschborn, Germany: GTZ, 62 pp.

Flores-Flores, V. and Tekelenburg, A. (1995). Dacti (*Dactylopius coccus* Costa) dye production. In *Agro-ecology, Cultivation and Uses of Cactus Pear*, ed. G. Barbera, P. Inglese, E. Pimienta-Barrios and E. de J. Arias-Jimenez, FAO Plant Production and Protection Paper 132. Rome, Italy: FAO, pp. 167–185.

Gibson, A. C. and Nobel, P. S. (1986). *The Cactus Primer*. Cambridge, MA: Harvard University Press.

Githure, C. W., Zimmermann, H. G. and Hoffmann, J. H. (1999). Host specificity of biotypes of *Dactylopius opuntiae* (Cockerell) (Hemiptera: Dactylopiidae): prospects for biological control of *Opuntia stricta* (Haworth) Haworth (Cactaceae) in Africa. *African Entomology*, **7**, 43–48.

Goeden, R. D., Fleshner, C. A. and Ricker, D. W. (1967). Biological control of prickly pear cacti on Santa Cruz Island, California. *Hilgardia*, **38**, 579–606.

Green, E. E. (1912). On the cultivated and wild forms of cochineal insects. *Journal of Economic Biology*, **7**, 79–93.

Greenfield, A. B. (2006). *A perfect red: empire, espionage, and the quest for the color of desire*. New York: Harper Collins. 338 pp.

Haile, M., Belay, T. and Zimmermann, H. G. (2002). Current and potential use of cactus in Tigray, Northern Ethiopia. *Acta Horticulturae*, **581**, 75–86.

Hoffmann, J. H. (1988). Pre-release assessment of *Nanaia* sp. (Lepidoptera: Phycitidae) from *Opuntia pascoensis* for biological control of *Opuntia aurantiaca* (Cactaceae). *Entomophaga*, **33**, 81–86.

Hoffmann, J. H. (1995). Biological control of weeds: the way forward, a South African perspective. In *Weeds in a Changing World*. ed. C. C. Stirton, BCPC Symposium Proceedings 64. Farnham, UK: BCPC, pp. 77–89.

Hoffmann, J. H. and Moran, V. C. (2008). Assigning success in biological weed control: what do we really mean? In *Proceedings of the XII International Symposium on Biological Control of Weeds*, held in La Grande Motte, France, 22–27 April 2007, ed. M. H. Julien *et al.* Wallingford, UK: CAB International, pp. 685–690.

Hoffmann, J. H., Impson, F. A. C. and Volchansky, C. R. (2002). Biological control of cactus weeds: implications of hybridization between control agent biotypes. *Journal of Applied Ecology*, **39**, 900–908.

Hoffmann, J. H., Moran, V. C. and Zeller, D. A. (1998a). Evaluation of *Cactoblastis cactorum* (Lepidoptera: Phycitidae) as a biological control agent of *Opuntia stricta* (Cactaceae) in the Kruger National Park, South Africa. *Biological Control*, **11**, 20–24.

Hoffmann, J. H., Moran, V. C. and Zeller, D. A.. (1998b). Long-term population studies and the development of an integrated management program for control of *Opuntia stricta* in Kruger National Park, South Africa. *Journal of Applied Ecology*, **35**, 156–160.

Hoffmann, J. H., Moran, V. C. and Zimmermann, H. G. (1999). Integrated management of *Opuntia stricta* (Haworth) Haworth (Cactaceae) in South Africa: an enhanced role for two, renowned insect agents. In *Biological Control of Weeds in South Africa (1990–1998)*, ed. T. Olckers and M. P. Hill. *African Entomology Memoir*, **1**, 15–20.

Hokkanen, H. M. T. and Pimentel, D. (1989). New associations in biological control: theory and practice. *Canadian Entomologist*, **121**, 829–840.

Hosking, J. R. and Deighton, P. J. (1981). Tiger pear is a continuing problem. *Agricultural Gazette of New South Wales*, **92**, 43–45.

Hosking, J. R. and Zimmermann, H. G. (1996). Integrated control of *Opuntia aurantiaca* in Australia and South Africa, the shift in emphasis from herbicidal to biological control. In *Proceedings of the IXth International Symposium on Biological Control of Weeds*, eds. V. C. Moran and J. H. Hoffmann. Stellenbosch, South Africa. University of Cape Town, 480 pp.

Julien, M. H. and Griffiths, M. W., eds. (1998). *Biological Control of Weeds. A World Catalogue of Agents and Their Target Weeds*. Fourth edn. Wallingford, UK: CABI Publishing. 223 pp.

Klein, H. (1999). Biological control of three cactaceous weeds, *Pereskia aculeata* Miller, *Harrisia martinii* (Labouret) Britton and *Cereus jamacaru* De Candolle in South Africa. In *Biological Control of Weeds in South Africa (1990–1998)*, ed. T. Olckers and M. P. Hill: *African Entomology Memoir*, **1**, 3–14.

Lounsbury, C. P. (1915). Plant killing insects: the Indian cochineal. *Agricultural Journal of South Africa*, **1**, 537–543.

Mann, J. (1969). *Cactus-feeding Insects and Mites*. Bulletin 256. Washington, DC: United States National Museum, 158 pp.

Mann, J. (1970). *Cacti naturalized in Australia and their control*. Brisbane, Queensland: Department of Lands, 128 pp.

McEvoy, P. B. and Coombs, E. M. (1999). Why things bite back: unintentional consequences of biological control. In *Non-target Effects of Biological Control*. ed. P. A. Follett and J. J. Duan. Boston, MA: Kluwer Academic Publishers, pp. 167–194.

McFadyen, R. E. and Tomley, A. J. (1981a). The successful biological control of Harrisia cactus (*Eriocereus martinii*) in Queensland. *Proceedings of the Sixth*

Australian Weeds Conference. Brisbane, Australia: Weed Society of Queensland, pp. 139–143.

McFadyen, R. E. and Tomley, A. J. (1981b). Biological control of Harrisia cactus, *Eriocereus martinii*, in Queensland by the mealy bug *Hypogeococcus festerianus*. In *Proceedings of the V International Symposium on Biological Control of Weeds*, ed. E. S. Delfosse. Melbourne, Australia: CSIRO, pp. 589–594.

Middleton, K. (1999). Who killed "Malagasy Cactus"? Science, environment and colonialism in southern Madagascar (1924–1930). *Journal of Southern African Studies*, **25**, 215–248.

Middleton, K. (2002). Opportunities and risks: a cactus pear in Madagascar. *Acta Horticulturae*, **581**, 63–73.

Moran, V. C. 1980. Interactions between phytophagous insects and their *Opuntia* hosts. *Ecological Entomology*, **5**, 153–167.

Moran, V. C. and Annecke, D. P. (1979). Critical reviews of biological control in South Africa. 3. The jointed cactus, *Opuntia aurantiaca* Lindley. *Journal of the Entomological Society of Southern Africa*, **42**, 299–329.

Moran, V. C. and Zimmermann, H. G. (1984). The biological control of cactus weeds: achievements and prospects. *Biocontrol News and Information* **5**, 297–320.

Moran, V. C. and Zimmermann, H. G. (1991). Biological control of cactus weeds of minor importance in South Africa. *Agriculture, Ecosystems and Environment*, **37**, 37–55.

Moran, V. C., Annecke, D. P. and Zimmermann, H. G. (1976). The identity and distribution of *Opuntia aurantiaca* Lindley. *Taxon*, **25**, 281–287.

Moran, V. C., Hoffmann, J. H. and Zimmermann, H. G. (2005). Biological control of invasive alien plants in South Africa: necessity, circumspection, and success. *Frontiers in Ecology and the Environment*, **3**, 77–83.

Nerd, A., Tel-Zur, N. and Mizrahi, Y. (2002). Fruits of vine and columnar cacti. In *Cacti: Biology and Uses*, ed. P. S. Nobel. Berkeley, CA: California University Press, pp. 185–198.

Olckers, T. and Hill, M. P., eds. (1999). Appendix to: *Biological control of weeds in South Africa (1990–1998)*. African Entomology Memoir 1, pp. 175–182.

Page, A. R. and Lacey, K. L. (2006). Economic impact assessment of Australian weed biological control. Technical Series 10. Adelaide, Australia: CRC for Australian Weed Management Systems, 150 pp.

Pemberton, R. W. (1995). *Cactoblastis cactorum* (Lepidoptera: Pyralidae) in the United States. An immigrant biological control agent or an introduction of the nursery industry? *American Entomologist*, **41**, 230–232.

Pettey, F. W. (1948). The biological control of prickly-pear in South Africa. *Science Bulletin, Department of Agriculture of the Union of South Africa*, **271**, 1–163.

Pretorius, M. W. and van Ark, H. (1992). Further insecticide trials for the control of *Cactoblastis cactorum* (Lepidoptera: Pyralidae) as well as *Dactylopius opuntiae* (Homoptera: Dactylopiidae) on spineless cactus. *Phytophylactica*, **24**, 229–233.

Pretorius, M. W., van Ark, H. and Smit, C. (1986). Insecticide trials for the control of *Cactoblastis cactorum* (Lepidoptera: Pyralidae) on spineless cactus. *Phytophylactica*, **18**, 121–125.

Ramakrishna Ayyar, T. V. (1931). The Coccidae of the prickly pear in South India and their economic importance. *Agriculture and Livestock in India*, **1**, 229–237.

Rao, V. P., Ghani, M. A., Sankaran, T. and Mathur, K. C. (1971). A review of biological control of insects and other pests in South-East Asia and the Pacific Region.

Technical Communication 6. Slough, UK: Commonwealth Institute of Biological Control, 149 pp.

Robertson, H. G. (1988). Spatial and temporal patterns of predation by ants on eggs of *Cactoblastis cactorum*. *Ecological Entomology*, **13**, 207–214.

Robertson, H. G. and Hoffmann, J. H. (1989). Mortality and life-tables of *Cactoblastis cactorum* (Berg.) (Lepidoptera: Pyralidae) compared on two host plant species. *Bulletin of Entomological Research*, **79**, 7–17.

Sheppard, A. W., Hill, R., DeClerck-Floate, R. A., *et al.* (2003). A global review of risk-benefit-cost analysis for the introduction of classical biological control agents against weeds: a crisis in the making? *Biocontrol News and Information*, **24**, 91–108.

Simberloff, D. and Stiling, P. (1996). How risky is biological control? *Ecology*, **77**, 1965–1974.

Simmonds, F. J. and Bennett, F. D. (1966). Biological control of *Opuntia* spp. by *Cactoblastis cactorum* in the Leeward Islands (West Indies). *Entomophaga*, **11**, 183–189.

Tomley, A. J. and McFadyen, R. E. (1985). Biological control of Harrisia cactus, *Eriocereus martinii*, in central Queensland by the mealybug, *Hypogeococcus festerianus*, nine years after release. In *Proceedings of the VI International Symposium on Biological Control of Weeds*, ed. E. S. Delfosse. Vancouver, Canada: Agriculture Canada, pp. 843–847.

Tryon, H. (1910). The "wild cochineal insects", with reference to its injurious action on prickly pear (*Opuntia* spp.) in India etc. and to its availability for the subjugation of this plant in Queensland and elsewhere. *Queensland Agricultural Journal*, **25**, 188–197.

Van Sittert, L. (2002). Our irrepressible fellow colonists: the biological invasion of prickly pear (*Opuntia ficus-indica*) in the Eastern Cape Colony c. 1870–1910. In *South Africa's Environmental History: Cases and Comparisons*. ed. R. W. Dovers, R. Edgecombe and B. Guest. Athens, OH: Ohio University Press and Cape Town South Africa: David Philip, pp. 139–159.

Van Wilgen, B. W., de Wit, M. P., Anderson, D. H. J., *et al.* (2004). Costs and benefits of biological control of invasive alien plants: case studies from South Africa. *South African Journal of Science*, **100**, 113–122.

Vigueras, A. L. and Portillo, L. (2001). Uses of *Opuntia* species and the potential impact of *Cactoblastis cactorum* (Lepidoptera: Pyralidae) in Mexico. *Florida Entomologist*, **84**, 493–498.

Volchansky, C. R., Hoffmann, J. H. and Zimmermann. H. G. (1999). Host-plant affinities of two biotypes of *Dactylopius opuntiae* (Homoptera: Dactylopiidae): enhanced prospects for biological control of *Opuntia stricta* (Cactaceae) in South Africa. *Journal of Applied Ecology*, **36**, 85–91.

Wallace, R. S. and Gibson, A. C. (2002). Evolution and systematics. In *Cacti: Biology and Uses*, ed. P. S. Nobel. Berkeley, CA: California University Press, pp. 1–22.

Zimmermann, H. G. (1979). Herbicidal control in relation to distribution of *Opuntia aurantiaca* Lindley and effects on cochineal populations. *Weed Research*, **19**, 89–93.

Zimmermann, H. G. (1989). *Control of Prickly Pear*. Farming in South Africa, Weeds Bulletin B.1.1/1998. Pretoria, South Africa: Goverment Printer, 2 pp.

Zimmermann, H. G. (1997). Opuntia invaders in Africa: solving conflicts of interest with biological control. In *Proceedings of International Workshop: 'Opuntia in Ethiopia: State of Knowledge in Opuntia Research,'* ed. M. Behailu and F. Tegegne. Tigray, Ethiopia: Mekelle University College, pp. 180–193.

Zimmermann, H. G. and Granata, G. (2002). Insect pests and diseases. In *Cacti: biology and Uses*, ed. P. S. Nobel. Berkeley, CA: California University Press, pp. 231–254.

Zimmermann, H. G. and Malan, D. E. (1980). A modified technique for the herbicidal control of jointed cactus, *Opuntia aurantiaca* Lindley, in South Africa. *Agroplantae*, **12**, 65–67.

Zimmermann, H. G. and Malan, D. E. (1981). The role of imported natural enemies in suppressing re-growth of prickly pear, *Opuntia ficus-indica*, in South Africa. In *Proceedings of the V International Symposium on Biological Control of Weeds*, ed. E. S. Delfosse. Melbourne, Australia: CSIRO, pp. 375–381.

Zimmermann, H. G. and Moran, V. C. (1982). Ecology and management of cactus weeds in South Africa. *South African Journal of Science*, **78**, 314–320.

Zimmermann, H. G. and Zimmermann, H. E. (1987). A novel use of a declared weed: young prickly pear leaves for human consumption. *Farming in South Africa, Weeds Bulletin B.1.2/1987*. Pretoria, South Africa: Goverment Printer, pp. 1–4.

Zimmermann, H. G., Burger, W. A. and Annecke, D. P. (1974). The biological control of jointed cactus in South Africa. In *Proceedings of the First National Weeds Conference*, held in Pretoria, 13–14 August 1974. Pretoria, South Africa: National Programme for Environmental Science, CSIR, pp. 204–211.

Zimmermann, H. G., Bloem, S. and Klein, H. (2004). *Biology, History, Threat, Surveillance and Control of the Cactus Moth*, Cactoblastis cactorum. Programme of Nuclear Techniques in Food and Agriculture. Vienna, Austria: IAEA/FAO, pp. 1–40.

Zimmermann, H. G., Moran, V. C. and Hoffmann, J. H. (2001). The renowned cactus moth, *Cactoblastis cactorum* (Lepidoptera: Pyralidae): its natural history and threat to native *Opuntia* floras in Mexico and the United States. *Diversity and Distributions*, **6**, 259–269.

Zimmermann, H. G., Perez S. C. M. and Rivera, A. B. (2005). *The Status of Cactoblastis cactorum (Lepidoptera: Pyralidae) in the Caribbean and the likelihood of its spread to Mexico*. Report to IAEA. Vienna, Austria: IAEA. Project TC MEX/5/029.

8

Chromolaena odorata (L.) King and Robinson (Asteraceae)

C. Zachariades, M. Day, R. Muniappan, and G. V. P. Reddy

8.1 Introduction

Chromolaena odorata (L.) King and Robinson (Asteraceae), formerly known as *Eupatorium odoratum* L., is a weedy pioneering shrub native to the Americas from southern USA to northern Argentina (Gautier, 1992). *Chromolaena odorata* has become one of the worst terrestrial invasive plants in the humid tropics and subtropics of the Old World over the past century (Holm *et al.*, 1977; Gautier, 1992). From its original point of introduction as an ornamental plant in northeastern India in the mid nineteenth century, it has spread throughout Southeast Asia, into parts of Oceania (Muniappan and Marutani, 1988; McFadyen, 1989; Waterhouse, 1994a), and into West and Central Africa (Gautier, 1992; Prasad *et al.*, 1996). A different form of *C. odorata* (see below), first recorded as naturalized in the 1940s (Hilliard, 1977), has invaded a large part of the subtropics of southern Africa (Goodall and Erasmus, 1996).

Individual *C. odorata* plants are easily controlled by chemical and/or mechanical means. However, as it is a weed mainly of the tropics and subtropics, many of the countries in which it is a problem do not have the resources to implement comprehensive control programs using conventional methods. Consequently, biological control has become an important management tool (Goodall and Erasmus, 1996; McFadyen, 1996a).

Research into the potential of biological control for *C. odorata* was initiated in the 1960s, when a survey of phytophagous insects on *C. odorata*, and the host-specificity testing of selected species, was conducted in Trinidad by the Commonwealth Institute for Biological Control, financed by the Nigerian Institute for Oil Palm Research (Cruttwell, 1972, 1974).

8.2 Taxonomy

Chromolaena odorata belongs to the Asteraceae (Compositae), a large, well-defined and highly evolved family (Toelken, 1983; Bremer, 1994; APG II, 2003). Members of the Asteraceae occur throughout the world, and are particularly abundant in the Americas, where the family may have originated (Bremer, 1994). There are many ornamentals and only a few crop plants in the Asteraceae (Toelken, 1983).

Biological Control of Tropical Weeds using Arthropods, ed. R. Muniappan, G. V. P. Reddy, and A. Raman. Published by Cambridge University Press. © Cambridge University Press, 2009.

Chromolaena odorata is a member of the Eupatorieae, a well-defined, largely New World tribe within the subfamily Asteroideae (King and Robinson, 1987). *Eupatorium* included over 1200 species before it was split by King and Robinson (1970), following which *Chromolaena* now includes more than 165 species, all from South and Central America and the West Indies (King and Robinson, 1987). Within the genus, *C. odorata*, *C. ivaefolia* (L.) King & Robinson and *C. laevigata* (Lam.) King & Robinson are widespread and occasionally weedy in the Americas, but only *C. odorata* has become a serious weed in the Old World (McFadyen, 1989). A minor infestation of *C. squalida* (DC.) King & Robinson from Brazil was recently found in northern Australia (Waterhouse, 2003), but this species does not appear to be particularly invasive.

In addition to *C. odorata*, several other invasive species were transferred from *Eupatorium*: *Ageratina adenophora* (Spreng.) King & Robinson, *A. riparia* (Regel) King & Robinson, *Fleischmannia microstemon* (Cass.) King & Robinson, *Austroeupatorium inulaefolium* (H.B.K.) King & Robinson, *Campuloclinium macrocephalum* (Less.) DC. and *Praxelis clematidea* (Griseb.) King & Robinson (Anonymous, 1983; King and Robinson, 1987; Julien and Griffiths, 1998; Henderson, 2001; Waterhouse, 2003). *Mikania micrantha* Kunth., *A. adenophora* and *Ageratum conyzoides* L., invasive weeds in the Old World, are also in the Eupatorieae.

The two main invasive forms of *C. odorata*, that occurring in Asia and West Africa and that occurring in southern Africa, differ from one another in morphology, biology, and ecology, and there is little variation within each form (Kluge, 1990; Lanaud *et al.*, 1991; Scott *et al.*, 1998; von Senger *et al.*, 2002; Ye *et al.*, 2004). Thus, they are functionally distinct entities, and have been characterized as biotypes (Zachariades *et al.*, 2004). It is important that these differences be considered when comparing studies on one invasive biotype with the other.

8.3 Morphology, biology, and phenology

Chromolaena odorata is a scrambling perennial shrub, with straight, pithy, brittle stems which branch readily, bear three-veined, ovate-triangular leaves placed oppositely, and with a shallow, fibrous root system (Holm *et al.*, 1977; Henderson, 2001). Capitula are borne in panicles at the ends of the branches and are devoid of ray florets. The corollas of the florets vary between plants from white to pale blue or lilac. Achenes are black with a pale pappus (Holm *et al.*, 1977; McFadyen, 1989). In open-land situations, *C. odorata* grows to 2–3 m in height, but it can reach up to 5–10 m when supported by other vegetation.

Within its native range, *C. odorata* shows marked morphological variability in terms of flower color, leaf shape and hairiness, smell of the crushed leaves, and plant architecture. In some regions, several forms and their intermediates co-occur, while in others, the population appears homogeneous; the basis for this variability presently remains unexplained (Zachariades *et al.*, 2004).

In contrast, the biotype invasive in Asia, Oceania, and West Africa is uniform in its morphological features, and has pale blue–lilac flowers, fairly hairy, dull-green leaves and stems, and a lax habit. The southern African biotype of *C. odorata* is distinct from this more widespread biotype, having glabrous stems and leaves, which in consequence are bright yellow-green when young. The plant has a more upright growth habit, white flowers and the smell emitted by the crushed leaves is sharp when compared with that of the Asian–West African biotype.

The tropical and subtropical areas in which *C. odorata* grows as either a native or an invasive species are generally characterized by a dry season with shorter days and a rainy season with longer days. At the start of the wet season, established plants generate new shoots from the crown or from higher, undamaged axillary buds, while seeds in the soil, produced during the previous dry season, germinate (McFadyen, 1988, 1989). In its invasive range, several thousand *C. odorata* seeds may germinate per square meter, but subsequent high mortality of seedlings (Yadav and Tripathi, 1981; Epp, 1987) and self-thinning of older plants (Witkowski and Wilson, 2001) occurs. Under moist conditions, *C. odorata* branches trailing on the ground may root (Gautier, 1993), but vegetative reproduction does not contribute significantly to its pest status.

Plants grow vigorously throughout the wet season and flowering is initiated by a decrease in both day length and rainfall (Sajise *et al.*, 1974; Gautier, 1993). Flowering peaks in December–January in the northern hemisphere and June–July in the southern hemisphere. The onset of flowering coincides with the cessation of vegetative growth. Flowering is often prolific, and in both the native and invasive ranges, fertile seed is produced without pollination, as the species is apomictic (Coleman, 1989; Rambuda and Johnson, 2004). However, the flowers of *C. odorata* are also visited by insects (Ghazoul, 2004; Ramírez, 2004) and *C. odorata* is an important honey plant in Thailand (Thapa and Wongsiri, 1997).

Large numbers of seeds (in South Africa, one plant produced over one million seeds in a single season; Liggitt, 1983) are set in less than two months after flowering (Erasmus, 1985). In South Africa, seed production increased until plants were about 10 years old and declined dramatically after 15 years, with senescence (Witkowski and Wilson, 2001). Seeds are dispersed by wind, as well as via animal fur, clothing, and vehicles (Gautier, 1992, 1993; Blackmore, 1998). Germinability in the southern African biotype is initially low, but increases over a 6-month period (Erasmus and van Staden, 1987). Most seed loses its viability after a year, although a small proportion of the seed persists in the soil for several years, allowing for rapid recolonization after the removal of a parent population (Witkowski and Wilson, 2001). Once plants have flowered and seeded, the terminal, flower-bearing parts of the stems die, and in seasonally drier areas, the leaves wither and fall.

8.4 Distribution

8.4.1 Chromolaena odorata *in the neotropics*

In the Americas, *C. odorata* occurs from USA (southern Florida and Texas) (30° N) to north-western Argentina (about 30° S) (Fig. 8.1) (Gautier, 1992; Kriticos *et al.*, 2005), almost continuously wherever the habitat and climate are suitable. It is present on all the

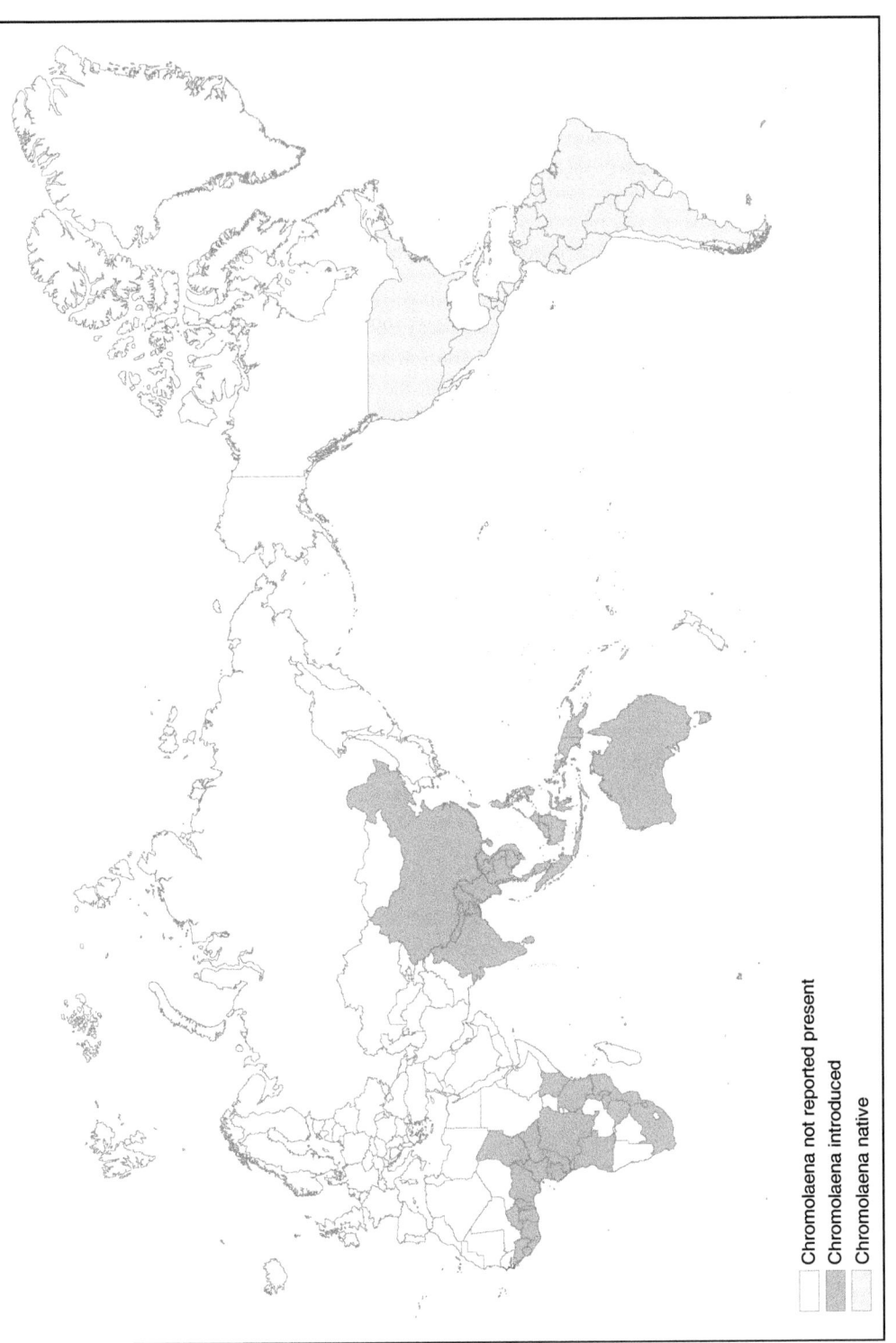

Fig. 8.1 Native and reported introduced range of *Chromolaena odorata*, based on presence records for each country.

Chromolaena not reported present

Chromolaena introduced

Chromolaena native

islands of the Caribbean. To the west of the Andes, its range extends as far south as northern Peru (Gautier, 1992). Plants that are similar to the southern African biotype have only been found on the northern Caribbean islands, particularly Jamaica and Cuba (Zachariades *et al.*, 2004).

8.4.2 Chromolaena odorata *in the Old World*

Chromolaena odorata is present in the majority of countries in the humid tropics and subtropics of the Old World (Fig. 8.1). Gautier (1992) and McFadyen (1989, 1996b) have discussed the invasive pathways of *C. odorata* in the Old World. The first record of naturalization was in the 1870s in Dacca and the Ganges floodplain (present-day India and Bangladesh). However, the plant was probably introduced into Asia as an ornamental plant in the early 1840s, through the Botanical Gardens in Kolkotta (Calcutta, India). By the early twentieth century, it was widespread in Assam, Bengal (India and Bangladesh), Myanmar (Burma) (Rao, 1920) and present in Thailand (Gautier, 1992). The plant was recorded in Vietnam and Laos in the 1930s (Gautier, 1992).

It was introduced as an ornamental to Peradeniya, Sri Lanka, in 1884 and it had naturalized in Sri Lanka by the 1930s (Grierson, 1980). *Chromolaena odorata* spread eastwards into China and south into Indonesia, especially during World War II (McFadyen, 2002), although it was present in both of these countries in the 1930s. Within Indonesia, *C. odorata* spread was aided by the transmigration program in the 1960s (McFadyen, 2002). The weed was first recorded in Nepal in the 1950s, the Philippines and Papua New Guinea in the 1960s (Henty and Pritchard, 1973) and in Timor and Taiwan in the 1980s (Wu *et al.*, 2004; Lai *et al.*, 2006). It was introduced to Guam in the 1960s (Stone, 1966) and it had spread to most of the Micronesian islands by 2000 (Muniappan *et al.*, 2004) (Fig. 8.1).

In Australia, *C. odorata* was found along the Tully River and at Bingil Bay in northern Queensland in 1994 (Waterhouse, 1994a) and at several other localities subsequently (McFadyen, 2004a). Two morphological forms were found in Australia. The more widespread form matched genetically with the Asian–West African biotype, whereas the more localized form matched with material from southern Brazil (Scott *et al.*, 1998). Another species, *C. squalida* (Waterhouse, 2003) was also found in the vicinity, implying that all three taxa may have been imported with fodder seed from Brazil (Waterhouse and Zeimer, 2002).

Chromolaena odorata appeared in West Africa much later than in Asia. It was probably accidentally introduced into Nigeria in 1937, through imported seeds of *Gmelina arborea* Roxb. (Verbenaceae) from Sri Lanka (Ivens, 1974). It was recorded in Côte d'Ivoire in the early 1950s and these two nodes of infestation coalesced. Although it is unclear whether separate nodes of infestation appeared in Cameroon and the Central African Republic in the 1930s or whether the plant spread to these countries from Nigeria in the 1960s, the infestation throughout West and Central Africa is represented by the Asian–West African biotype alone. By the mid 1990s, *C. odorata* had been recorded from

Guinea in the west, south into northern Angola, and east into the central parts of the Democratic Republic of the Congo (Hoevers and M'Boob, 1996). Recent records include Chad (Timbilla, 1998), Burkina Faso and The Gambia (CAB International, 2004) and most recently, it has been reported to be common in western Kenya and present in Tanzania (B. le Ru, and Q. Mann, personal communications).

In the mid nineteenth century, *C. odorata* from Jamaica appeared in a list of plants growing in the Cape Town Botanic Gardens, South Africa, and it was found naturalized around Durban in the late 1940s (Hilliard, 1977; Zachariades *et al.*, 2004). From there, it spread rapidly along the coast and is now present from Port St. Johns in the south, where it has probably reached its ecological limit, up into southern Mozambique, and inland through Swaziland into northern South Africa (Goodall and Erasmus, 1996; Macdonald *et al.*, 2003).

A single specimen collected from northern Zimbabwe in the late 1960s (Gautier, 1992) appears identical to the Asian–West African biotype (C. Zachariades, unpublished data) and may be linked to an unconfirmed *C. odorata* infestation in northern Mozambique and southern Malawi (J. Findlay, personal communication). If this is true, Mozambique is the only country with both the southern African and Asian–West African biotypes of *C. odorata*. The weed was introduced to Mauritius before 1949, probably from Asia.

8.4.3 Ecology and impacts

In its native range, *C. odorata* is a common species, growing from the coast to an altitude of 1000–1500 m (McFadyen, 1988, 1991), although it has been collected at a maximum altitude of almost 3000 m (Gautier, 1992). In Central and South America, *C. odorata* is common in areas with a rainfall exceeding 1500 mm per annum (McFadyen, 1991). In Trinidad, where the only native-range ecological studies were conducted, the plant prefers well-drained soils, tending to die under waterlogged conditions (Cruttwell, 1972). It grows best in sunny, open areas such as roadsides, abandoned fields, pastures, and disturbed forests, but it tolerates semishade conditions as well. It does not thrive under the shaded conditions of undisturbed forest or in closely planted, well-established orchards.

In the Americas, *C. odorata* is only weedy on occasion, presumably because its natural enemies keep it under control. It acts as a pioneer plant, growing to high densities in recently disturbed (e.g. slashed, overgrazed) areas, but it is soon outcompeted by successional vegetation and disappears after a few years (Cruttwell, 1972; McFadyen, 1988, 1989).

The Asian–West African biotype was predicted to invade areas in the Old World with a minimum annual rainfall of 1200 mm (McFadyen, 1989) but has a considerably lower limit (Gautier, 1992; Timbilla, 1998; Kriticos *et al.*, 2005). It does not thrive in areas with extremely high, year-round rainfall, such as the coastal areas of Myanmar (Burma) (Kriticos *et al.*, 2005). It is common up to 1000 m asl in Asia (McFadyen, 1989) and has been recorded at 2000 m in Cameroon (Timbilla, 1998).

The southern African biotype is more cool tolerant (Kriticos *et al.*, 2005), occurring up to the 31st parallel in frost-free zones with an annual rainfall of 500–1500 mm, although at the lower rainfall limits it is restricted to drainage lines and watercourses (Goodall and Erasmus, 1996). In Swaziland, it was found as high as 850 m asl (Goodall *et al.*, 1994). However, it continues to spread northwards in Africa, and thus may also be tolerant of more tropical conditions.

In the Old World, *C. odorata* is an invasive transformer species (*sensu* Richardson *et al.*, 2000), at least partly because it lacks natural enemies. It grows rapidly and often forms a dense scrambling thicket that grows through and over the existing vegetation. It most readily invades areas of natural or human-induced disturbance, but can invade undisturbed land. *Chromolaena odorata* affects both subsistence and commercial agriculture, including crops and plantations (e.g. oil palm, rubber, coffee, cacao, coconuts, cashews, cassava, yam, banana, plantain), grazing lands, and silviculture (e.g. teak, pines, eucalypts).

Chromolaena odorata invades a wide range of natural vegetation types, from grassland through savanna, bush, open woodland and forest margins and gaps (Goodall and Erasmus, 1996; Prasad *et al.*, 1996). Natural forests are not usually invaded by *C. odorata* due to its high light requirements, but forest degradation allows the weed to establish (Norbu, 2004), suppressing the recruitment of trees. Forest gaps that naturally develop through tree-fall are colonized rapidly by *C. odorata* (Epp, 1987; Goodall and Erasmus, 1996). The plant scrambles up through the surrounding trees and emerges on top of the canopy, eventually causing its collapse (Goodall and Erasmus, 1996). However, removal of *C. odorata* allows rapid regeneration of indigenous forest (Honu and Dang, 2000). In cleared fields infested by *C. odorata*, indigenous forest was able to regenerate and outcompete the weed if an adequate seed bank of tree species remained (de Rouw, 1991).

Chromolaena odorata impacts on biodiversity, the conservation thereof, and eco-tourism (Macdonald and Frame, 1988; Goodman, 2003). Thickets of the weed prevent the free movement of livestock and wildlife. In South Africa, growth of *C. odorata* along riverbanks interfered with the egg-laying of Nile crocodile and altered the sex ratio in the progeny through shading of nests (Leslie and Spotila, 2001). In Thailand, *C. odorata* attracted butterflies away from an indigenous tree, resulting in reduced butterfly pollination (Ghazoul, 2004).

The weed affects human livelihoods, both through its impacts on agriculture and, in areas with a distinct dry season, because it is a fire hazard (Holm *et al.*, 1977; Liggitt, 1983; Macdonald, 1983; Muniappan and Marutani, 1988; Goodall and Erasmus, 1996; Hoevers and M'Boob, 1996; McWilliam, 2000). The dry pithy stems and leaves are rich in oils (Moni and Subramoniam, 1960) and burn readily (Hoevers and M'Boob, 1996; McFadyen, 1989), although the plant's flammability is contested (Goodall and Erasmus, 1996). Dense *C. odorata* infestations often represent an increased fuel load compared with the native vegetation, resulting in fires of increased intensity (McFadyen, 2004b). These cause considerable damage to the surrounding native vegetation and give the resprouting *C. odorata* plants a further competitive advantage. In South Africa, the native vegetation growing along forest margins is of low flammability and thus protects the forest interior; once this vegetation is replaced by *C. odorata*, fire is able to penetrate the forest margin, causing progressive erosion of forest edges (Macdonald, 1983).

Regular burning of grassland and savanna reduces the establishment of both the Asian–West African and the southern African biotypes, although the Asian–West African biotype appears more resistant to fire than the southern African one (Macdonald and Frame, 1988; Gautier, 1996; Goodall, 2000). Both biotypes are reported to resprout from the crown after fire (Macdonald, 1983; Muniappan and Marutani, 1988), but fire-induced mortality rates for southern African plants are likely to be higher.

Chromolaena odorata contains a diverse range of secondary chemicals, including flavonoids, terpenoids, and alkaloids (Talapatra *et al.*, 1974; Biller *et al.*, 1994). The plant is generally not eaten by vertebrate herbivores, and the high nitrate levels in tender foliage could be the cause of livestock death (Sajise *et al.*, 1974). Pyrrolizidine alkaloids in the flowers killed goats which ate the flowers (McFadyen, 2004b). Despite the toxic properties of some plant parts, *C. odorata* leaves appear to be of high nutritive value (Apori *et al.*, 2000).

The allelopathic properties of the weed (Sahid and Sugau, 1993) aid it in gaining dominance in vegetation, and in replacing other aggressive invaders such as *Lantana camara* L. (Verbenaceae) (R. Muniappan, personal observation) and *Imperata cylindrica* (L.) Beauv. (Poaceae) (Eussen and de Groot, 1974; Ivens, 1974) in Asia and Africa.

8.5 *Neotropical arthropods released on* C. odorata *in its invasive range*

Cruttwell (1972, 1974) recorded approximately 225 phytophagous insects and mites on *C. odorata*, mainly in Trinidad but also in other parts of its native range. Of these, a few were identified as being restricted to *C. odorata*, or to a few close relatives. Following host range testing in Trinidad, *Pareuchaetes pseudoinsulata* Rego Barros (Lepidoptera: Arctiidae), *Mescinia* nr. *parvula* (Zeller) (Lepidoptera: Pyralidae), *Apion brunneonigrum* Beguin-Billecoq (Coleoptera: Curculionidae) and the mite *Acalitus adoratus* Keifer (Acarina: Eriophyidae) were sufficiently host specific to be recommended as biological control agents (Bennett and Cruttwell, 1973; Cruttwell, 1973a, 1977a, b), whereas *Dichomeris* n. sp. (Lepidoptera: Gelechiidae) had a wide host range (Cruttwell, 1973b). Following Gagne (1977), Cock (1984) recommended that some of the Cecidomyiidae could be considered as biological control agents. Reviews of agents released and established on *C. odorata* worldwide are available in Waterhouse (1994b), Julien and Griffiths (1998) and Muniappan *et al.* (2005), while McFadyen (1996a) reviewed the biological control programs that have been implemented against *C. odorata*. Seven species of biological control agents are currently established, either intentionally or by accident, in the Old World (Fig. 8.2).

In exploratory trips to the Americas from the late 1980s to the present, personnel of the South African Agricultural Research Council's Plant Protection Research Institute (ARC-PPRI) found several more insect species feeding on *C. odorata*, some of which were imported into quarantine in South Africa for rearing and host-range testing (Zachariades *et al.*, 1999; Strathie and Zachariades, 2004). In this section, we consider those species of

Distribution of agents

⬜ No agents present

⬜ A. adoratus

⬛ C. connexa

▨ P. pseudoinsulata

▨ P. insulata and C. eupatorivora

▨ A. adoratus and C. connexa (Palau not visible at map scale)

▨ A. adoratus, C. connexa, P. pseudoinsulata, A. anteas and A. thalia pyrrha

▨ A. adoratus and P. pseudoinsulata (including Tinian Is, CNMI not visible at map scale)

▨ A. adoratus, C. connexa and P. pseudoinsulata (including FSM, Roti & Saipan Is, CNMI & Guam not visible at map scale)

Fig. 8.2 Countries within the introduced range of *Chromolaena odorata* with biological control agents established either intentionally or accidentally

arthropods that have become established on *C. odorata* outside its native range, through either deliberate introduction as biological control agents or accidental introduction.

8.6 *Established biological control agents*

8.6.1 **Pareuchaetes pseudoinsulata** *Rego Barros* **(Lepidoptera: Arctiidae)**

The native range of *P. pseudoinsulata* is small, consisting of eastern Venezuela and Trinidad (Cock and Holloway, 1982). Initially identified as *Ammalo insulata* (Walk.), this butter-yellow moth lays its eggs in batches on the undersides of *C. odorata* leaves. The hairy caterpillars, which are black with red and white markings, feed on leaves, initially gregariously but later solitarily. Feeding damage by older larvae is characteristic, often leaving only the midrib. The larvae are nocturnal and the larger caterpillars shelter during the day in leaf litter under the plant, where they eventually pupate (Cruttwell, 1972).

Until recently, *P. pseudoinsulata* was the most widely released and established biological control agent on *C. odorata* (Waterhouse, 1994b; Julien and Griffiths, 1998; Muniappan *et al.*, 2005). In the early 1970s, it was released in Ghana, Nigeria, India, Sri Lanka, and Malaysia (Sabah). However, it established only in Sri Lanka and Malaysia and was effective only in Sri Lanka, where it inflicted widespread but sporadic defoliation (Dharmadhikari *et al.*, 1977; Waterhouse, 1994b).

In the 1980s, the moth was rereleased in India, where it established in the southwest, causing sporadic damage (Joy *et al.*, 1993; Waterhouse, 1994b). It was also released in Guam, where it established and controlled *C. odorata* to such a degree that the status of the weed could be downgraded (Muniappan *et al.*, 1989; Seibert, 1989). It was released on the islands within the Commonwealth of the Northern Marianas (Rota, Tinian, Saipan, and Aguijan), where it established and caused substantial damage. It failed to establish in Vietnam, Thailand, and South Africa (Julien and Griffiths, 1998). The moth was discovered in Brunei and on several islands in the Philippines in the 1980s. Because it had not been released there, it may have spread from Malaysia (Waterhouse, 1994b).

In the late 1980s and early 1990s, *P. pseudoinsulata* was released and established on the islands of Yap, Pohnpei, and Kosrae within the Federated States of Micronesia (FSM). In the 1990s, following successes in other countries, it was rereleased in Ghana (1991–1993), where it established, spread and inflicted substantial damage (Timbilla, 1996, 1998) and in Côte d'Ivoire, where it again failed to establish (Timbilla *et al.*, 2003), despite initial positive results (Julien and Griffiths, 1998). Releases in Indonesia resulted in establishment in Sumatra, Kalimantan (although this population may have moved from Sabah), and Sulawesi, causing substantial but sporadic damage, but it did not establish in Java or West Timor (Julien and Griffiths, 1998; Desmier de Chenon *et al.*, 2002a). It was released in Papua New Guinea (PNG), where it established in only the Markham Valley (Bofeng *et al.*, 2004) and in South Africa, where it again failed to establish (Strathie and Zachariades, 2002). In the 2000s, it was established on Chuuk (FSM) and released in Palau (Muniappan *et al.*, 2005).

Pareuchaetes pseudoinsulata has proved a difficult and unpredictable biological control agent, in terms of both establishment ease and subsequent effectiveness in controlling the weed. In some instances, the insect established after small releases; just over 2000 larvae in Sri Lanka (Dharmadhikari *et al.*, 1977) and 500 mated adults on Guam (Seibert, 1989), whereas in others, extremely large releases were necessary: 125 000 at one site in Ghana (Timbilla, 1996). In South Africa, about 350 000 larvae were released at two sites within a year, but the insect failed to establish at either site (Strathie and Zachariades, 2002).

Only in a few countries, such as Guam, was there a significant long-term reduction in the population of *C. odorata*. In many of the countries or regions where *P. pseudoinsulata* did establish, including Sri Lanka, the Marianas, India, Ghana, and Sumatra, initial establishment was followed by a spectacular population build-up and widespread, complete defoliation of *C. odorata* thickets, and high mortality rates of these plants (Dharmadhikari *et al.*, 1977; Seibert, 1989; Timbilla, 1996; Singh, 1998). However, after 1–2 years, the populations of the moth often declined dramatically and the weed recovered. Subsequent sporadic outbreaks resulting in defoliation occurred, but these were often less spectacular and were unpredictable over time and space.

Such population dynamics are typical of many "outbreak" or "epidemic" insect species (Wallner, 1987). *Tyria jacobaeae* (L.) (Lepidoptera: Arctiidae), which feeds on *Senecio jacobaea* L. (Asteraceae), is one of only two other arctiid species established as a weed biological control agent (Julien and Griffiths, 1998). It appears to have a similar impact on its target weed as do *Pareuchaetes* spp. on *C. odorata*, with periodic outbreaks causing massive defoliation but an overall unsatisfactory effect (Julien and Griffiths, 1998).

Native populations of *P. pseudoinsulata* in Trinidad (Cruttwell, 1972) were low in most areas, with outbreaks only in localized areas during the wet season. The population dynamics also varied from year to year. An initial rapid rise in population at one of these "outbreak sites" was followed by a population plateau and then a slow decline, coinciding with the appearance of parasitoids and disease. However, Cruttwell (1972) found a correlation between numbers of adults caught in light traps and rainfall levels 16 weeks earlier, and believed that the drop in population each year was due to a loss of plant quality due to cessation of rains, rather than parasitism or disease. Similar population dynamics have been reported for *T. jacobaeae* within its native range (Dempster, 1971).

A number of possible factors, including predation and parasitism, climate, and biotype matching, may explain why establishment of *P. pseudoinsulata* has been erratic. In areas where *P. pseudoinsulata* has established, factors affecting its abundance may include weather patterns, natural enemies (predators, parasitoids, disease) and herbivory-induced changes in host-plant quality (Cock and Holloway, 1982; Waterhouse, 1994b; Marutani and Muniappan, 1991). Such factors have been implicated in the population dynamics of other outbreak species (Wallner, 1987).

In India and Ghana, predators (particularly ants) were implicated in the non-establishment of *P. pseudoinsulata* in the 1970s, although evidence for this was anecdotal (Cock and Holloway, 1982). Ants and other invertebrate predators removed a high proportion of *P. pseudoinsulata* eggs placed in the field in South Africa (Kluge, 1994). A lack of predation may partly explain the relatively high success, both in terms of establishment and subsequent population increase, of releases of smaller numbers of *P. pseudoinsulata* on islands, where predation pressures are often lower (MacArthur and Wilson, 1967; Seibert, 1989). Although predation of *P. pseudoinsulata* by both invertebrates and vertebrates (toads, skinks, birds) was observed in Guam, Sri Lanka, and Pohnpei (Seibert, 1989; Waterhouse, 1994b), the effects of this on the moth's population were not assessed. The success of releases in some areas on continents illustrates that if Allee effects exacerbated by predation and dispersal are a factor in poor establishment, they can be overcome by the release of sufficient insects over a sufficient duration of time.

Levels of parasitism of *P. pseudoinsulata* remain low in its introduced range, even after many years in the field, compared with levels of parasitism in its native range (McFadyen, 1997), although a parasitoid caused up to 30% mortality on Guam (Seibert, 1989). An unidentified tachinid has been found depositing eggs on larvae in PNG, but has not prevented outbreak populations of *P. pseudoinsulata* occurring on a seasonal basis.

Pareuchaetes pseudoinsulata is highly prone to diseases such as microsporidia under laboratory conditions, and the release of unhealthy insects may have prevented their establishment in some instances (Singh, 1998; Strathie and Zachariades, 2002). A nuclear polyhedrosis virus caused substantial mortality in Trinidad (Cruttwell, 1972), but the culture was cleared of this on importation into India in the early 1970s (Waterhouse, 1994b). A culture of *P. pseudoinsulata* from Sri Lanka was released in India and appeared to be doing well until it died out due to a virus (Waterhouse, 1994b).

Climate has been viewed as affecting the establishment of many biological control agents (Cock and Holloway, 1982; Day *et al.*, 2003; Zalucki and van Klinken, 2006). A long, severe dry season may have prevented establishment of *P. pseudoinsulata* in some areas (Cock and Holloway, 1982; Seibert, 1989). The only sites in PNG where *P. pseudoinsulata* established were in the relatively dry areas of the Markham Valley (Bofeng *et al.*, 2004). South Africa is considerably cooler than Trinidad (Parasram, 2003) or other areas where *P. pseudoinsulata* has established in the Old World, and this may have contributed to the lack of establishment there.

8.6.2 Pareuchaetes insulata *(Walker) (Lepidoptera: Arctiidae)*

Pareuchaetes insulata occurs from Colombia through Central America to Texas, Florida, and on the larger Caribbean islands (Cock and Holloway, 1982). A culture of *P. insulata*, collected from *C. odorata* in Florida, was introduced into quarantine in South Africa in 1989 and was shown to be adequately host specific for release in South Africa (Kluge and Caldwell, 1993a). This species was eventually released from 2001 onwards in KwaZulu-Natal (KZN) province (Strathie and Zachariades, 2004), and establishment was confirmed in 2004 at one site (Zachariades and Strathie, 2006). The insect has subsequently spread along the wetter coastal belt and is causing high levels of localized damage to *C. odorata*. Its biology is similar to that of *P. pseudoinsulata*. It is thought that this species succeeded in South Africa where *P. pseudoinsulata* did not because of better climate matching (Byrne *et al.*, 2004). Although the form of *C. odorata* from which *P. insulata* was collected in Florida differs from the southern African biotype, the host range of the insect was sufficiently broad to allow establishment in South Africa.

8.6.3 Cecidochares connexa *Macquart (Diptera: Tephritidae)*

This tephritid fly was recorded on *C. odorata* in Trinidad, Mexico, and Bolivia (Cruttwell, 1974). It lays eggs in the shoot tips of the plant and the larvae develop within galls at the nodes. A strain of fly compatible with the Asian–West African *C. odorata* biotype was imported from Colombia into quarantine at the Indonesian Oil Palm Research Institute in North Sumatra under an Australian Centre for International Agricultural Research program in 1993, and host-range testing indicated its high specificity to *C. odorata* (McFadyen *et al.*, 2003). The fly was released in North Sumatra in 1995, and later on other Indonesian islands (Tjitrosemito, 1998; Desmier de Chenon *et al.*, 2002a; Wilson and Widayanto, 2004) where it established readily.

Cecidochares connexa was subsequently released and established in PNG (Orapa and Bofeng, 2004), Palau (Esguerra, 2002), Guam, Saipan, Rota and Pohnpei (Muniappan *et al.*, 2005; Cruz *et al.*, 2006), Chuuk, Kosrae and Yap (Muniappan *et al.*, 2007), East Timor (T. Paul, Charles Darwin University, personal communication), and in India (Bhumannavar and Ramani, 2007). Although it was released in Thailand (McFadyen *et al.*, 2003), it did not establish (Muniappan *et al.*, 2005).

Attempts to culture it in quarantine on the southern African *C. odorata* biotype failed, probably because of insect–plant incompatibility (Zachariades *et al.*, 1999). *Cecidochares connexa* was sent to Ghana in the late 1990s but no culture was established in quarantine (J. A. Timbilla, personal communication). The establishment of *C. connexa* in West Africa remains a priority, as it will enhance the effects of *P. pseudoinsulata* there.

The fly can be established easily from small founder populations. About 100 pairs were released at each of several sites across Indonesia (Wilson and Widayanto, 1998, 2002; Tjitrosemito, 1998), but as few as 26 individuals on Palau (Esguerra, 2002). Once established, fly populations can rapidly build up and spread (Desmier de Chenon *et al.*, 2002a; Wilson and Widayanto, 2002). In Sumatra, the fly spread over 200 km within 5 years following release (Desmier de Chenon *et al.*, 2002a), while in PNG it spread over 100 km within 4 years (Day and Bofeng, 2007).

The fly has been so effective, particularly in areas at lower elevations with a short dry season, that McFadyen *et al.* (2003) and Day and Bofeng (2007) have found successful biological control of *C. odorata* in many areas. Hundreds of galls per plant (>10 galls/m of stem) have sometimes been recorded, causing plants to become prematurely moribund with few flowers or seeds, or to die back (Desmier de Chenon *et al.*, 2002a; Wilson and Widayanto, 2002; McFadyen *et al.*, 2003; Day and Bofeng, 2007). The galls have been shown to act as a resource sink, and to slow the growth of the plant (Cruz *et al.*, 2006). Plant density has subsequently decreased on Lombok, in Irian Jaya (Wilson and

Widayanto, 2002) and in most PNG provinces where *C. odorata* is present (Day and Bofeng, 2007).

The fly appears less effective in higher-elevation, cooler areas and those with a longer dry season, as fewer generations are produced per annum and therefore populations build up slowly. Areas with a long, intense dry season are often burned at that time of the year, eliminating the fly population locally. Initially, there was concern that parasitism could reduce the effectiveness of *C. connexa*, as happened with *Procecidochares utilis* Stone (Diptera: Tephritidae) on *A. adenophora* in many areas where it was released (Julien and Griffiths, 1998). However, parasitism and predation levels have generally remained low, although up to 50% parasitism was recorded at one site in Java (McFadyen *et al.*, 2003).

8.6.4 **Actinote** *species (Lepidoptera: Nymphalidae)*

Larvae of *Actinote* feed on the leaves of species largely within the Asteraceae in Central and South America. The host plants of several *Actinote* species are in the taxa *Chromolaena* and *Mikania*, and some *Actinote* species feed on species within both genera (Ackery, 1988). Identification of *Actinote* species is difficult and the taxonomy of the group is uncertain (Francini and Penz, 2006). Cruttwell (1974) recorded *A. anteas* (Doubleday & Hewitson) from *C. odorata* in Trinidad, and it is also recorded from *Ageratum* and *Mikania* there (Desmier de Chenon *et al.*, 2002b). The host range of a culture of *A. anteas* from Costa Rica was tested in quarantine in South Africa, but the culture soon died out (Caldwell and Kluge, 1993). The same species, also from Costa Rica, was introduced into Indonesia in 1996 where it was subsequently approved for release following host-specificity studies (Desmier de Chenon *et al.*, 2002b).

A species initially identified as *A. parapheles* Jordan and later as *A. thalia pyrrha* Fabr. was imported from northeastern Brazil into South Africa in 1995. Host-specificity tests indicated that it fed to the same extent on two species of *Mikania* indigenous to South Africa as it did on *C. odorata*, and it was therefore not released (Zachariades *et al.*, 2002). A culture of this species was forwarded to Sumatra in 1999 to replace the culture of *A. anteas* that had been approved for release there, and it was released in the same year (Desmier de Chenon *et al.*, 2002b).

In Indonesia, the neotropical vine *M. micrantha* is as great a problem as *C. odorata*, and the oligophagous feeding habits of *Actinote* species were considered desirable, despite the presence of the native *M. cordata* (Burm. f.) B. L. Rob. *Actinote thalia pyrrha* is now widely established and common in Sumatra, where it not only feeds on *C. odorata* and *M. micrantha* but also the invasive alien *Austroeupatorium inulaefolium* (Desmier de Chenon *et al.*, 2002a, b; Zhigang *et al.*, 2004; R. Desmier de Chenon, A. Simamora and Nirwanto, Indonesian Oil Palm Research Institute, personal information, 2006).

A second species, initially identified as *A. thalia thalia* L., was collected in Venezuela and tested in South Africa, but it also fed on *Mikania* indigenous to South Africa and so was not approved for release. It was also forwarded to Sumatra where it was released (Desmier de Chenon *et al.*, 2002b). However, it has only established at two sites in

Indonesia and is considered not as successful as *A. thalia pyrrha* (R. Desmier de Chenon, A. Simamora, and Nirwanto, personal information, 2006).

8.6.5 Calycomyza eupatorivora *Spencer (Diptera: Agromyzidae)*

The larvae of this agromyzid fly form blotch mines on the leaves of *C. odorata*. Cruttwell (1974) recorded it in Trinidad, but at that stage it was considered the same species as *Calycomyza flavinotum* Frick, a nearctic species with a wide host range. *Calycomyza eupatorivora* has also been recorded from Jamaica, Hispaniola, Argentina, Venezuela, and Guadeloupe (Martinez *et al.*, 1993), and it or similar species are widespread in the neotropics (ARC-PPRI, unpublished data). A culture from Jamaica was first reared in quarantine in South Africa in 1997, and host range testing was completed in 2001, showing the fly to be highly host specific (Zachariades *et al.*, 2002).

Releases were initiated in 2003 and establishment confirmed the following year at one site on the KGN coast (Zachariades and Strathie, 2006). The insect has since spread at least 40 km along the coast, and has now been widely released on *C. odorata* in South Africa. However, it is unlikely that *C. eupatorivora* is having much effect on *C. odorata* yet. The fly displays a preference for young plants in shady conditions, and is likely to be effective on these plants only. It is common in Jamaica and a pest under laboratory conditions, but it is rare on the South American mainland. In South Africa, a congeneric leaf miner on lantana is heavily parasitized, which may contribute to its lack of effectiveness there (Baars and Neser, 1999).

8.6.6 Acalitus adoratus *Keifer (Acarina: Eriophyidae)*

Feeding by this eriophyid mite on the leaves of *C. odorata* induces abnormal growth of the epidermal hairs, resulting in formation of a white erineum patch (Cruttwell, 1977a). It has been recorded from Brazil, Bolivia, and Trinidad (Cruttwell, 1974). Studies in Trinidad indicated that it was host specific and could be damaging, and it was recommended for release as a biological control agent (Cruttwell, 1977a; Cock, 1984). It was never introduced deliberately, but was found in the field in the Philippines in 1987, and shortly thereafter in Thailand, Malaysia, and Indonesia (McFadyen, 1995). McFadyen (1995) suggested that it may have been accidentally released into the field in Sabah, Malaysia, through phoresy on *A. brunneonigrum* adults imported from Trinidad and released directly into the field. It now occurs through much of the range of *C. odorata* in Asia and Oceania (McFadyen, 1995; Muniappan *et al.*, 2005). It is thought not to cause any significant damage to *C. odorata*, although this has not been quantified.

8.7 *Other arthropod agents*

A number of other species of arthropods have been released on *C. odorata* but failed to establish. Some are outlined below.

8.7.1 Apion brunneonigrum *Beguin-Billecoq (Curculionidae)*

The larvae of this weevil develop in the flowers of only *C. odorata* and *C. ivaefolia* (Cruttwell, 1973a), and it has been recorded from Trinidad, Venezuela, and Argentina (Cruttwell, 1974). Adults were collected in Trinidad and released in Nigeria, Ghana, Sri Lanka, and Malaysia (Sabah) in the 1970s, India between 1972 and 1983, and in Guam in 1984. They persisted in Sabah for one year (Julien and Griffiths, 1998). However, *A. brunneonigrum* has not established anywhere and the reasons for its nonestablishment are unclear.

8.7.2 Mescinia *nr.* parvula *(Zeller) (Pyralidae)* (= Phestinia costella *Hampson*)

The larvae of this moth were collected from *C. odorata* and *C. ivaefolia* in Trinidad, and similar larvae have been collected from *C. odorata* in Mexico, Brazil, Jamaica, Cuba, Florida, and Venezuela (Cruttwell, 1974; Zachariades *et al.*, 1999; Strathie and Zachariades, 2004), and from *C. hookeriana* (Griseb.) King & Robinson in Argentina (Cruttwell, 1974). Solis *et al.* (2008) re-identified specimens of this moth from Jamaica, Trinidad, Guatemala, and Puerto Rico as *Phestinia costella*, and expect to occur throughout the native range of *C. odorata*. Cruttwell (1977b) showed that the moth was highly host specific, but was unable to rear it in the laboratory. Several other attempts to do so were also unsuccessful (Singh, 1998; Zachariades *et al.*, 1999, 2007). Low numbers of insects collected in the field in Trinidad were released in Guam in 1984, but the moth did not establish (Julien and Griffiths, 1998).

8.7.3 Pareuchaetes aurata aurata *(Butler) (Arctiidae)*

This moth has a range covering southern Brazil, Paraguay, Bolivia, and northern Argentina, and was recorded on *C. hookeriana* in Argentina (Cock and Holloway, 1982). A culture collected from *C. hookeriana* in Argentina was imported into South Africa, where it was shown to be sufficiently host specific to be safely released there (Kluge and Caldwell, 1993b). It was released in South Africa in the early 1990s but did not establish (Zachariades *et al.*, 1999).

In addition, several other species have been tested over the years but were found to have too broad a host range for release (Cruttwell 1973b; Zachariades *et al.*, 1999; Strathie and Zachariades, 2002), and several promising agents are currently being investigated or await permission for release (Zachariades and Strathie, 2006). There is still a need for additional agents, particularly in drier parts of the invasive range of *C. odorata* where the two *Pareuchaetes* species, *C. connexa* and *C. eupatorivora* cannot persist or are less effective.

At present, only ARC-PPRI in South Africa is conducting research on new potential agents. This research is focused on (i) agents from the center of origin of the southern

African *C. odorata* biotype, in order to ensure insect–plant compatibility and (ii) agents with a biology (a distinct diapause and/or a soil-dwelling stage) that will allow them to be effective in seasonally dry, fire-prone parts of the invasive range of *C. odorata* (Zachariades and Strathie, 2006).

8.8 Ecological interactions between the plant and arthropods

Few detailed studies have been undertaken so far on interactions between *C. odorata* and established biological control agents. Only *P. pseudoinsulata* has been established for any length of time, and many countries using biological control for *C. odorata* do not have the resources or expertise to conduct such studies. However, the reaction of *C. odorata* to feeding by *P. pseudoinsulata* has been well studied in Guam (Marutani and Muniappan, 1991; Raman *et al.*, 2006) and noted in India (Singh, 1998), Indonesia (Desmier de Chenon *et al.*, 2002a), and Ghana (Timbilla, 1996).

Typically, after two to three weeks' heavy feeding on a plant by *P. pseudoinsulata* larvae (i.e. under outbreak conditions), all the remaining leaves, both damaged and undamaged, turn from green to yellow and their toughness increases. This reaction is not induced by mechanical damage or feces, but by the feeding action of the larvae (Marutani and Muniappan, 1991), and/or the larval saliva.

Chlorophyll in these yellow leaves was much lower than green leaves, while nitrate-nitrogen levels and water soluble proteins increased. The subcellular physiology of this chlorosis mechanism has recently been described in detail, and involves the accumulation of lipidic materials which act as precursors of compounds that defend leaf cells from the impact of feeding (Raman *et al.*, 2006).

In laboratory trials, larvae preferred green leaves over yellow leaves. Larvae only survived on yellow leaves from the third instar onwards, but even then they grew at a slower rate and had a longer development time than those fed on green leaves. In the field, older larvae feeding on plants with yellow leaves remained on the upper parts of the plant during the day, rather than hiding in the leaf litter as usual. This was probably because they were obtaining insufficient nutrition from the leaves and needed to continue feeding, or to conserve energy, rather than move to the leaf litter each day. This behavior is likely to increase their risk of being preyed upon. About two to three weeks after larval feeding ceased, the leaves returned to their original green color (Marutani and Muniappan, 1991).

Yellowing of *C. odorata* bushes has also been recorded after feeding by *A. thalia pyrrha* larvae (R. Desmier de Chenon, A. Simamora and Nirwanto, personal information, 2006). In South Africa, the same response by the southern African *C. odorata* biotype to *P. insulata* feeding has recently been recorded (C. Zachariades, personal observation).

8.9 Economics of biological control efforts

Costs incurred by classical biological control of weeds are reasonably easy to calculate or estimate. These costs include the research stage (identification of the origin of the weed;

surveys for and selection of agents; rearing and testing of agents; release and establishment of agents; measuring their impact) and implementation (mass-rearing and redistribution of agents that have already established at certain sites). The cost-benefit ratio of biological control for a given weed, however, is more difficult to measure. It is dependent on (i) how much damage the weed is causing economically, socially and/or in terms of biodiversity, (ii) how effective the agent or suite of agents is, and (iii) the costs and practicalities of other control methods, which is especially relevant in developing countries, where resources are scarce for conventional control and the political will for coordinated, sustained, comprehensive weed control programs is often lacking. The cost of not introducing biological control, in terms of increasing future losses as the weed spreads and increases in density, versus the cost of introducing biological control, is a useful means of estimating the benefits of biological control (van Wilgen *et al.*, 2004).

We consider two case studies where some costs and benefits have been documented: *C. odorata* in Indonesia and Papua New Guinea (PNG), and in South Africa.

An ACIAR-funded project for biological control of *C. odorata* in Indonesia and the Philippines began in 1993 (McFadyen, 1996a) and in 1998, PNG replaced the Philippines as a partner. The project in PNG was extended twice and ended in June 2007, while the Indonesian project ended in 2002. The total cost of the project was conservatively about AU$2 million. During this time, *P. pseudoinsulata*, *C. connexa*, and the two *Actinote* spp. were introduced into Indonesia and *P. pseudoinsulata*, *C. connexa* and *C. eupatorivora* into PNG. Three of the four agents introduced into Indonesia are having an impact, with the *Actinote* spp. having an additional benefit by helping control *M. micrantha* and *A. inulaefolium* (R. Desmier de Chenon, A. Simamora and Nirwanto, personal information, 2006). Desmier de Chenon *et al.* (2002a) estimated that in plantations in Sumatra, the combined effects of *P. pseudoinsulata* and *C. connexa* reduced the cost of maintenance by 75%.

In PNG, where a conservative estimate of about AU$1 million was spent, *C. odorata* is present in 13 provinces (Day and Bofeng, 2007). *Calycomyza eupatorivora* failed to establish and *P. pseudoinsulata* is aiding control in only Morobe Province. However, within six years, *C. connexa* is already controlling *C. odorata* in most provinces. In several provinces, where *C. odorata* has been controlled and food gardens re-established, there is already a doubling of income and a 75% reduction in labour costs or time spent weeding. Villagers at Musu in Sandaun Province have reported that income from newly revitalized food gardens has increased by 50 Kina (AU$25) a week for a 1-ha block. In New Ireland and Manus provinces, similar figures have been reported (M. Day and I. Bofeng, PNG National Agricultural Research Institute, unpublished data). While the total area of infestations in PNG is not known, if just 100 ha of food gardens can be revitalized in the 13 provinces following the control of *C. odorata*, then the cost of the project in PNG can be recovered within eight years. These figures do not include savings in time spent weeding gardens or the social benefits of controlling *C. odorata* (Day and Bofeng, 2007; M. Day and I. Bofeng, unpublished data).

In South Africa, *C. odorata* is mainly a problem for biodiversity. Invasive alien plants have been identified as the major threat to protected areas in KZN, and *C. odorata* ranks

high amongst these (Goodman, 2003). Versveld *et al.* (1998) estimated that 326 139 ha
have been invaded by *C. odorata* in KZN (Marais *et al.*, 2004). In the Kube Yeni Game
Reserve in northern KZN, *C. odorata* infestations were estimated to have reduced the
carrying capacity from about 6 ha per large stock unit (LSU = 450 kg animal) to more
than 15 ha (Goodall and Morley, 1995). In the 2002/3 financial year, the *Working for
Water* program (WfW) spent about ZAR6.15 million nationally on clearing 3600 con-
densed hectares of *C. odorata*, of which just under 50% of the cost was in initial clearing
and the remainder in follow-up costs (Marais *et al.*, 2004).

Marais *et al.* (2004) estimated that at the current rates of clearing, it will take 15 years
to treat *C. odorata* at a national level. At costs adjusted to 2006 figures using South
African Production Price Index (PPI) (www.statssa.gov.za/keyindicators/ppi.asp), this
translates to almost ZAR105 million. However, Marais *et al.* (2004) acknowledged that
this is likely to be an underestimate. Among other assumptions they make, they base their
figures on only one follow-up treatment being needed. In addition, van Gils *et al.* (2004)
suggested that the efficiency of clearing *C. odorata* in an indigenous South African forest
was low, and suggested that only the use of an alternative plant cover after removal of
C. odorata reduced reinfestation substantially.

A *C. odorata* clearing programme that was initiated in the Hluhluwe-iMfolozi Game
Reserve in the 2003/4 financial year (Zachariades and Strathie, 2006) has cost almost
ZAR38 million (adjusted to 2006 PPI values = ZAR40 million) and has cleared 35 000 ha
in the park and surrounds (C. Terblanche, Ezemvelo KZN Wildlife, unpublished data).
When we consider the estimated 325 000 ha invaded in 1998 (Versveld *et al.*, 1998), it
seems likely that the real clearing costs of *C. odorata* nationwide will be closer to
ZAR400 million. These costs do not take into account the costs that infestations cause to
agriculture, ecotourism, or biodiversity.

The South African research programme on *C. odorata* biological control, initiated by
ARC-PPRI in 1988, has cost about ZAR16 million, adjusted to 2006 PPI figures. If the
research program continues until 2015, with two additional personnel, as is planned, it
would cost an additional ZAR14 million in 2006 figures. Therefore, the final cost will be
about ZAR30 million over 28 years. Mass-rearing and release of *C. odorata* biological
control agents (mainly the three *Pareuchaetes* species) have been conducted by the
South African Sugarcane Research Institute (SASRI) and WfW, at an estimated cost of
ZAR5–6 million, bringing the total cost of the biological control project to ZAR36
million.

Even if the final cost for biological control of *C. odorata* in South Africa runs to
ZAR40 million, this represents only 10% of costs for clearing 1998-level infestations of
C. odorata nationally. The cost-benefit ratio ultimately depends on the effectiveness of
biological control. Estimated clearing costs cited above are for reducing *C. odorata* levels
to <5%. Many cost-benefit studies of biological control have underestimated benefits of
biological control because they were conducted too early (McFadyen, 1998) or did not
look at the scenario with biological control versus without (van Wilgen *et al.*, 2004).
Versveld *et al.* (1998) estimated that, overall, weed biological control programs in South

Africa had already allowed a 20% saving of up to ZAR1.4 billion on the costs necessary for clearance, and that potential savings would eventually be more than 40%.

8.10 Measures of efficiency of biological control

A highly efficient biological control program would involve a quick and accurate determination of the origin and biotype of the weed species; a thorough survey of natural enemies in the area of origin, followed by importation, successful breeding and testing of selected potential biological control agents; the successful, easy establishment and spread of the agents, followed by a good level of control of the weed. Factors negatively affecting the success of programs may be intrinsic to the biological system (McFadyen, 1998) (e.g. not many potential agents available, native or valuable plants closely related to the target weed in the country of introduction, poor establishment or impact) and/or extrinsic (e.g. inexperienced researchers, few resources, political).

The efficiency of *C. odorata* biological control efforts globally has been substantially enhanced by the *IOBC Working Group on Chromolaena odorata*, through its regular international workshops, in terms of exchange of information, technology, and biological control agents (Boller *et al.*, 2006). It has also been enhanced by sustained funding at a national level in Ghana (in the 1990s), South Africa, and Micronesia and by international funding by ACIAR for Indonesia, the Philippines, East Timor, and PNG. However, not all international interventions have proved successful. A project funded by the European Economic Community in 1990–1992 produced limited results due to its short duration, and a UN Food and Agriculture Organization project in West Africa was blocked due to the controversy surrounding the usefulness of *C. odorata* as a fallow crop (McFadyen, 1996a; Prasad *et al.*, 1996).

However, despite these collaborative efforts, the efficiency of biological control of *C. odorata* worldwide has been variable and this can be attributed to both intrinsic and extrinsic factors. We consider two programmes as case studies below.

8.10.1 Biological control of C. odorata in PNG

Overall, the biological control of *Chromolaena* program in PNG has been very efficient, both because the program has been relatively cheap and because the main damaging agent, *C. connexa*, is easy to establish and very effective. *Cecidochares connexa* was introduced into PNG in 2001, following testing elsewhere which showed that the agent was host specific. It was only reared in the laboratory in PNG for about two years after initial release, until field populations were sufficiently high to warrant field collection for redistribution. During the rearing stage of the program, nearly 13 000 galls were released, compared with over 50 000 in the subsequent two years, when the agent was being field collected for redistribution (Day and Bofeng, 2007). Field release and establishment was very successful, with populations establishing at over 97% of release sites. As few as 50 galls could be released to achieve establishment, so 1 000 field-collected galls could be released over 20 new sites (Day and Bofeng, 2007). This meant that large numbers of new sites could be targeted. As the agent had been tested elsewhere, was easy and cheap to collect and release, and was damaging to *C. odorata*, there were substantial benefits for every dollar spent.

8.10.2 Biological control of C. odorata in South Africa

The South African biological control program has faced several obstacles not experienced by other countries where chromolaena is a problem. Firstly, the biotype of chromolaena in South Africa is different from that in other countries and where chromolaena has been controlled. Secondly, much of South Africa where chromolaena is found is cooler than where chromolaena is a problem in other countries so the agents may not be as damaging, or cannot establish due to the different climatic conditions.

Cecidochares connexa, which was highly successful in other countries, could not be used in South Africa, due to incompatibility with the southern African *C. odorata* biotype, and agent–plant compatibility problems have also resulted in the failure of the pathogen project to date. Large numbers of the three *Pareuchaetes* species were mass-reared and released. However, only one of these (*P. insulata*) established.

Several other insect species, most importantly *A. thalia pyrrha* and *Longitarsus* sp. (Coleoptera: Chrysomelidae), were tested, found to have broad host ranges and were rejected. The program spent time and resources identifying the small native range of the southern African biotype (the Greater Antilles) and has since begun exploration there for more suitable agents. Some of the recommended agents, for example *Melanagromyza eupatoriella* Spencer (Diptera: Agromyzidae) and *P. costella*, have proved difficult to maintain in quarantine for a long enough period for host-range tests to be conducted.

One of the current constraints on the program in South Africa is that no insect species with a distinct diapause or soil-dwelling stage have been found in the Greater Antilles (Strathie and Zachariades, 2004). Given the importance of such species for the seasonally dry areas of southern Africa invaded by *C. odorata*, it has thus been necessary to continue research on potential agents with a suitable biology from regions of South America that are climatically similar to southern Africa. Because the southern African *C. odorata* biotype is absent from these areas, it is also necessary to ensure that the host range of these insects is broad enough to allow them to establish on infestations of this biotype in South Africa.

The difference in the efficiencies of the *C. connexa* program in PNG and the South African program can be explained, in part, through serendipity, as the success of selected species as agents in classical biological control remains highly unpredictable (McFadyen, 1998). In addition, the South African program essentially needed to start *de novo* because biotype differences make *C. odorata* in southern Africa a different entity from that which had been studied elsewhere since the 1960s. In addition, South Africa has a cooler climate than many of the areas or regions where *C. odorata* occurs and has been controlled. As a result, agents that have worked in the tropics may not work in the cooler regions of South Africa or be suitable to control the southern African biotype.

8.11 Issues of sustainability of the identified biological control efforts

Ideally, once an agent is well established and distributed, no further effort should be necessary in redistributing it. The ideal biological control agent would provide fairly consistent control over time and would be able to locate isolated and small infestations of the target weed as they arise. *Cecidochares connexa* has shown itself to give highly sustainable control, because it spreads quickly and attacks plants consistently. It is good at locating scattered plants and can sustain a population even where *C. odorata* populations are low. In contrast, *P. pseudoinsulata* is an outbreak species which is highly unpredictable in terms of where and when it will control *C. odorata* to adequate levels. In some areas where it was released, initial outbreaks were not matched again, and the insect persists at a low population level. *Pareuchaetes pseudoinsulata* also tends to do better in large, dense infestations of *C. odorata*, and is less able to control *C. odorata* in scattered populations.

8.12 Conclusions

Chromolaena odorata is one of the worst weeds in the world, affecting agriculture and biodiversity in the tropical and subtropical regions of the Old World. Its earliest introduction seems to have been in the mid nineteenth century into India, and it has since spread to much of its suitable habitat in Asia, Africa, and Oceania. Some research has been conducted on its biology and ecology, as well as a good number of studies on its allelopathic properties, population dynamics, and impacts on indigenous vegetation. However, significant research gaps remain. Until recently, studies on the biotype of *C. odorata* invading Asia, Oceania, and West Africa have been equated with the more localized biotype invading southern Africa. However, the two appear to have different attributes and should functionally be regarded as separate entities.

Research on biological control of *C. odorata* was initiated in the 1960s. The first insects, *Pareuchaetes pseudoinsulata* and *Apion brunneonigrum*, were released in India, Malaysia, Sri Lanka, and West Africa in the early 1970s, but these releases were largely unsuccessful. Releases in various countries continued through the 1970s and 1980s, and *P. pseudoinsulata* became established in several countries, causing significant damage in some.

In the early 1990s, the *IOBC Working Group on Chromolaena odorata* was formed, and since then, regular international workshops have been held to discuss the management and biological control of the weed. The ACIAR program in Southeast Asia in the 1990s resulted in the establishment of the highly successful gall fly, *Cecidochares connexa*, throughout Southeast Asia and parts of Oceania. The agent was subsequently released in other countries, including India. The introduction and establishment of *C. connexa* has allowed a downgrading of the weediness status of *C. odorata* in many areas with high rainfall.

In South Africa, biological control research was initiated in the late 1980s and has involved much exploratory work in the Americas. *Pareuchaetes insulata* and *Calycomyza eupatorivora* were established in South Africa in the 2000s and are now spreading. South Africa continues to conduct research on several promising insect agents, including species which have a biology which should enable them to persist in areas with severe seasonal drought and fires. These agents should also be useful for the drier parts of Southeast Asia such as East Timor. After 40 years of research, *C. odorata* control is being achieved through the use of biological control, and these gains will be consolidated over the coming decade.

Acknowledgments

We thank Ezemvelo KZN Wildlife for permission to use unpublished data on clearing costs in Hluhluwe-iMfolozi park. The first author thanks the ARC and WfW for funding his time writing this chapter. Jimaima Le Grand (Queensland Department of Primary Industries and Fisheries) is thanked for generating the maps.

References

Ackery, P. R. (1988). Hostplants and classification: a review of nymphalid butterflies. *Biological Journal of the Linnean Society*, **33**, 95–203.

Anonymous (1983). *Important Weeds of the World*. Third edn. Leverkusen, Germany: Bayer AG.

APG II (2003). An update of the Angiosperm Phylogeny Group classification for the orders and families of flowering plants: APG II. *Botanical Journal of the Linnean Society*, **141**, 399–436.

Apori, S. O., Long, R. J., Castro, F. B. and Ørskov, E. R. (2000). Chemical composition and nutritive value of leaves and stems of tropical weed *Chromolaena odorata*. *Grass and Forage Science*, **55**, 77–81.

Baars, J.-R. and Neser, S. (1999). Past and present initiatives on the biological control of *Lantana camara* (Verbenaceae) in South Africa. *African Entomology Memoir*, **1**, 21–33.

Bennett, F. D. and Cruttwell, R. E. (1973). Insects attacking *Eupatorium odoratum* in the neotropics. 1. *Ammalo insulata* (Walk.) (Lep., Arctiidae), a potential biotic agent for the control of *Eupatorium odoratum* L. (Compositae). *Technical Bulletin of the Commonwealth Institute of Biological Control*, **16**, 105–115.

Bhumannavar, B. S. and Ramani, S. (2007) Introduction and establishment of *Cecidochares connexa* (Macquart) (Diptera: Tephritidae) for the biological suppression of *Chromolaena odorata* in India. In *Proceedings of the Seventh International Workshop on Biological Control and Management of Chromolaena and Mikania*, ed. P.-Y. Lai, G. V. P. Reddy and R. Muniappan. Taiwan: National Pingtung University, pp. 38–48.

Biller, A., Boppré, M., Witte, L. and Hartmann, T. (1994). Pyrrolizidine alkaloids in *Chromolaena odorata*. Chemical and chemoecological aspects. *Phytochemistry*, **35**, 615–619.

Blackmore, A. C. (1998). Seed dispersal of *Chromolaena odorata* reconsidered. In *Proceedings of the Fourth International Workshop on the Biological Control and*

Management of Chromolaena odorata, ed. P. Ferrar, R. Muniappan and K. P. Jayanth. *Agricultural Experiment Station, University of Guam, Publication* 216, 16–21.

Bofeng, I., Donnelly, G., Orapa, W. and Day, M. (2004). Biological control of *Chromolaena odorata* in Papua New Guinea. In *Proceedings of the Sixth International Workshop on Biological Control and Management of Chromolaena*, ed. M. D. Day and R. E. McFadyen. ACIAR Technical Reports. Canberra, Australia: ACIAR, pp. 55 14–16.

Boller, E. F., van Lanteren, J. C. and Delucchi, V. (2006). International Organization for Biological Control of Noxious Animals and Plants. *History of the First 50 Years (1956–2006)*. IOBC, Zurich, Switzerland, 275 pp.

Bremer, K. (1994). *Asteraceae – Cladistics and Classification*. Portland, OR: Timber Press.

Byrne, M. J., Coetzee, J., McConnachie, A. J., Parasram, W. and Hill, M. P. (2004). Predicting climate compatibility of biological control agents in their region of introduction. In *Proceedings of the XI International Symposium on Biological Control of Weeds*, ed. J. M. Cullen, D. T. Briese, D. J. Kriticos, *et al.* Canberra, Australia: CSIRO Entomology, pp. 28–35.

CAB International (2004). *Prevention and Management of Invasive Alien Species: Forging Cooperation throughout West Africa*. Proceedings of a workshop held in Accra, Ghana, 9–11 March 2004. Nairobi, Kenya: CAB International.

Caldwell, P. M. and Kluge, R. L. (1993). Failure of the introduction of *Actinote anteas* (Lep.: Acraeidae) from Costa Rica as a biological control candidate for *Chromolaena odorata* (Asteraceae) in South Africa. *Entomophaga*, **38**, 475–478.

Cock, M. J. W. (1984). Possibilities for biological control of *Chromolaena odorata*. *Tropical Pest Management*, **30**, 7–13.

Cock, M. J. W. and Holloway, J. D. (1982). The history of, and prospects for, the biological control of *Chromolaena odorata* (Compositae) by *Pareuchaetes pseudoinsulata* Rego Barros and allies (Lepidoptera: Arctiidae). *Bulletin of Entomological Research*, **72**, 193–205.

Coleman, J. R. (1989). Embryology and cytogenetics of apomictic hexaploid *Eupatorium odoratum* L. (Compositae). *Revista Brasileira de Genética*, **12**, 803–817.

Cruttwell, R. E. (1972). The insects of *Eupatorium odoratum* L. in Trinidad and their potential as agents for biological control. Unpublished Ph.D. thesis, University of the West Indies, Trinidad.

Cruttwell, R. E. (1973a). Insects attacking *Eupatorium odoratum* in the neotropics. 2. Studies of the seed weevil *Apion brunneonigrum* B.B., and its potential use to control *E. odoratum* L. *Technical Bulletin of the Commonwealth Institute of Biological Control*, **16**, 117–124.

Cruttwell, R. E. (1973b). Insects attacking *Eupatorium odoratum* in the neotropics. 3. *Dichomeris* sp. nov. (=*Trichotaphe* sp. nr *eupatoriella*) (Lep.: Gelechiidae), a leaf-roller on *Eupatorium odoratum* L. (Compositae). *Technical Bulletin of the Commonwealth Institute of Biological Control*, **16**, 125–134.

Cruttwell, R. E. (1974). Insects and mites attacking *Eupatorium odoratum* in the neotropics. 4. An annotated list of the insects and mites recorded from *Eupatorium odoratum* L., with a key to the types of damage found in Trinidad. *Technical Bulletin of the Commonwealth Institute of Biological Control*, **17**, 87–125.

Cruttwell, R. E. (1977a). Insects and mites attacking *Eupatorium odoratum* L. in the neotropics. 6. Two eriophyid mites, *Acalitus adoratus* Keifer and *Phyllocoptes cruttwellae* Keifer. *Technical Bulletin of the Commonwealth Institute of Biological Control*, **18**, 59–63.

Cruttwell, R. E. (1977b). Insects attacking *Eupatorium odoratum* L. in the neotropics. 5. *Mescinia* sp. nr. *parvula* (Zeller). *Technical Bulletin of the Commonwealth Institute of Biological Control*, **18**, 49–58.

Cruz, Z. T., Muniappan, R. and Reddy, G. V. P. (2006). Establishment of *Cecidochares connexa* (Diptera: Tephritidae) in Guam and its effect on the growth of *Chromolaena odorata* (Asteraceae). *Annals of the Entomological Society of America*, **99**, 845–850.

Day, M. and Bofeng, I. (2007). The status of biocontrol of *Chromolaena odorata* in Papua New Guinea. In *Proceedings of the Seventh International Workshop on Biological Control and Management of Chromolaena and Mikania*, ed. P.-Y. Lai, G. V. P. Reddy and R. Muniappan. Taiwan: National Pingtung University, pp. 53–67.

Day, M. D., Wiley, C. J., Playford, J. and Zalucki, M. P. (2003). *Lantana: Current Management Status and Future Prospects*. Monograph 102. Canberra, Australia: ACIAR, 1–128.

Dempster, J. P. (1971). The population ecology of the cinnabar moth, *Tyria jacobaeae* L. (Lepidoptera: Arctiidae). *Oecologia*, **7**, 26–67.

de Rouw, A. (1991). The invasion of *Chromolaena odorata* (L.) King & Robinson (ex *Eupatorium odoratum*), and competition with the native flora, in a rain forest zone, south-west Côte d'Ivoire. *Journal of Biogeography*, **18**, 13–23.

Desmier de Chenon, R., Sipayung, A. and Sudharto, P. (2002a). A decade of biological control against *Chromolaena odorata* at the Indonesian Oil Palm Research Institute in Marihat. In *Proceedings of the Fifth International Workshop on Biological Control and Management of* Chromolaena odorata, ed. C. Zachariades, R. Muniappan and L. W. Strathie. Pretoria, South Africa: ARC-PPRI, pp. 46–52.

Desmier de Chenon, R., Sipayung, A. and Sudharto, A. (2002b). A new biological agent, *Actinote anteas*, introduced into Indonesia from South America for the control of *Chromolaena odorata*. In *Proceedings of the Fifth International Workshop on Biological Control and Management of* Chromolaena odorata, ed. C. Zachariades, R. Muniappan and L. W. Strathie. Pretoria, South Africa: ARC-PPRI, pp. 170–176.

Dharmadhikari, P. R., Perera, P. A. C. R. and Hassen, T. M. F. (1977). The introduction of *Ammalo insulata* for the control of *Eupatorium odoratum* in Sri Lanka. *Technical Bulletin of the Commonwealth Institute of Biological Control*, **18**, 129–135.

Epp, G. A. (1987). The seed bank of *Eupatorium odoratum* along a successional gradient in a tropical rain forest in Ghana. *Journal of Tropical Ecology*, **3**, 139–149.

Erasmus, D. J. (1985) Achene biology and the chemical control of *Chromolaena odorata*. Unpublished Ph.D. thesis, University of Natal, South Africa.

Erasmus, D. J. and van Staden, J. (1987). Germination of *Chromolaena odorata* (L.) K. & R. achenes: effect of storage, harvest locality and the pericarp. *Weed Research*, **27**, 113–118.

Esguerra, N. M. (2002). Introduction and establishment of the tephritid gall fly *Cecidochares connexa* on Siam weed, *Chromolaena odorata*, in the Republic of Palau. In *Proceedings of the Fifth International Workshop on Biological Control and Management of* Chromolaena odorata, ed. C. Zachariades, R. Muniappan and L. W. Strathie. Pretoria, South Africa: ARC-PPRI, pp. 148–151.

Eussen, J. H. H. and de Groot, W. (1974). Control of *Imperata cylindrica* (L.) Beauv. in Indonesia. *Mededelingen Fakulteit Landbouw-Wetenschappen Gent*, **39**, 451–464.

Francini, R. B. and Penz, C. M. (2006). An illustrated key to male *Actinote* from Southeastern Brazil (Lepidoptera: Nymphalidae). *Biota Neotropica*, **6** (http://www.biotaneotropica.org.br/v6n1/pt/abstract?identification-key+bn00606012006).

Gagne, R. J. (1977). The Cecidomyiidae (Diptera) associated with *Chromolaena odorata* (L.) K. & R. (Compositae) in the Neotropical Region. *Brenesia*, **12/13**, 113–131.

Gautier, L. (1992). Taxonomy and distribution of a tropical weed, *Chromolaena odorata* (L.) R. King and H. Robinson. *Candollea*, **47**, 645–662.

Gautier, L. (1993). Reproduction of a pantropical weed: *Chromolaena odorata* (L.) R. King & H. Robinson. *Candollea*, **48**, 179–193.

Gautier, L. (1996). Establishment of *Chromolaena odorata* in a savanna protected from fire: an example from Lamto, central Côte d'Ivoire. In *Proceedings of the Third International Workshop on Biological Control and Management of* Chromolaena odorata, ed. U. K. Prasad, R. Muniappan, P. Ferrar, J. P. Aeschliman and H. de Foresta. Publication 202. Mangilao, Guam: Agricultural Experiment Station, University of Guam, 54–67.

Ghazoul, J. (2004). Alien abduction: disruption of native plant-pollinator interactions by invasive species. *Biotropica*, **36**, 156–164.

Goodall, J. M. (2000). Monitoring serial changes in coastal grasslands invaded by *Chromolaena odorata* (L.) R. M. King and Robinson. Unpublished M.Sc. thesis, University of Natal, South Africa.

Goodall, J. M. and Erasmus, D. J. (1996). Review of the status and integrated control of the invasive alien weed, *Chromolaena odorata*, in South Africa. *Agriculture, Ecosystems and Environment*, **56**, 151–164.

Goodall, J. G. and Morley, T. A. (1995). Ntambanana vegetation survey and veld improvement plan. Unpublished report submitted to the Mpendle Ntambanana Agricultural Company (Pty.) Ltd.

Goodall, J. M., Zimmermann, H. G. and Zeller, D. (1994). The distribution of *Chromolaena odorata* in Swaziland and implications for further spread. Unpublished report submitted to the Standing Committee of the 17th Meeting of the Southern African Regional Commission for the Conservation and Utilisation of the Soil (SARCCUS), Lesotho and to the FAO/UN, Rome.

Goodman, P. S. (2003). Assessing management effectiveness and setting priorities in protected areas in KwaZulu-Natal. *BioScience*, **53**, 843–850.

Grierson, A. J. C. (1980). Compositae. In *A Revised Handbook of the Flora of Ceylon*, Vol. 1, ed. M. D. Dassanayake and F. R. Fosberg. Rotterdam, The Netherlands: A. A. Balkema, pp. 111–278.

Henderson, L. (2001). *Alien Weeds and Invasive Plants*. Handbook No. 12. Pretoria, South Africa: ARC-PPRI.

Henty, E. E. and Pritchard, P. H. (1973). *Weeds of New Guinea and Their Control*. Botany Bulletin 7. Lae, Papua New Guinea: Department of Forests.

Hilliard, O. (1977). *Compositae in Natal*. Pietermaritzburg, South Africa: University of Natal Press.

Hoevers, R. and M'Boob, S. S. (1996). The status of *Chromolaena odorata* (L.) R. M. King and H. Robinson in West and Central Africa. In *Proceedings of the Third International Workshop on Biological Control and Management of* Chromolaena odorata, ed. U. K. Prasad, R. Muniappan, P. Ferrar, J. P. Aeschliman and H. de Foresta. Publication 202. Mangilao, Guam: Agricultural Experiment Station, University of Guam, pp. 1–5.

Holm, L. G., Plucknett, D. L., Pancho, J. V. and Herberger, P. D. (1977). *The World's Worst Weeds: Distribution and Biology*. Honolulu, HI: University Press of Hawaii.

Honu, Y. A. K. and Dang, O. L. (2000). Response of tree seedlings to the removal of *Chromolaena odorata* Linn. in a degraded forest in Ghana. *Forest Ecology and Management*, **137**, 75–82.

Ivens, G. W. (1974). The problem of *Eupatorium odoratum* L. in Nigeria. *Pest Articles News Summaries*, **20**, 76–82.

Joy, P. J., Lyla, K. R. and Satheesan, N. V. (1993). Biological control of *Chromolaena odorata* in Kerala (India). *Chromolaena odorata Newsletter*, **7**, 1–3.

Julien, M. H. and Griffiths, M. W. (1998). *Biological Control of Weeds: A World Catalogue of Agents and their Target Weeds*. Fourth edn. Wallingford, UK: CAB International Publishing.

King, R. M. and Robinson, H. (1970). Studies in the Eupatorieae (Compositae). XXIX. The genus *Chromolaena*. *Phytologia*, **20**, 196–209.

King, R. M. and Robinson, H. (1987). The genera of the Eupatorieae (Asteraceae). *Monographs in Systematic Botany from the Missouri Botanical Garden*, **22**, 1–581.

Kluge, R. L. (1990). Prospects for the biological control of triffid weed, *Chromolaena odorata*, in southern Africa. *South African Journal of Science*, **86**, 229–230.

Kluge, R. L. (1994). Ant predation and the establishment of *Pareuchaetes pseudoinsulata* Rego Barros (Lepidoptera: Arctiidae) for biological control of triffid weed, *Chromolaena odorata* (L.) King & Robinson, in South Africa. *African Entomology*, **2**, 71–72.

Kluge, R. L. and Caldwell, P. M. (1993a). Host specificity of *Pareuchaetes insulata* (Lep.: Arctiidae), a biological control agent for *Chromolaena odorata* (Compositae). *Entomophaga*, **38**, 451–457.

Kluge, R. L. and Caldwell, P. M. (1993b). The biology and host specificity of *Pareuchaetes aurata aurata* (Lepidoptera: Arctiidae), a "new association" biological control agent for *Chromolaena odorata* (Compositae). *Bulletin of Entomological Research*, **83**, 87–94.

Kriticos, D. J., Yonow, T. and McFadyen, R. E. (2005). The potential distribution of *Chromolaena odorata* (Siam weed) in relation to climate. *Weed Research*, **45**, 246–254.

Lai, P.-Y., Muniappan, R., Wang, T.-H. and Wu, C.-J. (2006). Distribution of *Chromolaena odorata* and its biological control in Taiwan. *Proceedings of Hawaiian Entomological Society*, **38**, 119–122.

Lanaud, C., Jolivot, M. P. and Deat, M. (1991). Preliminary results on the enzymatic diversity in *Chromolaena odorata* (L.) R. M. King and H. Robinson (Asteraceae). In *Proceedings of the Second International Workshop on Biological Control of Chromolaena odorata*, ed. R. Muniappan and P. Ferrar. *BIOTROP Special Publication* **44**, 71–77.

Leslie, A. J. and Spotila, J. R. (2001) Alien plant threatens Nile crocodile (*Crocodylus niloticus*) breeding in Lake St. Lucia, South Africa. *Biological Conservation*, **98**, 347–355.

Liggitt, B. (1983). *The Invasive Plant Chromolaena odorata, With Regard to its Status and Control in Natal*. Rural Studies Series, Monograph 2. Pietermaritzburg, South Africa: Institute of Natural Resources, University of Natal, pp. 1–41.

MacArthur, R. H. and Wilson, E. C. (1967). *The Theory of Island Biogeography*. Princeton, NJ: Princeton University Press.

Macdonald, I. A. W. (1983). Alien trees, shrubs and creepers invading indigenous vegetation in the Hluhluwe-Umfolozi Game Reserve Complex in Natal. *Bothalia*, **14**, 949–959.

Macdonald, I. A. W. and Frame, G. W. (1988). The invasion of introduced species into nature reserves in tropical savannas and dry woodlands. *Biological Conservation*, **44**, 67–93.

Macdonald, I. A. W., Reaser, J. K., Bright, C., *et al.*, eds. (2003). *Invasive alien species in southern Africa: national reports and directory of resources*. Cape Town: Global Invasive Species Programme.

Marais, C., van Wilgen, B. W. and Stevens, D. (2004). The clearing of invasive alien plants in South Africa: a preliminary assessment of costs and progress. *South African Journal of Science*, **100**, 97–103.

Martinez, M., Etienne, J., Abud-Antun, A. and Reyes, M. (1993). Les Agromyzidae de la République Dominicaine (Diptera). *Bulletin de la Société entomologique de France*, **98**, 165–179.

Marutani, M. and Muniappan, R. (1991). Interactions between *Chromolaena odorata* (Asteraceae) and *Pareuchaetes pseudoinsulata* (Lepidoptera: Arctiidae). *Annals of Applied Biology*, **119**, 227–237.

McFadyen, R. E. C. (1988) Ecology of *Chromolaena odorata* in the neotropics. In *Proceedings of the First International Workshop on Biological Control of* Chromolaena odorata, ed. U. Prasad, R. Muniappan, E. Ferrar J. P. Aeschliman and H. de Foresta. Mangilao, Guam: Agricultural Experiment Station, University of Guam, pp. 13–20.

McFadyen, R. E. C. (1989). Siam weed: a new threat to Australia's north. *Plant Protection Quarterly*, **4**, 3–7.

McFadyen, R. E. C. (1991). The ecology of *Chromolaena odorata* in the neotropics. In *Proceedings of the Second International Workshop on Biological Control of* Chromolaena odorata, ed. R. Muniappan and P. Ferrar. *BIOTROP Special Publication*, **44**, 1–9.

McFadyen, R. E. C. (1995) The accidental introduction of the *Chromolaena* mite, *Acalitus adoratus*, into South-East Asia. In *Proceedings of the VIII International Symposium on Biological Control of Weeds*, ed. E. S. Delfosse and R. R. Scott. Melbourne: DSIR/CSIRO, pp. 649–652.

McFadyen, R. C. (1996a). Biocontrol of *Chromolaena odorata*: divided we fail. In *Proceedings of the Ninth International Symposium on Biological Control of Weeds*, ed. V. C. Moran and J. H. Hoffmann. Cape Town: University of Cape Town Press, pp. 455–459.

McFadyen, R. C. (1996b). National report from Australia and the Pacific. In *Proceedings of the Third International Workshop on Biological Control and Management of* Chromolaena odorata, ed. U. K. Prasad, R. Muniappan, P. Ferrar, J. P. Aeschliman and H. de Foresta. Publication, 202. Mangilao, Guam: Agricultural Experiment Station, University of Guam, 39–44.

McFadyen, R. E. (1997). Parasitoids of the arctiid moth *Pareuchaetes pseudoinsulata* (Lep.: Arctiidae), an introduced biocontrol agent against the weed *Chromolaena odorata* (Asteraceae), in Asia and Africa. *Entomophaga*, **42**, 467–470.

McFadyen, R. E. C. (1998). Biological control of weeds. *Annual Review of Entomology*, **43**, 369–393.

McFadyen, R. E. C. (2002). Chromolaena in Asia and the Pacific: spread continues but control prospects improve. In *Proceedings of the Fifth International Workshop on Biological Control and Management of* Chromolaena odorata, ed. C. Zachariades, R. Muniappan and L. W. Strathie. Pretoria, South Africa: ARC-PPRI, pp. 13–18.

McFadyen, R. C. (2004a). Chromolaena in Australia: new developments. *Chromolaena odorata Newsletter*, **16**, 1–2.

McFadyen, R. C. (2004b). Chromolaena in East Timor: history, extent and control. In *Proceedings of the Sixth International Workshop on Biological Control and Management of Chromolaena*, ed. M. D. Day and R. E. McFadyen. ACIAR Technical Reports 55. Canberra, Australia: ACIAR, pp. 8–10.

McFadyen, R. E. C., Desmier de Chenon, R. and Sipayung, A. (2003). Biology and host specificity of the chromolaena stem gall fly, *Cecidochares connexa* (Macquart) (Diptera: Tephritidae). *Australian Journal of Entomology*, **42**, 294–297.

McWilliam, A. (2000). A plague on your house? Some impacts of *Chromolaena odorata* on Timorese livelihoods. *Human Ecology*, **28**, 451–469.

Moni, N. S. and Subramoniam, R. (1960). Essential oil from *Eupatorium odoratum* – a common weed in Kerala. *Indian Forester*, **86**, 209.

Muniappan, R. and Marutani, M. (1988). Ecology and distribution of *C. odorata* in Asia and the Pacific. In *Proceedings of the First International Workshop on Biological Control of* Chromolaena odorata, ed. R. Muniappan. Mangilao, Guam: Agricultural Experiment Station, University of Guam, pp. 21–24.

Muniappan, R., Englberger, K., Bamba, J. and Reddy, G. V. P. (2004). Biological control of chromolaena in Micronesia. In *Proceedings of the Sixth International Workshop on Biological Control and Management of Chromolaena*, ed. M. D. Day and R. E. McFadyen. ACIAR Technical Reports 55. Canberra, Australia: ACIAR, pp. 11–12.

Muniappan, R., Englberger, K. and Reddy, G. V. P. (2007). Biological control of *Chromolaena odorata* in the American Pacific Micronesian Islands. In *Proceedings of the Seventh International Workshop on Biological Control and Management of Chromolaena and Mikania*, ed. P.-Y. Lai, G. V. P. Reddy and R. Muniappan. Taiwan: National Pingtung University, pp. 49–52.

Muniappan, R., Reddy, G. V. P. and Lai, P. Y. (2005). Distribution and biological control of *Chromolaena odorata*. In *Invasive Plants: Ecological and Agricultural Aspects*, ed. Inderjit. Basel, Switzerland: Birkhauser Verlag, pp. 223–233.

Muniappan, R., Sundaramurthy, V. T. and Viraktamath, C. A. (1989). Distribution of *Chromolaena odorata* (Asteraceae) and bionomics and consumption and utilization of food by *Pareuchaetes pseudoinsulata* (Lepidoptera: Arctiidae) in India. In *Proceedings of the VII International Symposium on Biological Control of Weeds*, ed. E. S. Delfosse. Rome: Istituto Sperimentale per la Patologia Vegetale, Ministero dell'Agricoltura e delle Foreste, pp. 401–409.

Norbu, N. (2004). Invasion success of *Chromolaena odorata* in the Terai of Nepal. Unpublished M.Sc. thesis, International Institute for Geo-information Science and Earth Observation, Enschede, Netherlands.

Orapa, W. and Bofeng, I. (2004). Mass production, establishment and impact of *Cecidochares connexa* on chromolaena in Papua New Guinea. In *Proceedings of the Sixth International Workshop on Biological Control and Management of Chromolaena*, ed. M. D. Day and R. E. McFadyen. ACIAR Technical Reports 55. Canberra, Australia: ACIAR, pp. 30–35.

Parasram, W. A. (2003). The role of climate in optimal release strategy design for *Pareuchaetes insulata*: a new control agent for triffid weed (*Chromolaena odorata*). *Proceedings of the South African Sugar Technologists' Association*, **77**, 210–215.

Prasad, U. K., Muniappan, R., Ferrar, P., Aeschliman, J. P. and de Foresta, H., eds. (1996). *Proceedings of the Third International Workshop on Biological Control and*

Management of Chromolaena odorata. Publication 202. *Mangilao, Guam*: Agricultural Experiment Station, University of Guam.

Raman, A., Muniappan, R. N., Silva-Krott, I. U. and Reddy, G. V. P. (2006). Induced-defense responses in the leaves of *Chromolaena odorata* consequent to infestation by *Pareuchaetes pseudoinsulata* (Lepidoptera: Arctiidae). *Journal of Plant Disease and Protection*, **113**, 234–239.

Rambuda, T. D. and Johnson, S. D. (2004). Breeding systems of invasive alien plants in South Africa: does Baker's rule apply? *Diversity and Distributions*, **10**, 409–416.

Ramírez, N. (2004) Ecology of pollination in a tropical Venezuelan savanna. *Plant Ecology*, **173**, 171–189.

Rao, Y. R. (1920). Lantana insects in India. *Memoirs, Department of Agriculture in India, Entomology Series, Calcutta*, **5**, 239–314.

Richardson, D. M., Pysek, P., Rejmánek, M., *et al.* (2000). Naturalization and invasion of alien plants: concepts and definitions. *Diversity and Distributions*, **6**, 93–107.

Sahid, I. B. and Sugau, J. B. (1993). Allelopathic effects of lantana (*Lantana camara*) and Siam weed (*Chromolaena odorata*) on selected crops. *Weed Science*, **41**, 303–308.

Sajise, P. E., Palis, R. K., Norcio, N. V. and Lales, J. S. (1974). The biology of *Chromolaena odorata* (L.) R. M. King and H. Robinson. I. Flowering behavior, pattern of growth and nitrate metabolism. *Philippine Weed Science Bulletin*, **1**, 17–24.

Scott, L. J., Lange, C. L., Graham, G. C. and Yeates, D. K. (1998). Genetic diversity and origin of Siam weed (*Chromolaena odorata*) in Australia. *Weed Technology*, **12**, 27–31.

Seibert, T. F. (1989). Biological control of the weed, *Chromolaena odorata* (Asteraceae), by *Pareuchaetes pseudoinsulata* (Lepidoptera: Arctiidae) on Guam and the Northern Mariana Islands. *Entomophaga*, **35**, 531–539.

Singh, S. P. (1998). A review of biological suppression of *Chromolaena odorata* (Linnaeus) King and Robinson in India. In *Proceedings of the Fourth International Workshop on the Biological Control and Management of* Chromolaena odorata, ed. P. Ferrar, R. Muniappan and K. P. Jayanth. Publication 216. Mangilao, Guam: Agricultural Experiment Station, University of Guam, pp. 86–92.

Solis, M. A., Metz, M. A. and Zachariades, C. (2008). Identity and generic placement of *Phestinia costella* Hampson (Lepidoptera: Pyralidae: Phycitinae) reared on the invasive plant *Chromolaena odorata* (L.) R. M. King and H. Rob (Asteraceae). *Proceedings of the Entomological Society of Washington*, **110**, 292–301.

Stone, B. C. (1966). Further additions to the flora of Guam. 3. *Micronesica*, **2**, 133–141.

Strathie, L. W. and Zachariades, C. (2002). Biological control of *Chromolaena odorata* in South Africa: developments in research and implementation. In *Proceedings of the Fifth International Workshop on Biological Control and Management of* Chromolaena odorata, ed. C. Zachariades, R. Muniappan and L. W. Strathie. Pretoria, South Africa: ARC-PPRI, pp. 74–79.

Strathie, L. W. and Zachariades, C. (2004). Insects for the biological control of *Chromolaena odorata*: surveys in the northern Caribbean and efforts undertaken in South Africa. In *Proceedings of the Sixth International Workshop on Biological Control and Management of Chromolaena*, ed. M. D. Day and R. E. McFadyen. ACIAR Technical Reports 55. Canberra, Australia: ACIAR, pp. 45–52.

Talapatra, S. K., Bhar, D. S. and Talapatra, B. (1974). Flavonoid and terpenoid constituents of *Eupatorium odoratum*. *Phytochemistry*, **13**, 284–285.

Thapa, R. and Wongsiri, S. (1997). *Eupatorium odoratum*: a honey plant for beekeepers in Thailand. *Bee World*, **78**, 175–178.

Timbilla, J. A. (1996). Status of *Chromolaena odorata* biological control using *Pareuchaetes pseudoinsulata*, in Ghana. In *Proceedings of the Ninth International Symposium on Biological Control of Weeds*, ed. V. C. Moran and J. H. Hoffmann. Cape Town, South Africa: University of Cape Town Press, pp. 327–331.

Timbilla, J. A. (1998). Effect of biological control of *Chromolaena odorata* on biodiversity: a case study in the Ashanti region of Ghana. In *Proceedings of the Fourth International Workshop on the Biological Control and Management of Chromolaena odorata*, ed. P. Ferrar, R. Muniappan and K. P. Jayanth. Publication 216. Mangilao, Guam: Agricultural Experiment Station, University of Guam, pp. 97–101.

Timbilla, J. A., Zachariades, C. and Braimah, H. (2003). Biological control and management of the alien invasive shrub *Chromolaena odorata* in Africa. In *Biological Control in IPM Systems in Africa*, ed. P. Neuenschwander, C. Borgemeister and J. Langewald. Wallingford, UK: CABI Publishing, pp. 145–160.

Tjitrosemito, S. (1998). Introduction of *Procecidochares connexa* (Diptera: Tephritidae) to Java island to control Chromolaena odorata. In *Proceedings of the Fourth International Workshop on the Biological Control and Management of* Chromolaena odorata, ed. P. Ferrar, R. Muniappan and K. P. Jayanth. Publication 216. Mangilao, Guam: Agricultural Experiment Station, University of Guam, pp. 66–72.

Toelken, H. R. (1983). Compositae. In *Flowering Plants in Australia*, ed. B. D. Morley and H. R. Toelken. Adelaide, Australia: Rigby, pp. 300–314.

van Gils, H., Delfino, J., Rugege, D. and Janssen, L. (2004). Efficacy of *Chromolaena odorata* control in a South African conservation forest. *South African Journal of Science*, **100**, 251–253.

van Wilgen, B. W., de Wit, M. P., Anderson, H. J., *et al.* (2004). Costs and benefits of biological control of invasive alien plants: case studies from South Africa. *South African Journal of Science*, **100**, 113–122.

Versveld, D. B., Le Maitre, D. C. and Chapman, R. A. (1998). Alien invading plants and water resources in South Africa: a preliminary assessment. Water Research Commission Report No. TT 99/98, C.S.I.R. No. ENV/S-C 97154.

von Senger. I., Barker, N. P. and Zachariades, C. (2002). Preliminary phylogeography of Chromolaena odorata: finding the origin of South African weed. In *Proceedings of the Fifth International Workshop on Biological Control and Management of Chromolaena odorata*, ed. C. Zachariades, R. Muniappan and L. W. Strathie. Pretoria, South Africa: ARC-PPRI, pp. 90–99.

Wallner, W. E. (1987). Factors affecting insect population dynamics: differences between outbreak and non-outbreak species. *Annual Review of Entomology*, **32**, 317–340.

Waterhouse, B. M. (1994a). Discovery of *Chromolaena odorata* in northern Queensland, Australia. *Chromolaena odorata Newsletter*, **9**, 1–2.

Waterhouse, B. M. (2003). Know your enemy: recent records of potentially serious weeds in northern Australia, Papua New Guinea and Papua (Indonesia). *Telopea*, **10**, 477–485.

Waterhouse, B. M. and Zeimer, O. (2002). 'On the brink': the status of *Chromolaena odorata* in northern Australia. In *Proceedings of the Fifth International Workshop on Biological Control and Management of* Chromolaena odorata, ed. C. Zachariades, R. Muniappan and L. W. Strathie. Pretoria, South Africa: ARC-PPRI, pp. 29–33.

Waterhouse, D. F. (1994b). Biological Control of Weeds: Southeast Asian Prospects. *ACIAR Monograph*, **26**, 1–302.

Wilson, C. G. and Widayanto, E. B. (1998). A technique for spreading the *Chromolaena* gall-fly, *Procecidochares connexa*, to remote locations. In *Proceedings of the Fourth International Workshop on the Biological Control and Management of* Chromolaena odorata, ed. P. Ferrar, R. Muniappan and K. P. Jayanth. Publication 216. Mangilao, Guam: Agricultural Experiment Station, University of Guam, pp. 63–65.

Wilson, C. G. and Widayanto, E. B. (2002). The biological control programme against *Chromolaena odorata* in eastern Indonesia. In *Proceedings of the Fifth International Workshop on Biological Control and Management of* Chromolaena odorata, ed. C. Zachariades, R. Muniappan and L. W. Strathie. Pretoria, South Africa: ARC-PPRI, pp. 53–57.

Wilson, C. G. and Widayanto, E. B. (2004). Establishment and spread of *Cecidochares connexa* in eastern Indonesia. In *Proceedings of the Sixth International Workshop on Biological Control and Management of Chromolaena*, ed. M. D. Day and R. E. McFadyen. ACIAR Technical Reports 55. Canberra, Australia: ACIAR, pp. 39–44.

Witkowski, E. T. F. and Wilson, M. (2001). Changes in density, biomass, seed production and soil seed banks of the non-native invasive plant, *Chromolaena odorata*, along a 15 year chronosequence. *Plant Ecology*, **152**, 13–27.

Wu, S.-H., Hsieh, C.-F. and Rejmánek, M. (2004). Catalogue of the naturalized flora of Taiwan. *Taiwania*, **49**, 16–31.

Yadav, A. S. and Tripathi, R. S. (1981). Population dynamics of the ruderal weed *Eupatorium odoratum* and its natural regulation. *Oikos*, **36**, 355–361.

Ye, W. H., Mu, H. P., Cao, H. L. and Ge, X. J. (2004). Genetic structure of the invasive *Chromolaena odorata* in China. *Weed Research*, **44**, 129–135.

Zachariades, C., Strathie-Korrûbel, L. W. and Kluge, R. L. (1999). The South African programme on the biological control of *Chromolaena odorata* (L.) King & Robinson using insects. *African Entomology Memoir*, **1**, 89–102.

Zachariades, C., Strathie, L. W. and Kluge, R. L. (2002). Biology, host specificity and effectiveness of insects for the biocontrol of *Chromolaena odorata* in South Africa. In *Proceedings of the Fifth International Workshop on Biological Control and Management of* Chromolaena odorata, ed. C. Zachariades, R. Muniappan and L. W. Strathie. Pretoria, South Africa: ARC-PPRI, pp. 160–166.

Zachariades, C. and Strathie, L. W. (2006). Biocontrol of chromolaena in South Africa: recent activities in research and implementation. *Biocontrol News and Information*, **27**, 10N–15N.

Zachariades, C., Strathie, L., Delgado, O. and Retief, E. (2007). Pre-release research on biocontrol agents for chromolaena in South Africa. In *Proceedings of the Seventh International Workshop on Biological Control and Management of Chromolaena and Mikania*, ed. P.-Y. Lai, G. V. P. Reddy and R. Muniappan. Taiwan: National Pingtung University, pp. 68–80.

Zachariades, C., von Senger, I. and Barker, N. P. (2004). Evidence for a northern Caribbean origin for the southern African biotype of *Chromolaena odorata*. In *Proceedings of the Sixth International Workshop on Biological Control and Management of Chromolaena*, ed. M. D. Day and R. E. McFadyen. ACIAR Technical Reports 55. Canberra, Australia: ACIAR, pp. 25–27.

Zalucki, M. P. and van Klinken, R. D. (2006). Predicting population dynamics of weed biological control agents: science or gazing into crystal balls? *Australian Journal of Entomology*, **45**, 330–343.

Zhigang, L., Schichou, H., Mingfang G., *et al.* (2004). Rearing *Actinote thalia pyrrha* (Fabricius) and *Actinote anteas* (Doubleday and Hewitson) with cutting and potted *Mikania micrantha* Kunth. In *Proceedings of the Sixth International Workshop on Biological Control and Management of Chromolaena*, ed. M. D. Day and R. E. McFadyen. ACIAR Technical Reports, 55, Canberra, Australia: ACIAR, pp. 36–38.

9

Clidemia hirta (L.) D. Don (Melastomataceae)

Patrick Conant

9.1 Introduction

Clidemia hirta (L.) D. Don (Melastomataceae) commonly known as soapbush or Koster's curse, is a neotropical plant ranging from southern Mexico to northern Argentina and east to the islands of the West Indies (Wester and Wood, 1977). It is a densely branched shrub, which grows up to four meters in height. The stems are covered with red bristles. The leaves are opposite, simple, and petiolate. Five to seven major veins originate at the base of the leaf and extend to the apex. The inflorescence is a panicle that can be subterminal or axillary. The calyx has five hairy linear lobes atop a long urceolate hypanthium. The corolla consists of five to seven small white petals. Fruits are borne in clusters of dark blue berries, hairy ovoid, 6 to 9 mm long and can have well over 100 seeds per fruit (Gleason, 1939; Wagner *et al.*, 1990). The seeds are 0.5 mm in diameter and are readily dispersed by frugivorous birds (Simmonds, 1933; Garrison, 2003), and probably other vertebrates (including humans via footwear) such as mongoose. Soil disturbance by wild pigs in Hawaii facilitates invasion of *C. hirta* into native forest (Smith, 1992).

In its native range, *C. hirta* occupies forest edges, streams, trails, roadsides, and disturbed sites. It occurs as scattered plants, occasionally as thickets, which flourish for a few years and succumb to competition or diseases and insects. It tolerates a wide range of soil conditions as long as moisture is adequate and flourishes where annual rainfall ranges from about 1200 to 4000 mm. It grows from near sea level to 1500 m. It is rarely found either outside forests or in deep shade (Wester and Wood, 1977). In Trinidad, the plant prefers modest shade and has not been observed growing in the open by itself, nor in original forest land (Cook 1929). *Clidemia hirta* does not occur in old growth forests in Costa Rica (De Walt *et al.*, 2004).

As an alien invasive species in the absence of its natural enemies and fast-growing competitors, it can form dense, monotypic thickets under forest canopies where it will shade out most of the vegetation beneath them (Binggeli, 1997; Smith, 1992). Peters (2001) referred to *C. hirta* as a treefall gap specialist in Pasoh Forest Reserve in

Biological Control of Tropical Weeds using Arthropods, ed. R. Muniappan, G. V. P. Reddy, and A. Raman. Published by Cambridge University Press. © Cambridge University Press, 2009.

Peninsular Malaysia. In Hawaii it invades open sites and shaded understory in both native and alien dominated forests. De Walt *et al.*, (2004) attributed this expansion of habitat tolerance to release from natural enemy attack that confines the plant in its native range. The rapid growth of *C. hirta*, its prolific seed production and lack of natural enemies can explain its invasiveness and often dominance in suitable habitats.

Between 1880 and 1886, Koster accidentally introduced seeds of *C. hirta* to Fiji in coffee nursery stock, where its invasiveness was first noticed around 1920 (Paine, 1934; Simmonds, 1937). In Hawaii, it was first observed on Oahu in 1941. Since then, it has spread to the Islands of Hawaii in 1972, Molokai in 1973, Maui in 1977, Kauai in 1982 (Nakahara *et al.*, 1992), and Lanai in 1988 (Smith, 1992). By 1989, about 40 500 ha were infested on Oahu alone. It has been accidentally introduced to Singapore (Corlett, 1992), Peninsular Malaysia (Wester and Wood, 1977; Peters, 2001), Thailand (Napompeth, 2004), Indonesia (Whittaker *et al.*, 1995), Taiwan (Yang, 2001), American Samoa, Vanuatu, Wallis, and Futuna, Solomon Islands (Waterhouse and Norris, 1987; Space and Flynn, 2000), Palau (Schreiner, 1989), Ascension Island (Duffy, 1964), Seychelle Islands (Fleischmann, 1997), Sri Lanka (Ashton *et al.*, 2001), Madagascar (Brown and Gurevitch, 2004), Reunion Island (Mascarene Islands) (Baret *et al.*, 2006), and eastern Africa (Tanzania) (Sheil, 1994). The World Wildlife Fund Australia (2003) listed *C. hirta* in Julatten, Queensland, in 2001. The infested area there is estimated to be 316 ha and the government has made the weed a target for eradication. Although it appears the weed is being dispersed by birds, progress toward eradication has been good to date (P. Maher, personal communication).

9.2 Economic importance

A few positive attributes of this species in its native habitat are that it helps to revegetate disturbed areas and serves as a food source for wildlife in its native range. The threats this weed poses to places where it has become invasive far outweigh its benefits. In Fiji it invaded pastures, and rubber and coconut plantations, increasing costs of production due to the necessity of weed control (Simmonds, 1933). Sheep have been shown to control most weeds in plantations but will not eat *C. hirta* (Chee and Faiz, 2002). Goats suffer toxicity from hydrolyzable tannin when fed the plant (Murdiati *et al.*, 1990). Complaints have come in to the Hawaii Department of Agriculture (HDOA) from at least one rancher on the Hamakua coast (Hawaii Island) about the weed invading rangeland. Pastures in the wetter parts of Maui, Kauai, and Molokai are likewise invaded and forage plants are displaced.

9.3 *Clidemia hirta* as a natural area pest

It is a serious threat to understory plant species in tropical island ecosystems (Pacific Island Ecosystems at Risk, website, www.hear.org/pier). In Tau Island of American Samoa, this weed made up one half of the ground cover in some areas in the National Park (Cook, 2001).

Clidemia hirta is one of the invasive plants in the evergreen forests of the East Usambara Mountains in Tanzania that has spread from Amani and throughout the main range of the mountains. It was found mixed with *Lantana camara* L. (Verbenaceae) or in pure stands. It was one of the weeds that occupied the forest edge or gaps and it influences successional processes either by itself or in combination with *L. camara* (Sheil, 1994).

9.4 Control methods

9.4.1 Chemical/mechanical control

In managed natural areas in Hawaii, chemical and mechanical control costs are high, if the weed is not already uncontrollable in the area (Kawelo, personal communication, U.S. Army Directorate of Public Works Environmental Division., Schofield Barracks, Hawaii). Mowing is usually ineffective as the weed grows back from the stumps. Spraying with broadleaf herbicides is the most widely used control method (Motooka *et al.*, 2003), and is supplemented by hand pulling in the rubber plantations in Malaysia (Rubber Research Institute of Malaya, 1973). Both chemical and mechanical methods provide temporary control and are expensive and labor intensive. Even though this weed has spread to Asia and Africa and the islands in the Indian Ocean in recent years, no effort has been made to suppress the weed except for manual weeding or spraying with herbicides in the plantation.

9.4.2 Biological control

Waterhouse and Norris (1987) reviewed biological control efforts of *C. hirta* for the South Pacific islands and Nakahara *et al.* (1992) and Conant (2002) did the same for Hawaii. Out of 17 agents tested for host specificity in Hawaii, six arthropods and one fungal agent were field released (Table 9.1).

Liothrips urichi *Karny (Thysanoptera: Phlaeothripidae)*

Biology and host-specificity studies of this thrips were conducted in 1927–28 in Trinidad (Simmonds, 1933; Cook, 1929). Potted *C. hirta* infested with the thrips were shipped in cold storage to Fiji in 1930. About 20 000 thrips were received and most of them were transferred directly to plants in the field and some were kept in the laboratory for further multiplication and release. By 1932–1933 several hundred hectares of thrips-stunted *C. hirta* had been overgrown by plant competitors of greater forage value. Shaded and greatly weakened by thrips attack, these plants were soon defoliated and killed. Regrowth was readily located and attacked by the thrips. By 1937 the competitive ability of the weeds was permanently impaired by continued thrips attack except in a few shaded and wet areas, and successful biological control was attained (Simmonds, 1937; Rao *et al.*, 1971; Julien and Griffiths, 1998).

Liothrips urichi was introduced to the Solomon Islands from Fiji but appears not to have established even after three releases made in 1938, 1973, and 1975 (Julien and Griffiths,

Table 9.1 *Releases of natural enemies of* Clidemia hirta *in the Pacific islands*

Natural enemy	Origin	Year first released	Established	Plant parts attacked
Liothrips urichi Karny (Phlaeothripidae)	Trinidad, West Indies	Fiji 1930	Yes	Shoots
		Solomons 1938	No	
		Hawaii 1953	Yes	
		Palau 1960	Yes (Babeldaob)	
		Samoa 1974	Yes (Tutuila, Tay)	
Lius poseidon Napp (Buprestidae)	Trinidad, West Indies	Hawaii 1988	Yes	Adult–leaf feeder, Larva–leaf miner
Colletotrichum gloeosporioides (Penz.) Sacc. f. sp. *Clidemiae* Trujillo (Melanconiaceae)	Panama	Hawaii 1986	Yes	Shoots
Ategumia ebulealis Guenée (Pyralidae)	Trinidad, West Indies	Hawaii 1970	Yes (Oahu)	Leaves
	Puerto Rico	Palau 1972	No	
Antiblemma acclinalis Hübner (Noctuidae)	Tobago,West Indies	Hawaii 1995	Yes (Oahu), (no recovery, Kauai)	Leaves
Mompha trithalama Meyrick (Momphidae)	Tobago, West Indies	Hawaii 1995	Yes	Flowers and fruits
Carposina bullata Meyrick (Carposinidae)	Tobago, West Indies	Hawaii 1995	?	Flowers and fruits

1998). It was shipped from Fiji to American Samoa and established on Tutuila Island in 1974 (Tauiliili and Vargo, 1993). Recently it was introduced to Tau Island in American Samoa (Space and Flynn, 2000). Cook (2001) observed that *L. urichi* does appear to exert control on the weed on Tutuila Island. In 1972, *L. urichi* was introduced into Palau, and established. It has proven effective only in open and sunny areas (Schreiner, 1989).

The rest of the biological control of *C. hirta* work has taken place in the Hawaiian Islands (Conant, 2002) except for the release of the thrips and *Ategumia ebulealis* (Guenee) (Lepidoptera: Pyralidae) in Palau (Schreiner, 1989).

Carposina bullata *Meyrick (Lepidoptera: Carposinidae)*

The eggs of this moth are laid on flower buds. After emerging, larvae enter near the apical end of the buds or occasionally enter fruit. The larva attaches webbing to nearby plant parts and may damage up to five or six flowers. Damaged flowers do not produce viable seed. Pupation takes place on the ground. At 26 °C. the duration from egg to adult took 35 to 45 days (Burkhart, 1994). Unfortunately, emergence was very poor with shipments from Tobago to Hawaii. It was first field released on Oahu in 1995 and again in 1998, but fewer than five adults were released at each of two sites. A total of 129 adult moths were released on Hawaii Island starting in November of 1998 through January of 2000 at two sites in the lower Puna district. It was recovered once in 2002 in the lower Stainback Road area in Puna but has not been found since, in spite of several attempts. Its status remains uncertain.

Mompha trithalama *Meyrick (Lepidoptera: Momphidae)*

The moth lays its eggs on flowers and fruits. Larvae enter the flower or an immature green berry and feed around the core inside. Feeding is confined to one fruit. Pupation occurs at the juncture of the stem and flower stalk or between two primary veins of a leaf. In the outdoors, the life cycle is completed in 30 days, but during dry conditions, emergence from pupae may be delayed up to 100 days (Burkhart, 1994). It was released in 1995 on Oahu; Hawaii in 1999; and Maui, Molokai, and Kauai in 2002 and it has been recovered on all these islands. The characteristic red and white annulated late instar larvae are easy to find in mature green *C. hirta* fruits where the moth has been released.

Antiblemma acclinalis *Hubner (Lepidoptera: Noctuidae)*

Eggs are laid on the undersides of leaves. Young larvae make characteristic rectangular holes while late instars eat large irregular portions of the leaves and may consume more than one leaf. Third instar and older larvae migrate off the plant during the day. Pupation takes place on the ground. The life cycle is completed in about 36 days in the field. The insect prefers humid, shady environments (Burkhart, 1987). It was released in 1995 on Oahu and Kauai but no recovery effort has been made on the latter. It was recovered on Oahu but has not been collected there recently and may have suffered from the parasitization and predation that many lepidopteran weed natural enemies in Hawaiian Islands have.

Ategumia matutinalis (Blepharomastix ebulealis) *(Guenée)*
(Lepidoptera: Pyralidae)

This leaf-rolling caterpillar, obtained from Puerto Rico and Trinidad in 1969, was released on Oahu in 1970 and Hawaii in 1972. The life cycle is roughly five weeks under laboratory conditions. More than one leaf may be connected with webbing as larvae feed on the lower surface of leaves. Mature larvae fold leaves as shelter and can consume entire leaves. The first recovery on Oahu was made in 1974, but it was never recovered on Hawaii Island. Parasitoids are probably reducing the effectiveness of this moth (Reimer and Beardsley, 1986). It was also released in Palau in 1972 but did not establish (Schreiner, 1989).

Lius poseidon *Napp. (Coleoptera: Buprestidae)*

This beetle was introduced from Trinidad to Oahu and Kauai in the Hawaiian chain in 1988. It has since been released on Maui, Hawaii, and Molokai. Eggs are usually laid near the edge of a leaf not far from the petiole. Larvae bore into the leaf and mine within, pupating in a "blotch" at the end of the mine. The life cycle is completed in about 40 days. Adults feed on leaves, producing a characteristic ragged feeding "channel" from the leaf edge inward. Conant (2001) reared adults of this beetle from the leaves of another weed, *Tibouchina herbacea* (Melastomataceae), and this is the only other known host plant in Hawaii.

Collectorichum gloeosporioides f. sp. clidemiae *Trujillo*

An isolate of the fungus *C. gloeosporioides*, discovered in Panama, was introduced into Hawaii and it has proven to be an effective control agent under conditions of moderate to high humidity. These conditions prevail in *C. hirta* infested zones in Hawaii (Trujillo *et al.*, 1986; Norman and Trujillo, 1995). The fungus causes periodic defoliation over contiguous areas, typically less than a few hundred square meters. However, plants will often recover by resprouting from the stem at ground level. Terminals on foliated plants may die from infection. It has established on all the Hawaiian Islands except on Lanai.

9.5 Biotic interference

Biotic interference has probably made at least two of the natural enemies released in Hawaii less effective (Reimer and Beardsley, 1986; Reimer, 1988). Reimer (1988) reported biotic interference of *L. urichi* by two species of predators, the bigheaded ant, *Pheidole megacephala* (F.) (Hymenoptera: Formicidae) and the bug *Montandoniola morguesi* (Puton) (Hemiptera: Anthocoridae). *P. megacephala* carried away larval stages of the thrips and the bug fed on eggs, larvae, and pupae. Reimer and Beardsley (1986) reported *A. matutinalis* to be parasitized by four species of hymenoptera including *Trathla flavororbitalis* (Cameron) (Hymenoptera: Ichneumonidae), *Brachymeria obscurata* (Walker) (Hymenoptera: Chalcididae), *Meteorus laphygmae* Viereck (Hymenoptera: Braconidae), and *Casinaria infesta* (Cresson) (Hymenoptera: Ichneumonidae). They also reared out *Trichogramma* sp. near *higai* Oatman and Platner (Hymenoptera: Trichogrammatidae) from a single egg of *A. matutinalis* but they did not investigate egg parasitism rates. *Trichogramma* spp. in Hawaii are known to attack a wide range of lepidopteran eggs and probably also have caused biotic interference with the other moths used against *C. hirta*. They also stated that combined levels of parasitism by the larval parasitoids alone was high, and in combination with unknown levels of attack by *Trichogramma* spp., would explain the low numbers of *A. matutinalis* seen in the field.

It is more likely that all the lepidoptera released against this weed are affected by biotic interference, since Hawaii now has a significant fauna of generalist alien hymenopteran parasitoids (Henneman and Memmott, 2001). Lepidoptera in Hawaii, whether purposely introduced or adventives, are known to be parasitized by several species of hymenoptera and at least one tachinid fly species that are already present in the isles (Funasaki *et al.* 1988). Populations of newly arrived Lepidoptera typically either explode initially and subsequently decrease or they never reach levels where they inflict population level damage to their host plants. *Antiblemma acclinalis* became established at release sites and was recovered but it has become rare on the island of Oahu and presumably also on Kauai where it was also released. Recently, an unidentified pteromalid wasp was found parasitizing the larvae of *M. trithalama* in the immature fruits of *C. hirta* on the Island of Hawaii.

9.6 Current status of *C. hirta*

Clidemia hirta is mentioned in several publications as being invasive in native habitat in countries with tropical or semitropical zones. These include American Samoa (Cook, 2001) and Singapore (Corlett, 1992). In the Seychelle Islands, Fleischmann (1997) mentioned that *C. hirta* has been on Silhouette Island since 1987 and now forms "dense often monospecific stands on disturbed sites" and is together with *Merremia peltata* considered a potentially serious invader of native plant communities." Gerlach (2004) listed *C. hirta* as one of the most invasive plants in the Seychelles, particularly at low elevation. However, at a mid-elevation site (550 m), abundance actually decreased significantly over a 10-year period. He attributed this to the canopy of forest trees closing in over gaps that had been invaded by *C. hirta*. This observation is encouraging but seems to be atypical of tropical natural areas where it is an invasive alien plant.

Peters (2001) attributed the observation of *C. hirta* becoming established in Pasoh Forest Reserve in Penninsular Malaysia in the early 1990s to M. S. Ashton (personal communication, Yale School of Forestry and Environmental Studies). Peters referred to the weed as a treefall gap specialist in that forest. He found that the presence of the weed was significantly correlated with past pig disturbance in light gaps, suggesting that light availability confined infestations to high light environments. He found *C. hirta* biomass increased steadily, seed production was steady, and recruitment and establishment were observed with no mortality. His results suggested that *C. hirta* may be influencing forest succession via competition with native species. As long as disturbance continues there, the prognosis for the Pasoh forest seemed ominous. Peters' (2005) later study suggested that native herbivores of native Melastomataceae reduced the invasiveness of *C. hirta* (compared with the lack of effects of natural enemies found by De Walt *et al.* (2004) in Hawaii). Although the weed was confined to high light environments in forests, such canopy openings are also needed by the native dipterocarps to regenerate and maintain the native canopy.

C. hirta in Hawaii remains one of the most invasive alien shrubs in wet native forests from sea level to 1500 m, forming dense monospecific stands in some areas of the Hawaiian Islands (excluding Lanai where it is incipient). Next to *Miconia calvescens* (Melastomataceae), Loope *et al.* (2004) listed *C. hirta* as one of the most serious threats to Maui's rain forests. *C. hirta* is mentioned specifically as a threat in numerous endangered plant recovery plans of the U.S. Fish and Wildlife Service in Hawaii as a threat (C. Russel, personal communication).

9.7 Efficacy of natural enemies

No quantitative efficacy studies have yet been done in Hawaii on any of the natural enemies released. *Clidemia hirta* is not considered a high priority agricultural weed by the Hawaii Department of Agriculture. Consequently (and unfortunately), very little effort has been put into efficacy studies of natural enemies released. Since more classical biological control work is still much needed in natural areas of Hawaii, HDOA should, in the future, fund staff dedicated to evaluation work (or work with other cooperating agencies) to evaluate new natural enemies released as well as those of the past. Funding is now routinely appropriated by HDOA for post-release evaluation of new natural enemies proposed for control of both weed and insect pests.

Although efficacy of individual species of natural enemies has not been directly studied in Hawaii, De Walt *et al.* (2004) did quantify growth of *C. hirta* with natural enemies either present or excluded at four field sites on the Island of Hawaii. Their work indicates that any natural enemies present had little effect on the growth of wild plants. Plants in Hawaii sprayed with either insecticide or fungicide or both (to kill the natural enemies) did not show significantly higher survival than those left untreated. In fact, survival was close to 100% regardless of treatment. However, the same experiments in Costa Rica produced significantly higher survival with any of the three same treatments on plants in the understory. Survival was increased by 41% with the combination treatment. Relative growth rate increased significantly on fungicide-sprayed plants in Costa Rica, but there was no effect on such plants in Hawaii.

De Walt (2006) using matrix projection models showed that biocontrol agents that reduce survival across all vegetative stages are more likely to cause declines than those that attack seeds or seedlings only. Fortunately, Nakahara *et al.* (1992) listed several insect agents already found in the West Indies that attack different parts of the plant and some could have potential to cause the kind of damage to different stages of the plant that De Walt (2006) said was needed. However, in Hawaii, it should be noted that diurnal leaf-feeding caterpillars would probably be subject to significant biotic interference. Also, even the combined attack of a complex of seed feeders might not be adequate since De Walt (2006) predicted that almost 100% of seedling recruitment per adult plant would need to be thwarted. De Walt (2004) also mentioned a probable gall-forming cecidomyid fly, weevils, and stem borers she observed in Costa Rica as having future potential for biocontrol in countries where the weed is an alien species. We now know from her work that effective potential biocontrol agents do seem to be present in the neotropics. The job of tropical biological control practitioners is to find them.

9.8 Environmental and economic sustainability

Clidemia hirta is spreading throughout the Pacific and Southeast Asia and making inroads into the Old World tropics. It has great potential to adversely alter natural mesic and hydric habitats on a large scale, both agricultural and natural. Both cattle rangeland and tree plantations of any kind are at risk of being choked with this weed in any high rainfall areas outside the neotropics where it is native. Its toxicity to at least goats is worrisome and it could therefore possibly be toxic to other livestock. Chemical control would be a considerable added cost for farmers and ranchers in areas where invasive shrubs are not already a problem. In many tropical areas where agricultural profits are marginal, a single species of highly invasive shrub could make land economically unproductive.

Not only are tropical agricultural habitats at risk from this weed, the sustainability of natural tropical biodiversity outside of the neotropics is threatened by this weed. An agricultural weed can often be managed at some scale if there is the will and the effort. Invasive species in natural ecosystems, however, by definition have few enemies. Invasion by a single species of plant can be on a landscape level and affect an entire country, particularly the island nations that populate the tropical Pacific and Indian Oceans. *Lantana camara* is a good example of such a weed that has dominated many mesic and xeric tropical landscapes around the world (Day *et al.*, 2003). *Clidemia hirta* could prove to be a mesic-hydric habitat equivalent. Cooperation among tropical countries in the science of biological control of weeds is the only practical solution to the looming global loss of biodiversity and weed-caused loss of agricultural production.

References

Ashton, M. S., Gunatilleke, C. V. S., Singhakumara, B. M. P. and Gunatilleke, I. A. U. N. (2001). Restoration pathways for rain forest in southwest Sri Lanka: a review of concepts and models. *Forest Ecology and Management*, **154**, 409–430.

Baret, S., Rouget, M., Richardson, D. M., *et al.* (2006). Current distribution and potential extent of the most invasive alien plant species on La Reunion (Indian Ocean, Mascarene Islands). *Australian Ecology*, **31**, 747–758.

Binggeli, P. (1997). *An Overview of Invasive Woody Plants in the Tropics*. Invasive Woody Plants in the Tropics research group (http://www.bangor.ac.uk/~afs101/iwpt/welcome.shtml).

Brown, K. A. and Gurevitch, J. (2004). Long-term impacts of logging on forest diversity in Madagascar. *Proceedings of the National Academy of Sciences of the USA*, **101**, 6045–6049.

Burkhart, R. M. (1987). Supplemental report on the host range and life history of *Antiblemma acclinalis* Hubner (Lepidoptera: Noctuidae). Unpublished report to PPC Banch, Hawaii Department of Agriculture, 13 pp.

Burkhart, R. M. (1994). *Carposina bullata* Meyrick (Lepidoptera: Carposinidae) and *Mompha trithalama* Meyrick (Lepidoptera:Momphidae): potential biocontrol agents for the weed *Clidemia hirta* in Hawaii. Unpublished report to PPC Banch, Hawaii Department of Agriculture, 24 pp.

Chee, Y. K. and Faiz, A. (2002). *Sheep Grazing Reduces Chemical Weed Control in Rubber*. Carberra, Australia: Australian Centre for International Agriculture Research (http://www.aciar.gov.au/publications/proceedings/32/paper26.pdf), 4 pp.

Conant, P. (2001). A new host record for *Lius poseidon* Napp (Coleoptera: Buprestidae). *Proceedings of the Hawaiian Entomological Society*, **35**, 147.

Conant, P. (2002). Classical biological control of *Clidemia hirta* (Melastomataceae) in Hawaii using multiple strategies. In *Proceedings of a Workshop on Biological Control of Invasive Plants in Native Hawaiian Ecosystems*, ed. C. W. Smith, J. Denslow and S. Hight. Technical Report 129. Honolulu, HI: Hawaii Pacific Cooperative Studies Unit, University of Hawaii, pp. 13–20.

Cook, B. A. (1929). Some notes on the plant associates and habitat of *Clidemia hirta* (L.) D. Don in Trinidad. *Fiji Agricultural Journal*, **2**, 92–93.

Cook, R. P. (2001). Specificity of *Liothrips urichi* (Thysanoptera: Phlaeothripidae) for *Clidemia hirta* in American Samoa. *Proceedings of the Hawaiian Entomological Society*, **35**, 143–144.

Corlett, R. T. (1992). The ecological transformation of Singapore. *Journal of Biogeography*, **19**, 411–420.

Day, M., Wiley, C. J., Playford, J. and Zalucki, M. P. (2003). *Lantana: Current Management Status and Future Prospects*. Monograph 102. Canberra, Australia: Australian Centre for International Agricultural Research, 30 pp.

De Walt, S. J. (2006). Population dynamics and potential for biological control of an exotic invasive shrub in Hawaiian rain forests. *Biological Invasions*, **8**, 1145–1158.

De Walt, S. J., Denslow, J. S. and Ickes, K. (2004). Natural-enemy release facilitates habitat expansion of the invasive tropical shrub *Clidemia hirta*. *Ecology*, **85**, 471–483.

Duffy, E. (1964). The terrestrial ecology of Ascension Island. *Journal of Applied Ecology*, **1**, 219–251.

Fleischmann, K. (1997). Invasion of alien woody plants on the islands of Mahe and Silhouette, Seychelles. *Journal of Vegetation Science*, **8**, 5–12.

Funasaki, G. Y., Lai, P., Nakahara, L. M., Beardsley, J. W. and Ohta, A. K. (1988). A review of biological control introductions in Hawaii: 1890–1985. *Proceedings of the Hawaiian Entomological Society*, **28**, 105–160.

Garrison, J. S. E. (2003). The role of alien tree plantations and avian seed dispersers in native dry forest restoration in Hawaii. Ph.D. dissertation (unpublished), University of Hawaii at Manoa. 370 pp.

Gerlach, G. (2004). A 10-year study of changes in forest vegetation on Silhouette island, Seychelles. *Journal for Nature Conservation*, **12**, 149–155.

Gleason, H. A. (1939). The genus *Clidemia* in Mexico and Central America. *Brittonia*, **3**, 97–140.

Henneman, M. L. and Memmott, J. (2001). Infiltration of a Hawaiian community by introduced biological control agents. *Science*, **293**, 1314–1416.

Julien, M. H. and Griffith, M. W. (1998). *Biological Control of Weeds. A World Catalog of Agents and Their Target Weeds*. Wallingford, UK: CAB International, 223 pp.

Loope, L. L., Starr, F. and Starr, K. (2004). Protecting endangered plant species from displacement by alien plants on Maui, Hawaii. *Weed Technology*, **18**, 1472–1474.

Motooka, P., Castro, L., Nelson, D., Nagai, G. and Ching, L. (2003). *Weeds of Hawaii's pastures and natural areas*. Honolulu, HI: Hawaii College of Tropical Agriculture and Human Resources, University of Hawaii, 184 pp.

Murdiati, T. B., McSweeney, C. S., Campbell, R. S. F. and Stoltz, D. S. (1990). Prevention of hydrolysable tannin toxicity in goats fed *Clidemia hirta* by calcium hydroxide supplementation. *Journal of Applied Toxicology*, **10**, 325–331.

Nakahara, L. M., Burkhart, R. M. and Funasaki, G. Y. (1992). Review and status of biological control of Clidemia in Hawaii. In *Alien Plant Invasions in Native*

Ecosystems of Hawaii, Management and Research, ed. C. P. Stone, C. W. Smith and J. T. Tunison. Honolulu, HI: Cooperative National Parks Resources Studies Unit, University of Hawaii, pp. 452–465.

Napompeth, B. (2004). Management of invasive species in Thailand. Extension Bulletin 544. Taiwan: Food and Fertilizer Technology Center, 11 pp (http://www.agnet.org/library/article/eb544.html).

Norman, D. J. and Trujillo, E. E. (1995). Development of *Colletotrichum gloeosporioides* f.sp. *clidemiae* and *Septoria passiflorae* into two mycoherbicides with extended viability. *Plant Disease*, **79**, 1029–1032.

Paine, R. W. (1934). The control of Koster's curse (*Clidemia hirta*) on Taveuni. *Fiji Agricultural Journal*, **7**, 10–21.

Peters, H. A. (2001). *Clidemia hirta* invasion at the Pasoh Forest Reserve: an unexpected plant invasion in an undisturbed tropical forest. *Biotropica*, **33**, 60–68.

Peters, H. A. (2005). Distributional constraints on an invasive neotropical shrub. *Clidemia hirta*, in Malaysian dipterocarp forest. *Ecotropicos*, **18**, 65–72.

Rao, V. P., Ghani, M. A., Sankaran, R. and Mathur, K. C. (1971). *A Review of Biological Control of Insects and Other Pests in South-east Asia and the Pacific Region*. Technical Communication 6. Slough, UK: Commonwealth of Biological Control, 149 pp.

Reimer, N. J. (1988). Predation on *Liothrips urichi* Karny (Thysanoptera: Phaleothripidae): a case of biotic interference. *Environmental Entomology*, **17**, 132–134.

Reimer, N. J. and Beardsley, J. W. (1986). Some notes on parasitization of *Blepharomastix ebulealis* (Gunee) (Lepidoptera: Pyralidae) in Oahu forests. *Proceedings of the Hawaiian Entomology Society*, **27**, 91–93.

Rubber Research Institute of Malaya. (1973). *Clidemia hirta* in South Johore. *Planter's Bulletin*, **128**, 140–144.

Schreiner, I. (1989). Biological control introductions in the Caroline and Marshall Islands. *Proceedings of the Hawaiian Entomological Society*, **29**, 57–69.

Sheil, D. (1994). Naturalized and invasive species in the evergreen forests of the East Usambara Mountains, Tanzania, Africa. *African Journal of Ecology*, **32**, 66–71.

Simmonds, H. W. (1933). Biological control of the need *Clidemia hirta*. *Bulletin of Entomological Research*, **24**, 345–348.

Simmonds, H. W. (1937). The biological control of the weed *Clidemia hirta* commonly known in Fiji as 'The Curse'. *Fiji Agricultural Journal*, **8**, 37–39.

Smith, C. W. (1992). Distribution, status, phenology, rate of spread and management of *Clidemia* in Hawaii. In *Alien Plant Invasions in Native Ecosystems of Hawaii, Management and Research*, ed. C. P. Stone, C. W. Smith and J. T. Tunison. Honolulu, HI: Cooperative National Parks Resources Studies Unit, University of Hawaii, pp. 241–253.

Space, J. C. and Flynn, T. (2000). *Observations on Invasive Plant Species in American Samoa* (www.hear.org/pier/reports/asreport.htm).

Tauiliili, P. and Vargo, A. M. (1993). History of biological control in American Samoa. *Micronesica*, (Suppl.) **4**, 57–60.

Trujillo, E. E., Latterell, F. M. and Rossi, A. E. (1986). *Colletotrichum gloeosporioides*, a possible control agent for *Clidemia hirta* in Hawaiian forests. *Plant Disease*, **70**, 974–976.

Wagner, W. L., Herbst, D. R. and Sohmer, S. H. (1990). *Manual of the flowering plants of Hawaii*. Honolulu, HI: University of Hawaii Press and Bishop Museum Press, 1853 pp.

Waterhouse, D. F. and Norris, K. R. (1987). *Biological Control: Pacific Prospects.* Melbourne, Australia: Inkata Press, 454 pp.

Wester, L. L. and Wood, H. B. (1977). Koster's curse (*Clidemia hirta*), a weed pest in Hawaiian forests. *Environmental Conservation*, **4**, 35–41.

Whittaker, R. J., Partomihardjo, T. and Riswan, S. (1995). Surface and buried seed banks from Krakatau, Indonesia: Implications for the sterilization hypothesis. *Biotropica*, **27**, 346–354.

World Wildlife Fund Australia. (2003). *Weeds and Pests: Eradicating the Invasive Threat.* Position Paper 03/01. Sydney, Australia: WWF Australia.

Yang, S. (2001). A new record and invasive species in Taiwan: *Clidemia hirta* (L.) D. Don. *Taiwania*, **46**, 232–237.

10

Coccinia grandis (L.) Voigt (Cucurbitaceae)

R. Muniappan, G. V. P. Reddy, and A. Raman

10.1 Introduction

Coccinia grandis (L.) Voigt (= *C. indica* Wight et Arnold, *Coccinia cordifolia* (Auct.)) (Cucurbitaceae, Violales) commonly known as ivy gourd, scarlet gourd, tindori, tindola, or kovai kai, is native to north-central East Africa (Chun, 2001), but it is also found wild in the Indo-Malayan region (Singh, 1990). *Coccinia* includes 29 additional species and they are found only in tropical Africa (Singh, 1990). *Coccinia grandis* was introduced by humans mostly as a food crop to several countries in Asia Australia, Pacific Islands, the Caribbean, and southern United States (Jeffrey, 1967; Linney, 1986; Nagata, 1988; Singh, 1990; Telford, 1990). It has become naturalized in these parts of the world because it is capable of thriving well in warm, humid, tropical regions. In Fiji, it occurs as a naturalized weed in degraded land, cane fields, and road sides (Smith, 1981). Of these introductions, only in Hawaii (Murai *et al.*, 1998) and the Mariana Islands (McConnell and Muniappan, 1991) did it become invasive in the 1980s.

Coccinia grandis is a dioecious, perennial, and herbaceous climber, with glabrous stems, tuberous roots, and axillary tendrils. Leaves are alternate and simple. Fruit is a smooth, bright red, ovoid to ellipsoid berry 2.5–6 cm (Whistler, 1995). It is a smothering, aggressive vine, with an extensive tuberous root system. In Hawaii, *C. grandis* is highly naturalized and spreads rapidly in disturbed sites, 0–245 m in elevation (Wagner *et al.*, 1999). It usually covers trees, understory vegetation, fences, power poles, and other human-made structures in residential neighborhoods and agricultural areas. When stems of *C. grandis* touch soil, they strike roots readily at the nodes (Chun, 2001). Because the fruit is edible to several birds, they act as dispersers of seeds (Ciesla, 2002). Shoot tips and immature fruits are used in Asian cooking, and therefore, long-range dispersal has resulted due to trade and movement of people who carried either the cuttings or the seeds for planting in new locations. Two varieties of *C. grandis* are recognized; tender fruits are bitter in one variety and not bitter in another, and the latter is used in Asian cooking (Ramachandran and Subramaniam, 1983; Kunkel, 1984; Manandhar, 2002). Morphologically no difference is evident between them, however; both varieties are invasive and are found to grow close to each other.

Biological Control of Tropical Weeds using Arthropods, ed. R. Muniappan, G. V. P. Reddy, and A. Raman. Published by Cambridge University Press. © Cambridge University Press, 2009.

Coccinia grandis hosts several insects such as *Diaphania indica* (Saunders) (Lepidoptera: Pyralidae), *Aulacophora* spp. (Coleoptera: Chrysomelidae), *Bactrocera cucurbitae* (Coquillett) (Diptera: Tephritidae), *Aphis gossypii* Glover (Hemiptera: Aphididae), *Liriomyza* spp. (Diptera: Agromyzidae), *Leptoglossus australis* (Fabricius) (Hemiptera: Coreidae), and *Bemisia* spp. (Hemiptera: Aleyrodidae) that attack several commercially important species of Cucurbitaceae (Horner, 2003). By harboring these insects *C. grandis* facilitates increases in their populations and represents a threat to the production of Cucurbitaceae crop species. In its native habitat it is not a serious weed, because it is kept in check by competing plants and natural enemies, as in the case for most invasive species.

10.2 Control methods

Cutting and slashing with hand held-tools and equipment have been practiced as temporary means of control to a limited extent in household areas and public parks, but this technique has no impact on the population of *C. grandis* as it readily grows back from the left-over stems and stubbles. Vines growing over the neighboring vegetation make mechanized removal difficult. Further, proper disposal of the vines is necessary otherwise the cut stems in contact with the soil will strike roots and aid the spread of the weed. Chemical control is also difficult to adopt because of the nature of the vines growing intermingled with or over other vegetation. Basal bark applications of 2,4-D or triclopyr has been recommended in Hawaii; however, finding basal stems is difficult in dense stands. (Motooka *et al.*, 2002). In the Mariana Islands attempts to adopt chemical control were given up as it either proved ineffective or costly (R. Muniappan and G. V. P. Reddy, personal observation).

10.3 Biological control

When *C. grandis* became an invasive weed in the Hawaiian islands, the Hawaii Department of Agriculture declared it a noxious weed because of its impact on agriculture and biodiversity, and they initiated a biological control program to manage it (Chun, 2001). To identify natural enemies of *C. grandis* in its area of origin, Robert Burkhart (Hawaii Department of Agriculture, Honolulu) surveyed in Kenya in 1992 (Chun, 2001). *Coccinia grandis* was found mostly along the coastal region adjoining the Indian Ocean, along the south of Mombasa and around the inland Lake Victoria basin. Over 30 species of insects were recovered from *C. grandis* on the basis that they were found feeding on it. Preliminary host-range tests were conducted on most of the commercially grown economic cucurbitaceous plants grown in Hawaii at a temporary base set up at Diani Beach on the coast south of Mombasa (Chun, 2001). Only four insect species were found to be host specific and were sent to the quarantine facility at the Hawaii Department of Agriculture. However, tests in Hawaii showed that one hemipteran was not specific and hence destroyed. The three remaining natural enemies were subjected to further host-specificity testing for a greater range of plant species. These were the stem-boring moth *Melittia oedipus* (Lepidoptera: Sesiidae), described by Oberthur from specimens collected in Zanzibar in 1878 (cited in Eichlin, 1995) and two leaf-mining weevils, *Acythopeus cocciniae* O'Brien and Pakaluk (Coleoptera: Curculionidae), and *Acythopeus burkhartorum* O'Brien and Pakaluk (Coleoptera: Curculionidae) (O'Brien and Pakaluk, 1998).

10.3.1 Melittia oedipus *Oberthur (Lepidoptera: Sesiidae)*

The life history of *M. oedipus* has been described by Chun (2001). It is diurnal; adults emerge from the pupal cases by mid morning and mate the same day. Eggs are laid within a short period after mating and they are laid on different parts of *C. grandis*, including leaves, stems, and tendrils. Each female adult is capable of laying 60–140 eggs (Chun, 2001). Eggs were reddish brown and dome shaped. Eggs hatched in about 10 days. The neonate larvae bored into the stems and the pupation took place within the stem. Larval duration ranged from 29 to 57 days and the pupal duration was 14–25 days.

Host-specificity studies

In Hawaii, 19 species of Cucurbitaceae and nine species from other families, including vines that grow adjacent to *C. grandis* and some species of the families that belong to Vilolales following centrifugal phylogenetic method (Wapshere, 1974), were subjected to both oviposition and larval feeding tests (Chun, 2001). These tests proved that no plant other than *C. grandis* was suitable as a host for oviposition, with the exception of cucumber on which a few larvae developed, but the adults that emerged from them were weak and did not mate and produced no eggs (Chun, 2001). Based on these studies a USDA–APHIS permit was issued and this agent was field released in 1996 (Hennessey, 1996). In March 2005, a culture of *M. oedipus* was brought from Hilo, Hawaii, to the Containment Laboratory at the University of Guam. Since most of the economic and related species of plants were already tested in Hawaii, it was decided to test only the plant *Zehneria guamensis* (Merr.) Fosberg (Cucurbitaceae), which is endemic to the Mariana Islands (Stone 1970). Both "choice" experiments, growing plants of both *C. grandis* and *Z. guamensis* in the same cage, and "no choice" experiments, growing plants of either *C. grandis* or *Z. guamensis* in one cage, were conducted by releasing 10 pairs of adult moths for egg laying and by attaching 10 eggs for larval development in each cage. Based on the results of these studies, USDA–APHIS has granted permits to release this agent in Guam and Saipan (the islands invaded by *C. coccinia* in the Marianas) (USDA, 2007). Field releases were made on Guam and Saipan beginning July and August 2007, respectively. Investigations on the field establishment and its efficacy are being investigated.

10.3.2 Acythopeus cocciniae *O'Brien and Pakaluk (Coleoptera: Curculionidae)*

Adults of *Acythopeus cocciniae* feed on younger leaves of *C. grandis*, mostly between veins making small holes. The adult female weevil digs a pit on the leaf to lay an egg and covers it with a fluid. Eggs hatch in about eight days and yellow neonate larvae mine

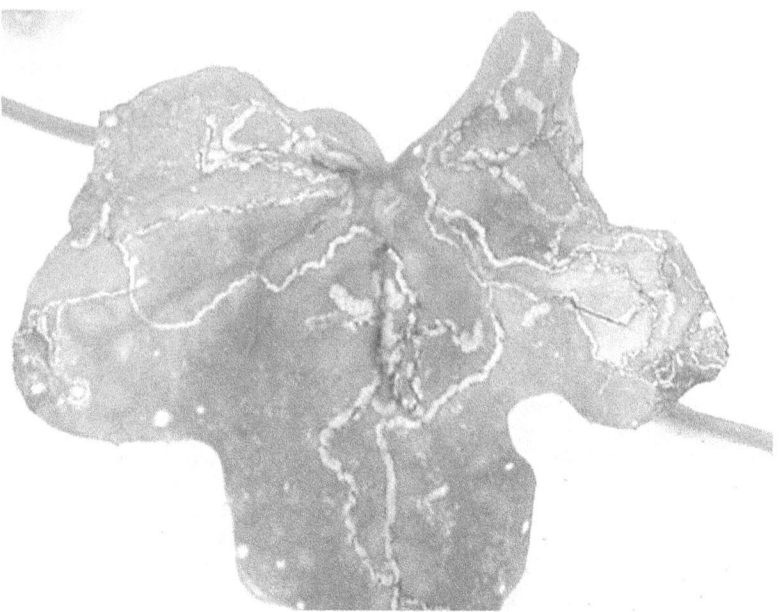

Fig. 10.1 *Coccinia grandis* leaf mined by *Acythopeus cocciniae.*

leaves (Fig. 10.1). Larval duration is 9–10 days. Pupation takes place at the end of the tunnel and the pupa is black in color. Pupal duration is about 15 days. According to Murai *et al.* (1998) adults live up to 203 days in Hawaii but one adult lived up to 480 days on Guam (J. P. Bamba, personal communication).

Host-specificity studies

In Hawaii, 38 species of plants belonging to 17 families were tested for host specificity of *A. cocciniae* at the Hawaii Department of Agriculture and the weevil was found to be specific to *C. grandis*. Based on results from these studies, USDA–APHIS granted a permit for field releases in 1999 (Broda-Hydorn, 1999). In 2002 a culture of *A. cocciniae* was brought to Guam from the Hawaii Department of Agriculture to the Quarantine Laboratory at the University of Guam. In discussions with the Guam and federal officials, it was decided that *A. cocinae* specificity tests should be conducted on *Z. guamensis*, a cucurbitaceous plant endemic to Guam. No adult feeding or larval mining was observed on *Z. guamensis* in "choice" tests. Some adult feeding and larval mining occurred on *Z. guamensis* in "no-choice" tests (J. P. Bamba, unpublished data; Horner, 2003). In 2003, USDA–APHIS issued a permit to field release this agent in Guam and Saipan. Accordingly, field releases were made on Guam and Saipan in May 2003 and subsequent surveys have revealed that *A. cocciniae* has established in Guam and Saipan. Figs. 10.2 and 10.3 show lush *C. grandis* vines before the release of *A. cocciniae* and defoliated vines one year after the release.

Fig. 10.2 *Coccinia grandis* stand before release of *Acythopeus cocciniae*.

Fig. 10.3 *Coccinia grandis* stand 12 months after release of *Acythopeus cocciniae*.

10.3.3 Acythopeus burkhartorum *O'Brien and Pakaluk*
(Coleoptera: Curculionidae)

This is a black, medium-sized weevil. Adults live up to two years feeding on the leaves of *C. grandis*. Eggs are inserted singly in tender petioles or tendrils or stems and they hatch in 7–10 days. Larvae induce galls; while living within galls, they undergo five developmental instars. Larval duration is about 25 days. Pupation takes place in the middle portion of the gall severed on either side in the form of a barrel which drops to the ground. Adults emerge in 25–30 days (Raman *et al.*, 2007).

Host-Specificity studies

In Hawaii 38 species of plants belonging to 17 families were tested for host specificity of *A. burkhartorum* at the Hawaii Department of Agriculture and it was found to be specific to *C. grandis*. Based on these studies, USDA–APHIS granted a permit to field release this agent in 1999 (Broda-Hydorn, 1999). Consequently, field releases were made on Guam and Saipan in October 2004 and February 2005, respectively. Field establishment of *A. burkhartorum* is yet to be confirmed in both the islands. Host-specificity tests conducted in Guam revealed that *A. burkhartorum* is specific to *C. grandis*. No feeding hole or gall development occurred on *Z. guamensis* either in the "choice" or in the "no-choice" tests (Raman *et al.*, 2007; USDA, 2004).

10.4 Effect of natural enemies in Hawaii, Guam and Saipan on survival of introduced biocontrol agents

In Hawaii, a few male parasitoids of *Eupelmus* sp. (Hymenoptera: Eupelmidae) were observed from the field-collected eggs of *M. oeidipus* (Chun, 2001). However, it has not reduced the efficacy of this agent. Also, rats were found to gnaw plants and remove larvae from the stems of *C. grandis*. A local hymenopteran parasitoid has been observed to attack the pupae of *A. cocciniae* in Guam and Saipan. Up to 60% parasitism has been observed in some parts of Guam. Ants have been observed to predate on the pupae of *A. burkhartorum* in the laboratory in Guam.

10.5 Efficacy of the agents

The release of the stem-boring moth *M. oeidipus* and the leaf-mining weevil *A. cocciniae* have had a significant impact on the population of *C. grandis* in Hawaii. *Acythopeus burkhartorum* has established only in some shady areas in Hawaii and its contribution to the suppression of *C. grandis* has been minimal or nil. The release of *A. cocciniae* in Guam and Saipan has resulted in defoliation of *C. grandis* in some areas and parasitism has probably reduced its efficacy. The establishment of *A. burhartorum* and *M. oedipus* in Guam and Saipan is yet to be confirmed.

10.6 Conclusion

Coccinia grandis, a native of East Africa, has been introduced and naturalized in different parts of tropical Asia, Pacific, and Americas, but it has become invasive only in the Hawaiian and Mariana Islands of the Pacific. Chemical and mechanical methods of control proved to be ineffective, uneconomical, not feasible, and unsustainable. The classical biological control approach adopted in Hawaii by introducing the natural enemies *M. oedipus*, *A. cocciniae*, and *A. burkhartorum* has resulted in suppression of *C. grandis*. It has saved the government of Hawaii and the public thousands of dollars spent in mechanical and chemical methods to control this weed. Introduction of *A. cocciniae* in Guam and Saipan in 2003 has already given encouraging results in suppression of *C. grandis* in spite of parasitism. The recent release of *M. oedipus* is expected to supplement the effect of *A. cocciniae*. Suppression of this weed has resulted in the reduction of fruits and foliage available for multiplication of melon fly and other pests of cucurbitaceous crops; it is therefore expected to enhance the prospects of eradicating the melon fly in the northern Mariana Islands. Biological control has been the most sustainable method available for *C. grandis* than all the other options explored thus far.

References

Broda-Hydorn, S. (1999). *Field release of* Acythopeus burkhartorum and A. cocciniae *(Coleoptera: Curculionidae) Nonindigenous Weevils For Biological Control of Ivy Gourd,* Coccinia grandis *(Cucurbitaceae), in Hawaii*. Environmental Assessment. Riverdale, MD: USDA-APHIS, 9 pp.

Chun, M. E. (2001). Biology and host specificity of *Melittia oedipus* (Lepidoptera: Sesiidae), a biological control agent of *Coccinia grandis* (Cucurbitaceae). *Proceedings of the Hawaiian Entomological Society*, **35**, 85–93.

Ciesla, W. M. (2002). Invasive insects, pathogens and plants in Western and Pacific Island forests. Report prepared for Western Forestry Leadership Coalition, 2580 Youngfield Street, Lakewood, CO 80215, USA, 120 pp.

Eichlin, T. D. (1995). New data and a redescription for *Melittia oedipus*, an African vine borer (Lepidoptera: Sesiidae). *Tropical Lepidoptera*, **6**, 47–51.

Hennessey, R. D. (1996). *Field release of* Melittia oedipus *(Lepidoptera: Sesiidae) for Biological Control of Ivy Gourd,* Coccinia grandis *(Cucurbitaceae), in Hawaii*. Environmental Assessment. Riverdale, MD: USDA-APHIS, 12 pp.

Horner, T. A. (2003). *Field release of* Acythopeus cocciniae *(Coleoptera: Curculionidae), a Nonindigenous Leaf-mining Weevil for Control of* Coccinia grandis *(Cucurbitaceae), in Guam and Saipan*. Environmental Assessment. Riverdale, MD: Policy and Program Development, USDA-APHIS, 13 pp.

Jeffrey, C. (1967). Flora of tropical East Africa. In *Cucurbitaceae*, ed. E. Milne-Redhead and R. M. Polhill. London: Crown Agents for Oversea Governments, 156 pp.

Kunkel, G. (1984). *Plants for Human Consumption*. Koenigstein, Germany: Koeltz Scientific Books.

Linney, G. (1986). *Coccinia grandis* (L.) Voigt: a new cucurbitaceous weed in Hawaii. *Hawaiian Botanical Society Newsletter*, **25**, 79–84.

Manandhar, N. P. (2002). *Plants and People of Nepal.* Portland, OR: Timber Press

McConnell, J. and Muniappan, R. (1991). Introduced ornamental plants that have become weeds on Guam. *Micronesica,* Supplement, **3**, 47–49.

Motooka, P., Ching, L. and Nagai, G. (2002). Herbicidal weed control methods for pasture and natural areas of Hawaii. CTAHR publication WC-8. Honolulu, HI: Cooperative Extension Service, College of Tropical Agriculture and Human Resources, University of Hawai'i.

Murai, K., Chan, M. and Culliney, T. (1998). *Host range studies of two African* Acythopeus *spp. (Coleoptera: Curculionidae), potential biocontrol agents for* Coccinia grandis *(Cucurbitaceae).* Honolulu, HI: Plant Pest Control Branch, Hawaii Department of Agriculture. 19 pp.

Nagata, K. M. (1988). Notes on some introduced flora in Hawaii. *Bishop Museum Occasional Papers,* **28**, 79–84.

O'Brien, C. W. and Pakaluk, J. (1998). Two new species of *Acythopeus* Pascoe (Coleoptera: Curculionidae: Baridinae) from *Coccinia grandis* (L.) Voigt (Cucurbitaceae) in Kenya. *Proceedings of the Entomological Society of Washington,* **100**, 764–774.

Ramachandran, K. and Subramaniam, B. (1983). Scarlet gourd, *Coccinia grandis,* little-known tropical drug plant. *Economic Botany,* **37**, 380–383.

Raman, A., Cruz, Z. T., Muniappan, R. and Reddy, G. V. P. (2007). Biology, host-specificity, and field release of gall-inducing *Acythopeus burkhartorum* (Coleoptera: Curculionidae: Baridinae), a biological control agent for the invasive weed *Coccinia grandis* (Cucurbitaceae) in Guam and Saipan. *Tijdschrift voor Entomologie,* **150**, 181–191.

Singh, A. K. (1990). Cytogenetics and evolution in Cucurbitaceae. In *Biology and Utilization of the Cucurbitaceae,* ed. D. M. Bates, R. W. Robinson and C. Jeffrey. Ithaca, NY: Cornell University Press.

Smith, A. C. (1981). *Flora Vitiensis Nova: A New Flora of Fiji.* Vol. 2. Lawai, Kauai, HI: National Tropical Botanical Garden, 810 pp.

Stone, B. C. (1970). The flora of Guam. *Micronesica,* **6**, 1–659.

Telford, I. R. H. (1990). Cucurbitaceae. In *Manual of the Flowering Plants of Hawaii,* Vol. 1, ed. W. L. Wagner, D. R. Herbst and S. H. Sohmer. Honolulu, HI: University of Hawaii Press/Bishop Museum Press, pp. 568–581.

USDA [United States Department of Agriculture] (2004). *Field release of* Acythopeus burkhartorum *(Coleoptera: Curculionidae) a Non-indigenous Weevil for the Control of Ivy Gourd,* Coccinia grandis *(Cucurbitaceae), in Guam and Northern Mariana Islands.* Environmental Assessment 2004. Riverdale, MD: Animal and Plant Health Inspection Service (APHIS), 12 pp.

USDA [United States Department of Agriculture] (2007). *Field Release of* Melittia oedipus *(Lepidoptera: Sesiidae), a Non-indigenous Moth for Control of Ivy Gourd,* Coccinia grandis *(Cucurbitaceae), in Guam and the Northern Mariana Islands.* Environmental Assessment 2007. Riverdale, MD: Animal and Plant Health Inspection Service (APHIS), 13 pp.

Wagner, W. L., Herbst, D. R. and Sohmer, S. H. (1999). *Manual of the Flowering Plants of Hawaii.* 2 vols. Bishop Museum Special Publication 83, Honolulu, HI: University of Hawaii and Bishop Museum Press.

Wapshere, A. J. (1974). A strategy for evaluating the safety of organisms for biological weed control. *Annals of Applied Biology,* **77**, 201–211.

Whistler, W. A. (1995). *Wayside Plants of the Islands.* Hong Kong: Everbest Printing Company.

11

Eichhornia crassipes (Mart.) Solms–Laub. (Pontederiaceae)

J. A. Coetzee, M. P. Hill, M. H. Julien, T. D. Center, and H. A. Cordo

11.1 Introduction

Eichhornia crassipes (Mart.) Solms-Laub. (water hyacinth; Pontederiaceae), an erect free-floating herbaceous plant, is indigenous to tropical South America (Gopal, 1987), but has been spread throughout the world. In the absence of its original suite of natural enemies, and usually in nutrient-enriched waters, it quickly becomes invasive, and is now the most important aquatic weed worldwide (Center, 1994; Julien *et al.*, 1996). It colonizes still or slow moving waters, resulting in thick extensive mats, which impede water traffic, reduce water quality (Edwards and Musil, 1975), and alter social structures for human riparian communities, such as those living on the Sepik River of Papua New Guinea. Infestations continue to plague freshwater bodies, particularly in tropical Africa (Navarro and Phiri, 2000), India (Kathiresan, 2000), and China (Ding *et al.*, 2001), causing significant environmental, economic and social problems, particularly for communities reliant on water bodies for sustenance and survival (Fig. 11.1).

11.2 Taxonomy

Eichhornia crassipes is in the Pontederiaceae, a taxonomically problematic family, which has recently been included in the Commelinales (APG II, 2003; Strange *et al.*, 2004). Eight other genera occur in this family of predominantly neotropical, freshwater aquatics, and eight species in the genus *Eichhornia* (Cook, 1998), all of which originated in South America, except *E. natans* (P. Beauv.) which is native to tropical Africa (Gopal, 1987). Only *E. crassipes* is regarded as a pantropical aquatic weed.

The common names of *E. crassipes* are "water hyacinth", "waterhyacinth" or "water-hyacinth." The two-word spelling suggests that it is part of the true "hyacinth" family (Hyacinthaceae), therefore the Weed Science Society of America uses "waterhyacinth" as the standard spelling (WSSA, 1984), whereas no standardized usage exists elsewhere.

Biological Control of Tropical Weeds using Arthropods, ed. R. Muniappan, G. V. P. Reddy, and A. Raman. Published by Cambridge University Press. © Cambridge University Press, 2009.

Fig. 11.1 Imbuando Village on Imbuand Lagoon, Sepik River, Papua New Guinea. *Eichhornia crassipes* caused total disruption to floodplain village life before biological control by *Neochetina* species resolved the problem in the mid 1990s. (Photo by M. Julien, CSIRO.)

11.3 Description

Eichhornia crassipes is a free-floating aquatic macrophyte that displays two different morphologies with intermediates, dependent on the conditions in which it grows (Fig. 11.2). In dense stands, the petioles are elongated (up to 1 m in length in nutrient-rich waters devoid of herbivores) with circular leaves; but are short (<30 cm) and bulbous, with kidney-shaped leaves where the plants are not in dense mats, or along the edge of infestations (Center and Spencer, 1981). The 6–10 glabrous leaves are arranged in basal rosettes, each leaf lasting up to 6–8 weeks before senescence. Both the rhizome and the fibrous, feathery roots remain submerged. The root morphology is highly plastic and the plasticity is related to nutrient, particularly phosphorus(P), availability in the water. Lateral roots are generally longer and denser at low P levels than at high P levels (Xie and Yu, 2003). The root–shoot ratio varies inversely with nutrient, particularly nitrogen, availability.

11.4 Biology and ecology

Reproduction is both sexual and vegetative. The showy flowers are pale blue or violet, displaying a yellow central patch in the standard perianth lobe, and are borne in spikes. The Pontederiaceae is one of only two monocotyledonous families that display genetic polymorphism of tristyly, in which all flowers of an individual plant possess one of three distinct corresponding style and stamen length phenotypes (Eckenwalder and Barrett, 1986). The intermediate-style form of *E. crassipes* is prevalent in its introduced range, whereas the long-styled form occurs less frequently. The short-style forms predominate in areas of its native range in South America but have not been recorded in its introduced range (Barrett, 1977; Barrett and Forno, 1982). Flowers produce large numbers of long-lived seeds that can remain viable for up to 20 years in sediments (Matthews, 1967; Gopal, 1987). Sexual reproduction is limited by a scarcity of suitable pollinators and a lack of appropriate sites for germination and seedling establishment (Barrett, 1980).

The main mode of population increase is vegetative, via ramets (daughter plants) formed from axillary buds on stolons produced through elongation of internodes (Center and Spencer, 1981). Once the ramets have developed roots, the stolons either decay or break, separating from the parent plant. Thus *E. crassipes* populations increase rapidly, doubling under suitable conditions every 11–18 days (Edwards and Musil, 1975).

(a)

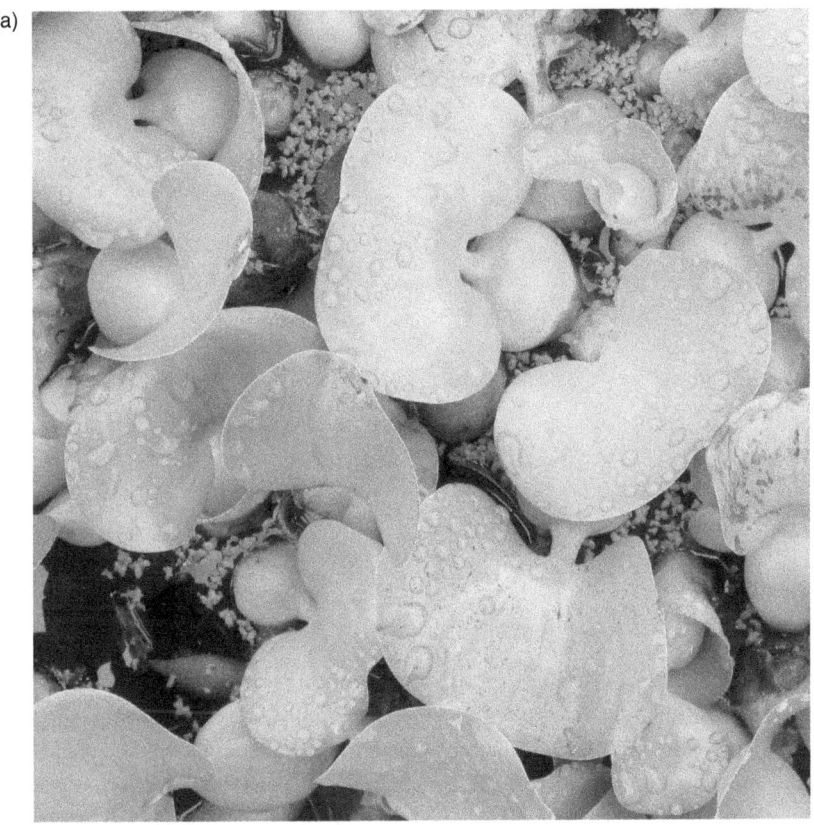

Fig. 11.2 Two *Eichhornia crassipes* morphologies. (a) Bulbous petiole morph (photo by J. Coetzee, Rhodes University), (b) attenuated petiole morph (photo courtesy of A. King, University of the Witwatersrand).

(b)

Fig. 11.2 (cont.)

Neutral pH favors *E. crassipes* proliferation, although the plant can tolerate pH levels from 4 to 10; high light intensities and nutrient-rich water also encourage population build-up. Growth is directly correlated with nutrient concentrations (Gopal, 1987) – as nitrogen and phosphorus increase in concentration, so too does *E. crassipes* biomass accumulation (Gossett and Norris, 1971; Reddy *et al.*, 1989, 1990). Consequently, eutrophication of water, or a steady flow of less enriched waters that provides a continuing supply of nutrients to the roots, leads to thick stands of *E. crassipes*. Optimal growth also occurs at temperatures of 28–30 °C, while growth ceases when water temperatures drop below 10 °C (Gopal, 1987). During these times of stress, stored carbohydrates from the stem are used as energy reserves (Owens and Madsen, 1995), but prolonged cold temperatures, below 5 °C, result in death of the plants, limiting the distribution of *E. crassipes* in high latitudes (Gopal, 1987; Owens and Madsen, 1995).

11.5 Distribution

C. F. P. von Martius first described *E. crassipes* from Brazil in 1823. It is indigenous to the New World tropics, and has its center of origin in Amazonia, Brazil (Barrett and Forno, 1982), with anthropogenic spread to other areas such as Venezuela, parts of central South America, and the larger Caribbean islands (Penfound and Earle, 1948; Edwards and Musil, 1975). The first authentic record of *E. crassipes* outside South America is from New Orleans in 1884 (Penfound and Earle, 1948). Afterwards, *E. crassipes* plants spread around the USA, and by the end of the nineteenth century were recorded in Egypt, India, Australia, and Java (Gopal, 1987). Its distribution is now mainly pantropical, but it also occurs in warm temperate regions of the world, limited to latitudes of 40° N and S (Gopal, 1987). Even though the first introduction of *E. crassipes* to the African continent was made in Egypt between 1879 and 1892 (Edwards and Musil, 1975), many invasions in Africa were first noticed only in the 1980s and it continues to invade many waterways of Africa, even though regional bans have been placed on its transport, and numerous control efforts have been implemented (Navarro and Phiri, 2000).

11.6 Impact

Eichhornia crassipes is recognized as the world's worst aquatic weed, because of the significant ecological impacts it has on the environment, and the associated cascading socioeconomic effects. Dense impenetrable mats restrict access to water, negatively impacting fisheries and related commercial activities, the effectiveness of irrigation canals, navigation and transport, hydroelectric programs, and tourism (Navarro and Phiri, 2000). Other problems include property damage during floods as a result of *E. crassipes* building up against bridges, fences, walls, obstructing water flow, and increasing flood levels. Arguably, the most affected are poverty-stricken communities in rural Africa, where the extent of these effects are yet to be fully measured. *Eichhornia crassipes* alters the livelihoods of any community with high dependence on freshwater waterways for food (subsistence or commercial), transport, and clean water. Ecologically, benthic and littoral diversity is reduced (Masifwa *et al.*, 2001; Toft *et al.*, 2003; Midgley *et al.*, 2006). For example, Midgley *et al.*, (2006) found that the benthic invertebrate community beneath *E. crassipes* mats was significantly less diverse than the community in open water on New Years Dam, South Africa, and similarly Masifwa *et al.* (2001) found a decrease in littoral macroinvertebrate diversity beneath dense *E. crassipes* mats on Lake Victoria in Uganda. Increases in the populations of vectors of human and animal diseases, such as bilharzia, malaria, and cholera, are also associated with *E. crassipes* infestation because these plants interfere with pesticide application (Harley *et al.*, 1996). Because of its rapid growth rate whereby it can double in number in suitable habitat every 11 to 18 days (Edwards and Musil, 1975), *E. crassipes* is able to outcompete native aquatic plants by utilizing the available nutrients in the water, and by successfully competing for space and sunlight (Cilliers, 1991).

11.7 Utilization

One hectare of *E. crassipes* may contain more than 2 million individual plants with a total wet mass of >300 t (Center and Spencer, 1981), and it is this sheer biomass of plant material that has provoked research into its utilization (Julien *et al.*, 1999; Lindsey and Hirt, 1999). Suggested uses of *E. crassipes* include biogas production (Harley, 1990), use as animal fodder, fertilizer, in the manufacture of paper and furniture, in waste water treatment, and in water quality management (Julien *et al.*, 1999). *Eichhornia crassipes* is a relatively cheap and environmentally friendly tool for the clarification of contaminated water because of its ability to absorb heavy metals (common pollutants) and its ability to grow rapidly (Muramoto and Oki, 1983; Zhu *et al.*, 1999). However, the problem of either storing or managing the harvested plant materials contaminated with toxic material remains unsolved.

The key factor mitigating against the utilization of *E. crassipes* is its nearly 95% water content (Harley, 1990). To gain 1 t of dry material, 9 t of fresh material have to be collected (Julien *et al.*, 1996), which makes the cost of drying for the paper and furniture industries commercially unviable (Julien *et al.*, 1999). In addition, *E. crassipes* as fodder for horses and cattle is of inferior quality, again due to its high water content, and it is also unpalatable due to the high potash and chlorine content (Edwards and Musil, 1975). Therefore, utilization of *E. crassipes* is not feasible as a control method due to the low market demand for water hyacinth based products, the inaccessibility of most *E. crassipes* infestations and the high cost of processing the raw material (Julien *et al.*, 1996). For these reasons, *E. crassipes* utilization does not appear to be commercially viable, and consideration of possible utilization of *E. crassipes* should therefore not prevent the execution of control programs against it. In addition, a reliance on *E. crassipes* would create a conflict of interest and lead to further spread of the weed, as has already been observed in some areas in Africa.

11.8 Control

Eichhornia crassipes exhibits many features of a successful weed – it is able to invade, to dominate, and to persist (Cousens and Mortimer, 1995). It has become such a serious problem in its introduced range because of the absence of natural enemies (Cilliers, 1991). Three control methods have been implemented against water hyacinth. Herbicidal control has been successful against small infestations accessible by land, air, or boat, but is relatively expensive, although it has the advantage of being quick and temporarily effective. *Eichhornia crassipes* is susceptible to herbicides such as 2,4-dichlorophenox-yacetic acid (2,4-D), diquat, paraquat, and glyphosate (Gopal, 1987), which have resulted in successful control in small, single-purpose water systems such as irrigation canals and dams (Wright and Purcell, 1995). In South Africa a severe *E. crassipes* infestation on the Hartebeespoort Dam was brought under control using the terbutryn herbicide Clarosan 500FW in the late 1970s (Ashton *et al.*, 1979). Despite its apparent success, herbicidal control provides only short-term relief and, subsequently, must be regularly and fre-quently reapplied (Center *et al.*, 1999). Furthermore, many *E. crassipes*-infested sites are used for potable water, washing, and fishing, and so the use of chemical sprays con-taminates these sites and threatens human health (Julien *et al.*, 1999).

The second control method is manual and mechanical removal. In developing coun-tries, manual removal with simple mechanical devices is still practiced (Julien *et al.*, 1999). This method of control is effective only for small infestations, as it is labor intensive; most problem infestations consist of large interwoven mats that are usually difficult to separate (Cilliers, 1991). Moreover, the long-lived seeds germinate and reinfest sites soon after exposure to light as a consequence of mechanical and chemical removal of the plant (Edwards and Musil, 1975).

Zimbabwe initiated a manual-removal program on Lake Chivero in the early 1980s (Chikwenhere and Phiri, 1999). The manual removal team consisted of 500 workers, working 8 h/day. Although almost 500 t of *E. crassipes* were removed, the rapid regeneration of the weed decelerated the effort and proved expensive, with no obvious impact 6 months later, which led to the decision of using a bulldozer, a boat, a conveyor, and dump trucks. Even though almost 2 ha of plants were cleared daily, neither manual removal nor mechanical harvesting effectively reduced the amount of *E. crassipes* in the lake (Chikwenhere and Phiri, 1999). Mechanical control has also been implemented around Port Bell and Owen Falls Dam on Lake Victoria with limited success (Mailu, 2001). Furthermore, the remoteness of many infestations makes chemical and mechanical control virtually impossible. Despite these problems, reasonably successful results were obtained in Mexico, where a combined chemical–mechanical program, using the herbi-cide 2,4-D and a triturator, was implemented to control *E. crassipes* on the Trigomil Dam (Gutiérrez *et al.*, 1996).

The third method of *E. crassipes* control is biological, which is the only one that offers economical and sustainable control of the weed (Harley *et al.*, 1996). Through plant damage caused by feeding, biological control agents can disrupt the competitive balance between plant species, in favor of native species (Van *et al.*, 1998). Biological control is advantageous because control persists with little ongoing cost and usually with no negative environmental impacts (Julien *et al.*, 1996). Research into the biological control of *E. crassipes* was initiated by the United States Department of Agriculture in 1961, and the first control agents against *E. crassipes* were released in Florida 11 years later (Harley, 1990).

The speed and efficiency of biological control of *E. crassipes* depends on several factors, but under ideal conditions, control in tropical areas can be expected in 3–5 years (Julien *et al.*, 1999).

11.8.1 Biological control agents – arthropods

To date, biological control agents against *E. crassipes* have been released in at least 33 countries (Julien and Griffiths, 1998). The most successful control agents against the weed have been *Neochetina bruchi* Hustache and *N. eichhorniae* Warner (Coleoptera: Curculionidae), and *Niphograpta albiguttalis* (Warren) (= *Sameodes albiguttalis* (Warren)) (Lepidoptera: Pyralidae), which have established throughout the world, wherever biological control against *E. crassipes* has been implemented (Julien and Griffiths, 1998; Julien *et al.*, 1999; Julien *et al.*, 2001).

Neochetina eichhorniae and N. bruchi *(Coleoptera: Curculionidae)*

Neochetina eichhorniae and *N. bruchi* complete their life cycles on *E. crassipes*. The adults are nocturnal, and lay eggs in leaf and petiole tissues, at a rate of 7.3 and 8.5 eggs/female/day, for *N. eichhorniae* and *N. bruchi*, respectively, in Argentina (DeLoach and Cordo, 1976), but these rates may vary with location. The eggs hatch in 7–10 days. The hatching larvae burrow down the leaf petiole to the crown, where the third instars inflict most damage (DeLoach and Cordo, 1976) by feeding on axillary buds (Center, 1994). Larval development ranges from 30 to 45 days, although *N. bruchi* develop slightly faster than *N. eichhorniae* (Center, 1994). The larvae move into the upper roots of the plants to pupate underwater in cocoons made of a parchment-like substance and root material. Pupation is limited, however, when the plants are rooted in mud as a result of fluctuating water levels. The adults emerge after about 7 days, or may remain in the cocoons for extended periods (Center, 1994). They begin to feed soon after emergence by stripping the mesophyll (DeLoach and Cordo, 1976). Generation time from egg to adult is faster in *N. bruchi* (96 days) than *N. eichhorniae* (120 days) (DeLoach and Cordo, 1976), and again, these rates vary with location; for example, a study in Uganda showed that *N. bruchi* took 74 days to complete development, compared with 93 days for *N. eichhorniae* (Njoka *et al.*, 2006). Both species can produce up to three generations per year in their native range (DeLoach and Cordo, 1976), but temperature and plant nutrient status influence generation time in their introduced ranges, and thus the level of control.

Generally, the two *Neochetina* species complement each other, and control of *E. crassipes* is enhanced when the two occur together (Julien *et al.*, 1999). *Neochetina bruchi* lays more eggs and its larvae develop faster than *N. eichhorniae* (96 days compared with 120 days). Further, *N. bruchi* prefers older bulbous leaves for oviposition, while *N. eichhorniae* prefers young central leaves (Harley, 1990). The weevils are therefore able to coexist because of a change in abundance of their preferred oviposition sites as a result of seasonal changes in the plant. In varying nutrient conditions, one or other of the two species may dominate. Plant nitrogen content has a large effect on the *E. crassipes*–weevil interaction, in that higher nitrogen levels lead to faster weevil population growth and increased damage, although *N. bruchi* are more dependent on better quality plant material than are *N. eichhorniae* (Heard and Winterton, 2000). Nonetheless, it appears that better control of water hyacinth occurs when both species are present at a particular site (Julien *et al.*, 1999).

Niphograpta albiguttalis *(Lepidoptera: Pyralidae)*

Adults of *Niphograpta albiguttalis* Warren lay their eggs in the aerenchyma of *E. crassipes* leaves, with a female laying an average of 370 eggs in her lifetime (Center, 1994). Eggs take 3–4 days to hatch. The five larval instars inflict damage by feeding within the petioles, damaging the growth meristems, but damage is not usually inflicted to the crown base. After two weeks of development, the larvae pupate inside healthy petioles, within silken cocoons, and adults emerge from the petioles after 7–10 days (DeLoach and Cordo, 1978). Adult longevity ranges from 4 to 9 days, and up to five generations per year are produced (Center, 1994). The morphological form of *E. crassipes* is important for successful establishment of the moths, which prefer smaller plants, with inflated (bulbous) leaf petioles, typical of the colonizing form of the plant (Center, 1984a).

The control agents *N. albiguttalis*, *N. eichhorniae*, and *N. bruchi* often do not kill *E. crassipes* shoots but inflict varying degrees of leaf mortality (Center, 1984b). Adult weevil feeding on the epidermis of the leaves and larval tunneling through the petiole and the meristematic tissue in the crown of the plant inflicts significant damage (Forno, 1981). The larval tunneling of *N. albiguttalis* through the petioles also causes considerable damage to the plants through injury of the growth tips (DeLoach and Cordo, 1978). However, moth populations are both spatially and temporally patchy. *Eichhornia crassipes* appears to tolerate these injuries by rapid leaf production, which replaces the leaves damaged by herbivory (Center, 1984b). However, the impact of weevil and moth herbivory can culminate in the death of the plant in a manner not readily attributed to their feeding activity, but through the disruption of *E. crassipes* leaf dynamics, when the rate of leaf mortality exceeds that of leaf production (Center and Van, 1989; Van and Center, 1994). Petiole damage hinders the ability of the plant to stay afloat, and the translocation of nutrients is reduced, adversely affecting the nutrient dynamics of the plant (Center and Van, 1989). As a result of tunneling by larvae in the crown and in the lower petioles, water enters the plants causing waterlogging, which encourages invasion by secondary fungi that rot tissues, contributing to the death of the plant and eventual sinking.

Orthogalumna terebrantis *(Acarina: Galumnidae)*

Due to the success of the *Neochetina* weevils and *N. albiguttalis*, the water hyacinth mite *Orthogalumna terebrantis* Wallwork has not been released in as many countries as the other three agents mentioned above. The shiny dark brown–black adults (0.5 mm long) are barely visible to the naked eye. Eggs are laid on the leaves, usually in damaged areas of the leaf, which provide ideal oviposition sites, and hatch after 7–8 days. One larval and three nymphal stages occur, which complete development after about 15 days (Cordo and DeLoach, 1975, 1976). Only the leaf blades are fed upon, and the mite larvae and nymphs produce characteristic feeding galleries extending towards the tip of the leaf, between veins. Adults emerge from exit holes at the end of the gallery, and live for up to 85 days. The ratio of adult to nymph to larva is 1:5:10 (Perkins, 1974). In Argentina, Cordo and DeLoach (1976) found that *O. terebrantis* population numbers varied considerably from year to year, but could produce up to three generations per year.

In high densities, *O. terebrantis* can produce large numbers of feeding galleries in *E. crassipes* leaves, which can be extremely damaging to the plants (Hill and Cilliers, 1999). The impact this damage has on populations of *E. crassipes* has not been fully quantified, so a postrelease evaluation is currently underway in South Africa.

Eccritotarsus catarinensis *(Hemiptera: Miridae)*

Eccritotarsus catarinensis (Carvalho), a leaf-sucking bug, is the most recent agent released against *E. crassipes*. It was first released in South Africa in 1996 (Hill *et al.*, 1999). It has long-lived, mobile adults (both sexes live approximately 50 days) that damage the plant. Eggs are laid in leaf tissue, on the undersurface of the leaf, and hatch after about nine days. The four nymphal instars and the adults feed gregariously, mainly on the undersurface of the *E. crassipes* leaves, inducing chlorosis and death of leaves due to loss of chlorophyll from palisade parenchyma (Hill *et al.*, 1999). This limits the overall growth rate of the plant because the loss of photosynthetic capacity results in reduced carbon fixation, and consequently reduced biomass production (Coetzee *et al.*, 2007).

The mirid has established in South Africa, and is proving to be damaging in sites that do not experience winter frost. Releases of the mirid have also been made in Malawi (Phiri *et al.*, 2001) and China (Ding *et al.*, 2001) in the late 1990s and early 2000s, but whether it has established in these countries is uncertain. It was also released in Benin in 1999, but failed to establish, although the reasons for this are unclear (Ajuonu *et al.*, 2007). The mirid was rejected for release in Australia because of nontarget feeding on native *Monochoria vaginalis* (Burman f.) Kunth. (*Pontederiaceae*) (Stanley and Julien, 1999), but is being considered for release in the USA and Thailand.

Xubida infusella *(Lepidoptera: Pyralidae)*

Larval feeding by the moth *Xubida infusella* (Walker) (= *Acigona infusella*) was found to inflict considerable damage to *E. crassipes* in its native range (DeLoach, 1975). Egg masses are laid in crevices formed by folded or touching leaves, and the hatching larvae tunnel into the petioles towards the base of the plant. Extreme damage is inflicted by extensive feeding in the petioles and rootstock. Prior to pupation, the larvae cut a window in the petiole, through which adults emerge. Development is completed in about two months.

The moth has only been released in Australia and Papua New Guinea, in 1996 and 1997, respectively, and it has established in both these countries (Julien and Stanley, 1999). It was also imported into South Africa for host-specificity testing, but the imported culture was terminated in 1998 as other higher priority agents required greater attention (Hill and Cilliers, 1999). The USA considered releasing *X. infusella*, but recent studies in Australia demonstrated greater damage to the North American native pickerelweed, *Pontederia cordata* L. (*Pontederiaceae*), than to *E. crassipes* (Stanley *et al.*, 2007), and so it has been rejected as a control agent in the USA. Despite being oligophagous, *X. infusella* could add to the control of *E. crassipes* in countries where pickerelweed does not occur (Stanley *et al.*, 2007).

11.8.2 Biological control agents – pathogens

Several fungal pathogens have been reported to attack *E. crassipes* in various parts of the world. Those with the most potential are discussed. *Acremonium zonatum* (Sawada) W. Gams (Ascomycotina) is a fungus that causes necrotic zonate leaf spots, distinguished by spreading lesions primarily on the upper leaf surface (Martyn and Freeman, 1978). *Cercospora rodmanii* Conway and *C. piaropi* Tharp have recently been merged into one species, *C. piaropi* (Tessman *et al.*, 2001), which is capable of decreasing *E. crassipes* biomass, and in some instances has caused substantial decline of *E. crassipes* populations (Freeman and Charudattan, 1984; Charudattan *et al.*, 1985; Martyn, 1985; Morris, 1990). It causes dark brown spots on *E. crassipes* leaves and petioles, which can become necrotic. Combined feeding by the *Neochetina* weevils and infection with *C. piaropi* has additive effects on the biological control of *E. crassipes* (Moran, 2005). *Alternaria eichhorniae* Nag Raj & Ponappa, first reported as a potential control agent for *E. crassipes* in India in 1970, is a highly virulent, host-specific pathogen that induces distinct necrotic spots surrounded by yellow halos on the leaves. Its potential as a mycoherbicide was investigated in Egypt (Shabana *et al.*, 1995, 1997, 2001).

11.8.3 New agents

Although many of the *E. crassipes* control agents have been successful in controlling populations of *E. crassipes* in many parts of the world, there remain numerous geographical regions where *E. crassipes* infestations still cause considerable problems. It has therefore been suggested that the herbivore pressure on *E. crassipes* may be further enhanced by introducing additional control agents (Stanley and Julien, 1999). Based on flower morphology, it has been suggested that the area of greatest genetic diversity of *E. crassipes* lies in the Amazon Basin, and so it is here that the greatest diversity of natural enemies would be expected. However the Upper Amazon Basin has never been surveyed for natural enemies. Explorations undertaken in 1999 and 2000 by the United States Department of Agriculture (USA and Argentina), Commonwealth Agricultural

Bureau International (UK) and Plant Protection Research Institute (South Africa) near Iquitos, Peru, at the confluence of the Marañon and Ucayali rivers showed that the abundance and diversity of natural enemies was greater than anywhere else surveyed on the continent. In this region, most of the arthropods previously known were found and over 50 fungal isolates have been delimited including several new species and even new genera (Evans and Reeder, 2001).

Several additional species of insects occurring on *E. crassipes* in South America could be considered for introduction as biological-control agents (Cordo, 1999). These include *Bellura densa* Walker (Lepidoptera: Noctuidae), *Cornops aquaticum* (Bruner) (Orthoptera: Acrididae), *Megamelus scutellaris* Berg (Hemiptera: Delphacidae), *Taosa inexacta* Walker (Hemiptera: Dictyopharidae), and several species of *Thrypticus* (Diptera: Dolichopodidae). Other insects that have been mentioned by explorers, for which basic information is not available, should be investigated to determine their field host plant ranges as a first step to assessing their potential for use in biologic control efforts. These include the petiole-mining flies *Eugaurax setigena* Sabrosky (Diptera: Chloropidae), *Hydrellia* sp. (Diptera: Ephydridae), and *Chironomus falvipilus* Rempel (Diptera: Chironomidae); the flower-feeding *Calleida* sp. (= *Brachinus* sp.) (Coleoptera: Carabidae) and *Flechtmannia eichhorniae* Keifer (Acarina: Eriophyidae) (Center *et al.*, 2002).

Bellura densa is an oligophagous moth, native to North America. Its natural host is *P. cordata*, but it is very damaging to *E. crassipes*, and also feeds on taro (*Colocasia esculenta* L. (*Araceae*)), an important crop plant in many tropical regions of the world. Based on this, it was rejected for release in South Africa (Hill and Cilliers, 1999), and despite extreme damage to *E. crassipes*, Center and Hill (2002) have recommended that it not be introduced outside of its native range.

Cornops aquaticum is also oligophagous, but it too is extremely damaging to *E. crassipes*. Results from the field and laboratory host-specificity testing showed that it accepts other pontederiads, *Commelina* sp. (*Commelinaceae*) and *Canna indica* L. (Cannaceae) (Silveira Guido and Perkins, 1975; Hill and Oberholzer, 2000; Oberholzer and Hill, 2001), rendering it unsuitable for release in the USA. It was, however, approved for release in July 2007 after more than 10 years in quarantine in South Africa, because the risk to native *Monochoria africana* (Solms-Laub.) N. E. Brown (*Pontederiaceae*) is considered negligible. Further testing is required to determine its impact on *E. crassipes* under different nutrient conditions, its thermal tolerance, and its interactions with the control agents already released in South Africa, and the rest of Africa. There is concern about the potential impacts *C. aquaticum* could have on the level of control already exerted by the *Neochetina* weevils, the most widely used and successful agents released in Africa, if it spreads from South Africa. We therefore need to establish whether it will have synergistic or antagonistic effects on *E. crassipes* biocontrol before it is released.

A more promising monophagous agent is the delphacid bug, *M. scutellaris*. Field data and laboratory host-specificity testing in its native range have confirmed its host specificity (Sosa *et al.*, 2007). It is currently being considered for release in South Africa and the USA and if, upon release and completion of subsequent impact studies, it is found to be significantly damaging, it should be considered for use elsewhere.

Recent studies have been conducted on the biology of various *Thrypticus* sp. in Argentina, investigating life histories and host specificity (Bickel and Hernández, 2004; Hernández *et al.*, 2007). Based on abundance and distribution, the most promising agents are *Thrypticus truncatus* Bickel and Hernandez and *T. sagittatus* Bickel and Hernandez, both specific to *E. crassipes*, and whose larvae cause damage by mining the petioles, auguring well for their use as biocontrol agents. On the other hand, the less known *T. circularis* Bickel and Hernandez that attacks colonizing, bulbous plants (Bickel and Hernandez, 2004), is perhaps equally specific and safe but potentially better for controlling expanding populations of *E. crassipes*.

The other promising agents that have yet to be completely evaluated are the dictyopharid *T. inexacta*, and the carabid *Brachinus* sp., which appear to be host specific and damaging in the field.

11.9 Limitations to successful biocontrol of *E. crassipes*

Biocontrol of *E. crassipes* has been met with varied success across the world (Fig. 11.3). Control has been very successful in tropical countries such as Papua New Guinea (Julien and Orapa, 1999, 2001), and Malawi (Phiri *et al.*, 2001), and it rapidly reduced the problem on Lake Victoria (Ogwang and Molo, 1999; Cock *et al.*, 2000; Albright *et al.*, 2004; Wilson *et al.*, 2007), where only *N. eichhorniae* and *N. bruchi* have been released. In Benin, control mainly by *N. eichhorniae* was substantial but not yet satisfactory (van Thielen *et al.*, 1994; Ajuonu *et al.*, 2003). On the other hand, five arthropod and one pathogen species of control agents have been released in South Africa since 1974, more than anywhere else in the world, yet the level of control does not meet that reached in tropical and subtropical regions (Hill and Olckers, 2001). *Eichhornia crassipes* also remains a significant problem in the south of China (Ding *et al.*, 2001; Chu *et al.*, 2006), India (Kathiresan, 2000), Mexico (Jiménez and Balandra, 2007), the southern USA, and some parts of Australia despite intensive implementation of biocontrol programmes.

Controlling *E. crassipes* without considering the reasons for its seemingly limitless potential for growth addresses only part of the predicament that infestations present, so it is imperative to consider these infestations as symptoms of much bigger problems. The success of biocontrol programs on *E. crassipes*, as exemplified by the impact of control agents on *E. crassipes*, is without a doubt affected by plant quality, which is in turn determined by the nutrient status of the water that it grows in (Heard and Winterton, 2000; Wilson *et al.*, 2006; Coetzee *et al.*, 2007). As a result of increased nutrient levels, eutrophic waters support denser stands of *E. crassipes* (Hill and Cilliers, 1999), which in turn affects the population growth rate of the control agents, and therefore damage to the weed (Julien *et al.*, 1996; Wilson *et al.*, 2007). In some locations the enriched nutrient status of the plants adversely affects the impact of the insect control agents on the plants

(a)

(b)

Fig. 11.3 New Years Dam, South Africa, where *Neochetina eichhorniae* was released in 1990, and by 2000 had reduced the infestation to 10% cover. (a) In 1997, the dam was more than 90% covered by *Eichhornia crassipes*, (b) In 2003, *E. crassipes* covered less than 10% of the dam.

as the insects are unable to suppress the growth rate adequately, due to the rate at which the plants proliferate (Hill and Cilliers, 1999; Coetzee *et al.*, 2007). For example, Hammarsdale Dam, in KwaZulu-Natal, South Africa, is a highly eutrophic system that receives runoff from a wastewater treatment plant, which collects effluent from textile factories and a chicken farm. Both *N. eichhorniae* and *E. catarinensis* have been released here, and despite having reached high population densities, they have had minimal impact on the *E. crassipes* infestation, presumably due to the high growth rate of the plants (Hill and Olckers, 2001). In contrast, within two and a half years after release, the two weevil species controlled *E. crassipes* growing on a nutrient-enriched lake that received sewage from treatment works outflow from Papua New Guinea's capital city, Port Moresby (Julien *et al.*, 1999; Julien and Orapa, 2001).

The role of contamination of waterways that cause eutrophication and toxicity that affect the biology and population dynamics of biological control agents remains unresolved. These and other issues limit effectiveness of the agents and the level of control that can be achieved (see Julien, 2001). Contamination from industrial waste has negatively affected biocontrol in China, where both species of weevil and *E. catarinensis* have been released (Ding *et al.*, 2001; Chu *et al.*, 2006).

Great success was obtained in controlling *E. crassipes* on Lake Victoria using the two *Neochetina* weevils. However, very lush *E. crassipes* that grows close to the wastewater outflows near Entebbe, Uganda, remains undamaged. This implies that close to the outlets the plants are taking up chemicals at levels toxic to the weevils or that the water is so toxic that pre-pupae cannot prepare their underwater cocoon or that pupae cannot survive. Whereas, just hundreds of meters away the insect populations thrived, the weed was destroyed, and open water exists.

Another factor affecting biological control of *E. crassipes* is the hydrology of the smaller water bodies where the weed is a problem (Hill and Olckers, 2001; Julien, 2001). Many of the worst *E. crassipes* infestations occur in small, shallow water bodies. These provide ideal growing conditions, under reduced physical stress from wave action, resulting in the proliferation of the plant mats. In larger, deeper water bodies, such as Lake Victoria, wind and wave action fragment the plant mats, and help to sink the insect-weakened plants (Hill and Olckers, 2001).

Periodic flooding and drought of nonimpounded water systems also cause variable results in control of *E. crassipes* (Hill and Cilliers, 1999; Julien, 2001). The intermittent removal of both the weed and its control agents results in resurgence of *E. crassipes* mats from seed banks or, in the absence of agents, allows proliferation to pre-biocontrol levels (Hill and Olckers, 2001). For this reason, *E. crassipes* biocontrol in India is compromised because the life cycles of the control agents are interrupted in hot summers by the complete drying up of water bodies, and further, by heavy rainfall during the monsoon seasons when plants and insects are washed away in flood waters (Kathiresan, 2000).

Climate may also have an effect on the level of control of *E. crassipes*, particularly in subtropical regions where winter frosts occur. Cold winter temperatures hinder successful biocontrol of *E. crassipes* in the more temperate regions of the USA, South Africa, and China (Fig. 11.4). The active growing season for *E. crassipes* and its agents is restricted to the warmer summer months (Hill and Cilliers, 1999), but both are assumed to remain dormant over winter (Hill and Olckers, 2001). *Eichhornia crassipes* regenerates during spring, while the control agent populations have to regenerate from low numbers due to cold-induced mortality, and reduced, if any, reproductive output. Therefore, the agent

(a)

Fig. 11.4 A high-altitude site in South Africa where biological control of *Eichhornia crassipes* is limited by cold winter temperatures. (a) The site in summer, (b) plants damaged by winter frost. (Photos courtesy of A. King, University of the Witwatersrand.)

(b)

Fig. 11.4 (cont.)

populations only reach significant levels during mid-summer (Hill and Cilliers, 1999; Julien, 2001). The synchrony of this phenology is vital for the control agents to increase their populations quickly enough to exert some level of control over the weed.

Possibly one of the biggest factors affecting successful biocontrol of *E. crassipes* is interference from herbicide operations (Center *et al.*, 1999; Hill and Olckers, 2001). Because biocontrol of *E. crassipes* is not immediate, it often does not meet management objectives, and so herbicides are used to obtain immediate results (Julien, 2001). Agent populations crash or disperse as a result of plant mortality, resulting in *E. crassipes* mats proliferating after regeneration from seed and isolated untreated plants, in the absence of control agents (Hill and Olckers, 2001) (Fig. 11.5).

(a)

(b)

Fig. 11.5 Mbozambo Swamp, South Africa, receives nutrient-rich effluent from a sugar mill resulting in lush *Eichhornia crassipes* growth (a). Both *Neochetina* spp. and *Eccritotarsus catarinensis* have established here, and had recently started causing significant damage to the plants towards the end of 2006 (b and c). However, managers of this system employed herbicides to reduce the infestation in January 2007, thereby interfering with the biological control program at this site (d). (Photos courtesy of A. King, University of the Witwatersrand.)

(c)

(d)

Fig. 11.5 (cont.)

11.10 Cost-benefit analysis

Defining "control" and the length of time taken to achieve it is a fundamental issue in quantifying the benefits of control. Complete control of *E. crassipes* is considered to be reached when *E. crassipes* populations are reduced below an ecologically or economically viable threshold, and are maintained at that threshold with no requirement of an additional intervention. Biological control is considered the most cost-effective method, but it takes a long time (3–5 years under ideal conditions), compared with manual control, which achieves instant success in a short period of time, but requires considerable human input to do so and the results are not sustainable. Herbicidal control also achieves success in a short time period, but is expensive, has negative environmental side effects, and requires considerable follow-up.

Benefits relating to biological control of aquatic weeds have been assessed for Australia (Page and Lacey, 2006). However, it was not possible to separate the benefits concerning biological control of *E. crassipes*, water lettuce, and salvinia, because these plants occur in water bodies in similar conditions. The combined costs of the biological control projects was approx. Au$5 million in 1974–1993. The *E. crassipes* project cost Au$636 000 in the period 1974–1991, and the combined cost-benefit ratio was 27.5:1.

A much higher cost-benefit ratio was achieved for the biological-control program in southern Benin, due to the direct economic effects on the local people. At its peak of infestation, *E. crassipes* reduced the annual income of approximately 200 000 people by about US$85 million, compared with the total cost of the control program of about US$2 million (in 1999 US$ accrued at 6% p.a., for a total duration of 20 years), yielding a cost-benefit ratio of 124:1 (De Groote *et al.*, 2003).

Although no complete cost-benefit analysis of *E. crassipes* control in South Africa has been undertaken, van Wyk and van Wilgen (2002) compared the costs of controlling *E. crassipes* under herbicide application, biological control, and integrated control. The most expensive method was herbicidal control (US$250/ha), while a biological control approach was much less expensive (US$44/ha), but the best return of investment was provided by integrated methods (US$39/ha).

11.11 Conclusions

Eichhornia crassipes impacts on all aspects of water-resource utilization and has led to widespread environmental degradation (Hill, 2003). The impacts of this weed have been most severe in Africa, where large rivers, lakes, and dams, vital for the economic development of the continent, have been rendered unutilizable. *Eichhornia crassipes* had a major negative economic and ecological impact on Lake Victoria, where 20 000 ha of the weed threatened the economy of the basin, which is estimated to be worth US$3–4 billion annually (Albright *et al.*, 2004). Despite being widely regarded as a major threat, very few studies have quantified the impact of this weed.

Nonetheless, *E. crassipes* does not pose the same threat to waterways as it did 10 years ago. The biological control programs have been successful, particularly in the tropics, and have been responsible for reducing *E. crassipes* infestations to levels where they are no longer considered problematic. Biological control has markedly reduced the threat of *E. crassipes* in many parts of the world and represents the highest return on investment of any of the control options.

Hill and Julien (2004) maintain that the key to success of any biological control program, but particularly in poorer rural countries, is appropriate transfer of technology and flexibility of the programs. Furthermore, political support is vital to the success of any *E. crassipes* control program, engendered through the publicizing of the success, where impacts can be observed at the landscape level and real impacts accrue to affected communities.

References

Ajuonu, O., Byrne, M., Hill, M., Neuenschwander, P. and Korie, S. (2007). Survival of the mirid *Eccritotarsus catarinensis* as influenced by *Neochetina eichhorniae* and *Neochetina bruchi* feeding scars on leaves of water hyacinth *Eichhornia crassipes*. *BioControl*, **52**, 193–205.

Ajuonu, O., Schade, V., Veltman, B., Sedjro, K. and Neuensch wander, P. (2003). Impact of the weevils *Neochetina eichhorniae* and *N. bruchi* (Coleoptera: Curculionidae) on water hyacinth, *Eichhornia crassipes* (Pontederiaceae), in Benin, West Africa. *African Entomology*, **11**, 153–161.

Albright, T. P., Moorhouse, T. G. and McNabb, T. J. (2004). The rise and fall of water hyacinth in Lake Victoria and the Kagera River Basin, 1989–2001. *Journal of Aquatic Plant Management*, **42**, 73–84.

APG II (2003). An update of the Angiosperm Phylogeny Group classification for the orders and families of flowering plants: APG II. *Botanical Journal of the Linnean Society*, **141**, 399–436.

Ashton, P. J., Scott, W. E., Sten, D. J. and Wells, R. J. (1979). The chemical control programme against the water hyacinth *Eichhornia crassipes* (Mart.) Solms on Hartebees4poort Dam: historical and practical aspects. *South African Journal of Science*, **75**, 303–306.

Barrett, S. C. H., (1977). Tristyly in *Eichhornia crassipes* (Mart.) Solms (water hyacinth). *Biotropica*, **9**, 230–238.

Barrett, S. C. H. (1980). Sexual reproduction in *Eichhornia crassipes* (water hyacinth). II. Seed production in natural populations. *Journal of Applied Ecology*, **17**, 113–124.

Barrett, S. C. H. and Forno, I. W. (1982). Style morph distribution in New World populations of *Eichhornia crassipes* (Mart.) Solms-Laubach (water hyacinth). *Aquatic Botany*, **13**, 299–306.

Bickel, D. J. and Hernández, M. C. (2004). Neotropical *Thrypticus* (Diptera: Dolichopodidae) reared from water hyacinth, *Eichhornia crassipes*, and other Pontederiaceae. *Annals of the Entomological Society of America*, **97**, 437–449.

Center, T. D. (1984a). Dispersal and variation in infestation intensities of water hyacinth moth, *Sameodes albiguttalis* (Lepidoptera: Pyralidae) populations in peninsular Florida. *Environmental Entomology*, **13**, 482–491.

Center, T. D. (1984b). Leaf life tables: a viable method for assessing sublethal effects of herbivory on waterhyacinth shoots. In *Proceedings of the IV International Symposium on Biological Control of Weeds*, ed. E. S. Delfosse, Vancouver, Canada: Agriculture Canada, pp. 511–524.

Center, T. D. (1994). Biological control of weeds: waterhyacinth and waterlettuce. In *Pest Management in the Subtropics: Biological Control – A Florida Perspective*, ed. D. Rosen, F. D. Bennett and J. L. Capinera. Andover, UK: Intercept Publishing Company, pp. 481–521.

Center, T. D. and Hill, M. P. (2002). Field efficacy and predicted host range of the pickerelweed borer, *Bellura densa*, a potential biological control agent of water hyacinth. *BioControl*, **47**, 231–243.

Center, T. D. and Spencer, N. R. (1981). The phenology and growth of water hyacinth (*Eichhornia crassipes* (Mart.) Solms) in a eutrophic north-central Florida lake. *Aquatic Botany*, **10**, 1–32.

Center, T. D. and Van, T. K. (1989). Alteration of water hyacinth (*Eichhornia crassipes* (Mart.) Solms) leaf dynamics and phytochemistry by insect damage and plant density. *Aquatic Botany*, **35**, 181–195.

Center, T. D., Dray, F. A., Jr., Jubinsky, G. P. and Grodowitz, M. J. (1999). Biological control of water hyacinth under conditions of maintenance management: can herbicides and insects be integrated? *Environmental Management*, **23**, 241–256.

Center, T. D., Hill, M. P., Cordo, H. and Julien, M. H. (2002). Waterhyacinth. In *Biological Control of Invasive Plants in the Eastern United States*, ed. R. G. van Driesche, S. Lyon, B. Blossey, M. S. Hoddle and R. Reardon. Morgantown, WV: USDA Forest Service, pp. 41–64.

Charudattan, R., Linda, S. B., Kluepfel, M. and Osman, Y. A. (1985). Biocontrol efficacy of *Cercospora rodmanii* on waterhyacinth. *Phytopathology*, **75**, 1263–1269.

Chikwenhere, G. P. and Phiri, G. (1999). History of water hyacinth and its control efforts on Lake Chivero in Zimbabwe. In *Proceedings of the First IOBC Global Working Group Meeting for the Biological and Integrated Control of Water Hyacinth*, ed. M. P. Hill, M. H. Julien, and T. D. Center. Pretoria, South Africa: Plant Protection Research Institute, pp. 91–97.

Chu, J., Ding, Y. and Zhuang, Q. (2006). Invasion and control of water hyacinth (*Eichhornia crassipes*) in China. *Journal of Zhejiang University Science*, **7**, 623–626.

Cilliers, C. J. (1991). Biological control of water hyacinth, *Eichhornia crassipes* (Pontederiaceae), in South Africa. *Agriculture, Ecosystems, and Environment*, **37**, 207–218.

Cock, M., Day, R., Herren, H., *et al.* (2000). Harvesters get that sinking feeling. *Biocontrol News and Information*, **21**, 1–8.

Coetzee, J. A., Byrne, M. J. and Hill, M. P. (2007). Impact of nutrients and herbivory by *Eccritotarsus catarinensis* on the biological control of water hyacinth, *Eichhornia crassipes*. *Aquatic Botany*, **86**, 179–186.

Cook, C. D. K. (1998). Pontederiaceae. In *Families and Genera of Vascular Plants*, Vol. IV, ed. K. Kubitzki. Berlin: Springer-Verlag, pp. 395–403.

Cordo, H. A. (1999). New agents for biological control of waterhyacinth. In *Proceedings of the First IOBC Global Working Group Meeting for the Biological and Integrated Control of Water Hyacinth*, ed. M. P. Hill, M. H. Julien and T. D. Center. Pretoria, South Africa: Plant Protection Research Institute, pp. 68–74.

Cordo, H. A. and DeLoach, C. J. (1975). Ovipositional specificity and feeding habits of the waterhyacinth mite, *Orthogalumna terebrantis*, in Argentina. *Environmental Entomology*, **4**, 561–565.

Cordo, H. A. and DeLoach, C. J. (1976). Biology of the waterhyacinth mite in Argentina. *Weed Science*, **24**, 245–249.

Cousens, R. and Mortimer, M. (1995). *Dynamics of Weed Populations*. Cambridge, UK: Cambridge University Press.

De Groote, H., Ajuonu, O., Attignon, S., Djessou, R. and Neuenschwander, P. (2003). Economic impact of biological control of water hyacinth in southern Benin. *Ecological Economics*, **45**, 105–117.

DeLoach, C. J. (1975). Evaluation of candidate arthropods for biological control of waterhyacinth: studies in Argentina. In *Proceedings of a Symposium on Water Quality Management through Biological Control*, ed. P. L. Brezonik and J. L. Fox. Gainesville, FL: University of Florida, pp. 45–50.

DeLoach, C. J. and Cordo, H. A. (1976). Life cycle and biology of *Neochetina bruchi*, a weevil attacking waterhyacinth in Argentina, with notes on *N. eichhorniae*. *Annals of the Entomological Society of America*, **69**, 643–652.

DeLoach, C. J. and Cordo, H. A. (1978). Life history and ecology of the moth *Sameodes albiguttalis*, a candidate for biological control of water hyacinth. *Environmental Entomology*, **7**, 309–321.

Ding, J., Wang, R., Fu, W. and Zhang, G. (2001). Water hyacinth in China: its distribution, problems and control status. In *Proceedings of the Second Meeting of the Global Working Group for the Biological and Integrated Control of Waterhyacinth*, held in Beijing, China, ed. M. H. Julien, M. P. Hill, T. D. Center and J. Ding. Canberra, Australia: Australian Centre for International Agricultural Research, pp. 29–32.

Eckenwalder, J. E. and Barrett, S. C. H. (1986). Phylogenetic systematics of Pontederiaceae. *Systematic Botany*, **11**, 373–391.

Edwards, D. and Musil, C. J. (1975). *Eichhornia crassipes* in South Africa – a general review. *Journal of the Limnological Society of Southern Africa*, **1**, 23–27.

Evans, H. C. and Reeder, R. H. (2001). Fungi associated with *Eichhornia crassipes* (water hyacinth) in the upper Amazon Basin and prospects for their use in biological control. In *Proceedings of the Second Meeting of the Global Working Group for the Biological and Integrated Control of Water Hyacinth*, held in Beijing, China, 9–12 October 2000, ed. M. H. Julien, M. P. Hill, T. D. Center and J. Ding. Canberra, Australia: Australian Centre for International Agricultural Research, pp. 62–70.

Forno, I. W. (1981). Effects of *Neochetina eichhorniae* on the growth of waterhyacinth. *Journal of Aquatic Plant Management*, **19**, 27–31.

Freeman, T. E. and Charudattan, R. (1984). *Cercospora rodmanii Conway, a Biocontrol Agent of Waterhyacinth*. Bulletin 842. Gainesville, FL: Agricultural Experiment Stations, Institute of Food and Agricultural Sciences, University of Florida, 18 pp.

Gopal, B. (1987). *Water Hyacinth*. Amsterdam: Elsevier.

Gossett, D. R. and Norris, W. E., Jr. (1971). Relationship between nutrient availability and content of nitrogen and phosphorous in tissues of the aquatic macrophyte, *Eichhornia crassipes* (Mart.) Solms. *Hydrobiologia*, **38**, 15–28.

Gutiérrez, E., Huerto, R. and Arreguin, F. (1996). Strategies for waterhyacinth (*Eichhornia crassipes*) control in Mexico. *Hydrobiologia*, **340**, 181–185.

Harley, K. L. S. (1990). The role of biological control in the management of water hyacinth, *Eichhornia crassipes*. *Biocontrol News and Information*, **11**, 11–22.

Harley, K. L. S., Julien, M. H. and Wright, A. D. (1996). Water hyacinth: a tropical worldwide problem and methods for its control. In *Proceedings of the Second International Weed Control Congress*, held in Copenhagen in June 1996, ed. H. Brown, G. W. Cussans, M. D. Devine, *et al.* Slagelse, Denmark: Department of Weed Control and Pesticide Ecology, pp. 639–644.

Heard, T. A. and Winterton, S. L. (2000). Interactions between nutrient status and weevil herbivory in the biological control of water hyacinth. *Journal of Applied Ecology*, **37**, 117–127.

Hernández, M. C., Pildain, M. B., Novas, M. V., Sacco, J. and Lopez, S. E. (2007). Mycobiota associated with larval mines of *Thrypticus truncatus* and *T. sagittatus* (Diptera, Dolichopodidae) on waterhyacinth, *Eichhornia crassipes*, in Argentina. *Biological Control*, **41**, 321–326.

Hill, M. P. (2003). The impact and control of alien aquatic vegetation in South African aquatic ecosystems. *African Journal of Aquatic Science*, **28**, 19–24.

Hill, M. P. and Cilliers, C. J. (1999). A review of the arthropod natural enemies, and factors that influence their efficacy, in the biological control of water hyacinth, *Eichhornia crassipes* (Mart.) Solms-Laubach (Pontederiaceae), in South Africa. In *Biological Control of Weeds in South Africa (1990–1998)*, ed. T. Olckers and M. P. Hill. African Entomology Memoir 1. Hatfield, South Africa: Entomological Society of Southern Africa, pp. 103–112.

Hill, M. P. and Julien, M. H. (2004). The transfer of appropriate technology: key to the successful biological control of five aquatic weeds in Africa. In *Proceedings of the XI International Symposium on Biological Control of Weeds*, ed. J. M. Cullen, D. T. Briese, D. J. Kriticos, *et al.* Canberra, Australia: CSIRO Entomology, pp. 370–374.

Hill, M. P. and Oberholzer, I. G. (2000). Host specificity of the grasshopper, *Cornops aquaticum*, a natural enemy of water hyacinth. In *Proceedings of the X International Symposium on the Biological Control of Weeds*, held 4–14 July 1999, Montana State University, Bozeman, ed. N. R. Spencer. Sidney, MT: USDA-ARS, pp. 349–356.

Hill, M. P. and Olckers, T. (2001). Biological control initiatives against water hyacinth in South Africa: constraining factors, success and new courses of action. In *Proceedings of the Second Meeting of the Global Working Group for the Biological and Integrated Control of Water Hyacinth*, held in Beijing, China, 9–12 October 2000, ed. M. H. Julien, M. P. Hill, T. D. Center and J. Ding. Canberra, Australia: Australian Centre for International Agricultural Research, pp. 33–38.

Hill, M. P., Cilliers, C. J. and Neser, S. (1999). Life history and laboratory host range of *Eccritotarsus catarinensis* (Carvalho) (Heteroptera: Miridae), a new potential natural enemy released on water hyacinth (*Eichhornia crassipes* (Mart.) Solms-Laub.) (Pontederiaceae) in South Africa. *Biological Control*, **14**, 127–133.

Jiménez, M. M. and Balandra, A. G. (2007). Integrated control of *Eichhornia crassipes* by using insects and plant pathogens in Mexico. *Crop Protection*, **26**, 1234–1238.

Julien, M. H. (2001). Biological control of water hyacinth with arthropods: a review to 2000. In *Proceedings of the Second Meeting of the Global Working Group for the Biological and Integrated Control of Water, Hyacinth* held in Beijing, China, 9–12 October 2000, ed. M. H. Julien, M. P. Hill, T. D Center and J. Ding. Canberra, Australia: Australian Centre for International Agricultural Research, pp. 8–20.

Julien, M. H. and Griffiths, M. W. (1998). *Biological Control of Weeds: A World Catalogue of Agents and their Target Weeds*. Fourth edn. Wallingford, UK: CABI Publishing.

Julien, M. H. and Orapa, W. (1999). Structure and management of a successful biological control project for water hyacinth. In *Proceedings of the First Meeting of the IOBC Global Working Group for the Biological and Integrated Control of Water Hyacinth*, held 16–19 November 1998, Zimbabwe, ed. M. P. Hill, M. H., Julien and T. D Center. Pretoria, South Africa: Plant Protection Research Institute, pp. 123–134.

Julien, M. H. and Orapa, W. (2001). Insects used for biological control of the aquatic weed water hyacinth in Papua New Guinea. *Papua New Guinea Journal of Agriculture, Forestry and Fisheries*, **44**, 49–60.

Julien, M. H. and Stanley, J. (1999). Recent research on biological control for water hyacinth in Australia. In *Proceedings of the First Meeting of the IOBC Global Working Group for the Biological and Integrated Control of Water Hyacinth*, held 16–19 November 1998, Zimbabwe, eds. M. P. Hill, M. H., Julien and T. D. Center. Pretoria, South Africa: Plant Protection Research Institute, pp. 52–61.

Julien, M. H., Griffiths, M. W. and Wright, A. D. (1999). *Biological Control of Water Hyacinth. The Weevils* Neochetina bruchi *and N.* eichhorniae: *Biologies, Host Ranges, and Rearing, Releasing and Monitoring Techniques for Biological Control of* Eichhornia crassipes. Monograph 60. Canberra, Australia: Australian Centre for International Agricultural Research (ACIAR), 87 pp.

Julien, M. H., Griffiths, M. W. and Stanley, J. N. (2001). *Biological Control of Water Hyacinth. The Moths* Niphograpta albiguttalis *and* Xubida infusellus: *Biologies, Host Ranges, and Rearing, Releasing and Monitoring Techniques for Biological Control of* Eichhornia crassipes. Monograph 79. Canberra, Australia: Australian Centre for International Agricultural Research (ACIAR), 79 pp.

Julien, M. H., Harley, K. L. S., Wright, A. D., *et al.* (1996). International co-operation and linkages in the management of water hyacinth with emphasis on biological control. In *Proceedings of the IX International Symposium on Biological Control of Weeds*, held 21–26 January 1996, Stellenbosch, South Africa, ed. V. C. Moran and J. H. Hoffman. Rondesbosch South Africa: University of Cape Town, pp. 273–282.

Kathiresan, R. M. (2000). Allelopathic potential of native plants against water hyacinth. *Crop Protection*, **19**, 705–708.

Lindsey, K. and Hirt, H. M. (1999). Use Water Hyacinth! A Practical Handbook of Uses for Water Hyacinth from Across the World. Winnenden, Germany: Anamed.

Mailu, A. M. (2001). Preliminary assessment of the social, economic and environmental impacts of water hyacinth in the Lake Victoria basin and the status of control. In *Proceedings of the Second Meeting of the Global Working Group for the Biological and Integrated Control of Water Hyacinth*, held in Beijing, China, 9–12 October 2000, ed. M. H. Julien, M. P. Hill, T. D Center and J. Ding. Canberra, Australia: Australian Centre for International Agricultural Research, pp. 130–139.

Martyn, R. D. (1985). Waterhyacinth decline in Texas caused by *Cercospora piaropi*. *Journal of Aquatic Plant Management*, **23**, 29–32.

Martyn, R. D. and Freeman, T. E. (1978). Evaluation of *Acremonium zonatum* as a potential biocontrol agent of waterhyacinth. *Plant Disease Reporter*, **62**, 604–608.

Masifwa, W. F., Twongo, T. and Denny, P. (2001). The impact of water hyacinth, *Eichhornia crassipes* (Mart) Solms on the abundance and diversity of aquatic macroinvertebrates along the shores of northern Lake Victoria, Uganda. *Hydrobiologia*, **452**, 79–88.

Matthews, L. J. (1967). Seedling establishment of water hyacinth. *PANS(C)*, **13**, 7–8.

Midgley, J. M., Hill, M. P. and Villet, M. H. (2006). The effect of water hyacinth, *Eichhornia crassipes* (Martius) Solms-Laubach (Pontederiaceae), on benthic biodiversity in two impoundments on the New Year's River, South Africa. *African Journal of Aquatic Science*, **31**, 25–30.

Moran, P.J. (2005). Leaf scarring by the weevils *Neochetina eichhorniae* and *N. bruchi* enhances infection by the fungus *Cercospora piaropi* on waterhyacinth, *Eichhornia crassipes*. *BioControl*, **50**, 511–524.

Morris, M.J. (1990). *Cercospora piaropi* recorded on the aquatic weed, *Eichhornia crassipes*, in South Africa. *Phytophylatica*, **22**, 255–256.

Muramoto, S. and Oki, Y. (1983). Removal of some heavy metals from polluted water by water hyacinth (*Eichhornia crassipes*). *Bulletin of Environmental Contamination and Toxicology*, **30**, 170–177.

Navarro, L. and Phiri, G. (2000). *Water Hyacinth in Africa and the Middle East. A Survey of Problems and Solutions*. Ottawa Canada: International Development Research Centre.

Njoka, S.W., Ochiel, G.R.S., Manyalar, J.O. and Okeyo-Owuor, J.B. (2006). The life history and survival of *Neochetina* in Lake Victoria basin: basis for biological weed control. In *Proceedings of the 11th World Lakes Conference,* Vol. 2, ed. E. Odada, D.O. Olago, W. Ochola, *et al.* Nairobi, Kenya: Ministry of Water and Irrigation. pp. 593–599.

Oberholzer, I.G. and Hill, M.P. (2001). How safe is the grasshopper *Cornops aquaticum* for release on water hyacinth in South Africa? In *Proceedings of the Second Meeting of the Global Working Group for the Biological and Integrated Control of Water Hyacinth*, held in Beijing, China, 9–12 October 2000, ed. M.H. Julien, M.P. Hill, T.D. Center and J. Ding. Canberra, Australia: Australian Centre for International Agricultural Research, pp. 82–88.

Ogwang, J.A. and Molo, R. (1999). Impact studies on *Neochetina bruchi* and *Neochetina eichhorniae* in Lake Kyoga, Uganda. In *Proceedings of the First Meeting of the IOBC Global Working Group for the Biological and Integrated Control of Water Hyacinth*, held 16–19 November 1998, Zimbabwe, ed. M.P. Hill, M.H. Julien and T.D. Center. Pretoria, South Africa: Plant Protection Research Institute, pp. 10–13.

Owens, C.S. and Madsen, J.D. (1995). Low temperature limits of waterhyacinth. *Journal of Aquatic Plant Management*, **33**, 63–68.

Page, A.R. and Lacey, K.L. (2006). Economic impact assessment of Australian weed biological control. *CRC for Australian Weed Management Technical Series*, **10**, 123–127.

Penfound, W.M.T. and Earle, T.T. (1948). The biology of the waterhyacinth. *Ecological Monographs*, **18**, 448–473.

Perkins, B.D. (1974). Arthropods that stress water hyacinth. *PANS* **20**, 304–314.

Phiri, P.M., Day, R.K., Chimatiro, S., *et al.* (2001). Progress with biological control of water hyacinth in Malawi. In *Proceedings of the Second Meeting of the Global Working Group for the Biological and Integrated Control of Water Hyacinth*, held in Beijing, China, 9–12 October 2000, ed. M.H. Julien, M.P. Hill, T.D. Center and J. Ding. Canberra, Australia: Australian Centre for International Agricultural Research, pp. 47–52.

Reddy, K.R., Agami, M. and Tucker, J.C. (1989). Influence of nitrogen supply rates on growth and nutrient storage by water hyacinth (*Eichhornia crassipes*) plants. *Aquatic Botany*, **36**, 33–43.

Reddy, K.R., Agami, M. and Tucker, J.C. (1990). Influence of phosphorus supply on growth and nutrient storage by water hyacinth (*Eichhornia crassipes*) plants. *Aquatic Botany*, **37**, 355–365.

Shabana, Y. M., Baka, Z. A. M., and Abdel-Fattah, G. M. (1997). *Alternaria eichhorniae*, a biological control agent for waterhyacinth: mycoherbicidal formulation and physiological and ultrastructural host responses. *European Journal of Plant Pathology*, **103**, 99–111.

Shabana, Y. M., Charudattan, R., and Elwakil, M. A. (1995). Identification, pathogenicity and safety of *Alternaria eichhorniae* from Egypt as a bioherbicide agent for waterhyacinth. *Biological Control*, **5**, 123–135.

Shabana, Y. M., Elwakil, M. A. and Charudattan, R. (2001). Biological control of water hyacinth by a mycoherbicide in Egypt. In *Biological and Integrated Control of Waterhyacinth,* Eichhornia crassipes: *Proceedings of the Second Meeting of the Global Working Group for the Biological and Integrated Control of Water Hyacinth* held in Beijing, China, 9–12 October 2000, ed. M. H. Julien, M. P. Hill, T. D. Center and J. Ding. Canberra, Australia: Australian Centre for International Agricultural Research, pp. 53–56.

Silveira Guido, A. and Perkins, B. D. (1975). Biology and host specificity of *Cornops aquaticum* (Bruner) (Orthoptera: Acrididae), a potential biological control agent for water hyacinth. *Environmental Entomology*, **4**, 400–404.

Sosa, A. J., Cordo, H. A. and Sacco, J. (2007). Preliminary evaluation of *Megamelus scutellaris* Berg (Hemiptera: Delphacidae), a candidate for biological control of waterhyacinth. *Biological Control*, **42**, 129–138.

Stanley, J. N. and Julien, M. H. (1999). The host range of *Eccritotarsus catarinensis* (Heteroptera: Miridae), a potential agent for the biological control of water hyacinth (*Eichhornia crassipes*). *Biological Control*, **14**, 134–140.

Stanley, J. N., Julien, M. H. and Center, T. D. (2007). Performance and impact of the biological control agent *Xubida infusella* (Lepidoptera; Pyralidae) on the target weed *Eichhornia crassipes* (waterhyacinth) and on a non-target plant, *Pontederia cordata* (pickerelweed) in two nutrient regimes. *Biological Control*, **40**, 298–305.

Strange, A., Rudall, P. J. and Prychid, C. J. (2004). Comparative floral anatomy of Pontederiaceae. *Botanical Journal of the Linnean Society*, **144**, 395–408.

Tessman, D. J., Charudattan, R., Kistler, H. C. and Rosskopf, E. N. (2001). A molecular characterization of *Cercospora* species pathogenic to water hyacinth and emendation of *C. piaropi*. *Mycologia*, **93**, 323–334.

Toft, J. D., Simenstad, C. A., Cordell, J. R. and Grimaldo, L. F. (2003). The effects of introduced water hyacinth on habitat structure, invertebrate assemblages, and fish diets. *Estuaries*, **26**, 746–758.

Van, T. K. and Center, T. D. (1994). Effect of paclobutrazol and waterhyacinth weevil (*Neochetina eichhorniae*) on plant growth and leaf dynamics of waterhyacinth (*Eichhornia crassipes*). *Weed Science*, **42**, 665–672.

Van, T. K., Wheeler, G. S. and Center, T. D. (1998). Competitive interactions between Hydrilla (*Hydrilla verticillata*) and Vallisneria (*Vallisneria americana*) as influenced by insect herbivory. *Biological Control*, **11**, 185–192.

van Thielen, R., Ajuonu, O., Schade, V., *et al.* (1994). Importation, release, and establishment of *Neochetina* spp. (Curculionidae) for the biological control of water hyacinth, *Eichhornia crassipes* (Lil.: Pontederiaceae), in Benin, West Africa. *Entomophaga*, **39**, 179–188.

van Wyk, E. and van Wilgen, B. W. (2002). The cost of water hyacinth control in South Africa: a case study of three options. *African Journal of Aquatic Science*, **27**, 141–149.

Weed Science Society of America (WSSA) (1984). Composite list of weeds. *Weed Science*, **32** (Suppl. 2) 1–137.

Wilson, J. R. U., Ajuonu, O., Center, T. D., *et al.* (2007). The decline of water hyacinth on Lake Victoria was due to biological control by *Neochetina* spp. *Aquatic Botany*, **87**, 90–93.

Wilson, J. R. U., Rees, M. and Ajuonu, O. (2006). Population regulation of a classical biological control agent: larval density dependence in *Neochetina eichhorniae* (Coleoptera: Curculionidae), a biological control agent of water hyacinth *Eichhornia crassipes*. *Bulletin of Entomological Research*, **96**, 145–152.

Wright, A. D. and Purcell, M. F. (1995). *Eichhornia crassipes* (Mart.) Solms-Laubach. In *The Biology of Australian Weeds,* Vol. 1, ed. R. H. Groves, and R. C. H. Shepherd. Melbourne, Australia: R. G. and F. J. Richardson, pp. 111–121.

Xie, Y. and Yu, D. (2003). The significance of lateral roots in phosphorus (P) acquisition of water hyacinth (*Eichhornia crassipes*). *Aquatic Botany*, **75**, 311–321.

Zhu, Y. L., Zayed, A. M., Qian, J. H., Souza, M. and Terry, N. (1999). Phytoaccumulation of trace elements by wetland plants. II. Water hyacinth. *Journal of Environmental Quality*, **28**, 339–344.

12

Lantana camara Linn. (Verbenaceae)

Michael D. Day and Myron P. Zalucki

12.1 Introduction

Lantana camara Linn. (Verbenaceae) (lantana) is a pantropical weed, affecting pastures, orchards, and native forests in about 70 countries worldwide (Day *et al.*, 2003b). *Lantana camara* (*sensu lato*) is a composite species and is thought to have originated from two or more lantana species from tropical America. Dutch explorers introduced the plant into the Netherlands in the 1600s from Brazil (Stirton, 1977). It was then hybridized in glasshouses in Europe prior to its introduction to other countries as an ornamental.

Lantana camara can grow as individual clumps or dense thickets, displacing desirable species. In disturbed natural forests, it can form the dominant understory, disrupting succession and decreasing biodiversity. Its allelopathic qualities can reduce vigor of plant species nearby and reduce productivity in orchards (Holm *et al.*, 1991). *Lantana camara* outcompetes native pastures, interferes with the mustering of cattle, and causes death of stock by poisoning (Swarbrick *et al.*, 1998). In Southeast Asia and the Pacific Island communities, it can reduce productivity in orchards and plantations and interferes with harvesting. It flowers prolifically and the seeds are dispersed by birds (Swarbrick *et al.*, 1998). *Lantana camara* has several uses, mainly as herbal medicines and in some areas as firewood and mulch (Sharma *et al.*, 1988; Sharma and Sharma, 1989).

Lantana camara can be controlled through the use of chemicals, mechanical removal, fire, and planting of competitive species. However, in many situations these methods are not feasible. *Lantana camara* growing on steep hillsides or along creeks is often inaccessible for either chemical application or mechanical removal, and fire is not an option in native forests or orchards and plantations. Therefore, biological control is seen as the only viable and sustainable long-term solution to managing *L. camara* (Day *et al.*, 2003b).

Efforts directed at biocontrol of *L. camara* started in 1902, and since then 41 agents have been released in 44 countries (Day *et al.*, 2003b). Despite intense efforts in several countries, biocontrol of *L. camara* has been only partially successful and the weed is rarely controlled totally anywhere within its introduced geographical range. Several factors appear to influence the success of biocontrol of *L. camara*, including intraspecific

Biological Control of Tropical Weeds using Arthropods, ed. R. Muniappan, G. V. P. Reddy, and A. Raman. Published by Cambridge University Press. © Cambridge University Press, 2009.

variation, climate, and biology and ecology of both the plant and agents (Broughton, 2000a; Day and Neser, 2000; Day *et al.*, 2003b; Zalucki *et al.*, 2007). This chapter reviews the biocontrol programs and assesses the effectiveness of the agents.

12.2 Taxonomy

Lantana Linn. belongs to the family Verbenaceae, within the order Lamiales and includes up to 150 species (Gujral and Vasudevan, 1983), but it has long been the subject of taxonomic uncertainty. It is separated into four sections: *Sarcolippia* and *Rhytocamara* (with a few species each); *Calliorheas*, more diverse and widespread than the first two sections and includes *L. montevidensis* – a weed in some countries, which has naturalized in Australia, Africa, and parts of India (Day *et al.*, 2003b); and *Camara*, consisting of three complexes based on *L. urticifolia*, *L. hirsuta* and *L. camara*. The *L. camara* complex in *Camara* section includes the weedy lantana, referred to as *L. camara* Linn. (*sensu lato*).

Lantana was first introduced into Europe in 1636 (Stirton, 1977) and was popularly cultivated in the second half of the nineteenth century (Swarbrick *et al.*, 1998). The material grown in Europe included in its lineage strains of several American taxa, and modern varieties were developed through long periods of hybridization and selection (Stirton, 1977). Following importation to other countries, these have subsequently become naturalized and continue to hybridize in the field (Cilliers and Neser, 1991) thus resulting in a highly variable polyploid and complex species. The resulting taxa have been variously referred to as either distinct species (White, 1929), forms (Parsons and Cuthbertson, 2001), cultivars (Howard, 1969), biotypes (Swarbrick, 1986), subspecies, or varieties (CSIR, 1962). Worldwide, more than 650 variety names exist and these "taxa" differ in flower color, spininess, leaf shape, toxicity, susceptibility to herbivore attack and ecology (Diatloff and Haseler, 1965; Howard, 1969; Smith and Smith, 1982).

It is widely recognized that the weedy form of *L. camara* is morphologically distinct in different regions of its naturalized range compared with *Lantana* spp. in its native range (Sanders, 2006) and this has critical implications in the search and collection of potential biological control agents. Some preliminary DNA studies have been conducted which suggests that weedy lantana in Australia at least, is most closely related to *L. urticifolia* in Mexico (Scott *et al.*, 2002). However, the relatedness of lantana elsewhere has not been conducted.

In this chapter, we address only the weedy "taxa" of *Lantana* [section *Camara*] that are the most widespread and, economically and environmentally, important and will refer to the various types of weedy lantana as varieties.

12.3 Distribution of *Lantana camara*

Lantana (section *Camara*) is native only to the Americas, with a distribution from Florida and Texas in the north to northern Argentina and Uruguay in the south (Fig. 12.1). *Lantana camara* (*s. s.*) occurs from Mexico, through Bahamas, Greater Antilles, and northwest South America (Sanders, 2006). The weedy taxa of lantana naturalized in the

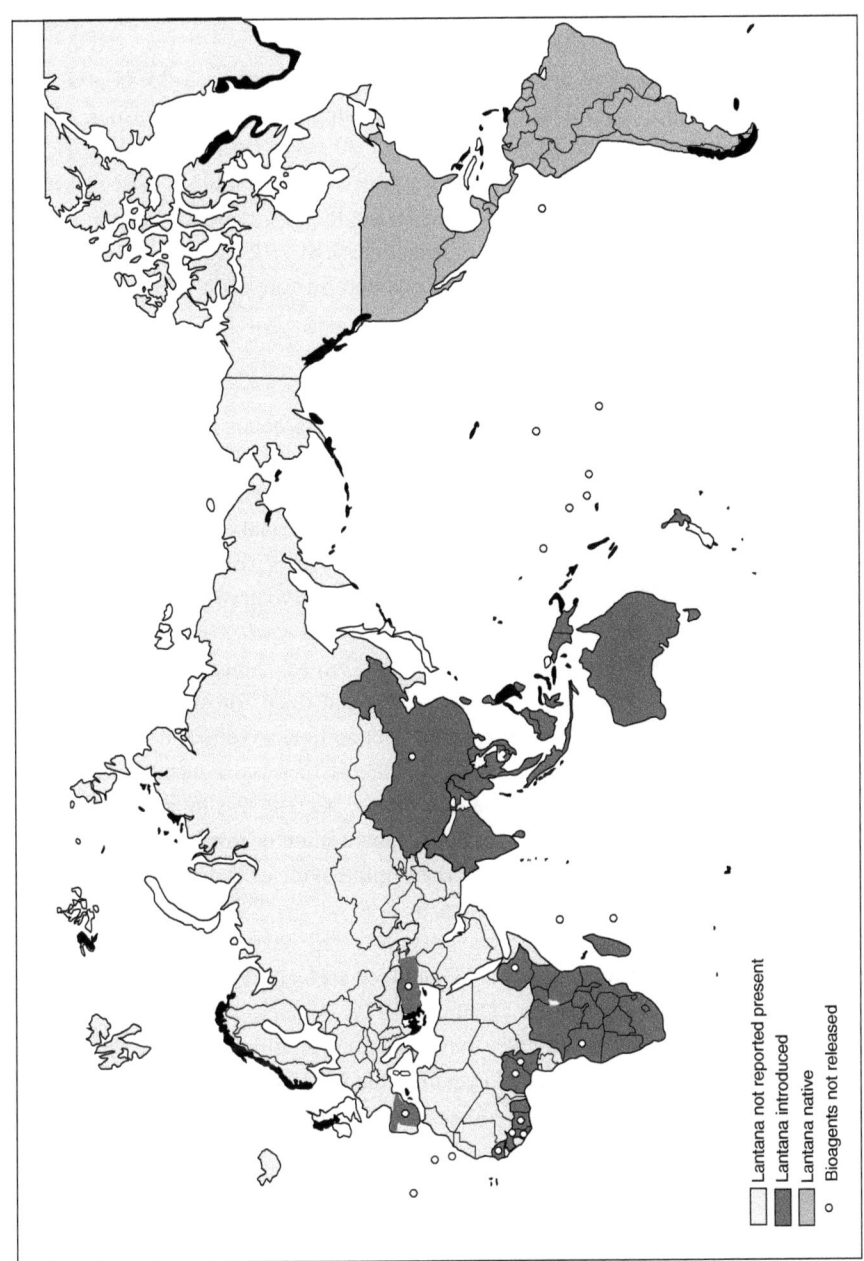

Fig. 12.1 Countries and/or island groups where the taxa of *Lantana* section *Camara* are native (light gray) and introduced or naturalized (dark gray). The map indicates the presence of lantana in a country and not its distribution within that country. Open circles indicate countries that are reported not to have any biological control agents.

Old World, being of hybrid origin, do not have a "native" range *per se*. The hybrids are almost certainly from various species within the section *Camara* and it is more likely that the various weedy varieties are derived from multiple parental species, so that different varieties may have progenitors with different geographic ranges.

Lantana is now naturalized in approximately 70 countries between 35° N and 35° S (Day *et al.*, 2003b) (Fig. 12.1). The distribution of lantana is still increasing, with it infesting many new countries and islands in the past 30 years (Waterhouse and Norris, 1987; Denton *et al.*, 1991; Harley, 1992). Even in areas such as South Africa and India, where lantana has been established since the mid 1800s, there is evidence that the weed is spreading, most likely facilitated by logging practices and other forms of habitat disturbance (Stirton, 1977; Sharma *et al.*, 1988; Wells and Stirton, 1988).

12.4 Habitats

Lantana tolerates a wide range of ecological and climatic factors, occurring in diverse habitats and on a variety of soil types. It generally grows best in open unshaded situations such as degraded lands and rain forest edges, and forests recovering from either fire or logging. Disturbed areas such as roadsides (edges), railway tracks and canals are also favorable for lantana (Thaman, 1974; Winder and Harley, 1983). Lantana grows well on rich volcanic soils (Humphries and Stanton, 1992) and benefits from the destructive grazing activities of pigs, cattle, goats, horses, sheep, and deer (Thaman, 1974; Denton *et al.*, 1991; Fensham *et al.*, 1994). It can grow at altitudes from sea level up to 2000 m asl (Matthew, 1971). Lantana tolerates modest shade and grows well in plantations and open forests (Humphries and Stanton, 1992), but it does not flower readily in these conditions (Wells and Stirton, 1988). Lantana infestations can expand into marginal habitats if there is reduced herbivory, as original habitat restrictions, such as climate and soil type, become less important.

Lantana does not appear to have either an upper temperature or rainfall limit. It usually occurs in tropical areas receiving 3000 mm/year rainfall with well-drained soils; infestations are usually restricted to riparian zones in drier areas (Swarbrick *et al.*, 1998). Lantana seldom occurs where temperatures frequently fall below 5 °C, although some varieties withstand minor, infrequent frosts. Prolonged freezing temperatures kill aerial woody branches and induce defoliation (Thaman, 1974; Winder, 1980; Graaff, 1986).

Lantana varieties and climatic tolerance have some correlation; only the "pink flowering" varieties occur at higher altitudes (>500 m in Australia; >1800 m in southern India), whereas the "orange" and "red flowering" varieties occur at lower altitudes (Matthew, 1971; Day *et al.*, 2003a).

12.5 Ecology

Lantana can flower as early as the second growing (summer) season. Plants flower year-round in most places if adequate moisture, temperature, and light are available (Gujral and Vasudevan, 1983; Graaff, 1986), with flowering peaking during the wet, summer months. In cooler or drier regions, flowering occurs only in the warmer or wetter months (Winder, 1980; Swarbrick *et al.*, 1998).

Lantana is pollinated by thrips (Mohan Ram and Mathur, 1984), Lepidoptera (Kugler, 1980; Hilje, 1985), and to a lesser extent, by sunbirds and hummingbirds (Winder, 1980). There are conflicting reports on lantana's ability to self-pollinate. Mohan Ram and Mathur (1984) considered lantana to be self-compatible, but insects were necessary for pollination. Neal (1999) found individual lantana flowers were capable of self-pollination. However, in laboratory experiments Barrows (1976) found that lantana flowers did not self-pollinate. Pollination results in 85% fruit set (Hilje, 1985) with each infructescence bearing about eight fruits (Barrows, 1976).

Seeds are widely dispersed, usually by birds but also mammals such as sheep, goats, cattle, foxes, and jackals (Sharma *et al.*, 1988; Wells and Stirton, 1988; Swarbrick *et al.*, 1998). On larger landmasses, indigenous bird species feed on the fruits, aiding in dispersal, whereas on smaller islands, exotic bird species such as the Chinese turtledove, *Streptopelia chinensis*, and the Indian mynah, *Acridotheres tristis*, aid in seed dispersal.

Seeds require intense light for germination and early growth (Duggin and Gentle, 1998), so seedlings are less likely to survive beneath parent bushes. Germination rate of lantana is 20–49% under both laboratory and field conditions (Duggin and Gentle, 1998; Swarbrick *et al.*, 1998). However, these rates increased when the fleshy seed pulp was manually removed. This higher germination rate is comparable to the seeds extracted from the feces of birds. Seeds germinate at any time of the year provided sufficient soil moisture (Parsons and Cuthbertson, 2001) is available. Low germination rates are offset by extremely low rates of seedling mortality (Sahu and Panda, 1998) and lantana's capabilities for vegetative propagation and high seed production. Prostrate stems can strike root at nodes when covered by moist soil, fallen leaves, or other debris and grow into flowering shoots (Neal, 1999).

Species of *Lantana* in tropical America generally occur in small clumps (1 m in diameter) and while they are common along roadsides and in open fields, they are not considered a weed (Palmer and Pullen, 1995). Within its naturalized range, lantana often forms dense monospecific thickets 1–4 m in height (Winder and Harley, 1983; Swarbrick *et al.*, 1998), while some varieties may grow up and over trees and reach heights of 8–15 m (Smith and Smith, 1982; Swarbrick *et al.*, 1998). In tropical areas, growth is continuous throughout the year, while in cooler climates, plants cease growing and undergo varying levels of defoliation in dry winter months (Winder, 1980). Rapid growth occurs in spring and early summer following rains.

While lantana infestations usually increase in wetter years, they do not recede during dry years. Mortality rate of mature lantana plants in their naturalized range is low (Sahu and Panda, 1998). In many regions, lantana is defoliated annually, by the complex of introduced biocontrol agents or during times of drought. Plants recover when the insect numbers have waned during the winter months and when early season rains commence (Greathead, 1971; Gupta and Pawar, 1984; Muniappan and Viraktamath, 1986; Baars and Neser, 1999; Day *et al.*, 2003a, b).

12.6 Impact of lantana

Under conditions of intense light, high soil moisture and soil nutrient regimens, lantana is an effective competitor against native colonizers. It has the potential to block succession and displace native species, inducing a reduction in biodiversity (Lamb, 1991; Loyn and French, 1991; Gentle and Duggin, 1998). One possible explanation for this reduction in biodiversity is that the allelopathic effects of lantana reduce seedling recruitment of almost all species under lantana and cause reduction in the girth growth of mature trees and shrubs (Gentle and Duggin, 1997). Allelopathy may explain why many invasive weeds such as lantana can survive secondary succession and become monospecific thickets (Hardin, 1960).

Lantana does not invade intact rain forests, but has been found along their margins. However, where forests have been disturbed because of logging, storms, or natural deaths, gaps are created, enabling lantana to encroach and establish. Further logging aggravates the condition and allows the lantana to spread or become thicker (Humphries and Stanton, 1992).

Lantana can alter fire regimes in natural areas by increasing fuel loads, producing fires intense enough to penetrate into the surrounding rain forest (Humphries and Stanton, 1992). Fuel loads provided by lantana have been implicated in destructive wildfires in northern Queensland rain forest margins (Fensham *et al.*, 1994).

Lantana is a major problem in most agricultural areas wherever it occurs. Once established in pastures, it forms large, impenetrable thickets, outcompeting valuable pasture species, blocking the movement of domestic stock to waterholes, poisoning stock, and interfering with mustering. In Australia, lantana costs the grazing industry alone Au$104 million per annum in terms of lost productivity and management expenses (AEC group, 2007).

Lantana has been implicated in the poisoning and death of a range of animals including cattle, buffalo, sheep, goats, horses, dogs, guinea pigs, and captive red kangaroos in numerous countries such as Australia, Brazil, Cuba, Fiji, Kenya, India, and Mexico (Day *et al.*, 2003b). The field cases of poisoning occur principally in young animals that have either been newly introduced into an area where lantana grows, or are without access to other fodder. Poisoning results in cholestasis, hepatotoxicity, and photosensitization: the early clinical signs being anorexia and severe constipation (Sharma, 1994).

In addition to its impact on grazing lands, lantana often causes a reduction in yield or impedes harvesting in plantations and perennial crops. It is a problem in coconut plantations in the Philippines (Cock and Godfray, 1985) and Fiji (Kamath, 1979), the Solomon Islands and Vanuatu (Harley, 1992); oil palms and rubber in Malaysia (A. A. Ismail, MARDI, personal communication); bananas in Australia and Samoa; copra in Vanuatu (Harley, 1992); citrus in Florida (Habeck, 1976); tea in India and Indonesia (Holm *et al.*, 1991); and timber plantations in Australia (Swarbrick *et al.*, 1998), South Africa (Graaff, 1986), Fiji (S. N. Lal, SPC, personal communication), Indonesia (CSIR, 1962), and India (Holm *et al.*, 1991).

Lantana causes a number of secondary impacts, especially in many tropical countries, where it shelters several serious pest arthropods, such as malaria-spreading mosquitoes in India (Gujral and Vasudevan, 1983) and tsetse flies in Rwanda, Tanzania, Uganda, and Kenya (Greathead, 1968; Katabazi 1983; Mbulamberi, 1990).

12.7 Uses of lantana

Lantana was originally introduced into most countries as a garden ornamental although in some countries it is planted as a hedge to keep out livestock (Ghisalberti, 2000). Today, lantana is seen as a pest in most countries in which it has naturalized. In spite of its pest status, it has several minor uses, mainly in herbal medicine. Extracts from the leaves exhibit antimicrobial, fungicidal, insecticidal, and nematicidal activity and have been used in folk medicine for the treatment of ulcers, tetanus, and malaria (CSIR, 1962; Sharma and Sharma, 1989; Ghisalberti, 2000).

The stems of lantana can be used to produce pulp for paper suitable for writing and printing (Gujral and Vasudevan, 1983), and fuel for cooking and heating (Sharma *et al.*, 1988), while the roots of lantana contain a substance that may be used in the rubber industry (Gujral and Vasudevan, 1983). Mixed with cattle dung, lantana has been used for biogas production, and the seeds have supplementary nutritive value when fed with wheat straw to sheep (Sharma *et al.*, 1988).

Although not planted for such uses, lantana has been used as a cover crop in deforested areas, helping to enrich the soil, increase nitrogen uptake in rice, and protect against erosion and surface cracking (CSIR, 1962; Greathead, 1968; Ghisalberti, 2000).

12.8 Biological control of lantana

In many countries, the infestations of lantana are either too large, or the land values too low, rendering conventional control of lantana with chemicals, machinery, or fire uneconomical. Therefore, biocontrol appears as the only viable, long-term solution to the management of this weed.

The first attempt at the biocontrol of lantana began in 1902, when 23 insect species were imported into Hawaii from Mexico. The "moderate" success experienced in Hawaii encouraged other countries to not only import insects that seemed safe and effective in Hawaii, but also to conduct explorations for new agents. Forty-one agents are now either deliberately or unintentionally released on lantana throughout the world and 27 of these have established in at least one country or island. Their country of origin, guild, host specificity, and status are listed in Table 12.1.

Only about 10 species contribute to any control of lantana, with the majority being of little or no assistance. Details of the most important agents are given below. The actual number of agents intentionally or accidentally introduced and their status in each country may vary from that presented in Table 12.1, as accurate and recent surveys have not been conducted for many countries or islands.

Table 12.1 *A list of countries and/or island groups where lantana is naturalized and the biological control agents introduced and their status in each country.*

Species	Family	Country	Feeding guild	Specificity	Ascension & St Helena Is. (UK)	Australia	Cape Verde Is.	Cook Is.	Federated States of Micronesia	Fiji	Ghana
A. compressa	Membracidae	Mexico	stem sucker	5		+					
A. championi	Cerambycidae	Mexico	stem borer	5		–					
A. parana	Chrysomelidae	Brazil	leaf feeder	1		–					
Apion sp. A	Apionidae	Mexico	flower feeder	1							
Apion sp B	Apionidae	Mexico	seed feeder	1							
A. illustrata	Noctuidae	Colombia	leaf feeder	2		–					
C. lantanae	Agromyzidae	Trinidad	leaf miner	3		+				+	
C. pygmaea	Chrysomelidae	Brazil	leaf feeder	2		–					–
C. lantanella	Gracillariidae	Mexico	leaf miner	1							
D. tigris	Noctuidae	Panama	leaf feeder	1	–	–			–	–	
E. garcia	Depressariidae	Brazil	leaf feeder	2		–					
E. lantana	Tortricidae	Mexico	flower feeder	1		+		+			
E. xanthochaeta	Tephritidae	Mexico	stem galler	3		–					
F. intermedia	Miridae	Jamaica	sap sucker	3		+					
H. laceratalis	Noctuidae	Kenya	leaf feeder	2	+	+	+		+	+	+
L. pusillidactyla	Pterophoridae	Mexico	flower feeder	3		+		+			
L. decora	Tingidae	Colombia, Peru	sap sucker	5		+		?	–	–	
M. lantanae	Hyphomycetes	Brazil	pathogen	1							
N. sunia	Noctuidae	USA	leaf feeder	1		+		–			
O. championi	Chrysomelidae	Costa Rica	leaf miner	2		+			–		
O. scabripennis	Chrysomelidae	Mexico	leaf miner	1		+		–		–	+
O. camarae	Agromyzidae	Florida	leaf miner	3		?					
O. lantanae	Agromyzidae	Mexico	seed feeder	1		+		?	?	+	+
O. insignis	Ortheziidae	Mexico	sap sucker	5	+		+				
P. xanthomelas	Cerambycidae	Mexico	stem borer	1							
P. parvus	Pseudococcidae	unknown	sap sucker	5		+		+			
P. spinipennis	Cerambycidae	Mexico	stem borer	1		–					
P. tuberculatum	Pucciniaceae	Brazil	pathogen	1		+					
P. santatalis	Pyralidae	Mexico	leaf feeder	1				–	–		
S. haemorrhoidalis	Pyralidae	Cuba, USA	leaf feeder	1		+				+	+
Septoria sp.	Sphaeropsidaceae	Ecuador	leaf pathogen	1							
S. bazochii	Lycaenidae	Mexico	flower feeder	5		–				+	
T. bifasciata	Tingidae	Trinidad	flower feeder	1							
T. elata	Tingidae	Brazil	leaf & flower feeder	1		–		–			
T. harleyi	Tingidae	Trinidad	flower feeder	1		–					
T. prolixa	Tingidae	Brazil	flower feeder	1		–					
T. scrupulosa	Tingidae	Mexico	sap sucker	5	+	+			+	+	+
T. echion	Lycaenidae	Mexico	flower feeder	5		–					
U. fulvopustulata	Chrysomelidae	Costa Rica	leaf miner	1		+			–		
U. girardi	Chrysomelidae	Argentina, Brazil	leaf miner	1	+	+		+	+	+	+
U. lantanae	Chrysomelidae	Brazil	leaf miner	1		–					

Hawaii	Hong Kong	India	Indonesia	Kenya	Madagascar	Malaysia	Marshall Is.	Mauritius	Myanmar	New Caledonia	New Zealand	Niue	Northern Mariana Is.	Palau	Papua New Guinea	Philippines	Republic of South Africa	Samoa	Solomon Is.	Sri Lanka	Swaziland	Taiwan	Tanzania	Thailand	Tonga	Uganda	Vanuatu	Vietnam	Zambia	Zimbabwe
−																	−													
−																														
−																	−													
		+			+									+	+	+	+			+	+	+	+			+	+			
+																	+													
−	−										+											−		−			−			
+	+									+				+			+					+								
+																	−													
																	+													
+									+	+		+		+	+	+	+				+									
+	+	+								+	+			+	+	+	+					+								−
+																	−								?					−
+																	+													
+											+						−													
−																	?													
+		+									+	?					+	?	?											
																	+													
+	?	+	+	+	+					+	+			+	+	+	+			?	+		+			+	+	+		+
+		+															+			?										
−																														
+												−					−													
−		−	−								+						+							−		+				?
+																														
+																														
−																														
																	−							−			−			
+		+	+	+	+			+		+		?	+		+	+		+	+		+		+		+	+	+		+	−
+																														
+		+						+		+	+		+	+	+	+		+	+				−	+	+		+			+

12.8.1 **Teleonemia scrupulosa** *Stål (Hemiptera: Tingidae)*

This tingid is found throughout Mexico and Central and South America (Waterhouse and Norris, 1987). Adults and nymphs feed in colonies, primarily on the undersurface of leaves where they suck the cell contents (Khan, 1945). However, they may also feed on flowers and shoot meristems (Fyfe, 1937). Adult and nymphal feeding causes chlorotic and necrotic lesions, leaf curling, and defoliation (Gupta and Pawar, 1984; Waterhouse and Norris, 1987). The occurrence of additional damage to plant parts removed from the feeding site suggests that salivary toxins may have a systemic effect (Khan, 1945; Harley and Kassulke, 1971). Eggs are partially inserted into the midrib and main veins on the undersides of leaves (Fyfe, 1937). The life cycle is short, taking about four weeks in summer conditions (Gupta and Pawar, 1984; Waterhouse and Norris, 1987).

Teleonemia scrupulosa was introduced into Hawaii in 1902 (Swezey, 1923) and has since been released in 31 countries, establishing in 29 (Julien and Griffiths, 1998, Day *et al.*, 2003b) (Table 12.1). When tingid populations are large, defoliation readily occurs, and when insect attack is combined with other environmental stresses, such as drought, plants may be killed (Harley and Kassulke, 1971). The stress to the plant caused by leaf damage or defoliation reduces flower and seed production significantly (Harley, 1970; Rao *et al.*, 1971; Muniappan *et al.*, 1996). The insect is more common in warm, drier areas and has caused defoliation to lantana infestations around central and southern Queensland, with the most damaging populations occurring in midsummer to autumn (Day *et al.*, 2003a). However, populations of the tingid can undergo rapid crashes once plants have become defoliated, or with the onset of adverse weather such as frost or heavy rain (Khan, 1945; Harley *et al.*, 1979).

Teleonemia scrupulosa would be a useful introduction into regions where it is not present. However, as *T. scrupulosa* has been found on several non-target species, host-specificity studies need to be undertaken in the target country prior to its importation.

12.8.2 **Ophiomyia lantanae** *(Froggatt) (Diptera: Agromyzidae)*

This fly is found from southern Brazil to southern USA (Winder and Harley, 1983; Palmer and Pullen, 1995). Adults feed on nectar from flowers and oviposit in immature fruits, usually one egg/fruit. The larvae feed mainly on the endosperm and in the pericarp of the fruit (Swezey, 1924; Harley, 1971), but do not damage the embryo. Thus the seed may be weakened, but not killed (Waterhouse and Norris, 1987). *Ophiomyia lantanae* has a life cycle of about 21 days.

Ophiomyia lantanae was introduced into Hawaii in 1902 (Swezey, 1923) and has since been introduced or naturally spread into 28 countries, establishing in 24 (Julien and Griffiths, 1998, Day *et al.*, 2003b; Table 12.1). It is possible that it was accidentally introduced in the shipments of lantana plants sent to some of the countries where lantana has become a weed (Sen-Sarma and Mishra, 1986; Day *et al.*, 2003b).

In the naturalized range of lantana, *O. lantanae* is frequently reported to infest high proportions (50–95%) of fruit (Swezey, 1924; Muniappan and Viraktamath, 1986; Denton *et al.*, 1991). However, there is dispute over the ability of the fly to reduce seed viability. Experimental studies examining the germination rates of infested versus uninfested fruit have revealed mixed results (Swezey, 1924; Broughton, 1999). Swezey's (1924) reported that 51% of infested berries had the embryo damaged, whereas Broughton (1999) examined dissected fruit and found that no embryos were damaged by the fly.

Vivian-Smith *et al.* (2006) found that seedling emergence of affected fruits was dependent on lantana variety. Damage to the seeds of the pink-edged red flowering variety resulted in lower seedling emergence rates than undamaged seeds, while emergence of seeds from the pink-flowering variety increased with damage. Irrespective of whether the fly may or may not reduce seed viability, there is strong evidence to suggest that infested fruits are less likely to be consumed by seed-dispersing birds (Denton *et al.*, 1991; Vivian-Smith *et al.*, 2006). Therefore, seeds from fruit damaged by *O. lantanae* are less likely to be dispersed and the long-distance spread of the weed can be slowed (Taylor, 1989).

12.8.3 Uroplata girardi *Pic (Coleoptera: Chrysomelidae)*

This beetle is found in Brazil, Paraguay and Argentina (Krauss, 1964; Winder and Harley, 1982). Adults feed and oviposit on the upper leaf surface. The larvae mine the leaves of lantana, feeding on the mesophyll layers and leaving the upper and lower epidermal layers intact. Usually one or two mines occur per leaf, with one larva in each mine. The life cycle takes about 40 days and there are normally about three generations/season. Adults may enter a facultative diapause during winter when plants are dry (Bennett and Maraj, 1967; Harley, 1969b).

Uroplata girardi has been introduced into 26 countries, establishing in 24 (Julien and Griffiths, 1998; Day *et al.*, 2003b, Table 12.1). Populations of *U. girardi* were slow to build up in some places, such as Hawaii, Uganda, India, and Micronesia (Greathead, 1971; Sen-Sarma and Mishra, 1986; Denton *et al.*, 1991), while in Australia and the Solomon Islands the populations built up rapidly following their introduction (Harley, 1969b; Scott, 1998). *Uroplata girardi* can perform well on lantana growing in semishade (Waterhouse and Norris, 1987; Denton *et al.*, 1991) and under these conditions it is better able to control lantana, which is less vigorous than elsewhere (Kamath, 1979). Damage caused by *U. girardi*, as with other leaf-feeding insects released on lantana, is insufficient to kill lantana bushes. However, *U. girardi* can cause severe defoliation in plants on a seasonal basis, resulting in a reduction in flowering and seed production (Day *et al.*, 2003a).

12.8.4 Octotoma scabripennis *Guérin-Méneville (Coleoptera: Chrysomelidae)*

This beetle is found from Mexico through to Nicaragua, as well as in parts of the Caribbean. Adults feed and oviposit on the upper surface of leaves. Larvae mine leaves and induce blotches. Development of egg through to adult takes 34–45 days, with a pre-oviposition period of 3–4 weeks. Usually three generations/year occur. Adults avoid seasonally unfavorable conditions by entering a facultative diapause (Harley, 1969b).

Octotoma scabripennis was introduced into Hawaii in 1953 and is now present in six countries (Julien and Griffiths, 1998; Day *et al.*, 2003b; Table 12.1). It has a predominantly subtropical distribution, preferring shady, wetter coastal areas (Baars and Neser, 1999; Day *et al.*, 2003a). It is not as damaging in India and New Caledonia as in Australia and South Africa (Sen-Sarma and Mishra, 1986; Julien and Griffiths, 1998).

Damage is most prominent in late spring and summer, when plants can become defoliated, reducing flowering and seed set (Cilliers, 1987; Baars and Neser, 1999; Day *et al.*, 2003a). Populations decline over winter, when temperatures are low and the plants are dry. Although the beetles may seasonally defoliate plants, reducing flowering and vigour, the plants do not die (Baars and Neser, 1999; Day *et al.*, 2003a).

12.8.5 *Calycomyza lantanae* (Frick) (Diptera: Agromyzidae)

This fly is found from Florida to Peru (Harley and Kassulke, 1974) and Brazil (Winder and Harley, 1983). Adults feed on flowers and larvae form blotch mines in the leaves. Larvae feed for 6–8 days and pupation occurs in the soil or leaf litter. Development from egg to adult takes about 25 days (Harley and Kassulke, 1974).

Calycomyza lantanae established in Australia in 1974 (Taylor, 1989), and was subsequently introduced or has spread naturally to 16 countries (Table 12.1). Most of these countries have not actively released lantana biocontrol agents, and in some, *C. lantanae* is the only leaf-feeding insect established (Julien and Griffiths, 1998). It is highly probable that *C. lantanae* will continue to spread to other countries. However, it is not as damaging as other agents released on lantana.

12.8.6 *Other agents*

Several other highly damaging agents that have a limited distribution or have established in only one or two countries or regions are also known. There are also several agents which have been released recently in Australia, Hawaii, and South Africa but have not been evaluated.

Two flower-feeding moths, *Epinotia lantana* (Busck) (Tortricidae) from Mexico and *Lantanophaga pusillidactyla* (Walker) (Pterophoridae) from Mexico and the Caribbean, have been reported to inflict damage on up to 80% of flowers and fruits of lantana in several Micronesian countries and islands (Denton *et al.*, 1991). *Epinotia lantana* oviposits in shoot tips and inflorescences. The larvae tunnel into new shoots or feed on the flowers, hollowing out the receptacles of the flower heads. Pupation occurs in the hollowed out receptacles or among the webbed remains of flowers (Harley, 1971). About 73% of inflorescences in Hawaii were infested with *E. lantana*, greatly reducing seed formation (Swezey, 1924). *Lantanophaga pusillidactyla* oviposits in flower heads and the larvae feed within the flowers or tunnel around the receptacle. The larvae feed for 7–10 days and pupate in inflorescences. Flowers within an inflorescence that are not eaten produce fruits (Swezey, 1924). Both moths have established in numerous countries, and may have been introduced via the importation of potted lantana plants.

Eutreta xanthochaeta Aldrich (Tephritidae) is found in Mexico (Koebele, 1903; Palmer and Pullen, 1995). Females oviposit in the growing tips of new shoots. The larvae bore into the stem and induce solitary, spheroid galls at the apical region of growing shoots. Each gall contains one larva. The length of the larval and pupal stages is 4–5 weeks and 2–3 weeks respectively. The fly shows a preference for new shoots, especially regrowth shoots, and high proportions of those shoots attacked, are killed (CSIRO unpublished records). In 1902, *Eutreta xanthochaeta* was introduced into Hawaii, where it has established on all islands (Swezey, 1924) and occurs throughout the year (Duan *et al.*, 1998). Harley and Kunimoto (1969) observed it attacking a large proportion of shoots produced beneath the girdles made by *Plagiohammus spinipennis* (Thomson). The fly can be damaging in drier parts of the islands, but lantana tends to outgrow the galls in wetter regions (M. Day, personal observation).

Leptobyrsa decora Drake (Tingidae) was collected near Lima, Peru, where it causes severe defoliation of lantana (Harley, 1971). It also occurs in Colombia and Ecuador. Adults and nymphs form colonies on the undersides of leaves, where they suck the cell contents. The life cycle takes 31 and 44 days in summer and winter, respectively (Harley and Kassulke, 1971), with adults surviving for 60–90 days (Misra, 1985; Mishra and Sen-Sarma, 1986). In heavy infestations, affected plants become leafless (Harley and Kassulke, 1971; Misra, 1985; Mishra and Sen-Sarma, 1986). *Leptobyrsa decora* has a high reproductive potential, is easy to rear in large numbers, and is relatively free of parasites (Harley and Kassulke, 1971). *Leptobyrsa decora* has established in only Australia and Hawaii, where it can cause severe defoliation on a seasonal basis in drier, high altitude areas (Day *et al.*, 2003a,b). The potential distribution of this species seems to be regulated by climatic conditions and it is unlikely to establish in subtropical or wet temperate regions.

Plagiohammus spinipennis Thomson (Cerambycidae) is found from Mexico to Peru. Adults feed mainly on the midrib and main veins of lantana leaves, although young shoots and stems are also eaten. Eggs are laid in an incision into the bark of lantana stems. The young larvae girdle the stems, before burrowing into the cambium (Harley, 1969a). They subsequently burrow into the xylem tissue and may burrow into the roots. *Plagiohammus spinipennis* is univoltine, with the larval stages lasting 8–9 months. Infested shoots wither when the larvae are two weeks old and branches are either weakened or killed by the actions of older larvae (Harley, 1969a). *Plagiohammus spinipennis* was introduced into Hawaii in 1953 and established at several localities, where the larvae girdled up to 97% of plants and 78% of stems. All attacked plants were severely damaged (Harley, 1969a). It did not establish in any other country in which it was introduced.

Falconia intermedia Distant (Miridae) is found in Mexico, Guatemala, and Honduras (Palmer and Pullen, 1998). Adults and nymphs feed on the intercellular tissues on the undersurface of leaves, causing severe chlorosis, defoliation, and a reduction in flowering. Adults live for about three weeks and lay 2–3 eggs/day. Eggs are laid on the undersides of leaves and nymphal development is completed in 20–25 days (Baars and Neser, 1999; Day and McAndrew, 2003). In 1999, *F. intermedia* was released in South Africa, where it established and damages lantana infestations at several release sites. It has established in Australia at only a few sites in north Queensland, but it is too early to confirm its impact (Day *et al.*, 2003b). *Falconia intermedia* shows considerable promise as a biocontrol agent due to its high reproductive and dispersal potential and its ability to cause substantial damage to lantana in its native range. *Falconia intermedia* appears to prefer areas that are warm and moist all year round. It is unlikely that it will perform well in areas where defoliation of lantana occurs in response to seasonal drought.

Ophiomyia camarae Spencer (Agromyzidae) is found from Florida to Venezuela and Brazil (Stegmaier, 1966; Winder and Harley, 1983; Palmer and Pullen, 1995). Adults either drink water or feed on nectar in lantana flowers and lay their eggs on the undersides of leaves (Simelane, 2002). Larvae tunnel along veins and enter the midrib. Late-instar larvae form herring-bone-shaped mines in the leaves, disrupting translocation and inducing leaves to abscise prematurely. There is usually only one mine per leaf but larger leaves can support 2–3 mines (Stegmaier, 1966; Simelane, 2002). Pupation occurs in the leaves and larvae in leaves that abscise prematurely can still complete development. The development time from egg to adult is approximately four weeks and adults live for about three weeks (Simelane, 2002). *Ophiomyia camarae* was released in South Africa in 2001 and has now spread throughout eastern South Africa and north to Swaziland and Mozambique (A. Urban, PPRI, personal communication). *Ophiomyia camarae* appears to prefer shady areas in the field. It has been reported to reduce stem height and diameter, leaf and flower density, and above-ground biomass by 19%, 28%, 73%, 99%, and 49%, respectively (Simelane and Phenye, 2005). *Ophiomyia camarae* has recently been approved for release in Australia and mines have been found at several sites in north and southeast Queensland.

Mycovellosiella lantanae var. *lantanae* (Chupp) Deighton (Mycosphaerellaceae) is widespread throughout the neotropics and is tolerant of a range of subtropical climatic zones, being found in Brazil (Barreto *et al.*, 1995) and Florida (Den Breeen and Morris, 2003). *Mycovellosiella lantanae* var. *lantanae* is a leaf-spot fungus, causing chlorotic, gray lesions of leaves and necrosis of flower buds and stalks. Damaged plants can become defoliated, reducing vigor and reproductive potential (Den Breeen and Morris, 2003). *Mycovellosiella lantanae* var. *lantanae* was approved for release in South Africa in 2001 but it is too early to determine the agent's impact on *L. camara* (Den Breeen and Morris, 2003).

Prospodium tuberculatum (Spegazzini) Arthur (Pucciniaceae) is found in Brazil, Ecuador, and Mexico (Barreto *et al.*, 1995). It is an autoecious rust, with a reduced life cycle. The main stage is the urediniospores, although teliospores can be found on lantana growing in high altitudes (Barreto *et al.*, 1995) or in temperate coastal areas (N. Riding and M. Day, personal observations). Leaf infections are in the form of dark purplish-brown lesions that can be irregular in shape. Severe lesions cause defoliation and infected plants are less vigorous and stunted (Tomley and Evans, 1992). In Brazil, *P. tuberculatum* can cause severe leaf necrosis, resulting in defoliation and leading to reduced vigor (Tomley and Evans, 1992; Barreto *et al.*, 1995). *Prospodium tuberculatum* was released in Australia in 2001 and is now present at over 60 sites and has spread upto 40 km. However, prolonged drought over most of eastern Australia has impeded its release and establishment in many areas.

Septoria sp. (Sphaeropsidaceae), a leaf-spot fungus, is found in Ibarra, Ecuador (Trujillo and Norman, 1995). Initial symptoms of chlorotic spots appear two weeks after inoculation, becoming necrotic lesions after four weeks. Defoliation can occur after six weeks (Trujillo and Norman, 1995). *Septoria* sp. was released in Hawaii in 1997, although the status of the pathogen on these islands has not been reported (Thomas and Ellison, 2000).

All other agents listed in Table 12.1 have either established in only a few countries, cause little damage to lantana, or have not established. Details of these species and more information on all the above species are available in Day *et al.* (2003b).

12.9 Ecological interactions and impact of agents

Biological control of weed species is based on the premise that insect herbivory can greatly reduce the fitness of attacked plants, and in some circumstances lead to a reduction in weed population density. Forty-one species of predominantly leaf and flower-feeding insects have been introduced to 44 countries or regions where lantana is considered a problem. Numbers of species released (or that have spread on their own) range from one (as in Thailand, New Zealand, and the Marshall Islands) to 31 (as in Australia). About a third of countries or regions (14) have released 2–3 agents. Seventeen countries or regions have released between 4 and 11 agents. Some agents have been introduced into many countries and have established widely; others have been introduced into only one country and have failed to establish (Table 12.1). In general, similar numbers of species have been introduced to oceanic islands: average 6.5, ranging from 25 (Hawaii) to 1 (Marshall Islands); and mainland countries: average 6.3 ranging from 31 (Australia) to 1 (Thailand). The proportion of agents that have established is slightly higher on islands (77%) than on mainlands (73%), while the average degree of control achieved (or at least claimed) is about double on islands (54%) than on mainlands (26%) (Zalucki *et al.*, 2007).

No relationship between the area infested with lantana and the proportion of species established is obvious, suggesting that success or failure has more to do with the selection, rearing, and release process for an agent than any other ecological predictor (Zalucki *et al.*, 2007). Not surprising though, as agents are generally selected from localities with climates similar to those where they are to be released, there was no relationship between

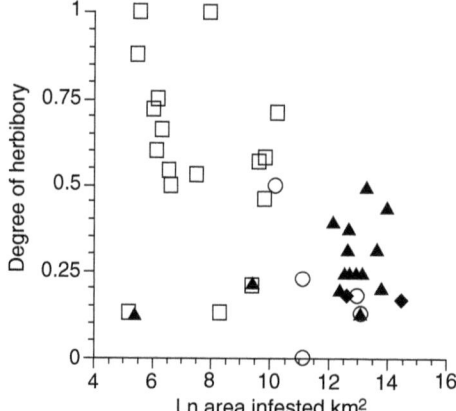

Fig. 12.2 Degree of herbivory of lantana achieved relative to the maximum possible (Zalucki *et al.*, 2007) plotted against the log normal area for Oceanic Islands (open squares), Archipelagos (filled diamond), Large Islands (open circles) and Mainland areas (filled triangles) infested with lantana.

index of suitability of a country (based on climate) for lantana and the proportion of species established.

The abundance of successfully introduced agents generally remains patchy but low (Day *et al.*, 2003b), so it is unlikely that high densities of already introduced agents are related to failure of new agents to establish. In addition, surveys in the native range of lantana indicate that potentially hundreds of species coexist, even within the same feeding guilds (Krauss, 1953; Winder and Harley, 1983; Palmer and Pullen, 1995).

Zalucki *et al.* (2007) found that the degree of herbivory (approximately equivalent to control) exerted in a country was not related to the number of species established. Overall, there was a weak negative but significant relationship for degree of control and area infested in a country (Fig. 12.2). This trend was due to the higher level of control reported on oceanic islands, which have generally smaller total infested areas (and possibly lower heterogeneity and number of habitats), relative to the level of control reported on mainlands (Fig. 12.2). When islands and mainlands were analyzed separately, the relationship was positive but not significant for mainland area (Fig. 12.3), with no relationship between total island area infested and degree of control.

12.10 Measures of efficiency of biological control

Measuring the efficiency of any biocontrol research program can be difficult and can often only be calculated once the program is complete. Biocontrol programs typically progress through a number of stages such as taxonomic, biological and ecological studies of the target weed, determining the distribution of the target weed in its native and

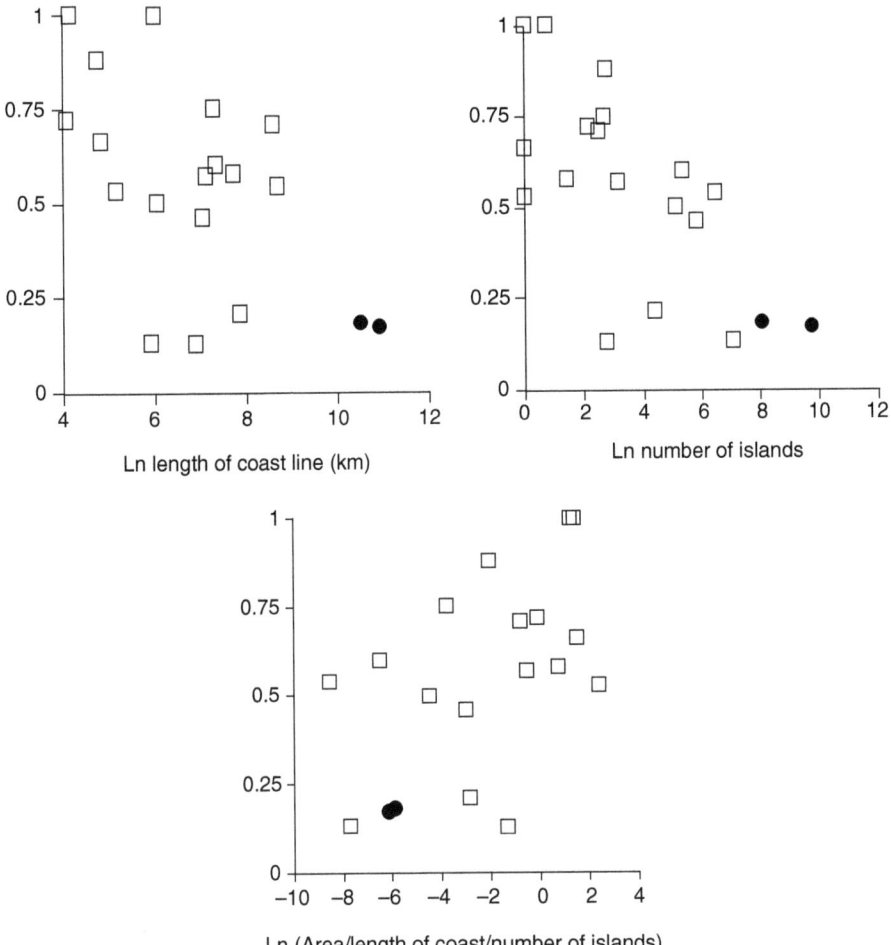

Fig. 12.3 Degree of herbivory for Oceanic Islands (and Archipelagos) plotted log normal of island area divided by the length of coastline and the number of islands, as a measure of the metapopulation structure of the landscape (Zalucki *et al.*, 2007).

introduced ranges, conducting exploration studies for potential agents within the native range of the target weed, host-specificity testing of priority agents, and finally the rearing and field release of approved agents. Highly efficient programs will result in quick transitions through each of the phases, hopefully resulting in successful biocontrol of the weed.

In reality, the process can be long, with typically several years between the time a weed is nominated as a target for biocontrol and the release of the first agent, and possibly many more years before control is achieved. Throughout the process, the program may stall. Taxonomic studies may reveal that there are too many closely related native species in the introduced range to suggest that any potential agents would be sufficiently host specific.

This has stalled several potential programs, for example the biocontrol of *Senna obtrusifolia* in Australia and several grass species. Exploration studies can be expensive and time consuming and may not result in any species being identified for further study, for example biocontrol of *L. montevidensis*.

Host-specificity testing may result in agents being rejected as they attack other plant species, and some agents following approval to field release may not establish. In terms of host specificity, few potential agents of lantana have been rejected in Australia, where there is a lack of closely related plant species to lantana. However, there are numerous examples of where potential agents have been discarded in South Africa, which has several native species of *Lantana* and the closely related genus *Lippia* (Day et al., 2003b).

Even if agents establish, there is no guarantee that they will have a significant impact on the target weed. Lantana biocontrol research has been conducted since 1902, resulting in the release of 41 agents in at least one country, with 27 species establishing. However, only 6–8 species are recorded as making any substantial contribution to lantana control. Therefore, for countries such as South Africa and Australia, biocontrol efficiency is less than desired. The problems and difficulties of achieving successful lantana biocontrol are discussed in the next section.

However, for many countries that utilize agents that have been previously researched and released elsewhere, there are significant efficiency gains. These countries can capitalize on the research conducted in other countries and target agents that are host specific as well as damaging. For example, *U. girardi* was imported into the Solomons from Hawaii. Consequently, lantana is virtually under control in the Solomon Islands. In addition, lantana is under control in Guam following the release of several agents that had been utilised elsewhere (Muniappan et al., 1996).

12.10.1 Factors influencing successful biological control of lantana

Factors found to limit the effectiveness of biocontrol agents (some were limited by more than one factor) were: climate (44% of cases), host incompatibility (33%), predators (22%), competition (12%), parasitoids (11%), and disease (8%) (Crawley, 1986). Several authors studying lantana biocontrol suggest that the *Lantana* species from which potential agents were collected, the variety of the target weedy lantana, climatic and geographical distribution of lantana, plant biology and ecology, release techniques or strategies, parasitism, and islands limit successful biocontrol of the weed (Broughton, 2000a; Day and Neser, 2000; Zalucki et al., 2007).

12.10.2 Lantana taxonomy

Sheppard (1992) suggests that genetically variable weeds that differ in their suitability to particular biocontrol agents are more difficult to control through biological means than weeds that are genetically homogeneous. The hybrid nature of lantana naturalized throughout the tropics poses major challenges for biocontrol programs. In most biocontrol programs, potential agents of a particular weed are found on the same species in its "natural range" and are therefore suited to the same plant in its weedy environment. As the weedy taxa of lantana are not indigenous anywhere, a major problem is in identifying the most suitable *Lantana* species on which to concentrate exploratory efforts in the native range. Potential agents collected from *Lantana* species in their native range which differs from the *Lantana* species in the introduced range may not be adapted to the new host and therefore fail to establish (Day and Neser, 2000). The interactions between various natural enemies and the different lantana varieties can be complex and difficult to predict (Baars and Neser, 1999).

Scott *et al.* (2002) suggest that *L. urticifolia* is one of the main sources of genetic material for the hybrid weedy lantana naturalized in Australia. Retrospective analyses in Australia show that a greater proportion of agents that were collected from *L. camara* and *L. urticifolia* established, compared with agents that were collected from other species of lantana (Table 12.1). In addition, a greater proportion of agents (20 out of 28) that were collected from Mexico and the Caribbean, thought to be the origin of the weedy form and the native range of *L. urticifolia*, established than agents collected elsewhere, where *L. urticifolia* was absent (Krauss, 1953; Winder and Harley, 1983; Palmer and Pullen, 1995). Fifteen (83%) of the agents that established were found on three or more lantana species, suggesting that agents that are oligophagous have a greater chance of establishing. Similar studies to determine relatedness of lantana in other countries or regions have not yet been undertaken.

12.10.3 Variety of target weedy lantana

There are over 650 named varieties of lantana worldwide (Howard, 1969), with different varieties possibly having different progenitors (Scott, 1998). Given that the different species of lantana have differing assemblages of insects associated with them in their native range, it is not surprising that some agents have been reported to show preference to, or perform better on, some varieties than others (Diatloff and Haseler, 1965; Harley and Kassulke, 1971; Harley *et al.*, 1979). A number of agents such as *P. spinipennis*, *E. xanthochaeta*, and *Strymon bazochii* (Godart) (Lycaenidae) have all established and are widespread in Hawaii, while all three have failed to establish elsewhere, despite several attempts. Lantana in Hawaii may have different progenitors from lantana in other countries and this may at least partly explain different establishment success (Day *et al.*, 2003b).

Even within a country, agents have shown differences in their preference for or performance on particular varieties. Ten of the 41 agents introduced to control lantana have shown some degree of preference for certain varieties within a country (Day *et al.*, 2003b). Many rusts are highly specific. The rust *P. tuberculatum* only affects the common pink flowering taxa in Australia (Tomley and Riding, 2002), while another rust, *Puccinia lantanae* Farlow (Pucciniaceae), attacks the common pink-edged red flowering lantana (C. Ellison, CABI, personal communication).

12.10.4 Climate

Climate is probably the single most important factor determining the distribution of insects and the effectiveness of biocontrol agents (Zalucki and van Klinken, 2006). Lantana occupies a wide range of habitats over a broad geographical distribution in many countries where it has been introduced. Consequently, climatic conditions vary widely throughout the naturalized range of lantana, affecting the distribution of biocontrol agents. In Australia, lantana is found from tropical areas in far north Queensland to temperate areas in southern New South Wales and only two agents, *L. pusillidactyla* and *O. lantanae*, are found in most areas. More often, agents are limited in their distribution; for instance, *L. decora* and *Uroplata fulvopustulata* Baly (Chrysomelidae) are found only in tropical north Queensland (Day *et al.*, 2003a). *Teleonemia scrupulosa* and *L. decora* are often found in dry areas or on north-facing slopes but are rarely found on south-facing slopes or on lantana growing under canopy, while *O. scabripennis* prefers the warm, moist coastal regions in South Africa and Fiji than the drier inland areas.

Not only do insect populations vary spatially and temporally according to climatic conditions, but the susceptibility of lantana to damage caused by these insects appears to be climatically dependent. Successful control of lantana has been reported in drier areas of some countries, where the combined stresses of drought and large populations of *T. scrupulosa* and other agents have been sufficient to kill mature plants (Swezey, 1924; Fullaway, 1959; Day *et al.* 2003a, b). Likewise, lantana growing beneath established pine plantations in Fiji has been largely controlled through the damage caused by *U. girardi*, in combination with reduced plant vigor associated with low light conditions (S. N. Lal, SPC, personal communication). However, climate can also alter plant characteristics that would otherwise make it suitable for herbivores. Frosts or seasonally dry conditions cause defoliation of plants making them unsuitable for leaf-feeding insects.

As lantana is frequently widespread and is not separated by major geographical barriers in its native range, it is probable that potential biocontrol agents are constrained to their geographic ranges by climate. If the climate in the naturalized range of lantana is greatly different from that where the agents occur naturally, it is less likely that they will establish successfully (Sutherst *et al.*, 1999).

In the long-term, climate change is likely to have a major impact on all aspects of insect–plant interactions, of which lantana is just an example. Climate change will affect lantana distribution directly, as well as the current agents and their effects on the plant. Some agents may become extinct; while others may increase in abundance, extend further south or remain unchanged. The net impact on the sustainability of management in this instance is likely to be small as impact on lantana is currently not large.

12.10.5 Plant biology

Leaf-feeding insects have been able to control many weeds, or at least severely retard plant growth and flowering such that they can limit the competitive ability of the weed and reduce its ability to spread. However, leaf-feeding insects rarely kill perennial weeds. Over half of the agents released on lantana have been leaf-feeding insects and it is clear that they have not been able to control lantana successfully in many areas. Insect populations tend to increase during summer when plants are healthy and decline during winter when temperatures decrease and plants are often without leaves. Any damage caused by agents, such as *T. scrupulosa*, *O. scabripennis*, or *U. girardi*, is only seasonal and the plant can recover. Even in the absence of natural enemies, lantana has the ability to survive defoliation when stressed as a result of dry winter months and/or frost, and reshoot and flower following spring rains and warmer temperatures.

Some insects such as *O. scabripennis* and *U. girardi* can survive winter by diapausing. For many others such as the leaf-feeding and flower-feeding lepidoptera, there is no diapause stage. Consequently, in the spring when lantana plants begin to recover, many of the agents are present only in low numbers or must colonize plants from elsewhere. Populations then slowly build up and by late summer reach levels that damage plants. However, the damage is not sustained as insect numbers again begin to decrease with the onset of winter. Therefore, as plant condition is linked to seasons and insect numbers tend to follow plant condition, it is unlikely that leaf-feeding agents in many countries will ever control lantana by themselves.

Seed and flower-feeding insects have also had limited impact on lantana. An individual lantana plant has the ability to produce thousands of flowers and seeds each season. Although there have been several flower and seed-feeding agents, such as *L. pusillidactyla*, *E. lantana* and *O. lantanae*, released on lantana and damaging up to 80% of flowers and/or fruit (Muniappan, 1989), large amounts of viable seed can still be produced, especially early in the season when the insects have yet to build up into damaging populations.

Only a few agents, such as *Aconophora compressa* Walker (Membracidae), *E. xanthochaeta* and *P. spinipennis*, which attack the stems, have been released on lantana and they have established in limited areas (Julien and Griffiths, 1998; Day *et al.*, 2003ab). The advantage of utilizing stem-attacking or root-feeding agents is that the agents do not require the plant to be in leaf all year round. Stem-boring or root-feeding insects attack the carbohydrate reserves of a plant and disrupt translocation. They often have life histories whereby adults emerge in summer when there is fresh leaf growth upon which to feed while the larvae feed in the stems or on the roots respectively during winter when the plant can be devoid of leaves.

To offset the rapid recovery of lantana from defoliation and the problems faced by insects, various plant pathogens, which have the benefit of rapid population growth, have been tried. Three species, *Septoria* sp. in Hawaii, *P. tuberculatum* in Australia and

M. lantanae in South Africa (Thomas and Ellison, 2000; Tomley and Riding, 2002; Trujillo and Norman, 1995; Den Breen and Morris, 2003) have been introduced. The advantages of using pathogens are that they have a short life cycle, a tremendous capacity to reproduce and disperse, and they have a resting stage to overcome unfavorable conditions. The impact of these three agents and whether they can overcome the intricacies of lantana's biology is still to be determined.

12.10.6 Effect of parasitism and predation on the effectiveness of agents

The importance of parasites and predators in reducing biocontrol agent populations has rarely been investigated but frequently alluded to as a cause of "failure." Newly introduced biocontrol agents may undergo rapid population explosions, causing severe defoliation to the target weed, only to suffer a subsequent population crash, after which the population never reaches the same size again (Fullaway, 1959; Gardner and Davis, 1982; Cilliers and Neser, 1991; Denton *et al.*, 1991).

Anecdotal reports have suggested up to 10 lantana agents are attacked by parasites, but in most cases, levels are low and do not appear to limit populations (Day *et al.*, 2003b). A series of studies undertaken by Duan and coworkers (1996, 1998) revealed that *E. xanthochaeta* was attacked by parasitoids introduced into Hawaii to combat fruit flies. However, parasitism rates in the wild were very low (Duan and Messing, 1996) and the gall fly larvae experienced high levels of mortality from reasons other than parasitism (Duan *et al.*, 1998). In addition, the levels of parasitism of *P. spinipennis* in Hawaii varied between sites and accounted for only 10% of the overall mortality at the site with the highest parasitism rates (Harley and Kunimoto, 1969).

Lantana insects collected from the Americas are generally not closely related to species occurring in the Old World and those species apparently free of parasites in their native range are rarely parasitised in their new environment. Therefore, parasites and predators attacking biocontrol insects in their new environment are likely to be generalist species, making it difficult to predict which biocontrol agents are likely to be parasitized in the target country.

12.10.7 Release techniques

Some biocontrol agents of lantana have, almost certainly, not established due to either the release of insufficient numbers or the use of inappropriate release techniques; an example is *Teleonemia harleyi* (Froeschner) (Tingidae) (Day and Neser, 2000). There are a number of recent papers proposing release methods to maximize establishment (e.g. Grevstad, 1996; Memmott *et al.*, 1996; Shea and Possingham, 2000; Day *et al.*, 2004). Release techniques should be based on the agent's biology, behavior, and the most suitable life stage for release. For most agents, adults are the most appropriate, as they are reasonably mobile and seek favorable feeding and/or oviposition sites (Day and Neser, 2000). For example, higher establishment rates were obtained when *Neogalea sunia* (Guenée)

(Noctuidae) was released as adults, compared with when larvae were released (Haseler, 1963).

12.10.8 Biogeography

Apart from Guam, the Solomons, and parts of Hawaii, lantana is generally not under adequate biocontrol. Even in regions where some degree of control has been achieved, such as in Hawaii (Davis et al., 1992), the plant remains a problem. The degree of control achieved with the agents released to date differs between island and mainlands, being generally higher in the former than the latter, presumably because there are (1) fewer natural enemies on islands, (2) lower climatic variation or more equable climates on islands, (3) "resource concentration" at least in small island groups (Fig. 12.2), and perhaps (4) fewer varieties of lantana to contend with, implying that if the local variety and climate are suitable for the agent, the agent will establish and exert control; if plant variety and climate are not suitable, the agent will not establish.

The degree of control is independent of the number of agents released for islands. Generally, fewer than five agents have been released and they either work or they do not. Nor is degree of control related to area of the infestations in island countries. However, how the area is arranged appears to be significant. Large numbers of islands and hence small islands, with necessarily lengthy coastlines, appear to be associated with lower control. We presume successful movement of agents amongst islands (patches) will be low when there are large numbers of small islands and the subsequent population of herbivores will be low (as will degree of control). Conversely, for large islands, with more contiguous distributions of lantana, degree of control is higher (Fig. 12.3). Unlike islands, degree of control for mainlands was positively related to the area of infestation and suggests that landscape level processes may be important to the abundance of established agents and degree of herbivory (Zalucki et al., 2007).

12.11 Economics of biological control efforts

To determine whether lantana biocontrol is economical, there needs to be data on the impact of the weed on agriculture and the environment, as well as the cost of control. While it is possible to measure this for the former, the latter is not so easy and for many countries, this information is not available. In Australia, lantana costs the grazing industry over $100 million annually in lost production and control (AEC group, 2007). However, costs to biodiversity and the environment are not available. A recent study found that the average person in Australia was prepared to pay $5/year to see lantana controlled in national parks (AEC group, 2003). As lantana is widespread in many countries, it is not feasible to continue with many conventional control methods which are costly and ongoing. Therefore, biocontrol is probably the only feasible long-term control option.

Biocontrol of lantana has been conducted for over 100 years. Countries such as USA (Hawaii), Australia, and South Africa have had active long-term projects resulting in at least 20 agents being released in each country. Total dollars spent on biocontrol is not available for any of these projects but conservative estimates would suggest in the order of tens of millions. Biocontrol, though, is at best only partially successful, initially suggesting that the program is not economical. However, a recent study in Australia suggested that even a reduction of 5% in lantana due to biocontrol would justify any biocontrol research, resulting in a benefit-cost ratio of Au\$9 per Au\$1 invested (AEC group, 2007). In other countries, which have introduced agents from Australia or elsewhere following tests demonstrating their specificity, projects would return even higher benefit-cost ratios, as costs are limited to importation, rearing and field releasing agents. For example, lantana is reported to be under control in the Solomon Islands and Guam, following the introduction of agents from Hawaii, suggesting that these programs have resulted in great economic benefits for both countries for the amount invested.

12.12 Issues of sustainability of the identified biological control efforts

Weed biocontrol by its nature is a long-term sustainable control strategy. Biocontrol agents once established continue to persist, although local populations may fluctuate and possibly even die out due to heatwaves, frosts, or fires. Agent populations are usually reestablished through natural dispersion from other areas. For instance, during surveys conducted on lantana in Australia over the past 10 years, various species were seasonally absent from some locations, yet in other years, the same agents were abundant. This is especially so with agents that have been present in a country for a long period of time.

Many lantana agents have attributes that help them overcome adverse conditions. Some agents can go into diapause while others have life cycles that are synchronous with the plant's seasonal phenology. The hispine beetles *Octotoma* spp. and *Uroplata* spp. all have adults that enter a facultative diapause during winter when temperatures are low and conditions are dry. Plants often lose their leaves during this time and it is advantageous for these insects to diapause when food quality and quantity are low.

12.13 Future lantana biocontrol

For most countries where lantana is a problem, increasing the number of species of biocontrol agents would be a priority. There are a number of countries where lantana is present, but are reported not to have any of the biocontrol agents, while many other countries have released only a few agents of the 41 that have been tried (Table 12.1). *Teleonemia scrupulosa*, *O. scabripennis*, *U. girardi*, and *O. lantanae* have proved to be damaging agents in a number of countries and could be introduced where they are not present. *Calycomyza lantanae*, *E. lantana*, and *L. pusillidactyla* are not as damaging as the aforementioned agents, but could assist in controlling lantana in countries where only a few agents are present. *Eutreta xanthochaeta*, *F. intermedia*, *L. decora*, and *O. camarae* are damaging but appear to have specialized climatic requirements. In addition, there are three pathogens, *M. lantanae* var. *lantanae*, *P. tuberculatum*, and *Septoria* sp., that have been recently released and could be tried in other countries once their impact on lantana has been assessed.

For countries such as Australia and South Africa that have imported many biocontrol agents, new and more effective agents need to be located in their host range and trialed. Particular characteristics worth considering when importing new agents are: the agents' ability to develop on the lantana variety being targeted, adaptation to the local climate in which agents will be released, and agents that attack the parts of the plant such as roots and stems, upon which few agents have been released.

12.13.1 Selecting future lantana biological control agents

There are a number of papers published on what makes a good biocontrol agent and some authors have offered a method for assessing agents or how exploration should be conducted (Harris, 1973; Goeden, 1983; Hokkanen and Pimentel, 1984; Wapshere *et al.*, 1989). Such methods consider guild, life history, and behavior of the agent and how they affect the plant in terms of biomass removal or the reduction of seed set (Harris, 1973; Winder and Harley, 1982). While these papers offer a guide, the intrinsic nature of the target weed will limit the effectiveness of any system. In addition, it is difficult to predict how a potential agent will perform once released (Zalucki and van Klinken, 2006). Factors such as climate, habitat, altitude, and the impact of predators and/or parasitoids will determine the effectiveness of an agent (Wapshere *et al.*, 1989).

Many of the insects first released in Hawaii attacked the fruits and/or flowers of lantana (Koebele, 1903). However, the contribution of flower- and fruit-feeding insects to seed loss appears to be limited because flower- and fruit-feeders are satiated when flowers and/or fruits are abundant, resulting in many seeds being unaffected. Conversely, seed losses caused by the agents are greatest when flowers and/or fruit are scarce (Crawley, 1989). Although flower- and fruit-feeding insects such as *E. lantana*, *L. pusillidactyla*, and *O. lantanae* have been effective in a few regions, such as Guam and some islands of Micronesia (Denton *et al.*, 1991), studies in Guam indicate that leaf-feeding insects account for greater reductions in seed set than the flower- and seed-feeders (Muniappan *et al.*, 1996).

There is currently a shift in the selection of agents, with insects that form galls, stem borers, root feeders, and pathogens preferred over leaf-feeding insects. This is because the activity of these agents is independent of the condition of the foliage and they have life cycles that are more suitable to seasonal variation and the condition of the plant or, in the case with pathogens, very short generation times. Unfortunately, few insects have been found to attack lantana stems or roots in the Americas (Koebele, 1903; Krauss, 1953; Winder and Harley, 1983; Palmer and Pullen, 1995). *Parevander xanthomelas* (Guérin-Méneville) (Cerambycidae), which has only been released in Hawaii (Julien and Griffiths, 1998), and *Longitarsus bethae* Savini and Escalona (Chrysomelidae), which is currently being studied in South Africa (Simelane, 2005), both attack the roots of lantana. Two stem-boring beetles, *P. spinipennis* and *Aerenicopsis championi* Bates (Cerambycidae), have been introduced to control lantana, but both were difficult to establish due to their long life cycles and problems associated with mass rearing.

Gall-forming agents can act as physiological sinks and can deplete important food reserves, causing the plant to die or become stunted and cease flowering. Gall-forming agents have been used successfully in other weed biocontrol programs, such as *Cecidochares connexa* (Macquart) (Tephritidae) on *Chromolaena odorata* (L.) King and Robinson (Asteraceae) (Day and Bofeng, 2007). Only one gall-forming insect, *E. xanthochaeta*, has been released on lantana and it has established only in Hawaii (Day *et al.*, 2003b).

The use of pathogens in weed biocontrol is a fairly recent development. Field evidence has shown that pathogens that have been utilized as biocontrol agents can be very damaging to weeds, for example *Maravalia cryptostegiae* (Cummins) Ono on rubber vine *Cryptostegia grandiflora* (Roxburgh) Brown (Asclepiadaceae) and *Puccinia xanthii* Schweinitz on Bathurst burr *Xanthium strumarium* L. (Asteraceae) (Julien and Griffiths, 1998). Surveys carried out in Brazil have identified several pathogens that are capable of causing significant damage to lantana (Tomley and Evans, 1992; Barreto *et al.*, 1995). These appear to be highly host specific, with damage not seen on closely related species of *Lantana* (Barreto *et al.*, 1995). By shifting the mix of agents from leaf feeders to gall formers, stem borers, root feeders, and pathogens, sustainable long-term management of the weed may be possible.

12.13.2 New agents currently being considered for release

While the actual number of insects and/or pathogens found attacking lantana is quite high (Koebele, 1903; Krauss, 1953; Winder and Harley, 1983; Barreto *et al.*, 1995; Palmer and Pullen, 1995), the number that are considered to be specific enough for further study or are climatically suited is much lower. Some of the potential agents that are currently being studied for importation or release are discussed below.

Aceria lantanae (Cook) (Acarina: Eriophyidae) causes galls on leaves and inflorescences, resulting in stunted plants in Florida, Mexico, the Caribbean, and Brazil (Flechtmann and Harley, 1974; Craemer and Neser, 1990). Galls have the potential to place huge physiological pressures on the plant such that the host will stop producing new shoots, flowers, or seeds. Galls are also known to carry viral plant diseases (Cromroy, 1976; Craemer and Neser, 1996). *Aceria lantanae* appears to be very host specific, inducing galls only in some lantana varieties but not attacking any of the native South African species of lantana (Urban *et al.*, 2001).

Adults of *Coelocephalapion camarae* Kissinger (Brentidae) feed on leaves and lay eggs in the petioles. Larvae bore into the petioles and induce small galls (Baars *et al.*, 2007). Galls on leaf petioles may cause leaves to desiccate and abscise. Research in South

Africa shows that *C. camarae* can disrupt the transport of essential solutes and cause a reduction in dry weight of roots and shoots. The adults are long-lived and diapause during winter when plants can lose their leaves. *Coelocephalapion camarae* is awaiting approval to be released (Baars *et al.*, 2007).

Longitarsus bethae is a root-feeding flea beetle found in Mexico (Simelane, 2005). Adults feed on leaves and lay their eggs in leaf litter. Larvae feed on the roots and pupate in the soil (Simelane, 2005). The insect has a number of generations per year and the adults diapause over winter when it is dry. *Longitarsus bethae* is considered a highly promising agent, as it is one of only a few root-feeding insects to be studied for the biocontrol of lantana. The beetle is waiting approval for release in South Africa (Simelane, 2005).

The rust fungus *Puccinia lantanae* Farlow (Pucciniaceae) is common on *L. camara* in tropical areas of Brazil, but is scarce on this plant in subtropical regions. *Puccinia lantanae* is of potential interest for classical weed biocontrol in warmer, more humid regions (Barreto *et al.*, 1995). Preliminary host testing and varietal susceptibility tests have found it has a narrow host range (C. Ellison, CABI, personal communication).

12.13.3 Classification and identification of naturalized taxa

The first step in any biocontrol program should be to correctly identify the target weed species and intraspecific taxa (Schroeder and Goeden, 1986). Work on the biocontrol of lantana has been conducted since 1902 and yet this vital step has not been fully addressed. Part of the problem stems from not fully realizing the complexity of the lantana group and its effect on biocontrol agents. In earlier exploration visits, agents were collected from plants morphologically similar to those in the naturalized range and many of these plants were collectively referred to as *L. camara*. Only recently, with DNA studies and a more thorough appreciation of the complexities of the group, we are now recognizing the affinities and taxonomic relationships of individual plants within this genus. Despite recent advances, there is still scope for further research and a need to look at particular characteristics of specimens more closely to separate groups, as variants within the taxon occur.

It is hoped that further DNA testing, combined with biochemical profiling and morphological studies, will enable a better understanding of the relationships of taxa within the lantana complex. Clarifying the taxonomy of the genus and, in particular the *Lantana* section *Camara*, is an essential prerequisite for successful biocontrol. In addition, by knowing the relatedness of the naturalized lantana between different countries, agents considered successful can be rereleased into countries that have suitable varieties of lantana for the agents.

12.13.4 Lantana biology and ecology

While numerous attempts to understand basic taxonomy and biology and many attempts at utilizing biocontrol agents have occurred, little progress has been made on what must be done to lantana to either kill the plant or at least reduce its vigor and seed set. Harris (1973) attempted to rate the different insect guilds in relation to their effectiveness as biocontrol agents for weeds in general, but such studies are of limited use when applied to lantana. Winder and van Emden (1980) and Broughton (2000b) studied various aspects of the impact of leaf-feeding insects, with both studies monitoring the effects of pruning plants at different levels and times. However, feeding by insects is a continuous process and quantitative studies that reflect this should be conducted in the field.

Finally, a better appreciation of the impact of each of the agents currently established is needed to determine their potential usefulness for other countries. So far, little information exists, apart from some earlier studies by Forno and Harley (1976), Winder (1980), Winder and Harley (1982) and some anecdotal reports. More recently, South African and Australian scientists have been trying to address this deficiency. Through field assessment of agents and manipulative experiments, it should be possible to make decisions on which guilds of agents are best to focus on in the future.

12.14 Conclusion

Given that lantana is not under control and that agent establishment is unrelated to the number of species introduced (but degree of control is), we would suggest further agents, particularly in the stem-boring, gall-forming, or root-feeding guilds, be considered. We also suggest any new campaigns to release agents against lantana take the opportunity to undertake experiments so as to improve the science behind the art of classical biocontrol. The prospects for sustainable lantana control may well then improve.

Acknowledgments

We thank W. Palmer, D. Panetta, and P. Paping (Department of Primary Industries and Fisheries, Australia) for providing information, photos, and maps and for suggesting amendments to the text.

References

AEC group (2003). *Economic Assessment of Environmental Weeds in Queensland*. Final Report for Queensland Department of Natural Resources and Mines, Brisbane, Australia, 96 pp.

AEC group. (2007). *Economic Impact of Lantana on the Australian Grazing Industry*. Final Report for Queensland Department of Natural Resources and Water, Brisbane, Australia, 39 p.

Baars, J. R. and Neser, S. (1999). Past and present initiatives on the biological control of *Lantana camara* (Verbenaceae) in South Africa. *African Entomology Memoir*, **1**, 21–33.

Baars, J.R., Hill, M. P., Heystek, F., Neser, S. and Urban, A. J. (2007). Biology, oviposition preference and impact in quarantine of the petiole-galling weevil,

Coelocephalapion camarae Kissinger, a promising candidate agent for biological control of *Lantana camara*. *Biological Control*, **40**, 187–195.

Barreto, R. W., Evans, H. C. and Ellison, C. A. (1995). The mycobiota of the weed *Lantana camara* in Brazil, with particular reference to biological control. *Mycological Research*, **99**, 769–782.

Barrows, E. M. (1976). Nectar robbing and pollination of *Lantana camara* (Verbenaceae). *Biotropica*, **8**, 132–135.

Bennett, F. D. and Maraj, S. (1967). Host specificity tests with *Uroplata girardi* Pic., a leaf-mining hispid from *Lantana camara* L. *Technical Bulletin of the Commonwealth Institute of Biological Control*, **9**, 53–60.

Broughton, S. (1999). Impact of the seed-fly, *Ophiomyia lantanae* (Froggatt) (Diptera: Agromyzidae), on the viability of lantana fruit in south-east Queensland, Australia. *Biological Control*, **15**, 168–172.

Broughton, S. (2000a). Review and evaluation of lantana biocontrol programs. *Biological Control*, **17**, 272–286.

Broughton, S. (2000b). Artificial defoliation effect on lantana growth and biomass. Ph.D. thesis. University of Queensland, pp. 117–132.

Cilliers, C. J. (1987). The evaluation of three insect natural enemies for biological control of the weed *Lantana camara* L. *Journal of the Entomological Society of Southern Africa*, **50**, 15–34.

Cilliers, C. J. and Neser, S. (1991). Biological control of *Lantana camara* (Verbenaceae) in South Africa. *Agriculture, Ecosystems and Environment*, **37**, 57–75.

Cock, M. J. W. and Godfray, H. C. J. (1985). Biological control of *Lantana camara* L. in the Philippines. *Journal of Plant Protection in the Tropics*, **2**, 61–63.

Craemer, C. and Neser, S. (1990). *Mites Imported Against Lantana*. Pretoria: South Africa Pretoria Weed Unit.

Craemer, C. and Neser, S. (1996). Eriophyoid mites (Acari: Eriophyoidea) as possible control agents of introduced plants in South Africa. In *Proceedings of the IX International Symposium on Biological Control of Weeds*, ed. V. C. Moran and J. H. Hoffman. Randebosch, South Africa: University of Cape Town p. 228.

Crawley, M. J. (1986). The population biology of invaders. *Philosophical Transactions of the Royal Society of London, B Biological Sciences*, **314**, 711–731.

Crawley, M. J. (1989). Insect herbivores and plant population dynamics. *Annual Review of Entomology*, **34**, 531–564.

Cromroy, H. L. (1976). The potential use of eriophyoid mites for the control of weeds. In *Proceedings of the IV International Symposium on Biological Control of Weeds*, ed. T. E. Freeman. Gainesville, FL: University of Florida, pp. 294–296.

CSIR (1962). *Lantana* Linn. (Verbenaceae). In *The Wealth of India: A Dictionary of Raw Materials and Industrial products, vol VI*. ed. B. N. Sastri. New Delhi: *Council of Scientific and Industrial Research*, pp. 31–34.

Davis, C. J., Yoshioka, E. and Kageler, D. (1992). Biological control of lantana, prickly pear, and *Hamakua pamakani* in Hawai'i: a review and update. In *Alien Plant Invasions in Native Ecosystems of Hawaii: Management and Research*, ed. C. P. Stone, C. W. Smith and J. T. Tunison. Honolulu, HI: University of Hawaii Press, pp. 411–431.

Day, M. D. and Bofeng, I. (2007). The status of biocontrol of *Chromolaena odorata* in Papua New Guinea. In *Proceedings of the 7th International Workshop on Biological*

Control and Management of Mikania micrantha *and* Chromolaena odorata. Taiwan: Kaohsiung University, pp. 53–67.

Day, M. D. and McAndrew, T. D. (2003). The biology and host range of *Falconia intermedia* (Distant) (Hemiptera: Miridae), a potential biological control agent for *Lantana camara* L. (Verbenaceae) in Australia. *Biocontrol Science and Technology*, **13**, 13–22.

Day, M. D. and Neser, S. (2000). Factors influencing the biological control of *Lantana camara* in Australia and South Africa. In *Proceedings of the X International Symposium on Biological Control of Weeds*, ed. N. R. Spencer. Sidney, MT: US DA-ARS, pp. 897–908.

Day, M. D., Briese, D. T., Grace, B. S., *et al.* (2004). Improving release strategies to increase the establishment rate of weed biocontrol agents. In *Proceedings of the 14th Australian Weeds Conference*, ed. B. M. Sindel and S. B. Johnson. Sydney, Australia: Weed Society of NSW, pp. 369–373.

Day, M. D., Broughton, S. and Hannan-Jones, M. A. (2003a). Current distribution and status of *Lantana camara* and its biological control agents in Australia, with recommendations for further biocontrol introductions into other countries. *Biocontrol News and Information*, **24**, 63N–76N.

Day, M. D., Wiley, C. J., Playford, J. and Zalucki, M. P. (2003b). *Lantana: Current Management Status and Future Prospects*. Canberra, Australia: ACIAR Monograph Series, 128 pp.

Den Breeen, A. and Morris, M. J. (2003). Pathogenicity and host specificity of *Mycovellosiella lantanae* var. *lantanae*, a potential biocontrol agent for *Lantana camara* in South Africa. *Biocontrol Science and Technology*, **13**, 313–322.

Denton, G. R. W., Muniappan, R. and Marutani, M. (1991). The distribution and biological control of *Lantana camara* in Micronesia. *Micronesica*, Suppl. **3**, 71–81.

Diatloff, G. and Haseler, W. H. (1965). Varietal differences in lantana. In *Australian Weeds Conference*, held in Toowoomba, Queensland. Brisbane, Arstralia: Weed Society of Queensland, pp. 11–12.

Duan, J. J. and Messing, R. H. (1996). Response of two opiine fruit fly parasitoids (Hymenoptera: Braconidae) to the lantana gall fly (Diptera: Tephritidae). *Environmental Entomology* **25**, 1428–1437.

Duan, J. J., Messing, R. H. and Purcell, M. F. (1998). Association of the opiine parasitoid *Diachasmimorpha tryoni* (Hymenoptera: Braconidae) with the lantana gall fly (Diptera: Tephritidae) on Kauai. *Environmental Entomology*, **27**, 419–426.

Duggin J. A. and Gentle, C. B. (1998). Experimental evidence on the importance of disturbance intensity for invasion of *Lantana camara* L. in dry rainforest–open forest ecotones in north-eastern NSW, Australia. *Forest Ecology and Management*, **109**, 279–292.

Fensham, R. J., Fairfax, R. J. and Cannell, R. J. (1994). The invasion of *Lantana camara* L. in Forty Mile Scrub National Park, north Queensland. *Australian Journal of Ecology*, **19**, 297–305.

Flechtmann, C. H. W. and Harley, K. L. S. (1974). Preliminary report on mites (Acari) associated with *Lantana camara* L. in the Neotropical region. *Anais da Sociedade Entomológica do Brasil*, **3**, 69–71.

Forno, I. W. and Harley, K. L. S. (1976). The evaluation of biocontrol agents with particular reference to two hispine beetles established on *Lantana camara* in Australia. In *Proceedings of the IV International Symposium on Biological Control of Weeds*, ed. T. E Freeman. Gainesville, FL: University of Florida, pp. 152–154.

Fullaway, D. T. (1959). *Biological Control of Lantana in Hawaii.* Biennial Report 1956–58 Honolulu, HI: Hawaiian Board of Agriculture and Forestry, 70–74.

Fyfe, R. V. (1937). The lantana bug, *Teleonemia lantanae* Distant. *Journal of the Council for Scientific and Industrial Research*, **10**, 181–186.

Gardner, D. E. and Davis, C. J. (1982). *The Prospects for Biological Control of Nonnative Plants in Hawaiian National Parks.* Technical Report 45. Honolulu, HI: University of Hawaii at Manoa, 53 pp.

Gentle, C. B. and Duggin, J. A. (1997). Allelopathy as a competitive strategy in persistent thickets of *Lantana camara* L. in three Australian forest communities. *Plant Ecology*, **132**, 85–95.

Gentle, C. B. and Duggin, J. A. (1998). Interference of *Choricarpia leptopetala* by *Lantana camara* with nutrient enrichment in mesic forests on the Central Coast of NSW. *Plant Ecology*, **136**, 205–211.

Ghisalberti, E. L. (2000). *Lantana camara* L. (Verbenaceae). *Fitoterapia*, **71**, 467–486.

Goeden, R. D. (1983). Critique and revision of Harris' scoring system for selection of insect agents in biological control of weeds. *Protection Ecology*, **5**, 287–301.

Graaff, J. L. (1986). *Lantana camara*, the plant and some methods for its control. *South African Forestry Journal*, **136**, 26–30.

Greathead, D. J. (1968). Biological control of *Lantana*. A review and discussion of recent developments in East Africa. *PANS (C)*, **14**, 167–175.

Greathead, D. J. (1971). Progress in the biological control of *Lantana camara* in East Africa and discussion of problems raised by the unexpected reaction of some of the more promising insects to *Sesamum indicum*. In *Proceedings of the II International Symposium on Biological Control of Weeds*, ed. P. H. Dunn. Slough, UK: Commonwealth Agricultural Bureaux, pp. 89–92.

Grevstad, F. S. (1996). Establishment of weed control agents under the influences of demographic stochasticity, environmental variability and Allee effects. In *Proceedings of the IX International Symposium on Biological Control of Weeds*, ed. V. C Moran and J. H. Hoffmann. Stellenbosch, South Africa: University of Cape Town, pp. 261–267.

Gujral, G. S. and Vasudevan, P. (1983). *Lantana camara* L., a problem weed. *Journal of Scientific and Industrial Research*, **42**, 281–286.

Gupta, M. and Pawar, A. D. (1984). Role of *Teleonemia scrupulosa* Stål in controlling *Lantana*. *Indian Journal of Weed Science*, **16**, 221–226.

Habeck, D. H. (1976). The case for biological control of lantana in Florida citrus groves. *Proceedings of the Florida State Horticultural Society*, **89**, 17–18.

Hardin, G. (1960). The competitive exclusion principle. *Science*, **131**, 1292–1297.

Harley, K. L. S. (1969a). Assessment of the suitability of *Plagiohammus spinipennis* (Thoms.) (Col., Cerambycidae) as an agent for the control of weeds of the genus *Lantana* (Verbenaceae). I. Life history and capacity to damage *L. camara* in Hawaii. *Bulletin of Entomological Research*, **58**, 567–577.

Harley, K. L. S. (1969b). The suitability of *Octotoma scabripennis* Guér and *Uroplata girardi* Pic (Col., Chrysomelidae) for the control of *Lantana* (Verbenaceae) in Australia. *Bulletin of Entomological Research*, **58**, 835–843.

Harley, K. L. S. (1970). Recent advances in biological control of *Lantana camara*. In *Proceedings of the Fourth Australian Weeds Conference*, held in Hobart, Tasmania, pp. 6–8.

Harley, K. L. S. (1971). Biological control of *Lantana*. *PANS (C)*, **17**, 433–437.

Harley, K. L. S. (1992). Report on surveys in Solomon Islands, Vanuatu and Fiji. Unpublished report, CSIRO Division of Entomology, Indooroopilly, Queensland, 31 pp.

Harley, K. L. S. and Kassulke, R. C. (1971). *Tingidae* for biological control of *Lantana camara (Verbenaceae)*. *Entomophaga*, **16**, 389–410.

Harley, K. L. S. and Kassulke, R. C. (1974). The suitability of *Phytobia lantanae* Frick for biological control of *Lantana camara* in Australia. *Journal of the Australian Entomological Society*, **13**, 229–233.

Harley, K. L. S. and Kunimoto, R. K. (1969). Assessment of the suitability of *Plagiohammus spinipennis* (Thoms.) (Col., Cerambycidae) as an agent for control of weeds of the genus *Lantana* (Verbenaceae). II. Host specificity. *Bulletin of Entomological Research*, **58**, 787–792.

Harley, K. L. S., Kerr, J. D. and Kassulke, R. C. (1979). Effects in S.E. Queensland during 1967–72 of insects introduced to control *Lantana camara*. *Entomophaga*, **24**, 65–72.

Harris, P. (1973). The selection of effective agents for the biological control of weeds. *The Canadian Entomologist*, **105**, 1494–1503.

Haseler, W. H. (1963). Progress in insect control of *Lantana*. *Queensland Agricultural Journal*, **89**, 65–68.

Hilje, L. (1985). Insectos visitadores y eficiencia reproductiva de *Lantana camara* L. (Verbenaceae). *Brenesia*, **23**, 293–300.

Hokkanen, H. and Pimentel, D. (1984). New approach for selecting biological control agents. *The Canadian Entomologist*, **116**, 1109–1121.

Holm, L. G., Plucknett, D. L., Pancho, J. V. and Herberger, J. P. (1991). *A Geographic Atlas of World Weeds*. Malabar, FL: Krieger Publishing Company, 391 pp.

Howard, R. A. (1969). A check list of cultivar names used in the genus *Lantana*. *Arnoldia*, **29**, 73–109.

Humphries, S. E. and Stanton, J. P. (1992). *Weed Assessment in the Wet Tropics World Heritage Area of North Queensland*. Report to The Wet Tropics Management Agency, Cairns, Australia, 75 pp.

Julien, M. H. and Griffiths, M. W. (1998). *Biological Control of Weeds: A World Catalogue of Agents and their Target Weeds*. 4th edn. Wallingford, UK: CABI Publishing, 223 pp.

Kamath, M. K. (1979). A review of biological control of insect pests and noxious weeds in Fiji (1969–1978). *Fiji Agricultural Journal*, **41**, 55–72.

Katabazi, B. K. (1983). The tsetse fly *Glossina fuscipes* in the sleeping sickness epidemic area of Busoga, Uganda. *East African Medical Journal*, **60**, 397–401.

Khan, A. H. (1945). On the lantana bug (*Teleonemia scrupulosa* Stål.). *Indian Journal of Entomology*, **6**, 149–161.

Koebele, A. (1903). Report on enemies of *Lantana camara* in Mexico, and their introduction into the Hawaiian Islands. In *The Introduction into Hawaii of Insects that Attack Lantana*, ed. R. C. L. Perkins and O. H. Swezey. *Bulletin of the Experiment Station of the Hawaiian Sugar Planters' Association*, **16**, 1–83.

Krauss, N. L. H. (1953). Notes on insects associated with lantana in Cuba. *Proceedings, Hawaiian Entomological Society*, **15**, 123–125.

Krauss, N. L. H. (1964). Some leaf-mining chrysomelids of lantana (Coleoptera). *The Coleopterists' Bulletin*, **18**, 92–94.

Kugler, H. (1980). On the pollination of *Lantana camara* L. *Flora*, **169**, 524–529.

Lamb, D. (1991). Forest regeneration research for reserve management: some questions deserving answers. In *Tropical Rainforest Research in Australia: Present Status*

and Future Directions for the Institute for Tropical Rainforest Studies,
ed. N. Goudberg, M. Bonell and D. Benzaken. Townsville, Australia: Institute
for Tropical Rainforest Studies, pp. 177–181.

Loyn, R. H. and French, K. (1991). Birds and environmental weeds in south-eastern
Australia. *Plant Protection Quarterly*, **6**, 137–149.

Matthew, K. M. (1971). The high altitude ecology of the lantana. *Indian Forester*, **97**,
170–171.

Mbulamberi, D. B. (1990). Recent outbreaks of human trypanosomiasis in Uganda. *Insect
Science and its Application*, **11**, 289–292.

Memmott, J., Fowler, S. V., Harman, H. M. and Hayes, L. M. (1996). How best to release
a biological control agent. In *Proceedings of the IX International Symposium for
Biological Control of Weeds*, ed. V. C. Moran. and J. H. Hoffmann, Rondebosch,
South Africa: University of Cape Town, pp. 291–296.

Mishra, S. C. and Sen-Sarma, P. K. (1986). Host specificity test and a note on life history
of *Leptobyrsa decora* Drake (Hemiptera: Tingidae) on teak. *Bulletin of Entomology*,
27, 81–86.

Misra, R. M. (1985). A note on *Leptobyrsa decora* Drake (Hemiptera: Tingidae). A
biocontrol agent of *Lantana camara* (Verbenaceae). *Indian Forester*,
111, 641–644.

Mohan Ram, H. Y. and Mathur, G. (1984). Flower-insect interaction in pollination.
Proceedings of the Indian Academy of Science (Animal Science), **93**, 359–363.

Muniappan, R. (1989). Biological control of *Lantana camara* L. in Yap. *Proceedings,
Hawaiian Entomological Society*, **29**, 195–196.

Muniappan, R. and Viraktamath, C. A. (1986). Status of biological control of the weed,
Lantana camara in India. *Tropical Pest Management*, **32**, 40–42.

Muniappan, R., Denton, G. R. W., Brown, J. W., *et al.* (1996). Effectiveness of the natural
enemies of *Lantana camara* on Guam: a site and seasonal evaluation. *Entomophaga*,
41, 167–182.

Neal, J. (1999). Assessing the sterility of ornamental lantana varieties: are we
exacerbating the weed problem? Honours thesis, Department of Botany, University
of Queensland, Brisbane.

Palmer, W. A. and Pullen, K. R. (1995). The phytophagous arthropods associated with
Lantana camara, L. *hirsuta*, L. *urticifolia*, and L. *urticoides* (Verbenaceae) in North
America. *Biological Control*, **5**, 54–72.

Palmer, W. A. and Pullen, K. R. (1998). The host range of *Falconia intermedia* (Distant)
(Hemiptera: Miridae): a potential biological control agent for *Lantana camara*
L. (Verbenaceae). *Proceedings of the Entomological Society of Washington*,
100, 633–635.

Parsons, W. T. and Cuthbertson, E. G. (2001). Common lantana. In *Noxious Weeds of
Australia*. Melbourne, Australia: CSIRO Publishing, pp. 627–632.

Rao, V. P., Ghani, M. A., Sankaran, T. and Mathur, K. C. (1971). *A Review of the
Biological Control of Insects and Other Pests in South-east Asia and the Pacific
Region*. Technical Communication 6., Slough, UK: Commonwealth Institute of
Biological Control, pp. 60–62.

Sahu, A. K. and Panda, S. (1998). Population dynamics of a few dominant plant species
around industrial complexes, in West Bengal, India. *Journal of Bombay Natural
History Society*, **95**, 15–18.

Sanders, R. W. (2006). Taxonomy of *Lantana* sect. *Lantana* (Verbenaceae). I. Correct
application of *Lantana camara* and associated names. *Sida*, **22**, 381–421.

Schroeder, D. and Goeden, R. D. (1986). The search for arthropod natural enemies of introduced weeds for biological control – in theory and practice. *Biocontrol News and Information*, **7**, 147–155.

Scott, L. J. (1998). *Identification of Lantana spp. Taxa in Australia and the South Pacific.* Final Report to ACIAR, Co-operative Research Centre for Tropical Pest Management, Brisbane, Australia, 26 pp.

Scott, L. J., Hannan-Jones, M. A. and Graham, G. C. (2002). Affinities of *Lantana camara* in the Australia-Pacific region. In *Proceedings of the 13th Australian Weeds Conference*, ed. H. Spafford-Jacob, J. Dodd and J. H. Moore. Perth, Australia: Plant Protection Society of Western Australia, pp. 471–474.

Sen-Sarma, P. K. and Mishra, S. C. (1986). Biological control of forest weeds in India – retrospect and prospects. *Indian Forester*, **112**, 1088–1093.

Sharma, O. P. (1994). Plant toxicosis in north-western India. In *Plant-associated Toxins: Agricultural, Phytochemical and Ecological Aspects*, ed. M. S. Colgate, and P. R. Dorling. Wallingford, UK: CAB International, pp. 19–24.

Sharma, O. P. and Sharma, P. D. (1989). Natural products of the lantana plant – the present and prospects. *Journal of Scientific and Industrial Research*, **48**, 471–478.

Sharma, O. P., Makkar, H. P. S. and Dawra, R. K. (1988). A review of the noxious plant *Lantana camara*. *Toxicon*, **26**, 975–987.

Shea, K. and Possingham, H. P. (2000). Optimal release strategies for biological control agents: an application of stochastic dynamic programming to population management. *Journal of Applied Ecology*, **37**, 77–86.

Sheppard, A. W. (1992). Predicting biological weed control. *Tree*, **7**, 290–291.

Simelane, D. O. (2002). Biology and host range of *Ophiomyia camarae*, a biological control agent for *Lantana camara* in South Africa. *BioControl*, **47**, 575–585.

Simelane, D. O. (2005). Biological control of *Lantana camara* in South Africa: targeting a different niche with a root-feeding agent, *Longitarsus* sp. *BioControl*, **50**, 375–387.

Simelane, D. O. and Phenye, M. S. (2005). Suppression of growth and reproductive capacity of the weed *Lantana camara* (Verbenaceae) by *Ophiomyia camarae* (Diptera: Agromyzidae) and *Teleonemia scrupulosa* (Heteroptera: Tingidae). *Biocontrol Science and Technology*, **15**, 153–163.

Smith, L. S. and Smith, D. A. (1982). The naturalised *Lantana camara* complex in Eastern Australia. *Queensland Botany Bulletin*, **1**, 1–26.

Stegmaier, C. E., Jr. (1966). A leaf miner on *Lantana* in Florida, *Ophiomyia camarae* (Diptera, Agromyzidae). *The Florida Entomologist*, **49**, 151–152.

Stirton, C. H. (1977). Some thoughts on the polyploidy complex *Lantana camara* L. (Verbenaceae). In *Proceedings of the Second National Weeds Conference of South Africa*, ed. D. P. Annecke, Cape Town, South Africa: A. A. Balkema, pp. 321–340.

Sutherst, R. W., Maywald, G. F., Yonow, T. and Stevens, P. M. (1999). *CLIMEX: Predicting the Effects of Climate on Plants and Animals, Version 1.1.* Melbourne, Australia: CSIRO.

Swarbrick, J. T. (1986). History of the lantanas in Australia and origins of the weedy biotypes. *Plant Protection Quarterly*, **1**, 115–121.

Swarbrick, J. T., Willson, B. W. and Hannan-Jones, M. A. (1998). *Lantana camara* L. In *The Biology of Australian Weeds*, ed. F. D. Panetta, R. H. Groves and R. C. H. Shepherd. Melbourne, Australia: R. G. and F. J. Richardson, pp. 119–140.

Swezey, O. H. (1923). Records of introduction of beneficial insects into the Hawaiian Islands. *Proceedings of the Hawaiian Entomological Society*, **5**, 299–304.

Swezey, O. H. (1924). Present status of lantana and its introduced insect enemies. In *The Introduction into Hawaii of Insects that Attack Lantana*, ed. R. C. L. Perkins and O. H. Swezey. *Bulletin of the Experiment Station of the Hawaiian Sugar Planters' Association*, **16**, 1–83.

Taylor, E. E. (1989). A history of biological control of *Lantana camara* in New South Wales. *Plant Protection Quarterly*, **4**, 61–65.

Thaman, R. R. (1974). *Lantana camara*: its introduction, dispersal and impact on islands of the tropical Pacific Ocean. *Micronesica*, **10**, 17–39.

Thomas, S. E. and Ellison, C. A. (2000). A century of classical biological control of *Lantana camara*: can pathogens make a significant difference? In *Proceedings of the X International Symposium on Biological Control of Weeds*, ed. N. R. Spencer. Sidney, MT: USDA-ARS, pp. 97–104.

Tomley, A. J. and Evans, H. C. (1992). Some problem weeds in tropical and sub-tropical Australia and prospects for biological control using fungal pathogens. In *Proceedings of the VIII International Symposium on Biological Control of Weeds*, ed. E. S. Delfosse and R. R. Scott. Melbourne, Australia: DSIR/CSIRO, pp. 477–482.

Tomley, A. J. and Riding, N. (2002). *Prospodium tuberculatum*, lantana rust, a new agent released for the biocontrol of the woody shrub *Lantana camara*. In *Proceedings of the 13th Australian Weeds Conference*, ed. H. Spafford-Jacob, J. Dodd and J. H. Moore. Perth, Australia: Plant Protection Society of Western Australia, pp. 389–390.

Trujillo, E. E. and Norman, D. J. (1995). *Septoria* leaf spot of lantana from Ecuador: a potential biological control for bush lantana in forests of Hawaii. *Plant Diseases*, **79**, 819–821.

Urban, A. J., Mpedi, P. F., Neser, S. and Craemer, C. (2001). Potential of the flower gall mite, *Aceria lantanae* (Cook) (Acari: Eriophyidae), for biocontrol of the noxious weed, *Lantana camara* L. (Verbenaceae). In *Proceedings of the 13th Entomological Congress*, ed. T. Olckers and D. J. Brothers. Pretoria, South Africa: Entomological Society of South Africa, pp. 67–68.

Vivian-Smith, G., Gosper, C. R., Wilson, A. and Hoad, K. (2006). *Lantana camara* and the fruit- and seed-damaging fly, *Ophiomyia lantanae* (Agromyzidae): seed predator, recruitment promoter or dispersal disruptor? *Biological Control*, **36**, 247–257.

Wapshere, A. J., Delfosse, E. S. and Cullen, J. M. (1989). Recent development in biological control of weeds. *Crop Protection*, **8**, 227–250.

Waterhouse, D. F. and Norris, K. R. (1987). *Biological Control: Pacific Prospects*. Melbourne: Inkata Press, 454 pp.

Wells, M. J. and Stirton, C. H. (1988). *Lantana camara*: a poisonous declared weed. *Farming in South Africa Weeds*, **A.27**, 1–4.

White, C. T. (1929). Weeds of Queensland: the correct botanical identity of the Lantanas naturalised in Queensland. *Queensland Agricultural Journal*, **April**, 294–296.

Winder, J. A. (1980). Factors affecting the growth of lantana in Brazil. Ph.D. thesis, Department of Agriculture and Horticulture, University of Reading, UK.

Winder, J. A. and Harley, K. L. S. (1982). The effects of natural enemies on the growth of *Lantana* in Brazil. *Bulletin of Entomological Research*, **72**, 599–616.

Winder, J. A. and Harley, K. L. S. (1983). The phytophagous insects on *Lantana* in Brazil and their potential for biological control in Australia. *Tropical Pest Management*, **29**, 346–362.

Winder, J. A. and van Emden, H. F. (1980). Selection of effective biological control agents from artificial defoliation/insect cage experiments. In *Proceedings of the V International Symposium on Biological Control of Weeds*, ed. E. S. Delfosse. Melbourne, Australia: CSIRO, pp. 415–439.

Zalucki, M. P., Day, M. D. and Playford, J. (2007). Will biological control of *Lantana camara* ever succeed? Patterns, processes and prospects. *Biological Control*, **42**, 251–261.

Zalucki, M. P. and van Klinken, R. (2006). Predicting population dynamics and abundance of introduced biological agents: science or gazing into crystal balls. *Australian Journal of Entomology*, **45**, 331–344.

13

Mimosa diplotricha C. Wright ex Sauvalle (Mimosaceae)

Lastus S. Kuniata

13.1 Introduction

Mimosa diplotricha C. Wright ex Sauvalle (= *Mimosa invisa* Mart. ex Colla) (Mimosaceae) is commonly known as the giant sensitive plant, creeping sensitive plant and nila grass; various local names also exist wherever it has been introduced (Waterhouse, 1994). It is a native of Central America to Brazil (Holm *et al.*, 1977) but has become a serious weed outside its natural range.

13.2 Distribution

It has been recorded as an invasive weed in American Samoa, Australia, Commonwealth of the Northern Mariana Islands, Cook Islands, Federated States of Micronesia, Fiji, French Polynesia, Guam, New Caledonia, Niue, Palau, Papua New Guinea, Samoa, Solomon Islands, Vanuatu, and Wallis and Futuna in the Pacific; Cambodia, China, India, Indonesia, Malaysia, the Philippines, Taiwan, Thailand, and Vietnam in Asia; and Mauritius, Nigeria, and Reunion in Africa (Holm *et al.*, 1977; Waterhouse and Norris, 1987; PIER, 2006; Invasive Species Specialist Group website, www.issg.org).

13.3 Ecology and biology

Mimosa includes 400–450 species, which are mostly native to Central and South America (Lewis and Elias, 1981). *Mimosa diplotricha* is widespread in South America, Central America, the West Indies, Mexico, Puerto Rico, parts of Africa, India, Southeast Asia, Australia, and the Pacific Islands (Waterhouse, 1994). It is considered a major weed in pasture, plantations, roadsides, and wet degraded lands and can also be a serious problem in crop areas (Waterhouse, 1994).

Mimosa diplotricha is a shrubby or sprawling annual that sometimes behaves as a perennial vine and forms a dense thicket (PIER, 2006). It has a strong root system, which is often woody at the decumbent base. Stems tend to bunch, often scramble over other

Biological Control of Tropical Weeds using Arthropods, ed. R. Muniappan, G. V. P. Reddy, and A. Raman. Published by Cambridge University Press. © Cambridge University Press, 2009.

plants, and are four-angled, the angles usually with a line of sharp prickles. Leaves are alternate, bright green, feathery and fern-like, each leaf divided into five to seven pairs of segments. Each segment carries about 20 pairs of tiny leaflets, which close when either disturbed or injured, and at night times. Small pale-pink flowers occur as round, fluffy globular umbels (12 mm across), on short stalks in the leaf axils. Numerous pods occur in clusters (each 25 × 6 mm when ripe). Covered by small prickles, these pods break later into 4–5 one-seeded pieces (Waterhouse and Norris, 1987; Waterhouse, 1994).

Mimosa diplotricha grows best in tropical regions with high moisture and in highly fertile soils (Swarbrick, 1997). It prefers open areas with a lot of sunlight. This plant has become a serious weed outside its natural habitat including the Pacific Rim, where it has been the context of several eradication programs. Early detection and control is usually recommended to prevent massive infestations. In agricultural situations, *M. diplotricha* was most likely introduced as a legume crop cover or through contaminated pasture seeds. *Mimosa diplotricha* seeds are transported by running water, vehicles, machinery, stock, and contaminated earth. To minimize spread of this weed, all heavy machinery and equipment must be properly cleaned before they are moved to other areas.

Mimosa diplotricha is an annual, which usually flowers and seeds from April (autumn) to the end of June (midwinter) in Australian conditions. In years when there has been little cold weather, plants seed from April to December and some plants of at least 10 cm height can set seeds.

This plant has the potential to produce 15–20 000 seeds/m^2 (Kuniata and Korowi, 2001). Seeds have been known to lie dormant for up to 50 years (Department of Natural Resources, Mines and Water, 2005). Seeds germinate once brought to the soil surface. These can germinate any time during the year, but sound establishment takes place during wet season. Fire can stimulate germination.

13.4 Impact – weed status

Mimosa diplotricha is a serious weed in the Pacific, Australia, Asia, and in some African countries (Waterhouse and Norris, 1987). It causes major problems in coconut, oil palm, tea, and rubber plantations, sugarcane and pineapple fields, cropland (cassava, tomatoes, upland rice, soybeans, maize, peanuts), and pasture land (Fig. 13.1). It is common along roadsides, wastelands, and in moist places restricting movement of humans, because of the nuisance of sharp prickles. Moreover, humans or stock bruised with *M. diplotricha* prickles can easily get infected. In a number of Pacific countries, *M. diplotricha* is considered amongst the top 10 worst weeds (Waterhouse and Norris, 1987, Waterhouse, 1997).

It is difficult to estimate economic losses and the cost of control of *M. diplotricha* country wide. However, Kuniata (1994) reported from cattle properties owned by Ramu Sugar Ltd. in the Ramu–Markham Valleys in Madang–Morobe Provinces, Papua New Guinea, that up to US$130 000 annually was spent on the chemical control and slashing of this weed. These costs could be higher in a more intensive cattle production system compared with the low input situations such as at Ramu Sugar which utilize natural

Fig. 13.1 Severe infestation of *Mimosa diplotricha* and *Sida* spp. in cattle property at Markham valley, Papua New Guinea.

pastures. On Ramu Sugar's sugarcane estate more than 80% of the cane fields were infested by *M. diplotricha*. It was estimated that up to three engine-hours downtime per day/harvester (approx. US$200 per harvester) was experienced as a result of *M. diplotricha* interference with normal sugarcane harvesting (green cane). Estimated total cost of lost time due the *M. diplotricha* interference was US$320 000 per year. In small village farmers' plots, the crops such as cassava, bananas, and sweet potatoes many not get harvested at all because of this weed. *Mimosa diplotricha* infestations cause erosion of endemic species such as *Imperata cylindrica* L. (Beauv.) (Poaceae: Gramineae) used for thatching roofs in remote villages in Papua New Guinea.

13.5 Control measures

Hand weeding is possible in smaller areas, but the prickles on the vines can cause serious sores. Slashing or rolling of the vines offers short-term reduction in vegetative growth. In a cassava crop, Alabi *et al.* (2004) showed that up to three activities of hand weeding–slashing are necessary for better yields.

A number of chemical products have been recommended for the control of *M. diplotricha*. Preemergent herbicides such as Balance (isoxaflutole), atrazine, and diuron can be used in seed beds, but only remain active for a few months and are too expensive for smallholders and for controlling the weed in degraded lands. Alabi *et al.* (2004) obtained better yields and returns in cassava crops with use of atrazine + metolachor for *M. diplotricha* control. Although Balance can effectively control this weed for 3–4 months in sugarcane (L. S. Kuniata, unpublished data), this treatment requires an excellent seed bed to be effective.

Paraquat + diuron applied postemergence has provided good control of *M. diplotricha* when applied at 2–5 leaf stage. 2,4-D + atrazine has been applied as an overall spray especially in pasture situations to control young to semimature stands (Swarbrick, 1997). Use of herbicides can be expensive and require repeated applications. Persistent herbicides used for *M. diplotricha* control are hazardous to humans handling them, contaminate the environment, and result in pesticide residues in animal products.

13.6 Biological control

The mechanical and chemical control of *M. diplotricha* requires sustained efforts and constant supply of resources which are often limited in developing countries such as in the South Pacific. The use of herbicides to control this weed can also lead to environmental degradation of ecologically sensitive sites such as on atolls. Biological control not only can provide sustainable control but also is safe to humans and the environment. However, as a prerequisite to a successful biocontrol program, the biology and ecology of both the agent and target weed species need to be thoroughly investigated. Modification of the environment (e.g. by the application of nitrogen) often helps to improve establishment by increasing host plant quality, which indirectly affects biological control agent performance. The effectiveness of this practice has been demonstrated with the aquatic weed *Salvina molesta* Mitchell (Salviniaceae) (Room and Thomas, 1985), *M. diplotricha* (Kuniata, 1994), and *Sida* spp. (Kuniata and Korowi, 2003). Nitrogen fertilizer applied to the plants indirectly increased insect numbers, resulting in a population explosion which subsequently caused severe damage to the target weeds. A number of biological control agents have been investigated to manage *M. diplotricha* (Julien and Griffiths, 1998). Up to 59 species of insects and three pathogens have been recorded on *M. diplotricha* in its native habitat (Garcia, 1982; Waterhouse and Norris, 1987).

13.6.1 Heteropsylla spinulosa *Muddiman, Hodkinson and Hollis* (Hemiptera: Psylloideae)

Heteropsylla includes about 15 described species mainly from Central America. Most species are recorded from the family Mimosaceae and appear to be host specific (Hodkinson and White, 1981). *Heteropsylla spinulosa* is a native of Central America and is confined to *M. diplotricha* as a host plant (Muddiman *et al.*, 1992). From about 100 plant species tested by Wilson and Garcia (1992), *H. spinulosa* developed successfully only on *M. diplotricha*, indicating its high specificity to this host. *Heteropsylla spinulosa* has been introduced for biological control of *M. diplotricha* in the Pacific Rim nations American Samoa, Australia, Cook Islands, Fiji, Palau, Papua New Guinea, Pohnpei, Samoa, Solomon Islands, and Yap (Kuniata, 1994; Muniappan and Schaefer, 1995; Esguerra *et al.*, 1997; Julien and Griffiths, 1998; DeMeo *et al.*, 2002). Feeding action of high populations of *H. spinulosa* adults and nymphs cause severe stunting and distortion of leaves and growing tips. Flowering is either greatly reduced or greatly prevented when young plants are attacked by *H. spinulosa.* Kuniata (1994) observed that application of nitrogen to stands of *M. diplotricha* indirectly increased the psyllid populations, subsequently inflicting severe damage on the plants (Fig. 13.2). The psyllid numbers remained low in plots that did not receive nitrogen fertilizer for the first two months after release and also the damage caused was slow. In contrast, the psyllid numbers were always high in plots receiving nitrogen fertilizer and the damage was quicker resulting in complete control within 2–3 months after release. In Papua New Guinea large stands of *M. diplotricha* were reduced to being an insignificant weed in pastures and other situations with 12 months of the psyllid releases. It took about four years for the psyllid to suppress the weed in long-term monitoring sites at Ramu Sugar in Papua New Guinea (Fig. 13.3). Similar results have been reported from all the countries wherein it was introduced.

Kuniata and Korowi (2003) discussed strategies of preserving the psyllid during dry seasons, such as application of nitrogen to the plants, raising of insects in irrigated plots, and, making releases when *M. diplotricha* is actively growing. In this way, the weed is controlled well before it becomes a problem in pastures and cropping situations.

Predatory arthropods such as the jumping spiders *Clynotis sp.* and *Cosmophasis sp.* (Salticidae) predatory bug, *Melanolestes* sp. (Hemiptera: Reduviidae) and a ladybeetle, *Eriopis connexa* Germar (Coleoptera: Coccinelidae) attack nymphs and adults of *H. spinulosa*. A predatory wasp, *Protonectarina sylveriae* De Saussure (Hymenoptera: Vespidae), also attacks the nymphs and adults of *H. spinulosa* (Garcia, 1985). Severe predation can occur but this normally happens late when most of the damage in the plant has occurred and the psyllid is migrating to new host plants.

13.6.2 Psigida (= Psilopigida) walkeri Grote (Lepidoptera: Citheroniidae)

Psigida walkeri is widespread in Brazil on *M. diplotricha*. The moth has also been evaluated in Australia and shown to cause severe damage on *Mimosa* spp. (Garcia, 1982; Vitelli *et al*, 2001). The larvae can cause considerable damage by feeding voraciously on the leaves, tender stems and branches including flower buds and tender seed pods. Attack by *P. walkeri* can prevent flowering and seed production. It was also shown that this biocontrol agent feeds on *Acacia* and *Neptunia* plants, thus the release of this agent was not done in Australia. In field conditions, the larvae of *P. walkeri* are parasitized by a tachinid fly *Lespesia* sp. (Waterhouse and Norris, 1987), which may limit *P. walkeri*'s efficacy as a biological control agent of *M. diplotricha*.

(a) Adults

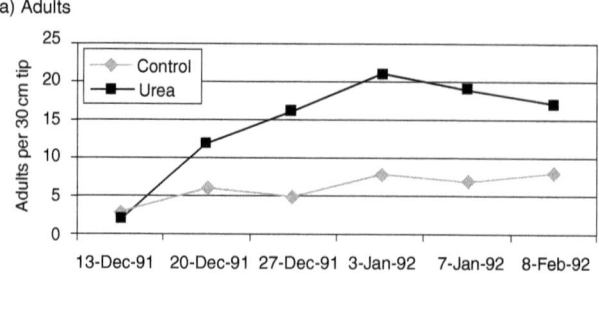

(b) Eggs

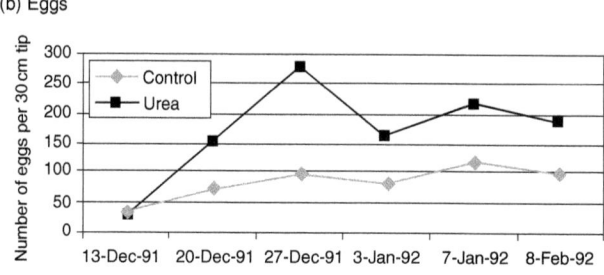

(c) Nymphs

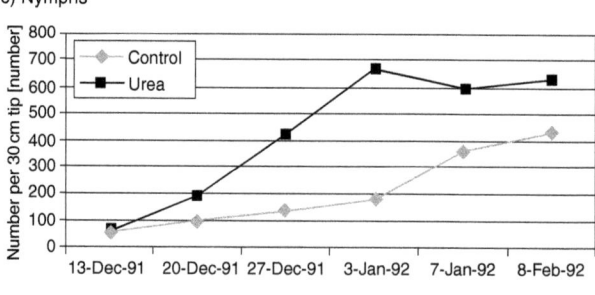

(d) Damage

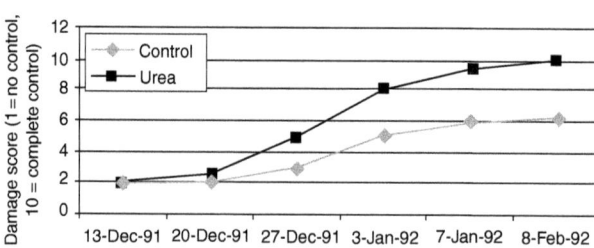

Fig. 13.2 Effect of nitrogen on *Heteropsylla spinulosa* (a) adults, (b) eggs, (c) nymphs and (d) damage in *Mimosa diplotricha* (Kuniata and Korowi, 2003).

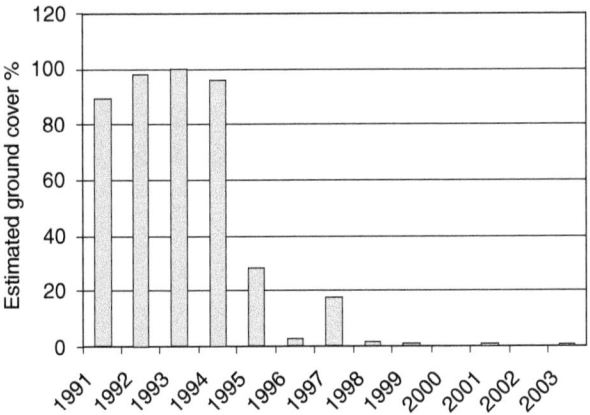

Fig. 13.3 Summary of *Mimosa diplotricha* cover in long-term monitoring site at Ramu Sugar plantation, Papua New Guinea (Kuniata and Korowi, 2003).

13.6.3 Other insects

Other insects that have been evaluated for the biological control of *M. diplotricha* included the larvae of the butterfly *Hemiargus hanno* Stoll (Lepidoptera: Lycaenidae), and the weevils *Promecops campanulicollis*, *Chalcodermus* nr *segnis* (Schonherr), and *Sibinia aspera* Marshall (Coleoptera: Curculionidae). The larvae of *H. hanno* feed on leaves, flowers, and seed pods of *M. diplotricha*. However, the larvae also feed on pigeon pea and other legumes, therefore this agent was not considered as a biological control agent. The initial work with the weevils has shown some potential and these could be further evaluated in the field. A number of insects have been recorded feeding on *M. diplotricha* in Papua New Guinea (Kuniata and Nagaraja, 1994). The most important are *Eurema hecabe* L. (Lepidoptera: Pieridae) and *Euproctis* sp. nr *trispila* Turner (Lepidoptera: Lymantriidae) the larvae of these butterflies can attack the tender shoots and leaves of *M. diplotricha* but they are not host specific. Although there may be other biological control agents that can be used for the control of *M. diplotricha*, *H. spinulosa* appears to be a better candidate to provide longer term control (Kuniata and Korowi, 2003), given its ability to breed fast in large numbers and disperse naturally by wind following releases.

13.7 Conclusion

Mimosa diplotricha, a native of Brazil and Central America has become an invasive weed in plantation crops, pastures, and wastelands in almost all of the Pacific countries, Asia, Australia, and Central Africa. Attempts to control this weed using physical and chemical means have given temporary results. Introduction and establishment of *H. spinulosa* into American Samoa, Australia, Cook Islands, Fiji, Palau, Papua New Guinea, Pohnpei,

Samoa, Solomon Islands, and Yap has provided a sustained suppression requiring little or no additional efforts to manage this weed. Most areas cleared of *M. diplotricha* by *H. spinulosa* in these island nations have been taken over by the native plants. India, Guam, and the Commonwealth of the Northern Mariana Islands are planning to introduce *H. spinulosa* for control of *M. diplotricha* in the near future (USDA, 2007).

References

Alabi, B. S., Ayeni, A. O. and Agboola, A. A. (2004). Economic assessment of manual and chemical control of thorny mimosa in cassava in Nigeria. Paper available from: http://www.cropscience.org.au/icsc2004/poster/2/4/1/447_alabibs.htm

DeMeo, R., Holm. T., Sengabau, F. and Miles, J. (2002). *Invasive Plants of Palau*. Republic of Palau: Palau Natural Resources Council, 34 pp.

Department of Natural Resources, Mines and Water (2005). Pest Series: giant sensitive plant: *Mimosa diplotricha (Mimosa invisa)*. Brisbane, Australia: The State of Queensland Government. Summary available from: http://www.dpi.qld.gov.au/cps/rde/dpi/hs.xsl/4790_7289_ENA_HTML.htm

Esguerra, N. M., William, J. D., Samuel, R. P. and Diopulos, K. J. (1997). Biological control of the weed, *Mimosa invisa* Von Martius on Pohnpei and Yap. *Micronesica*, **30**, 421–427.

Garcia, C. A. (1982). *A Survey of Insects Attacking Giant Sensitive Plant* (M. invisa) *in Brazil with Notes on its Distribution*. Brisbare, Australia: Queensland Department of Lands, 15 pp.

Garcia, C. A. (1985). *A Preliminary Report on the Host Specificity of* Heteropsylla *sp. (Hemiptera: Homoptera: Psyllidae) A Potential Agent for Biological Control of* Mimosa invisa *(Giant Sensitive Plant) in Queensland, Australia*. Report. Brisbone, Australia: Queensland Department of Lands, 10 pp. + 3 appendices.

Hodkinson, I. D and White, I. M. (1981). The Neotropical Psylloidea (Homoptera: Insecta): an annotated check list. *Journal of Natural History*, **15**, 491–523.

Holm, L. G., Plucknett, D. L., Pancho, J. V. and Herberger, J. P. (1977). *The World's Worst Weeds: Distribution and Biology*. Honolulu, HI: University Press of Hawaii, 609 pp.

Julien, M. H. and Griffiths, M. W. (1998). *Biological Control of Weeds: A World Catalogue of Agents and their Target Weeds*, 4th edn. Wallingford: CAB International. 223 pp.

Kuniata, L. S. (1994). Importation and establishment of *Heteropsylla spinulosa* (Homoptera: Psyllidae) for the biological control of *Mimosa invisa* in Papua New Guinea. *International Journal of Pest Management*, **40**, 64–65.

Kuniata, L. S. and Korowi, K. T. (2001). Biological control of the giant sensitive plant with *Heteropsylla spinulosa* (Homoptera: Psyllidae) in Papua New Guinea. In *Proceedings of the sixth Workshop on Tropical Agricultural Entomology*, held in Darwin, May 1998. Technical Bulletin 288. Darwin, Australia: Department of Primary Industry and Fisheries, pp. 145–151.

Kuniata, L. S. and Korowi, K. T. (2003). Bugs offer sustainable control of *Mimosa invisa* and *Sida* spp. in the Markham Valley, Papua New Guinea. *Proceedings of the XI International Symposium on Biological Control of Weeds*, ed. J. M. Cullen *et al.* Canberra, Australia: CSIRO Entomology, pp. 567–573.

Kuniata, L. S. and Nagaraja, H. (1994). Insects of the giant sensitive plant (*Mimosa invisa*) at Ramu, Papua New Guinea. *Papua New Guinea Journal of Agriculture, Forestry and Fisheries*, **37**, 36–39.

Lewis, G. P. and Elias, T. S. (1981). Mimoseae. In *Advances in Legume Systematics*, ed. R. M. Polhill and P. H. Raven. Kew: Royal Botanic Gardens, pp. 155–168.

Muddiman, S. B., Hodkinson, I. D. and Hollis, D. (1992). Legume-feeding psyllids of the genus *Heteropsylla* (Homoptera: Psyllidae). *Bulletin of Entomological Research*, **82**, 73–117.

Muniappan, R. and Schaefer, J. (1995). South Pacific Commission, German biological control project: a progress report. *Harvest*, **17**, 43–50.

Pacific Islands Ecosystems at Risk (PIER) (2006). http://www.hear.org/pier/species/mimosa_diplotricha.htm

Room, P. M. and Thomas, P. A. (1985). Nitrogen and establishment of a beetle for the biological control of the floating weed *Salvinia* in Papua New Guinea. *Journal of Applied Ecology*, **22**, 139–156.

Swarbrick, J. T. (1997). *Weeds of the Pacific Islands*. Technical paper No. 209. Noumea, New Caledonia: South Pacific Commission, 124 pp.

USDA [United States Department of Agriculture] (2007). *Field Release of* Hteropsylla spinulosa *(Homoptera: Psyllidae), a Non-indigenous Insect for Control of Giant Sensitive Plant*, Mimosa diplotricha *(Minosaceae), in Guam and the Commonwealth of the Northern Mariana Islands: Environmental Assessment 2007*. Riverdale, MD: Animal and Plant Health Inspection Service (APHIS), 17 pp.

Vitelli, M. P., Garcia, C., Lockett, C. J., West, G. M. and Willson, B. W. (2001). Host specificity and biology of the moth *Psigida walkeri* (Lepidoptera: Citheroniidae), a potential biological control agent for *Mimosa diplotricha* in Australia and the South Pacific. *Biological Control*, **22**, 1–8.

Waterhouse, D. F. (1994). *Biological Control of Weeds: Southeast Asian Prospects*. Monograph 26. Canberra, Australia: ACIAR, 302 pp.

Waterhouse, D. F (1997). *The Major Invertebrate Pests and Weeds of Agriculture and Plantation Forestry in the Southern and Western Pacific*. Monograph 44. Canberra, Australia: ACIAR, 99 pp.

Waterhouse, D. F. and Norris, K. R. (1987). *Biological Control: Pacific Prospects*. Melbourne, Australia: Inkata Press, 454 pp.

Wilson, B. W. and Garcia, C. A. (1992). Host specificity and biology of *Heteropsylla spinulosa* (Hom.: Psyllidae) introduced into Australia and Western Samoa for the biological control of *Mimosa invisa*. *Entomophaga*, **37**, 293–299.

14

Mimosa pigra L. (Leguminosae)

Tim A. Heard and Quentin Paynter

14.1 Introduction

Mimosa pigra L., a prickly, perennial, woody shrub native to tropical America from Mexico to Argentina, is listed in the Global Invasive Species Database as one of the *One Hundred of the World's Worst Invasive Alien Species*. We refer to *M. pigra* in the strict sense, which excludes *M. asperata*, a close relative which some authors have lumped with *M. pigra* as one species (Heard, 2004). Outside its native range, especially Southeast Asia and Australia, untreated infestations may double in area each year (Lonsdale, 1993; Triet *et al.*, 2004) and form dense thickets that affect both conservation areas and agricultural land (Samouth, 2004; Son *et al.*, 2004). These reduce the diversity of plants and animals and impact negatively on agriculture by competing with pasture species, hindering mustering of livestock, and restricting access to water by humans and livestock (Braithwaite *et al.*, 1989; Lonsdale *et al.*, 1989).

A biological control project in Australia against *M. pigra* has been active since 1979 with exploration work on natural enemies from bases in Brazil then Mexico. This has led to a steady stream of agents being released in Australia between 1983 and 2007 and in Asian countries between 1983 and 1998. A total of 13 insect species and two pathogenic fungi have been released against *M. pigra* in Australia (Table 14.1). Both pathogens failed to establish (Hennecke, 2004; Hennecke, 2006). Few biological control projects have continued for so long and released so many agents. To our knowledge, only *Lantana camara* has exceeded this number of total agents released. The release of a large number of species is not in itself a success. Some regard it as a failure in agent selection, a process which should have identified and released only the most effective agents. Success must be judged on the overall impact on the target organism. However, the wide range of agents provides a large pool of agents from which other countries have an option to choose.

Heard and Pettit (2005) estimated the biological control and integrated management components of the *M. pigra* project to cost US\$20 million over 24 years, with the 14 species released (at that time) costing an average of about US\$570 000/agent. The cost of introducing these agents to other countries will be less, as the exploration work and much

Biological Control of Tropical Weeds using Arthropods, ed. R. Muniappan, G. V. P. Reddy, and A. Raman. Published by Cambridge University Press. © Cambridge University Press, 2009.

Table 14.1 *The status of agents for biological control of* Mimosa pigra *in chronological order of release in Australia*

Agent	Plant part attacked	Where and when released	Where established
Acanthoscelides puniceus Johnson (Coleoptera, Bruchidae)	Mature hard seeds	Australia 1983 Thailand 1983 Malaysia 1991 Vietnam 1987 Myanmar, 1988	Australia Thailand Malaysia Vietnam Myanmar
Acanthoscelides quadridentatus (Schaeffer) (Coleoptera, Bruchidae)	Mature hard seeds	Australia 1983 Thailand 1983 Vietnam 1987 Myanmar, 1988	Thailand Vietnam Myanmar
Chlamisus mimosae Karren (Coleoptera, Chrysomelidae)	Leaves, stems	Australia 1985 Thailand 1985 Vietnam 1990	Australia
Neurostrota gunniella (Busck) (Lepidoptera, Gracillariidae)	Bores in pinnae and small stems	Australia 1989	Australia
Carmenta mimosa Eichlin & Passoa (Lepidoptera, Sesiidae)	Bores in large stems	Australia 1989 Thailand 1989 Vietnam 1996 Malaysia 1997 Indonesia 1998	Australia, Vietnam, Malaysia
Coelocephalapion aculeatum (Fall) (Coleoptera, Curculionidae)	Flower buds	Australia 1992 Thailand, 1991	
Coelocephalapion pigrae Kissinger (Coleoptera, Curculionidae)	Leaves and flower buds	Australia 1994	Australia
Phloeospora mimosae-pigrae H.C. Evans & G. Carrión (Fungus)	Leaves, stems and pods	Australia 1995	
Diabole cubensis L. (Fungus)	Leaves	Australia 1996	
Chalcodermus serripes Fahraeus (Coleoptera, Curculionidae)	Mature green seeds	Australia 1996	
Sibinia fastigiata Clark (Coleoptera, Curculionidae)	Young green seeds	Australia 1997	
Malacorhinus irregularis Jacoby (Coleoptera, Chrysomelidae)	Roots and leaves	Australia 2000	Australia
Macaria pallidata (Warren) (Lepidoptera, Geometridae)	Leaves	Australia 2002	Australia
Leuciris fimbriaria (Stoll) (Lepidoptera, Geometridae)	Leaves	Australia 2004	
Nesaecrepida infuscata (Coleoptera, Chrysomelidae)	Roots and leaves	Australia 2007	
Temnocerus debilis (Sharp) (Coleoptera, Rynchitidae)	Young leaves and tips	Potential future agent	

of the insect biology, rearing, and host-specificity testing have already been completed. These figures underline that biological control work is expensive at the beginning. Indeed, while Page and Lacey (2006) estimated the benefits to be an annual saving of AU$1.5 million (from decreased costs of chemical and mechanical control), the resulting benefit to cost ratio was just 0.8:1. This is low compared with outstanding examples such as 312 for prickly pear and 109 for rubbervine. However, Page and Lacey (2006) considered their *M. pigra* figure to be an underestimate as productivity gains and environmental, recreational, tourism, and cultural benefits were not considered. Furthermore, benefits should continue to increase, when the total impact of recently established agents improves biological control.

14.2 Native and cosmopolitan species that damage *M. pigra*

In Australia, more than 100 native species have been recorded on *M. pigra* (Wilson *et al.*, 1990). Some of these inflict conspicuous damage, although the impact is unknown. *Platyomopsis humeralis* White (Coleoptera: Cerambycidae) occasionally causes visible damage by girdling stems (Flanagan *et al.*, 1990) and efforts were made to redistribute this insect. *Mictis profana* (F.) (Hemiptera: Coreidae) breeds on *M. pigra*, causing stem-tip dieback, and is considered to make a positive, if small, contribution to biological control (Flanagan, 1994). Larvae of the moth *Maroga setiotricha* Meyrick (Xyloryctidae) feed in stems and cause a lot of damage per insect (Flanagan *et al.*, 1990). The widespread tropical plant pathogenic fungus *Lasiodiplodia theobromae* (Pat.) Griffon & Maubl. induces dieback of *M. pigra* in Australia in the dry season especially on plants stressed by drought and attacked by *Neurostrota gunniella* (Wilson and Pitkethley, 1992) or *Carmenta mimosa* (Blair Grace, personal communication). Napompeth (1982) listed the insects associated with *M. pigra* in Thailand and Indonesia. Cam *et al.* (1997) listed the insects on *M. pigra* in Vietnam. None of these insects was considered sufficiently damaging or safe for release outside their native range.

Risbecoma pigrae Rasplus (Hymenoptera: Eurytomidae) is a common seed-feeding wasp on *M. pigra* in Africa. It was first described from the Ivory Coast of Africa (Rasplus, 1988), and has been collected from northeastern Namibia, northeastern South Africa and Uganda (Stefan Neser, personal communication). The origin of this species is not known but because this type of seed-feeding chalcoid is generally host specific, it has potential for biological control purposes elsewhere. Extensive collecting over many years in Mexico and Central America has failed to recover this species (unpublished data), suggesting it originated in Africa.

14.3 Introduced biological control agents of *M. pigra*

Surveys for natural enemies conducted in seven countries by different collectors recorded approximately 420 species of insects that attack *M. pigra* (Harley *et al.*, 1995). Heard and Pettit (2005) showed that while large areas of the natural distribution were not surveyed, most natural enemies occur over the majority of the range of the host. Although the Isthmus of Panama is a barrier to many species, *M. pigra* was surveyed on either side of the Isthmus. Therefore, although many areas were not explored, the surveys were sufficiently thorough. Indeed, most species eventually released in Australia were discovered early in the exploration phase.

Fifteen agents (13 insects and two fungi) have been introduced into Australia. Several of these have been introduced to Asian countries also. Here we provide information on the origin, distribution, biology, interactions with host plant, host specificity, and parasites and predators, rearing and release, introduction to various countries, and the efficacy of the 13 insect agents released to date. Further information about these agents, and the species which were assessed but not released for various reasons, has been published elsewhere (Heard and Segura, 2004).

14.3.1 Stem borers

Neurostrota gunniella *(Busck) (Lepidoptera, Gracillariidae)*

Neurostrota gunniella is widespread wherever its hosts occur from southern Texas to Costa Rica and Cuba. Adults are frequently bred from *M. pigra, M. asperata*, and also from *Neptunia plena*, another legume native to the neotropics (Davis *et al.*, 1991). Eggs are laid singly on leaves; the first- and second-instar larvae mine the pinnules, and third-instar larvae enter the primary rachis and tunnel towards the stem where the remaining instars complete their development. The eighth (final) instar exits the stem, spins a cocoon between pinnules and then pupates. The total development time (egg–adult) is *c.* 30 days (at about 24 °C). Adult females lay a mean of 86 eggs (Davis *et al.*, 1991). Wilson *et al.* (1992) have described the rearing and release methods for *N. gunniella*.

No-choice host-specificity tests in the laboratory examined the oviposition preferences of adults and the ability of larvae to develop on test-plant species. Adults accepted two species of *Mimosa* (both weeds) and the four native Australian species of *Neptunia* for oviposition and the larvae completed development on all these species. The duration of larval development did not differ greatly among plant species but larval mortality on *Neptunia* spp. was higher than on *M. pigra*. Damage to *M. pigra* plants was much greater than to *Neptunia* species (Davis *et al.*, 1991). On the basis of these data, a decision was made to release it in Australia. However, Thailand did not approve release of this insect as laboratory trials showed that *N. gunniella* could reproduce on the introduced aquatic vegetable *Neptunia oleracea* (Forno *et al.*, 2000). It was predicted that, in Australia, *Neptunia* spp. would probably not support high populations and that nontarget attack to most *Neptunia* spp. was unlikely because only one species of *Neptunia* occurs in the same regions or habitats as *M. pigra*. However, this risk may have been understated because adult *N. gunniella* disperse rapidly (Wilson and Flanagan, 1990), while *M. pigra* has continued to spread. Nevertheless, Taylor *et al.* (2007) found that *Neptunia major* only suffered about 10% of the level of attack experienced by *M. pigra* in the field, and only plants growing in close proximity to *M. pigra* were affected. At the highest level of attack recorded on *N. major* in the field, tip death increased by 54% and plant height decreased by 6%. The effect of *N. gunniella* on the reproductive output of *N. major* could not be clearly resolved, but is likely to be small. Therefore, Taylor *et al.* (2007) concluded that the impact of *N. gunniella* on *N. major* is minor, confirming the predictions of the prerelease studies.

In contrast, *N. gunniella* damage significantly reduced both growth (14% reduction in radial growth in one season) and seed output (such that seed output was 60% lower than normal at the highest densities of the insect) of *M. pigra* plants (Lonsdale and Farrell, 1998), although these authors considered the damage insufficient to control the weed. By contrast, Paynter (2005) found that *N. gunniella* abundance was not correlated with seed set. Paynter (2005) confined sampling to within dense *M. pigra* thickets, and *N. gunniella* attack is greater on plants on the edge of stands compared with within stands (Smith and Wilson, 1995). Paynter (2006) hypothesized that a plant's capacity for compensatory growth (i.e. a density-dependent increase in growth and flowering of stems that escape attack) should be greater within stands, where *N. gunniella* attack is relatively low, compared with those at stand edges, where every stem may be heavily infested. Therefore, these contrasting results may indicate that spatial variation in *N. gunniella* abundance results in significant impacts (a greater than 50% reduction in seed production) on plants growing at the edges of *M. pigra* stands but not within dense stands.

Carmenta mimosa *Eichlin and Passoa (Lepidoptera: Sesiidae)*

Carmenta mimosa is known from southern Mexico, Honduras, and Nicaragua (Eichlin and Passoa, 1983). Eggs are laid on the plant and larvae enter the stem at the node. Large larvae sometimes girdle the stem and kill it. Pupation occurs within the stem. The adults live for 6–12 days and females lay an average of 295 eggs (Eichlin and Passoa, 1983). Host testing was confined to no-choice tests of larval development using transferred eggs. Larvae developed to adults on *M. pigra* only.

Carmenta mimosa was first released in Australia in 1989 (Forno *et al.*, 1991). The natural dispersal of *C. mimosa* is not rapid, presumably due to the long life cycle (mean: 98 days) and low rates of dispersal by adult moths. Two infestations in the Northern Territory increased by 5 and 10 times in density and 3 and 8 times in distribution in two years, indicating a rate of spread of about 2 km/year (Ostermeyer, 2000). Redistribution efforts were therefore continued and by 2004, *C. mimosa* was present on all river catchments with major *M. pigra* infestations and abundance at all sites was still increasing (Ostermeyer and Grace, 2007). Wilson *et al.* (1992) noted rearing and release methods for this agent. Ostermeyer *et al.* (2004) described a project for involving community groups and school children in the rearing and release of *C. mimosa* and *Macaria pallidata*.

Carmenta mimosa was predicted to be a useful agent because the larval feeding damage weakens stems, making them susceptible to breakage (Forno *et al.*, 1991). In Australia, this prediction has proved to be correct: Paynter (2005) showed that in three years, four out of eight stands where *C. mimosa* was absent expanded and none contracted. In contrast, none of nine stands where *C. mimosa* was present expanded and three contracted. Analysis of the age structure of *M. pigra* stands indicated that contracting stands were typically devoid of seedlings. *Mimosa pigra* seed rain was negatively correlated with *C. mimosa* damage and declined by more than 90% at the highest *C. mimosa* densities. Seed banks also declined with *C. mimosa* damage (Paynter, 2005). Furthermore, percentage cover of competing vegetation, mainly grasses and sedges, was significantly higher under stands defoliated by *C. mimosa*, probably because the dead and defoliated branches allowed light penetration to areas that would otherwise be heavily shaded. Competing vegetation both inhibits *M. pigra* seedling establishment (Lonsdale and Farrell, 1998) and apparently increased the suscep-tibility of *M. pigra* to fire, by increasing fuel loads beneath stands (Paynter, 2005).

Carmenta mimosa was also released in Thailand (1989), Vietnam (1996), Malaysia (1997), and Indonesia (1998). It did not establish in Thailand (Suasa-ard *et al.*, 2004) or Indonesia, where it was released at only one site (Soekisman Titrosoedirdjo, personal communication) but established in Vietnam, where it has spread and is being mass-reared and released (Son *et al.*, 2004). It appeared to establish in Malaysia (Soon and Chong, 1997), but recent information on its progress is lacking. According to the experience in Australia, we recommend an assessment of establishment and a renewal of redistribution efforts, especially as redistribution can be relatively cheap and easy (Ostermeyer *et al.*, 2004).

14.3.2 *Flower feeders*

Coelocephalapion aculeatum *(Fall) and* Coelocephalapion pigrae *Kissinger*
(Coleoptera: Curculionidae)

Coelocephalapion aculeatum was sourced from Mexico (it is also known from Texas, USA; Kissinger, 1992) and first released in Australia in 1992. *Coelocephalapion pigrae*, native to Venezuela and Brazil was discovered after the release of *C. aculeatum* and was first released in 1994 in Australia (Heard and Forno, 1996). Adults of both species feed on and oviposit into unopened flower buds, although *C. pigrae* also feeds on *M. pigra* leaves. Larvae of both species complete their development on the flowers. Host-specificity studies showed that both *C. aculeatum* and *C. pigrae* larvae could complete their development only on *M. pigra*, except for low levels of development on *Neptunia dimorphantha* Domin. (Forno *et al.*, 1994; Heard and Forno, 1996).

Although initially reported to have established in Australia, *C. aculeatum* has not been found in surveys for several years (Paynter, 2004a). *Coelocephalapion aculeatum* was also released in Thailand in 1991 where it also failed to establish (Suasa-ard *et al.*, 2004). In contrast, *C. pigrae* is now widespread in Australia and is still spreading to isolated infestations (Ostermeyer and Grace, 2007). *Coelocephalapion pigrae* was introduced into Malaysia for host testing but was not released (Soon and Chong, 1997).

Although *C. pigrae* prefers perfect flowers (i.e. containing female and male parts) over the male flowers, potentially allowing a greater impact on seed set (T. Heard, unpublished data), Paynter (2005, 2006) found no correlation between *C. pigrae* abundance and seed production in Australia. At one site in Australia, Paynter (2006) found *C. pigrae* abundance lagged behind sharp seasonal peaks in flower production and, while 55% of inflorescences were infested with *C. pigrae*, and an average of 2.7 beetles were reared per inflorescence, this level of infestation was estimated to destroy only 11% of flowers.

We speculate that the contrasting fortunes of *C. aculeatum* and *C. pigrae* in Australia are related to their ability to survive the dry season. Flowering almost ceases during the dry season in northern Australia (Lonsdale, 1988). *Coelocephalapion aculeatum* is an obligate flower feeder, whereas the adult *C. pigrae* also feed on *M. pigra* leaves, allowing them to survive in the absence of flowers. Both *Coelocephalapion* species may prove more effective in regions where flower production occurs throughout the year.

14.3.3 Seed feeders

Chalcodermus serripes *Fåhraeus (Coleoptera: Curculionidae)*

Chalcodermus serripes is a common seed predator through the entire native range of *M. pigra* (T. Heard, unpublished data). Adults lay eggs on seeds approaching their full size, but still green. Larvae develop inside the pod, each on a single seed. Testing focused on the oviposition and feeding preferences of adults, which proved to be entirely specific (Heard *et al.*, 1999).

Mass rearing this insect for release was difficult. Initially, permission was obtained to release field-collected adults following an intensive regimen process including holding of insects for two weeks, inspection for correct taxonomic identity and absence of parasites, washing in sodium hypochlorite solutions to eliminate adhering microorganisms, and internal inspection of a sample (5%) of beetles for pathogens. Eventually a method for mass rearing this species was developed and approximately 9300 field-collected or laboratory-reared adults were released between 1996 and 2000 (N. Graham, personal communication). Although a few adults were found at one site, one year after the last release had been made there, evidence is lacking that this species has persisted since then (Paynter, 2004a; Ostermeyer and Grace, 2007).

Sibinia fastigiata *Clark (Coleoptera: Curculionidae)*

Sibinia fastigiata is a common seed predator in Mexico, and Central and South America (Clark, 1984; T. Heard, unpublished data). Adults oviposit on small green seeds. Larvae develop inside the pod, each on a single seed. The first two larval instars feed on the perimeter of the seed allowing it to continue growing. Testing focused on the oviposition preferences of adults, which proved to be highly specific (Heard *et al.*, 1997). Adults feed nondestructively on pollen and possibly on nectar from open flowers.

Australian Quarantine and Inspection Service (AQIS) normally requires that imported insects be reared through one generation in quarantine before field release. Because *S. fastigiata* could not be reared in quarantine (as seeds do not grow in the low-light intensities), an exception was made by AQIS to release field-collected adults following an intensive regimen similar to that practiced on *C. serripes*. Approximately 25 importations from both Mexico and Brazil (1997–2002) resulted in the release of approximately 2450 field-collected adults. However, no evidence exists that *S. fastigiata* has established in Australia (Paynter, 2004a; Ostermeyer and Grace, 2007).

Acanthoscelides puniceus *Johnson and* Acanthoscelides quadridentatus *(Schaeffer) (Coleoptera: Bruchidae)*

Both *Acanthoscelides puniceus* and *A. quadridentatus* were imported from Mexico, although *A. quadridentatus* is widely distributed throughout the American tropics (Heard and Pettit, 2005). Seed feeders were chosen to target the dispersal of seed (Kassulke *et al.*, 1990) that was associated with rapid expansion of infestations (Lonsdale, 1993). In oviposition tests, conducted on a mixed colony, which was later shown to be only 10%, the eggs of *A. quadridentatus* were laid on 16 species but the resulting larvae only developed on *M. pigra*. Both *Acanthoscelides* spp. were released in Australia in 1983 using rearing and release methods described by Wilson *et al.* (1992).

Both species initially established in Australia although *A. puniceus* dominated from an early stage (Wilson and Flanagan, 1991) and only *A. puniceus* has been recovered in Australia in recent years (Paynter, 2004a). Ostermeyer and Grace (2007) showed that overall less than 2% of seeds were attacked and that the abundance of *A. puniceus* declined between 1997 and 2004, both in terms of number of sites where found and the abundance at those sites. They admit, however, that their surveys were done at times that were not ideal for recovering this species. Paynter (2005) found rates of attack up to *c.* 10%. While very high levels of seed predation are often assumed to be necessary for a seed feeder to control established infestations, if establishment of seedlings at the edge is seed limited then even the relatively low proportion of seeds consumed by *A. puniceus* may have an impact on the rate of invasion of *M. pigra* (Paynter, 2006). *Acanthoscelides puniceus* may prove more effective in regions where seed production is less strongly seasonal (Paynter, 2005).

Both species were also released in Thailand in 1983 (Julien and Griffiths, 1998) where rates of attack are sometimes much higher than in Australia, causing up to 87% damage to seeds (Suasa-ard *et al.*, 2004). The impacts on plant populations have not been quantified and opinions vary from "highly effective" (Napompeth, 1994) to "not enough to reduce *M. pigra* populations" (Suasa-ard *et al.*, 2004). *Acanthoscelides puniceus* was released in Malaysia in 1991, which although established, has not spread beyond the release site (Soon and Chong, 1997). Both species were released in Myanmar (Burma) from Thailand (Julien and Griffiths, 1998) and have established there (B. Napompeth, personal communication). These bruchids are also spreading naturally from Thailand to neighboring countries, and have been recovered in Laos, Malaysia, Singapore, Myanmar, and Indonesia (Napompeth, 1994).

14.3.4 Defoliators

Chlamisus mimosae *Karren (Coleoptera: Chrysomelidae)*

This species was described from a series obtained from a quarantine colony started from a collection made in Brazil in 1981 (Karren, 1989). Both larvae and adults feed on the epidermis of leaves and stems. Host specificity was tested using choice and no-choice tests of adult and first-instar larvae feeding. Only minor feeding damage by adults was detected and no larval feeding or development beyond first instar occurred on any of the species tested (R. C. Kassulke and K. L. S. Harley, unpublished data).

The first release in Australia was made in 1985, in Thailand in 1985, and in Vietnam in 1990 (Julien and Griffiths, 1998). It initially established in Thailand but then disappeared (B. Napompeth, personal communication). The Vietnamese release was a single attempt with a small number of individuals only and its fate is unknown. This species established in Australia in the Finnis River catchment where it occasionally reaches high numbers and causes moderate damage. Despite considerable efforts to redistribute it, *C. mimosae* has not persisted in other river systems (Ostermeyer and Grace, 2007). Even in the Finnis River system, *C. mimosae* numbers were low throughout the survey period 1997 to 2004. Ostermeyer and Grace (2007) suggested that it is unlikely to have a significant effect on *M. pigra* at such densities. Its long life cycle (3–4 months) makes it susceptible to predation, for example by green tree ants (T. Heard, personal observation).

Macaria pallidata Warren *(Lepidoptera: Geometridae)*

Macaria pallidata is a common and damaging species that is widespread across tropical America. Material for testing and eventually release into Australia was obtained from Mexico. Adults are short-lived, nocturnal and feed on nectar. Females deposit eggs directly onto leaves and stems. Larvae begin feeding by removing the top surface of the leaf. They feed, exposed without protection, on young and mature foliage completely stripping plants when larval densities are high. During the day, they rest on the tips of leaves in positions that imitate stems. They move very little from their original position until forced to look for more food after almost completely consuming pinnules. The larvae develop through five instars. Pupation is in the soil or among damaged plant tissue. Time from egg to adult under laboratory conditions at 25–27 °C is 25 days and there are several generations per year (Heard *et al.*, 2001).

The host specificity of *M. pallidata* was tested using laboratory larval development tests on 70 test plant species. Development to adult occurred on six species other than *M. pigra*. However, the survival rates were so low that these plants were considered unable to sustain a population of this insect species (Heard *et al.*, 2001).

Released in 2002, *M. pallidata* was found to have spread widely by 2004. It causes conspicuous larval feeding damage, and adults can be highly visible in the undergrowth around *M. pigra* populations (Routley and Wirf, 2006). Wirf (2006) demonstrated significant reductions in the growth of potted plants artificially infested with *M. pallidata* larvae, but field experiments have not yet quantified the impact of the agent under natural conditions. However, *M. pallidata* damage appears to be significant, contrary to predictions made before its release: Harley *et al.* (1995) listed *M. pallidata* as having no potential "because ectophagous larvae of Lepidoptera are often subjected to high levels of parasitism and predation." Grace (2005) showed that *M. pallidata* larvae are indeed subject to moderate levels of predation, and larval survival increased when ants were excluded. Larvae drop on a silken thread when disturbed (Heard *et al.*, 2001), perhaps providing a useful protection from predators. It will be interesting to see whether this species attracts more predators with time and, if so, whether they detract from its impact.

Leuciris fimbriaria *Stoll (Lepidoptera: Geometridae)*

Leuciris fimbriaria is an abundant and damaging defoliator of *M. pigra* in the native range of Central and South America. Larvae feed on leaves of all ages. Adults are nonfeeding, short-lived moths. Generation times are short and fecundity is high allowing rapid population increase. Culturing is easy allowing large releases to be made with minimal resources (Heard *et al.*, 2004).

The host specificity of this species was tested on colonies introduced from Mexico, using laboratory larval development tests on 69 plant species. Development to adult did not occur on any species other than *M. pigra* and *M. asperata*. Releases were made from 2004 but it has not yet been observed in the field (Ostermeyer and Grace, 2007).

14.3.5 Root feeders

Malacorhinus irregularis *Jacoby (Coleoptera: Chrysomelidae)*

Malacorhinus irregularis is known only from Mexico. The adults feed on leaves of the host and the larvae develop in the soil, feeding on small seedlings, imbibed seeds, roots, and root nodules. Larval feeding can kill seedlings (McIntyre *et al.*, 2007). Larvae prefer nitrogen-fixing root nodules to other plant parts and *M. pigra* relies on these nodules for nitrogen supply in low nutrient conditions, suggesting below-ground herbivory should impact on *M. pigra* vigor (McIntyre *et al.*, 2007). The larvae complete development from egg to adult in about 36 days and adults can live for up to six months (Heard *et al.*, 2005). Host-specificity tests were conducted to determine the suitability of seedlings and leaves for larval development, and suitability of leaves for adult feeding. No larval survival occurred on any plant species other than *M. pigra*. The extent of adult feeding on the test plants was negligible being less than 1% of that on *M. pigra* (Heard *et al.*, 2005).

This insect is not common in Mexico possibly due to the low abundance of *M. pigra* seedlings. It was predicted that, due to the availability of large numbers of seedlings, it could become common in Australia and damage seedlings and also defoliate mature plants, thereby complementing the impacts of *C. mimosa* and *N. gunniella*. The first release occurred in 2000. Establishment was confirmed in the Adelaide River system (Heard *et al.*, 2005), from where adults were collected and redistributed until 2005 with the release of a total of 35 500 beetles (Ostermeyer and Grace, 2007). Light traps have been used to confirm establishment at other sites in other river systems including one

Table 14.2 *Soil seed banks (seeds m^{-2}) of* Mimosa pigra *under dense* M. pigra *canopy in Australia (introduced range) and Mexico (native range)*

Year sampled	Soil seed bank	Site information	Reference
1986	8500–12 000	Australia, before biological control	Lonsdale *et al.* (1988)
2002	3710 (SE = 755)	Australia, after biological control	Paynter (2005)
2005	991	Australia, after biological control	Routley and Wirf (2006)
1985	117	Mexico, natural conditions	Lonsdale and Segura (1987)

site where considerable damage was visible (Routley and Wirf, 2006). Abundance of this species appears to be variable in space and time.

Nesaecrepida infuscata *(Schaeffer) (Coleoptera: Chrysomelidae)*

Nesaecrepida infuscata (Schaeffer) is among the most common insects on *M. pigra* in Mexico, Central America, and Venezuela. It was known as *Syphrea bibiana* in earlier publications, reports, and permits, but recently *N. infuscata* was shown to be a senior synonym of *S. bibiana* (Furth, 2006). The larvae feed on roots while the adults feed on leaves. Development time from egg to adult is 29 days with adults living a mean of 86 days and laying 4.4 eggs per day. As both roots and leaves of *M. pigra* are relatively undamaged in Australia, this species has potential to limit the growth and seed production by both leaf-feeding and root-feeding damage. Furthermore direct feeding on seedlings may increase their mortality. In host-specificity tests, larvae of *N. infuscata* did not develop on any plant species other than *M. pigra* and adult feeding on other plant species was minimal (Heard *et al.*, 2006). Permission for its release in Australia was gained in 2006. Colonies were sent to the Northern Territory for mass rearing and release in 2007.

14.3.6 *Overall impact of biological control agents of* M. pigra

A total of 13 insects and two fungi have been released in Australia to control *M. pigra* of which at least seven have established. Although it has proved difficult to separate the impacts of individual agents, it is clear that by 2003, both *C. mimosa* and *N. gunniella* were having a major impact on *M. pigra*, resulting in reduced seed rain, reduced seedling regeneration, and lower seed banks (Table 14.2). The impact of more recently released agents has not been measured, but appears to be substantial. Subsequent surveys, conducted after the release and establishment of *M. pallidata*, have indicated that *M. pigra* seed banks have continued to decline (Routley and Wirf, 2006), and are now approximately 10% of the prebiological control levels recorded by Lonsdale *et al.* (1988). However, they have to fall even further to reach the levels measured in the native range

(Lonsdale and Segura 1987). There has been little assessment of either establishment or impact of agents in Asia.

14.4 Integrated management

Prior to the introduction of damaging biological control agents, the available control measures were herbicide, fire, and mechanical control. These methods proved largely ineffective when used alone and Miller *et al.* (1992) suggested that an integrated approach that included biological control should provide the most effective management strategy. Although Paynter (2005) subsequently predicted that *Carmenta mimosa* should cause widespread reductions in *M. pigra* populations, modeling the impact of biological control showed that it might take decades to clear the huge existing patches of *M. pigra* (Paynter and Flanagan, 2002; Buckley *et al.*, 2004), indicating that these other control options will continue to be required.

Miller *et al.* (1992) envisaged that biological control would be used against large infestations that would be too costly to control by other means, while a range of techniques would be used to target outlying infestations to prevent weed spread. Buckley *et al.* (2004) predicted that biological control should enhance the impact of other control measures, implying that biological control could be integrated more widely. This prediction, however, relied on the assumption that biological control and techniques such as chemical weed control are compatible.

Paynter (2003) demonstrated that biological control could be compatible with herbicidal control because *N. gunniella* larvae were able to successfully complete their development before the death of plants that had been treated with herbicide. To investigate whether biological control could be integrated with other options, a large-scale (128-ha) split-plot experiment was performed to measure the impact of single and repeated applications of herbicide and crushing by bulldozer, either alone or in combination, on both *M. pigra* and five biological control agents that were abundant at the site (Paynter and Flanagan, 2004). In isolation, herbicide, bulldozing, and fire were not effective, but several combinations of techniques cleared *M. pigra* thickets and promoted establishment of competing vegetation that inhibited *M. pigra* regeneration from seed. Some treatment combinations that were effective were predicted by Buckley *et al.* (2004) to succeed only in combination with biological control. The most effective integrated weed management strategy was an application of herbicide in year 1, mechanical control followed by fire in year 2, and herbicide to target regrowth in year 3, with reduction of disturbance (such as feral pig rooting) where possible.

Depending on the species, biological control agent abundance on surviving *M. pigra* plants was either unchanged or increased following herbicide and/or bulldozing treatments and all agents recolonized regenerating *M. pigra* within one year of the fire treatment, and *N. gunniella* increased dramatically. The increased abundance of *N. gunniella* in response to all treatments was attributed to attack by this species being most common along stand edges. By reducing *M. pigra* populations from monocultures to smaller patches or individual plants, control treatments increased the ratio of "edge" plants to "thicket" plants and therefore the proportion of plants susceptible to *N. gunniella* attack. In contrast to *N. gunniella*, *C. mimosa* declined dramatically following the fire. Although *C. mimosa* is also most abundant at stand edges (Paynter, 2006) fire temporarily reduced the proportion of plants large enough to support *C. mimosa* larvae.

Paynter and Flanagan (2004) concluded that integrating control techniques can success-fully control dense *M. pigra* thickets and biological control integrates well with other control options and should lead to significant cost reductions for *M. pigra* management.

Paynter (2004b) investigated the value of revegetation, using native grasses and sedges, to accelerate the recovery of native vegetation and to prevent *M. pigra* regeneration from seed following fire. Sowing seed of several floodplain grasses and *Eliocharis dulcis* was unsuccessful. However, stolons of the native perennial grass *Hymenachne acutigluma* (Steud) Gilliland (Poaceae) established well when planted in wet mud and shallow water during the early dry season, as seasonal floodwaters subsided. Nevertheless, *H. acutigluma* stolons established and spread rather slowly and Paynter (2004b) con-cluded that revegetation should not be considered an alternative to the diligent control of *M. pigra* seedlings regenerating following control of *M. pigra* thickets.

B. Grace and Q. Paynter (unpublished data) showed that native wetland vegetation was just as effective in preventing regrowth of *M. pigra* as was the introduced pasture species para grass (*Urochloa mutica*), and that managing grazing pressure was more important in controlling regrowth of *M. pigra* than was the introduction of exotic pasture species. Grazing trials showed that after dense stands of *M. pigra* have been cleared, fewer *M. pigra* plants will regenerate in plots that had reduced grazing pressure, and there was no difference between plots with reduced grazing pressure and ungrazed plots (B. Grace and Q. Paynter, unpublished data).

14.5 Opportunities for the future

We propose the introduction of one final biological control agent if it proves sufficiently host specific. *Temnocerus* (= *Pselaphorhynchites) debilis* (Coleoptera: Rhynchitidae) is a common and damaging species in Mexico and other tropical American countries. These small beetles breed in leaf buds. Adults oviposit into buds and feed on the bud rachis. The larvae develop inside the dying ring-barked tips. Damage to young tips may provide the critical damage needed to provide a higher level of control. Limited success with rearing has been achieved, but oviposition and adult feeding tests using field-collected adults have shown a high level of specificity, so efforts are continuing to assess and develop this agent.

The future holds many opportunities for the introduction, reintroduction and redistri-bution in Asian and African countries of existing agents found in the native range of *M. pigra* and tested in Australia. This is especially true for those agents released in Australia after 1997 when the projects in Asia terminated. We believe that agents established in Australia, such as *Macaria pallidata* and *Malacorhinus irregularis*, will be the most promising candidates, but those that did not establish there should also be considered. In particular the green-seed feeders and pathogens offer potential in other regions. It may be that establishment did not occur in Australia for reasons that will not limit establishment in other countries. For example, the dry season in Australia is more severe than in the native range of *M. pigra* and in other areas of introduction. Although these agents may have failed to survive the dry season in Australia, they could establish in more benign climates.

Acknowledgments

We thank Bair Grace, Rieks van Klinken, Banpot Napompeth, Wendy Forno, and Raelene Kwong for commenting on the manuscript. We applaud the Australian Government's Department of Environment and ACIAR for its persistent funding of the *Mimosa pigra* project over several decades.

References

Braithwaite, R. W., Lonsdale, W. M. and Estbergs, J. A. (1989). Alien vegetation and native biota in tropical Australia: the spread and impact of *Mimosa pigra*. *Biological Conservation*, **48**, 189–210.

Buckley, Y. M., Paynter, Q. and Lonsdale, M. (2004). Modelling integrated weed management of an invasive shrub in tropical Australia. *Journal of Applied Ecology*, **41**, 547–560.

Cam, N. V., Lam, P. V., Dien, H. C., Huong, T. T. and Hien, N. T. (1997). Biological control of giant sensitive plant *Mimosa pigra* L. in Vietnam. Report of ACIAR Project 9313. Hanoi, Vietnam: VN Biological Control Research Center, National Institute for Plant Protection, 16 pp.

Clark, W. E. (1984). Species of *Sibinia* Germar (Coleoptera: Curculionidae) associated with *Mimosa pigra* L. *Proceedings of the Entomological Society of Washington*, **86**, 358–368.

Davis, D. R., Kassulke, R. C., Gillett, J. D. and Harley, K. L. S. (1991). Systematics, morphology, biology and host specificity of *Neurostrota gunniella* (Buesk) (*Lepidoptera: Gracillarid*), an agent for biological control of *Mimosa pigra* in Australia. *Entomological Society of Washington*, **93**, 16–44.

Eichlin, T. D. and Passoa, S. (1983). A new clearwing moth (Sesiidae) from Central America: a stem borer in *Mimosa pigra*. *Journal of the Lepidopterists Society*, **37**, 193–206.

Flanagan, G. J. (1994). The Australian distribution of *Mictis profana* and its life cycle on *Mimosa pigra*. *Journal of Australian Entomological Society*, **33**, 111–114.

Flanagan, G. J., Wilson, C. G. and Gillett, J. D. (1990). The abundance of native insects on the introduced weed *Mimosa pigra* in northern Australia. *Journal of Tropical Ecology*, **6**, 219–230.

Forno, I. W., Fichera, J. and Prior, S. (2000). Assessing the risk to *Neptunia oleracea* Lour. by the moth *Neurostrota gunniella* (Busck), a biological control agent for *Mimosa pigra* L. In *Proceedings of the X International Symposium on Biological Control of Weeds*, held 4–14 July 1999 in Bozeman, Montana, ed. N. R. Spencer. Sidney, MT: USDA-ARS, pp. 449–457.

Forno, I. W., Heard, T. A. and Day, M. D. (1994). Host specificity and aspects of the biology of *Coelocephalapion aculeatum* (Coleoptera: Apionidae), a potential biological control agent of *Mimosa pigra* (Mimosaceae). *Environmental Entomology*, **23**, 147–153.

Forno, I. W., Kassulke, R. C. and Day, M. D. (1991). Life-cycle and host testing procedures for *Carmenta mimosa* Eichlin and Passoa (Lepidoptera: Sesiidae), a biological control agent for *Mimosa pigra* L. (Mimosaceae) in Australia. *Biological Control*, **1**, 309–315.

Furth, D. G. (2006). The current status of knowledge of the Alticinae of Mexico (Coleoptera: Chrysomelidae). *Bonner Zoologische Beitrage*, **54**, 209–237.

Grace, B. (2005). Do predators affect the survival of *Macaria pallidata* larvae? Implications for the biological control of *Mimosa pigra* in the NT. *Northern Territory Naturalist*, **18**, 8–13.

Harley, K. L. S, Gillett, J. D, Winder, J., *et al.* (1995). Natural enemies of *Mimosa pigra* and *M. berlandieri* (Mimosaceae) and prospects for biological control of *M. pigra*. *Environmental Entomology*, **24**, 1664–1678.

Heard, T. A. (2004). The taxonomy of *Mimosa pigra*. In *Research and Management of* Mimosa pigra, ed. M. Julien, G. Flanagan, T. Heard, *et al.* Canberra, Australia: CSIRO Entomology, p. 10.

Heard, T. A. and Forno, I. W. (1996). Host selection and host range of the flower feeding weevil, *Coelocephalapion pigrae*, a potential biological control agent against *Mimosa pigra*. *Biological Control*, **6**, 83–95.

Heard, T. A. and Segura, R. (2004). Agents for biological control of *Mimosa pigra* in Australia: review and future prospects. In *Research and Management of* Mimosa pigra, ed. M. Julien, G. Flanagan, T. Heard, *et al.* Canberra, Australia: CSIRO Entomology, pp. 126–140.

Heard, T. A., Paynter, Q., Chan, R. and Mira, A. (2005). *Malacorhinus irregularis* for biological control of *Mimosa pigra*: host specificity, life cycle, and establishment in Australia. *Biological Control*, **32**, 252–262.

Heard, T. A., Forno, I. W. and Burcher, J. (1999). *Chalcodermus serripes* Coleoptera: Curculionidae, for biological control of *Mimosa pigra*: host relations and life cycle. *Biological Control*, **15**, 1–9.

Heard, T. A., Mira, A. and Zonneveld, R. (2001). Proposal for the release of *Macaria pallidata* for the biological control of *Mimosa pigra*. Unpublished report submitted to Australian Quarantine and Inspection Service, Canberra, Australia.

Heard, T. A., Mira, M., Chan, R. and Fichera, G. (2004). Application to release the defoliating Lepidoptera *Leuciris fimbriaria* (Geometridae) into Australia for biological control of the weed *Mimosa pigra*. Unpublished report submitted to Australian Quarantine and Inspection Service, Canberra, Australia.

Heard, T. A., Mira, A., Chan, R. and Fichera, G. (2006). Application to release *Neasecrepida infuscata* (Coleoptera: Chrysomelidae: Alticinae) into Australia for biological control of the weed *Mimosa pigra*. Unpublished report submitted to Australian Quarantine and Inspection Service, Canberra, Australia.

Heard, T. A. and Pettit, W. (2005). A review of the surveys for natural enemies of *Mimosa pigra*: what does it tell us about surveys for broadly distributed hosts? *Biological Control*, **34**, 247–254.

Heard, T. A., Segura, R., Martinez, M. and Forno, I. W. (1997). Biology and host range of the green seed weevil, *Sibinia fastigiata*, for biological control of *Mimosa pigra*. *Biocontrol Science and Technology*, **7**, 631–644.

Hennecke, B. (2004). The prospect of biological control of *Mimosa pigra* with fungal pathogens in Australia. In *Research and Management of* Mimosa pigra, ed. M. Julien, G. Flanagan, T. Heard, *et al.* Canberra, Australia: CSIRO Entomology, pp. 117–121.

Hennecke, B. R. (2006). Failure of *Diabole cubensis*, a promising classical biological control agent, to establish in Australia. *Biological Control*, **39**, 121–127.

Julien, M. H. and Griffiths, M. W. (1998). Biological Control Of Weeds: A World Catalogue of Agents and Their Target Weeds. Fourth edn. Wallingford, UK: CABI.

Kassulke, R. C., Harley, K. L. S. and Maynard, G. V. (1990). Host specificity of *Acanthoscelides quadridentatus* and *A. puniceus* (Col.: Bruchidae) for biological control of *Mimosa pigra* (with preliminary data on their biology). *Entomophaga*, **35**, 85–96.

Karren, J. B. (1989). *Chlamisus mimosae*, sp. nov. (Coleoptera: Chrysomelidae: Chlamisinae) from Brazil and imported into Australia and Thailand. *The Coleopterists Bulletin*, **43**, 355–358.

Kissinger, D. G. (1992). Apionidae from North and Central America. Part 4. Generic classification and introduction to the genus *Coelocephalapion* Wagner, with new species from Mexico and Venezuela (Coleoptera). *Insecta Mundi*, **6**, 65–77.

Lonsdale, W. M. (1988). Litterfall in an Australian population of *Mimosa pigra*, an invasive tropical shrub. *Journal of Tropical Ecology*, **4**, 381–392.

Lonsdale, W. M. (1993). Rates of spread of an invading species: *Mimosa pigra* in northern Australia. *Journal of Ecology*, **81**, 513–521.

Lonsdale, W. M. and Farrell, G. S. (1998). Testing the effects on *Mimosa pigra* of a biological control agent *Neurostrota gunniella* (Lepidoptera: Gracillaridae), plant competition and fungi under field conditions. *Biocontrol Science and Technology*, **8**, 485–500.

Lonsdale, W. M. and Segura, R. (1987). A demographic comparison of native and introduced populations of *Mimosa pigra*. In *Proceedings of the Eighth Australian Weeds Conference*. Sydney, Australia: Weed Society of New South Wales, pp 163–166.

Lonsdale, W. M., Harley, K. L. S. and Gillett, J. D. (1988). Seed bank dynamics of *Mimosa pigra*, an invasive tropical shrub. *Journal of Applied Ecology*, **25**, 963–976.

Lonsdale, W. M., Miller, I. L. and Forno, I. W. (1989). The biology of Australian weeds. 20. *Mimosa pigra* L. *Plant Protection Quarterly*, **4**, 119–131.

McIntyre, V., Grace, B. and Schmidt, S. (2007). Impacts of the biocontrol agent *Malacorhinus irregularis* (Coleoptera, Chrysomelidae) on *Mimosa pigra* seedlings and the importance of root nodules. *Biocontrol Science and Technology*, **17**, 365–374.

Miller, I. L., Napompeth, B., Forno, I. W. and Siriworakul, M. (1992) Strategies for the integrated management of *Mimosa pigra*. In *A Guide to the Management of* Mimosa pigra, ed. K. L. S. Harley. Canberra, Australia: CSIRO, pp. 110–115.

Napompeth, B. (1982). Preliminary screening of insects for biological control of *Mimosa pigra* L. in Thailand. In Mimosa pigra *Management, Proceedings of an International Symposium*, held in Chiang Mai, Thailand. Corvallis, OR: International Plant Protection Center. pp. 121–128.

Napompeth, B. (1994). Biological control of paddy and aquatic weeds in Thailand: integrated management of paddy and aquatic weeds in Asia. In *Proceedings of an International Seminar*, held in Tsukuba, Japan, 19–25 October 1992, ed. J. Bay-Petersen. Taipei: Food and Fertilizer Technology Center. pp 122–127.

Ostermeyer, N. (2000). Population density and distribution of the biological control agent *Carmenta mimosa* on *Mimosa pigra* in the Adelaide and Finnis River catchments of the Northern Territory. *Plant Protection Quarterly*, **15**, 46–49.

Ostermeyer, N. and Grace, B. S. (2007). Establishment, distribution and abundance of *Mimosa pigra* biological control agents in northern Australia: implications for biological control. *BioControl*, **52**, 703–720.

Ostermeyer, N., Grace, B., Paskins, M., McIntyre, V. and Routley, B. (2004). Biological control in the Top End: getting community involved. In *Proceedings of the 14th Australian Weeds Conference*, eds. B. M. Sindel and S. B. Johnson. Sydney, Australia: Weed Society of New South Wales, pp. 337–340.

Page, A. R. and Lacey, K. L (2006). *Economic Impact Assessment of Australian Weed Biological Control*. CRC for Australian Weed Management Technical Series 10. Glen Osmond, Australia: CRC for Australian Weed Management.

Paynter, Q. (2003). Integrated weed management: effect of herbicide choice and timing of application on the survival of a biological control agent of the tropical wetland weed, *Mimosa pigra*. *Biological Control*, **26**, 162–167.

Paynter, Q. (2004a). Evaluating *Mimosa pigra* biological control in Australia. In *Research and Management of* Mimosa pigra, ed. M. Julien, G. Flanagan, T. Heard, *et al.* Canberra, Australia: CSIRO Entomology, pp. 141–148.

Paynter, Q. (2004b). Revegetation of a wetland following control of the invasive woody weed, *Mimosa pigra*, in the Northern Territory, Australia. *Ecological Management and Restoration*, **5**, 191–198.

Paynter, Q. (2005). Evaluating the impact of biological control against *Mimosa pigra* in Australia. *Journal of Applied Ecology*, **42**, 1054–1062.

Paynter, Q. (2006). Evaluating the impact of biological control against *Mimosa pigra* in Australia: comparing litterfall before and after the introduction of biological control agents. *Biological Control*, **38**, 166–173.

Paynter, Q. and Flanagan, G. J. (2002). Integrated management of *Mimosa pigra*. In *Proceedings of the 13th Australian Weeds Conference*, held 8–13 September 2002, Perth, Western Australia, ed. H. Spafford-Jacob, J. Dodd and J. H. Moore. Perth, Australia: Plant Protection Society of Western Australia, pp. 165–168.

Paynter, Q. and Flanagan, G. J. (2004). Integrating herbicide and mechanical control treatments with fire and biological control to manage an invasive wetland shrub, *Mimosa pigra*. *Journal of Applied Ecology*, **41**, 615–629.

Rasplus, J. Y. (1988). Nouvelles espèces d'Eurytomidae principalment parasites de coléoptères séminivores de légumineuses en Côte d'Ivoire (Lamto) (Hymenoptera: Chalcidoidea). *Bollettino di Zoologia Agraria e di Bachicoltura*, **20**, Ser.ii, 89–114.

Routley, B. M. and Wirf, L. A. (2006). Advancements in biocontrol of *Mimosa pigra* in the Northern Territory. In *Proceedings of the 15th Australian Weeds Conference*, ed. C. Preston, J. H. Watts and N. D. Crossman. Adelaide, Australia: Weed Management Society of South Australia, pp. 561–564.

Samouth, C. (2004). *Mimosa pigra* infestations and the current threat to wetlands and floodplains in Cambodia. In *Research and Management of* Mimosa pigra, ed. M. Julien, G. Flanagan, T. Heard, *et al.* Canberra, Australia: CSIRO Entomology, pp. 29–32.

Smith, C. S. and Wilson, C. G. (1995). Close to the edge: microhabitat selection by *Neurostrota gunniella* Busck (Lepidoptera: Gracillariidae), a biological control agent for *Mimosa pigra* L. in Australia. *Journal of the Australian Entomological Society*, **34**, 177–180.

Son, N. H., Lam, P. V., Cam, N. V., *et al.* (2004). Preliminary studies on control of *Mimosa pigra* in Vietnam. In *Research and Management of* Mimosa pigra, ed. M. Julien, G. Flanagan, T. Heard, *et al.* Canberra, Australia: CSIRO Entomology, pp. 110–116.

Soon, L. G. and Chong, K. K. (1997). Biological control of giant sensitive plant (*Mimosa pigra*) in Southeast Asia. Unpublished Project Review, ACIAR, Canberra, Australia.

Suasa-ard, W., Sommartya, P. and Jaitui, S. (2004). Evaluation of seed feeding bruchids, *Acanthoscelides* species, as biological control agents for *Mimosa pigra* in Thailand. In *Research and Management of* Mimosa pigra, ed. M. Julien, G. Flanagan, T. Heard, *et al.* Canberra, Australia: CSIRO Entomology, pp. 122–125.

Taylor, D. B. J., Heard, T. A., Paynter, Q. and Spafford-Jacob, H. (2007). Non-target impact of a weed biological control agent on a native plant in northern Australia. *Biological Control*, **42**, 25–33.

Triet, T., Kiet, L. C., Thi, N. L. and Dan, P. Q. (2004). The invasion of *Mimosa pigra* in wetlands of the Mekong Delta, Vietnam. In *Research and Management of* Mimosa pigra, ed. M. Julien, G. Flanagan, T. Heard, *et al.* Canberra, Australia: CSIRO Entomology, pp. 45–51.

Wilson, C. G. and Flanagan, G. J. (1990). Establishment and spread of *Neurostrota gunniella* on *Mimosa pigra* in the Northern Territory. In *Proceedings of the Ninth Australian Weeds Conference*, held in Adelaide, July 1990. Adelaide, Australia: Weed Managment Society of Australia, pp. 505–507.

Wilson, C. G. and Flanagan, G. J. (1991). Establishment of *Acanthoscelides quadridentatus* (Schaffer) and *A. puniceus* Johnson (Coleoptera: Bruchidae) on *Mimosa pigra* L. *Journal of the Australian Entomological Society*, **30**, 279–280.

Wilson, C. G., Flanagan, G. J. and Gillett, J. D. (1990). The phytophagous insect fauna of the introduced shrub, *Mimosa pigra* in northern Australia and its relevance to biological control. *Environmental Entomology*, **19**, 776–784.

Wilson, C. G. and Pitkethley, R. N. (1992). *Botryodiplodia* dieback of *Mimosa pigra*, a noxious weed in northern Australia. *Plant Pathology*, **41**, 777–779.

Wilson, C. G., Forno, I. W., Smith, C. S. and Napompeth, B. (1992). Rearing and release methods for biological control agents. In *A Guide to the Management of* Mimosa pigra, ed. K. L. S. Harley. Canberra, Australia: CSIRO, pp. 49–62.

Wirf, L. A. (2006). The effect of manual defoliation and *Macaria pallidata* (Geometridae) herbivory on *Mimosa pigra*: implications for biological control. *Biological Control*, **37**, 346–353.

15

Parthenium hysterophorus L. (Asteraceae)

K. Dhileepan and L. Strathie

15.1 Introduction

Parthenium hysterophorus L. (Asteraceae), commonly known as parthenium, is a weed of global significance (Navie *et al.*, 1996; Fig. 15.1). Parthenium is a major weed in Australia and India. In Australia, parthenium was first identified in 1955, but was proclaimed as a noxious plant in 1975 (Auld *et al.*, 1983). Parthenium was accidentally introduced into India in 1955 (Rao, 1956), and has since spread to neighboring countries, including Pakistan (Javaid and Anjum, 2005; Shabbir and Bajwa, 2006), Sri Lanka (Jayasuriya, 2005), Bangladesh, and Nepal. Parthenium also occurs in southern China, Taiwan, Vietnam, and Israel in Asia (Nath, 1981; Joel and Liston, 1986), several Pacific islands including New Caledonia, Papua New Guinea, Seychelles, Vanuatu (Adkins *et al.*, 2005), and several African countries including Ethiopia, Kenya, Madagascar, Mozambique, South Africa, Somalia, Swaziland, and Zimbabwe (i.e. Wood, 1897; Hilliard, 1977; Njoroge, 1986, 1989, 1991; Nath, 1988; Frew *et al.*, 1996; Tamado and Milberg 2000; Tamado *et al.*, 2002a; MacDonald *et al.*, 2003; CABI, 2004; Da Silva *et al.*, 2004; Fessehaie *et al.*, 2005; Taye *et al.*, 2004b; Strathie *et al.*, 2005) (Fig. 15.1).

Parthenium is an annual herb with a deep-penetrating taproot system and an erect shoot system. Young plants form a rosette of leaves close to the soil surface. As it matures, the plant develops many branches on its upper half, and may eventually grow up to two meters (McFadyen, 1992). With good rainfall and warm temperature, parthenium has the ability to germinate and establish at any time of the year (e.g. Navie *et al.*, 1996; Tamado *et al.*, 2002b). Flowering usually commences 6–8 weeks after germination and soil moisture seems to be the major contributing factor to flowering (Navie *et al.*, 1996). Pollination is primarily by wind (Lewis *et al.*, 1988). Parthenium is a prolific seed producer and a fully-grown plant can produce more than 15 000 seeds in its lifetime (Haseler, 1976). Seeds persist and remain viable in the soil for reasonably long periods, with a seed bank half-life of approximately six years (Navie *et al.*, 1998a).

Parthenium can grow in a wide range of landscapes, including degraded and disturbed lands, degraded pastures, crops, and performing lands that include crops, orchards,

Biological Control of Tropical Weeds using Arthropods, ed. R. Muniappan, G. V. P. Reddy, and A. Raman. Published by Cambridge University Press. © Cambridge University Press, 2009.

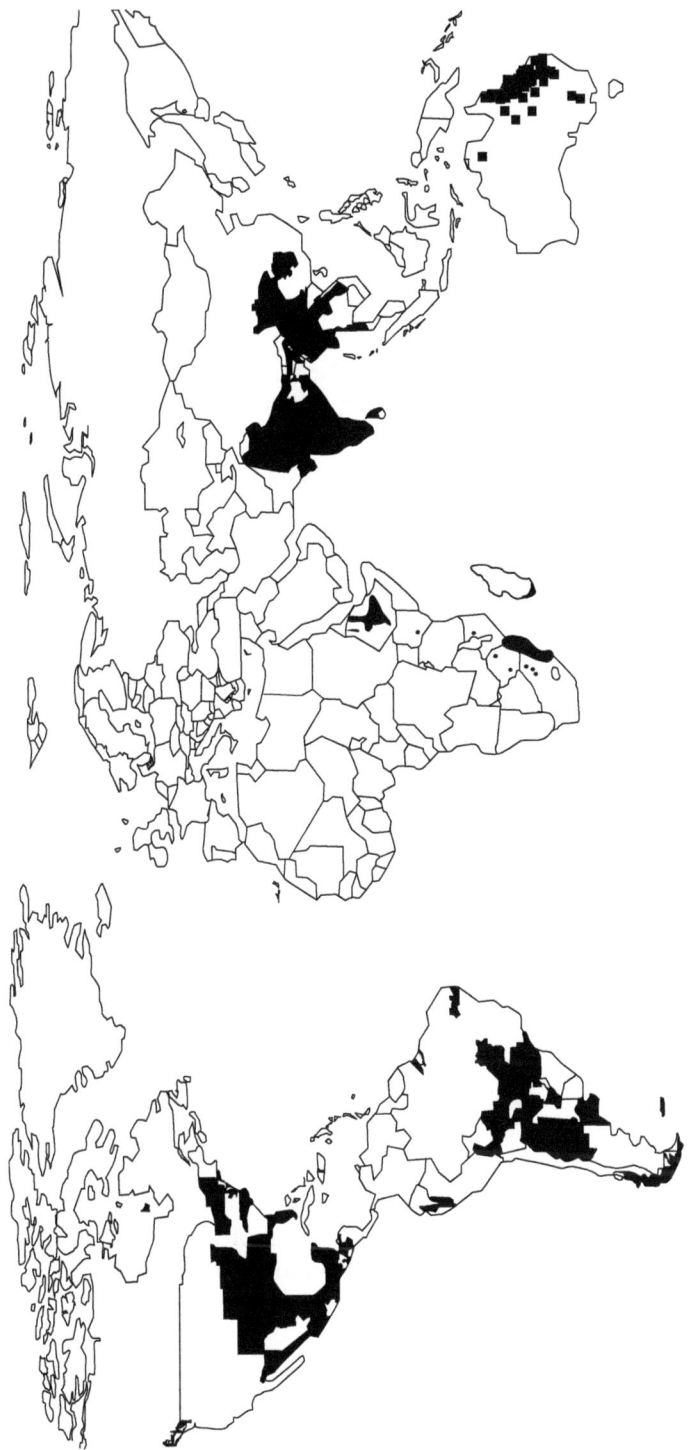

Fig. 15.1 Global distribution of *Parthenium hysterophorus*.

forests, and along railway tracks and roadsides, and streams and rivers (Navie *et al.*, 1996), across a wide range of habitats, ranging from hot, arid and semi-arid low altitude to humid high to mid-altitude areas (Taye, 2005).

15.2 Detrimental effects

Parthenium causes human (e.g. Subba Rao *et al.*, 1977; Wedner *et al.*, 1986; McFadyen, 1995; Kololgi *et al.*, 1997; Cheney, 1998) and animal health (Tudor *et al.*, 1982; Kadhane *et al.*, 1992) problems, agricultural losses (e.g. Navie *et al.*, 1996; Tamado and Milberg 2000; Tamado *et al.*, 2002a; Firehun and Tamado, 2006) as well as serious environmental problems (Chippendale and Panetta, 1994). Parthenium has reduced the richness and diversity of other plant species (Sridhara *et al.*, 2005) and their seed banks (Navie *et al.*, 2004). Parthenium also acts as a reservoir host for plant pathogens and insect pests of crop plants (Basappa, 2005; Govindappa *et al.*, 2005; Prasad Rao *et al.*, 2005). Parthenium and related genera contain sesquiterpene lactones (Picman and Towers, 1982), which induce contact dermatitis and other allergies in humans (Towers, 1981). Stock animals, especially horses, suffer from allergic skin reaction while grazing in fields infested by parthenium. Parthenium is generally unpalatable and toxic to cattle, buffalo, and sheep (e.g. Narasimham *et al.*, 1980; Kadhane *et al.*, 1992). Consumption of large quantities of parthenium taints mutton (Tudor *et al.*, 1982) and can even kill livestock.

In Australia, parthenium mainly occurs in Queensland, affecting 170 000 km^2 of prime grazing country (Fig. 15.2) (McFadyen, 1992; Chippendale and Panetta, 1994), and has the potential to spread throughout Australia (Adamson, 1996). Parthenium is a serious problem in perennial grasslands in central Queensland, where it reduces beef production by as much as Au\$16.5 m annually (Chippendale and Panetta, 1994).

In India, parthenium occurs in most states (e.g. Mahadevappa, 1997; Pandey and Dubey, 1989; Fig. 15.3) and is a weed of high relevance in cropping areas (Mahadevappa, 1997) inflicting yield losses of up to 40% in several crops (Khosla and Sobti, 1979). Parthenium in noncropping areas reduces forage production from 10% (Jayachandra, 1971) to 90% (Nath, 1981). Parthenium also occurs widely in Pakistan and Sri Lanka (Fig. 15.3).

In Ethiopia, parthenium is primarily a weed in sugarcane cropping areas and rangeland areas, and is ranked as the most serious weed by farmers (Tamado and Milberg, 2000; Firehun and Tamado, 2006; Fig. 15.4). Parthenium was first reported in Kenya in 1975, and since then it has rapidly spread throughout, affecting crops like coffee (Njoroge, 1986, 1989, 1991). In South Africa, parthenium occurs in the northeastern regions, with its distribution extending from the subtropical regions of KwaZulu-Natal province from around Durban, northwards to Mozambique and to Mpumalanga province and to the northwest and north of Pretoria (Fig. 15.4). Parthenium is a weed of sugarcane and banana plantations in South Africa. Parthenium has been reported in several of the national parks in Mpumalanga (e.g. Kruger National Park) and KwaZulu-Natal (e.g. Ndumo, Tembe, and Hluhluwe-iMfolozi parks) provinces. Parthenium occurs throughout Swaziland and in southern and central Mozambique (Fig. 15.4).

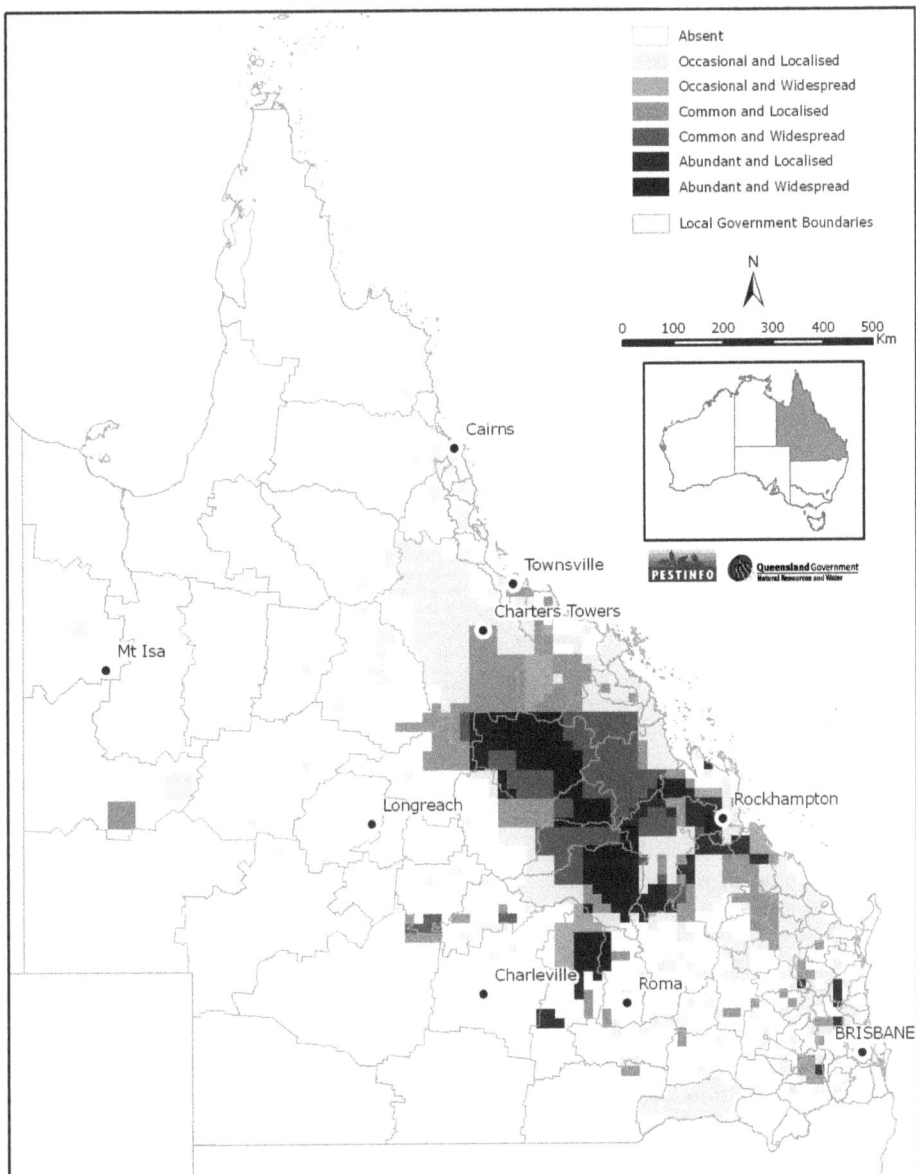

Fig. 15.2 Distribution and intensity of *Parthenium hysterophorus* infestation in Queensland, Australia in 2006.

15.3 Why biological control?

Herbicides provide effective control of parthenium (e.g. Navie *et al.*, 1996; Brooks *et al.*, 2004). Control using herbicides is the first line of defence (Holman, 1981), but high costs

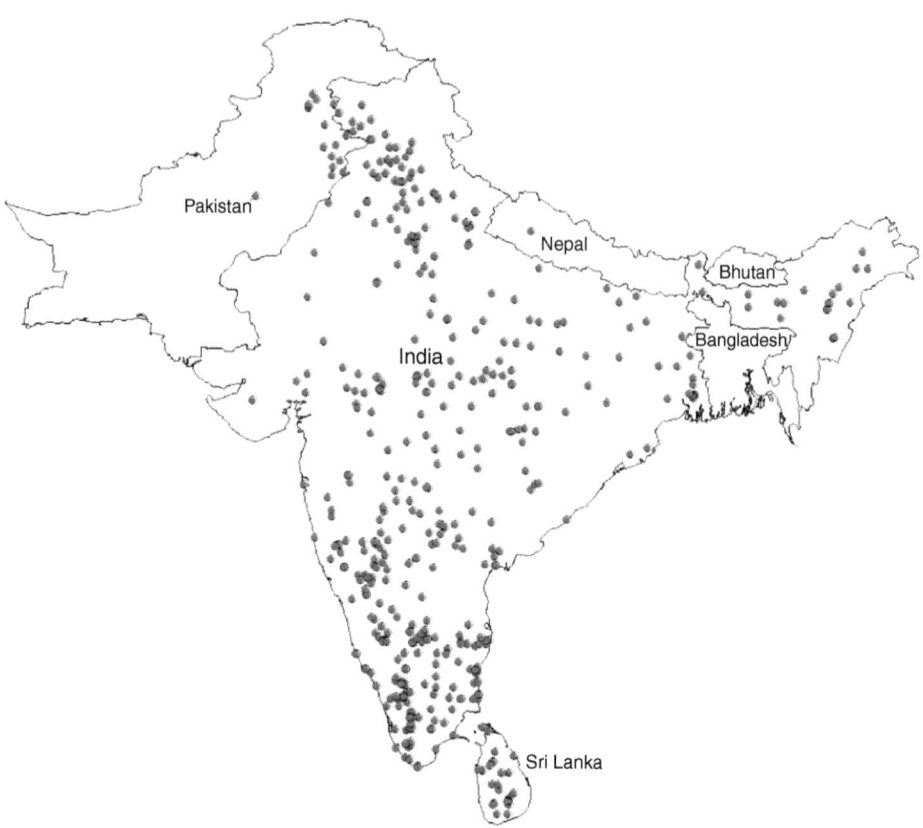

Fig. 15.3 Prevalence of parthenium in India, Pakistan, and Sri Lanka. (Dark circles represent documented locations with parthenium.)

of herbicides prohibit their long-term use for parthenium management in grazing areas, public and uncultivated areas, and forests. Herbicides are used to eradicate localized infestations, for roadside infestations, or when the weed is a problem in certain crops (Holman, 1981; Brooks *et al.*, 2004). In areas where the use of herbicides is neither economical nor effective (e.g. Njoroge, 1991), other options such as the use of competitive plants to displace parthenium (e.g. Joshi, 1991; Kandasamy and Sankaran 1997; O'Donnell and Adkins, 2005), allelopathy (e.g. Javaid *et al.*, 2006a; van der Laan, 2006), fire (Vogler *et al.*, 2002), and other physical methods including mulching green parthenium plants (e.g. in potato and rice crops in India and wheat, sorghum, and sunflower crops in Queensland, Australia) have been suggested as suitable options either individually or in combination.

Recent study has shown that fire does not reduce either the parthenium infestations or the soil seed bank, nor does smoke from such fires stimulate parthenium seed germination (Vogler *et al.*, 2002; Butler and Fairfax, 2003). In grazing areas, management

Fig. 15.4 Prevalence of parthenium in Africa. (Dark circles represent documented locations with parthenium).
Sources: USAID Integrated Pest Management Collaborative Research Support Program (IPM CRSP) project on "Management of the weed parthenium in Eastern and Southern Africa using Integrated Cultural and Biological Measures"; the *South African Plant Invaders Atlas*; and *Swaziland Alien Plants Database*.)

of parthenium can be achieved by maintaining acceptable levels of pasture grass growth to maximize competition against the weed. Pasture cover and composition are the key factors that influence the density of parthenium present in native pastures. Although various natural enemies are known to attack parthenium in countries where the weed has been introduced (e.g. Srikanth and Pushpalatha, 1991; Singh, 1997; Taye *et al.*, 2004c; Taye, 2005), most of them are either generalists (e.g. Farkya *et al.*, 1994; Mishra *et al.*, 1994; Taye *et al.*, 2002; Javaid *et al.*, 2006b), or crop pests and disease agents using parthenium as alternative hosts (e.g. Robertson and Kettle, 1994; Navie *et al.*, 1996; Evans, 1997a; Basappa, 2005; Govindappa *et al.*, 2005; Prasad Rao *et al.*, 2005), and none of them would exert any critical impact on parthenium. Hence, classical biological control has been seen as a better alternative than use of herbicides in perennial grasslands as well as in areas such as degraded lands and forests, where the use of herbicides is uneconomical.

15.4 Native range studies

The invasive parthenium (*P. hysterophorus*) is native to the landscape bordering the Gulf of Mexico, and has spread throughout southern USA, the Caribbean, and Brazil (Towers, 1981). Genotypic studies revealed that the parthenium genotypes in Australia, India, Mozambique, and South Africa originated from southern Texas, USA (Graham and Lang, 1998). A different race of *P. hysterophorus* with yellow flowers is native to Argentina, Bolivia, Chile, Paraguay, Peru, and Uruguay (Dale, 1981). Genotypic studies confirmed the existence of distinct North American and Central American populations (Graham and Lang, 1998). On the basis of the preliminary surveys conducted in Mexico and southern USA in North America, Brazil and Argentina in South America and the Caribbean islands (Bennett, 1976; McFadyen, 1976, 1979), Mexico was determined as the most suitable area for further explorations of natural enemies associated with parthenium.

Surveys conducted in North America, from bases at Monterrey and Cuernavaca in Mexico, and Temple, Texas, USA, yielded 262 phytophagous arthropod species and several fungal pathogens (Evans, 1983, 1997a, b; McClay, 1980; McClay *et al.*, 1995). Among them, at least 144 were found to feed on parthenium at some stage of their life cycle, of which 13 species were restricted to Ambrosiinae. Six insect species that were shown to be stenophagous and two rust fungi underwent host-range testing and were released in Australia from 1980 onwards (Table 15.1). Surveys in northwestern Parana, east Sao Paulo, and western Mato Grosso in Brazil, and northwestern Argentina yielded around 100 insect species (McFadyen, 1976, 1979), of which only three were nominated for further host-specificity tests (Table 15.1). None of the 15 species of insects recorded on parthenium in the Caribbean Islands (Bennett, 1976) was nominated for host-specificity tests.

15.5 Biological control agents

In Australia, biological control of parthenium was initiated in 1977 and since then, nine species of insects and two rust fungi have been introduced (McFadyen and McClay, 1981; McFadyen, 1985, 1992, 2000; McClay *et al.*, 1990; Wild *et al.*, 1992; Parker *et al.*, 1994; Dhileepan and McFadyen, 1997; Table 15.1). Among them, at least six species of insects and two rust fungi are known to be established in the field (Dhileepan *et al.*, 1996; Dhileepan and McFadyen, 1997; McFadyen, 1992, 2000). Leaf-feeding *Thecesternus hirsutus* Pierce (Coleoptera: Curculionidae) (the larvae feed and induce galls in the root) (McClay and Anderson, 1985), was imported into Australia for host-specificity tests in 1982 (McFadyen, 1992) and again in 1997. On both occasions no further progress was made as the insect could not be reared in quarantine (McFadyen, 1992; K. Dhileepan, unpublished data).

Table 15.1 *Biological control agents released against P. hysterophorus in Australia*

Insect/pathogen species (Common name)	Order: Family	Source country	Field release	Establishment	Distribution and abundance
Epiblema strenuana Walker (Stem-galling moth)	Lepidoptera: Tortricidae	Mexico	1982	Yes	Widespread and abundant
Zygogramma bicolorata Pallister (Leaf-feeding beetle)	Coleoptera: Chrysomelidae	Mexico	1980	Yes	Abundant, but seasonal and localized
Listronotus setosipennis Hustache (Stem-boring weevil)	Coleoptera: Curculionidae	Argentina and Brazil	1982	Yes	Widespread, but sporadic
Smicronyx lutulentus Dietz (Seed-feeding weevil)	Coleoptera: Curculionidae	Mexico	1981	Yes	Widespread, but sporadic
Conotrachelus albocinereus Fiedler (Stem-galling weevil)	Coleoptera: Curculionidae	Argentina	1995	Yes	Only in release sites, and occasional
Carmenta nr. *ithacae* (Beutenmüller) (Root-boring moth)	Lepidoptera: Sesiidae	Mexico	1998	Yes	Only in release sites, and occasional
Bucculatrix parthenica Bradley (Leaf-mining moth)	Lepidoptera: Bucculatricidae	Mexico	1984	Yes	Widespread, but less abundant
Puccinia abrupta var. *partheniicola* (Jackson) Parmelee (Winter rust)	Basidiomycotina: Uredinales	Mexico	1991	Yes	Localized and sporadic
Puccinia melampodii Dietel & Holway (Summer rust)	Basidiomycotina: Uredinales	Mexico	1999	Yes	Widespread, but highly seasonal
Platphalonidia mystica (Razowski & Becker) (Stem-boring moth)	Lepidoptera: Tortricidae	Argentina	1992	Unknown	
Stobaera concinna (Stål). Parthenium sap-feeding planthopper	Homoptera: Delphacidae	Mexico	1983	Unknown	

In India, a biological control program was initiated in 1983, and since then only the leaf-feeding *Zygogramma bicolorata* Pallister (Coleoptera: Chrysomelidae) has been introduced. In Sri Lanka, biological control efforts were initiated in 2003 with the importation of stem gall-inducing *Epiblema strenuana* Walker (Lepidoptera: Totricidae) and *Puccinia melampodii* Dietel and Holway (Uredinales) from Australia (Jayasuriya, 2005). Attempts to establish colonies of *Z. bicolorata* and *E. strenuana* in quarantine in Papua New Guinea have so far failed.

In South Africa, a biological control program on parthenium was initiated in 2003 with prioritization of *Z. bicolorata*, *Listronotus setosipennis* (Hustache) (Coleoptera: Curculionidae), *E. strenuana*, and *P. melampodii* (Strathie *et al.*, 2005; Ntushelo and Wood, 2008) chosen for their impact on parthenium and likely suitability for local climatic conditions within the invasive range of parthenium, which experiences a distinct dry season.

15.5.1 Epiblema strenuana *Walker (Lepidoptera: Totricidae)*

Epiblema strenuana is widely distributed in North America and parts of the Caribbean where it uses *Ambrosia* spp. and *Xanthium* spp. in the northern parts and *P. hysterophorus* in the southern parts of the range, which include Mexico, Virgin Islands, and Antigua (Bennett, 1976). This moth was introduced to Australia from Mexico in 1982 (McFadyen, 1992), after the necessary prerelease host-specificity tests (McClay, 1987). The moth became widespread within two years of introduction and now occurs in all partheniuminfested areas. In India, *E. strenuana* was not approved for field release due to oviposition and larval feeding on *Guizotia abyssinica* (L.f.) Cass. and *Helianthus annus* Linn. (Asteraceae) (Jayanth, 1987b; Singh, 1997). Host-specificity tests confirmed the suitability and safety of *E. strenuana* as a biological control agent for parthenium in Sri Lanka, and the moth was field released in 2004 (Jayasuriya, 2005). In South Africa, an attempt to establish a culture of *E. strenuana* in quarantine did not succeed, likely due to low humidity, but it will be imported from Australia again (Strathie *et al.*, 2005).

Epiblema strenuana is not being considered for introduction into Ethiopia, in view of its potential to feed on *G. abyssinica*, a major oil seed crop.

Adult moths have a life span of 7–11 days and females lay up to 1 000 eggs in their lifetime on young terminal leaves (McFadyen, 1992). The emerging larvae feed initially on the leaf, and then bore into the stem through the terminal or axillary meristem. Once larvae are inside the stem, their feeding induces a hollow, fusiform gall (Raman and Dhileepan, 1999). The larvae pupate inside the gall and in 4–6 weeks, the moth emerges through a hole in the gall, leaving the pupal skin extruded from the exit hole. Gall development can occur in all stages of plant growth (Dhileepan and McFadyen, 2001) and usually one larva occurs per gall (Raman and Dhileepan, 1999).

Epiblema strenuana causes visible symptoms on parthenium (McFadyen, 1992), and the impact becomes significant when the gall damage is initiated at early stages of plant

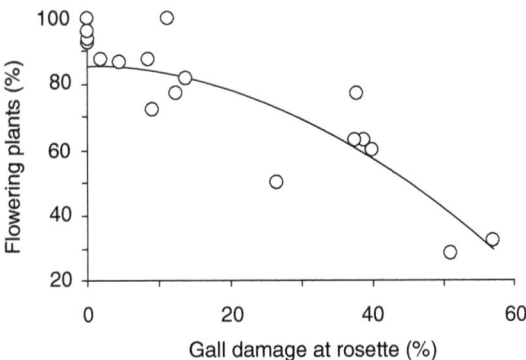

Fig. 15.5 Impact of galling by *Epiblema strenuana* at rosette stage on the proportion of plants attaining the flowering stage.

growth (Dhileepan and McFadyen, 1997; Navie *et al.*, 1998b). Gall damage at early stages of plant growth reduced plant height, main-stem height, flower production, leaf production, and shoot and root biomass (Dhileepan, 2001, 2003b, 2004; Dhileepan and McFadyen, 2001). In glasshouse conditions, all gall-bearing plants produced flowers irrespective of the growth stage at which the plants were galled, but produced fewer flowers than nongalled plants (Dhileepan and McFadyen, 2001). Gall induction during early growth stages in field cages prevented 30% of the plants from producing flowers (Fig. 15.5). Flower production per unit plant biomass was also lower in galled plants than in ungalled plants (Dhileepan and McFadyen, 2001), and the reduction was more intense when gall damage was initiated at early stages of plant growth. The negative impact of galling on parthenium is due to damage to phloem tissue, resulting in the disruption of the host plant's overall metabolism (Raman and Dhileepan, 1999; Florentine *et al.*, 2001, 2005; Raman *et al.*, 2006). Competition from grasses significantly increased the effectiveness of *E. strenuana* (Navie *et al.*, 1998b).

Gall damage is evident throughout the parthenium growing season on both rosette and flowering stages (Fig. 15.6), and the moth can have more than six generations each year (McFadyen, 1992). However, the levels of infestation by *E. strenuana* in Australia vary considerably (Dhileepan and McFadyen, 1997; Dhileepan, 2001, 2003b) due to lack of synchrony between parthenium germination and *E. strenuana* emergence.

15.5.2 Zygogramma bicolorata *Pallister (Coleoptera: Chrysomelidae)*

Zygogramma bicolorata was introduced from Mexico into Australia in 1980 (McFadyen and McClay, 1981). In Australia, evidence of *Z. bicolorata* activity on parthenium in the field was first noticed in 1990 (Dhileepan and McFadyen, 1997). Outbreaks of *Z. bicolorata* resulting in complete defoliation of small patches of parthenium were reported from within an area of 200 km^2 in central Queensland in 1993. Since then, due

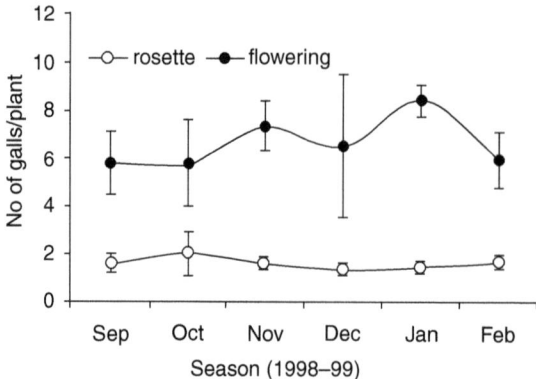

Fig. 15.6 Seasonal variation in *Epiblema strenuana* gall abundance in rosettes and flowering plants.

to both natural spread by the beetle and deliberate spread by farmers, the area with *Z. bicolorata* defoliation has increased to about 12 000 km², covering more than 50 properties in central Queensland (Dhileepan *et al.*, 2000a).

In India, *Z. bicolorata* was introduced from Mexico in 1984 in the vicinity of the city of Bangalore, and became established in the same year (Jayanth and Nagarkatti, 1987; Jayanth, 1987a; Jayanth and Bali, 1994a). However, its population levels attained damaging levels only after three years (Jayanth and Visalakshy, 1994a, 1996). Field releases continued in 15 states in India (Viraktamath *et al.*, 2004), and now after 20 years the beetle occurs in the majority of areas in India with parthenium infestations (Fig. 15.7), ranging from the tropical south to the sub-Himalayan regions in the north (Basappa, 1997; Maninder *et al.*, 1998; Susilkumar and Bhan, 1998; Pandey *et al.*, 2001; Uniyal *et al.*, 2001; Jadhav and Varma, 2001; Gupta and Sood, 2002, 2005; Gupta *et al.*, 2004; Bhatia *et al.*, 2005; Dhiman and Bhargava, 2005; Gautam *et al.*, 2005, 2006; Sarkate and Pawar, 2006; Sharma and Shujauddin, 2006), but not in the hot and dry arid northwestern region (i.e. Rajasthan State). It is unlikely that *Z. bicolorata* will survive here, as the summer temperature usually exceeds 45 °C, resulting in high rates of egg losses, and mortality among the larvae and diapausing adults (Jayanth and Bali, 1993a). Incidence of *Z. bicolorata* has also been reported from the Punjab region in Pakistan (Javaid and Shabbir, 2007; Fig. 15.7).

In South Africa, a *Z. bicolorata* colony was established in quarantine from adults collected in Australia, and host-specificity tests are progressing with results achieved so far indicating a strong likelihood of release (Strathie *et al.*, 2005). *Zygogramma bicolorata* has been introduced from South Africa into Ethiopian quarantine for further testing on native and economically important plant species in 2007. Attempts to establish *Z. bicolorata* colonies in quarantine in Sri Lanka and Papua New Guinea have not been successful.

Both adults and larvae feed on parthenium leaves (McFadyen and McClay, 1981), preferentially on younger leaves (Annadurai, 1989). Adults lay eggs either singly or in groups on the leaves, flower heads, stems, and on terminal and axillary buds (Jayanth,

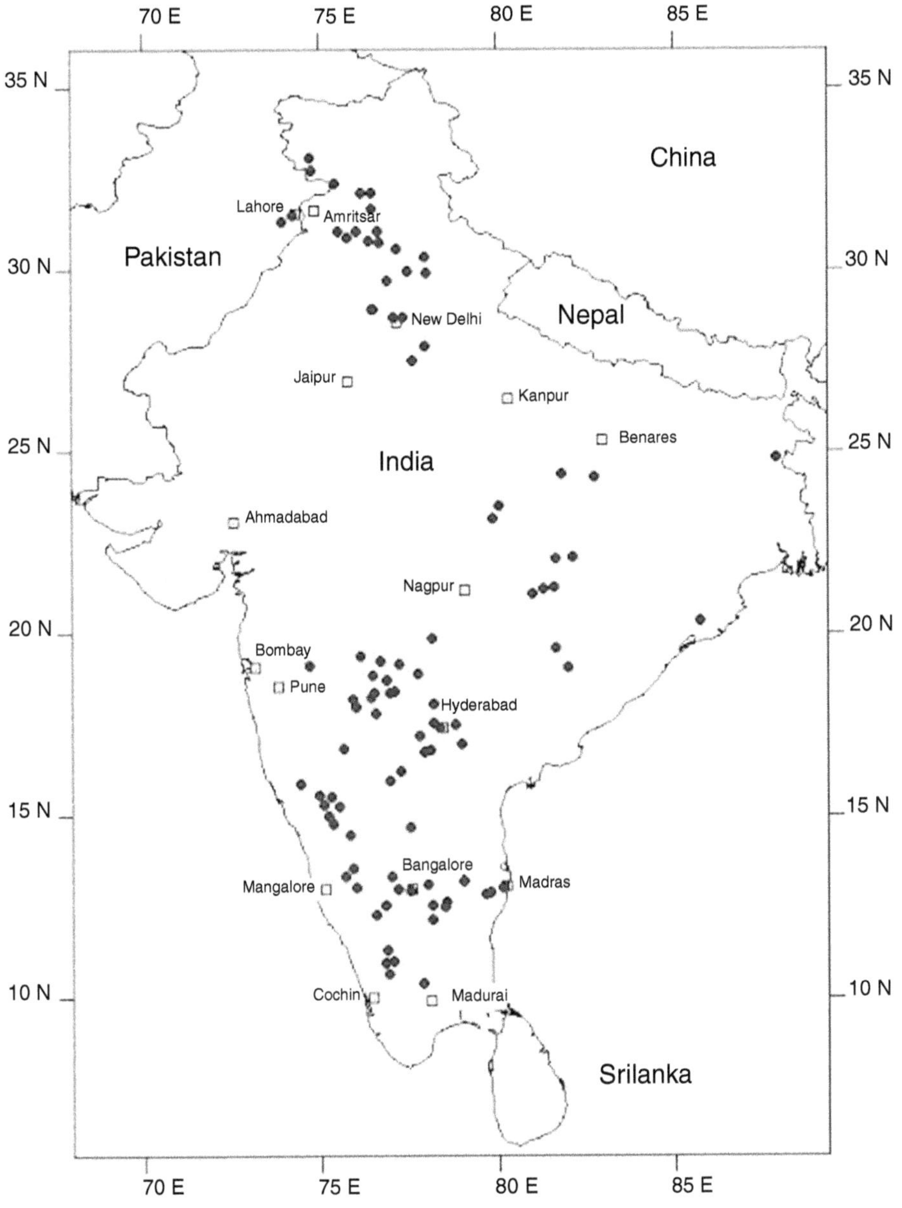

Fig. 15.7 Prevalence of *Zygogramma bicolorata* in India and Pakistan. (Dark circles represent documented locations with *Z. bicolorata*; empty squares represent towns.)

1987a). The emerging larvae feed voraciously on young leaves and the fully grown larvae burrow into the soil to pupate; the pupal stage lasts two weeks (McFadyen and McClay, 1981; Jayanth, 1987a). The life cycle is completed in 6–8 weeks and up to four generations per year occur depending on the rainfall and food availability (McFadyen, 1992). In Australia, adult beetles diapause in the soil in autumn (April–May) due to shorter days and cooler temperatures, and the adults emerge from soil in spring (September–November) responding to rainfall, increased temperature, and longer days. In India, adults undergo diapause in the dry season (November–April) and emerge with the commencement of monsoon rains (May–June) and higher temperature (Jayanth and Bali, 1993b). Adult beetles live up to two years and spend around six months diapausing in soil in autumn and winter in Australia (McFadyen, 1992). According to Indian data, males live longer than females (Jayanth and Bali, 1993c).

In greenhouse studies in Australia, defoliation by *Z. bicolorata* reduced plant height because of continuous feeding by *Z. bicolorata* on the vegetative apical meristems, resulting in reduced primary-stem height and altered branching pattern (Dhileepan *et al.*, 2000b). Sustained defoliation for three months reduced shoot and root biomass by 67% and 80%, respectively (Dhileepan *et al.*, 2000b). In totally defoliated plants, *Z. bicolorata* oviposited in the flower heads and the emerging larvae fed on the flowers, thus preventing the seed set.

In central Queensland, *Z. bicolorata* inflicted 91–100% defoliation, resulting in reductions in weed density by 32–93%, plant height by 18–65%, plant biomass by 55–89%, flower production by 75–100%, soil seed bank by 13–86% and seedling emergence in the following season by 73–90% (Dhileepan *et al.*, 2000a). At sites with continued outbreaks of *Z. bicolorata*, the existing soil seed bank is expected to drop, resulting in reduced density of parthenium in 6–7 years.

In India, *Z. bicolorata* caused 85–100% defoliation, resulting in nearly 100% reduction in parthenium weed density in Bangalore area (Jayanth and Bali, 1994a; Jayanth and Visalakshy, 1996). Similar impacts have been observed in other areas in India (e.g. Susilkumar, 2000; Dhiman and Bhargava, 2005; Jaipal, 2008). However, no significant effect on flower production in fully grown flowering parthenium plants because of defoliation by *Z. bicolorata* in India has been recorded (Jayanth and Bali, 1994a).

15.5.3 Listronotus setosipennis *(Hustache) (Coleoptera: Curculionidae)*

Listronotus setosipennis has a widespread distribution in northern Argentina and southern Brazil. It was introduced into Australia from these localities in 1982–1986 (McFadyen, 1985; Wild *et al.*, 1992) after host-specificity tests in Brazil (Wild, 1980) and Australia (Wild *et al.*, 1992). *Listronotus setosipennis*, immediately after field releases, established in Queensland in 1983 (Wild *et al.*, 1992), but its field incidence remained low and sporadic (McFadyen, 1992; Dhileepan *et al.*, 1996; Dhileepan and McFadyen, 1997; Dhileepan 2003a). In South Africa, a *L. setosipennis* colony was established in quarantine from adults collected from yellow-flowered parthenium in Santiago del Estero and outside Metán in Salta province in northwestern Argentina, and host-specificity tests are currently in progress (Strathie *et al.*, 2005).

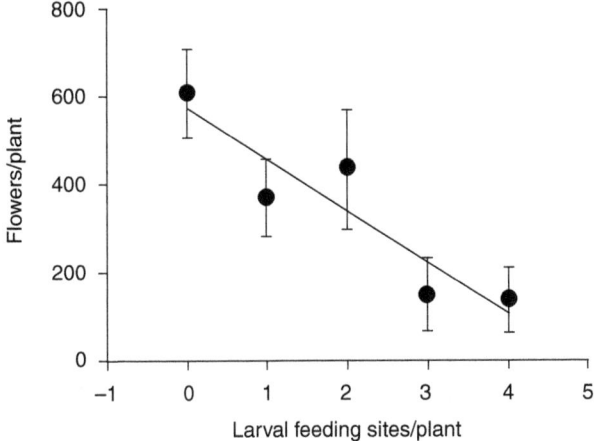

Fig. 15.8 Impact of *Listronotus setosipennis* larval density on flower production in glasshouse trials.

Listronotus setosipennis is a 5-mm-long nocturnal weevil, which feeds on parthenium leaves and flowers. Adults are winged, but are not known to fly. The adult female lays eggs singly in holes chewed in flowers, or leaf bases or on stem surfaces that are then covered by frass, and the newly emerged larvae tunnel into the stem. Five larval instars occur. Mature larvae may leave the stem and move to the root where they feed prior to pupation. The life cycle takes about 7 weeks and adult weevils live up to 8 months (Wild *et al.*, 1992).

Adult feeding and oviposition damage is negligible. Larval feeding has the ability to kill or prevent further development of parthenium seedlings (Wild *et al.*, 1992). In mature plants, higher larval density (> four larvae/plant) reduced the shoot biomass and number of seeds produced (Dhileepan *et al.*, 1996). In greenhouse studies, *L. setosipennis* reduced plant height by 51%, number of leaves by 78%, flower production by 63%, and plant biomass by 54% in rosette-stage plants (Dhileepan, 2003a). Flower production declined with increasing number of larvae per plant (Fig. 15.8), and it was estimated that a minimum of five larvae/plant is necessary to prevent the plant from flowering (Dhileepan, 2003a).

In field cage studies in Australia, damage by *L. setosipennis* in rosette-stage plants reduced primary stem height by 26% and flower production by 38%, but the impact on total plant height, basal stem width, root length, number of branches, root biomass, and total plant biomass was insignificant. In glasshouse and field cage studies, the impact of *L. setosipennis* on preflowering and flowering stages of parthenium was also insignificant (Dhileepan, 2003a). The negative impact of *L. setosipennis* on parthenium was more severe in the glasshouse than in field cages, due to lower population levels in the field cages (Dhileepan, 2003a).

In Australia, *L. setosipennis* was recorded in 48% of the parthenium-infested sites sampled in 1996–1998, but only 16% of the sites showed more than one larva/plant

(Dhileepan, 2003a). Incidence of *L. setosipennis* was more prevalent in plants growing on alluvial and black soils than on clay and sandy soils, possibly due to the soil quality appropriate for the construction of pupation chambers. *L. setosipennis* is a promising biological control agent in regions with prolonged dry periods and erratic rainfall pattern. The realized impact of *L. setosipennis* damage in the field in Australia has been less than the potential impact as estimated through glasshouse and field cage studies, due to the higher number of *L. setosipennis* larvae/plant utilized in the glasshouse and field cage trials (Dhileepan, 2003a).

15.5.4 Smicronyx lutulentus *Dietz (Coleoptera: Curculionidae)*

The seed-feeding weevil *Smicronyx lutulentus* occurs commonly on parthenium in Mexico and Texas, USA (Anderson, 1962; Bennett, 1976). This insect can also survive and reproduce on *Parthenium confertum* Gray, but its incidence was less than that on *P. hysterophorus*. Host-specificity studies confirmed that *S. lutulentus* is restricted to a few species within the taxon *Parthenium*, with *P. hysterophorus* being the most preferred host (McClay, 1979; McFadyen and McClay, 1981). The seed-feeding weevil was approved for field release in Australia in 1981 (McFadyen and McClay, 1981) and field releases were carried out between 1981 and 1983. Field establishment was confirmed only in 1996 (Dhileepan and McFadyen, 1997). Inability to establish *S. lutulentus* in quarantine in India using adults collected in Australia prevented further studies on the host specificity of this agent in India.

The adult is a 1.5 to 2.0 mm-long black weevil that feeds on flower buds and tender leaves. Adults live up to three months, and females lay an average of 237 eggs in their lifetime. Oviposition occurs in buds or freshly opened capitula, and emerging larvae feed on seeds; only one larva can feed per seed. Four larval instars occur. When the mature seed dehisces and falls to the ground, larvae burrow into the soil to pupate. An extended prepupal stage of 7–8 weeks occurs in the soil, which is influenced by temperature, usually at 30 °C (McFadyen and McClay, 1981). Emergence of adults is stimulated by rainfall.

Adults feed on young leaves, but inflict negligible feeding damage. Larval feeding causes significant reduction in seed output. In Mexico up to 30% seed destruction is attributed to *S. lutulentus* damage (McClay, 1985). In its native range, the seed-feeding weevil appears to have two generations per year, coinciding with autumn and summer rains.

In Australia, the weevil was established in central Queensland by 1996, and has since spread to northern Queensland (Dhileepan and McFadyen, 1997). However, the incidence of *S. lutulentus* is sporadic and localized with limited impact on seed production. In view of the enormous number of inflorescences produced/plant (mean $\pm$ standard error; 4963 $\pm$ 2192), a very high ($\geq$ 20 weevils/plant) *S. lutulentus* population is required to achieve even 50% reduction in seed production (Dhileepan *et al.*, 1996). With the current low infestation levels it will be difficult to estimate its impact on parthenium under field conditions.

15.5.5 Conotrachelus albocinereus *Fiedler (Coleoptera: Curculionidae)*

Conotrachelus albocinereus is native to Argentina, Brazil, and Colombia. In the native range, adults were collected only on parthenium and *Ambrosia artemisiifolia*, though sunflower is widely grown in the region.

The adults are usually 4–5 mm long, nocturnal and live up to three months (McFadyen, 2000). They feed on the stem tips and leaves, and lay eggs singly on the stem, mainly on the leaf axil (Jean-Marc, 1998). Newly emerged larvae feed on epidermal cells and burrow vertically into the stem to feed on the nutrient-rich parenchyma cells. Larval feeding induces elliptical galls on the main shoot axes. Larvae remain within the galls for about two months. The final larval instar chews a hole at the end of the tunnel and emerges through that to move to the soil for pupation. The adult emerges after three weeks.

Adult feeding damage is not significant, and causes only minor damage to young leaves (McFadyen, 2000). Damage is mainly due to larval feeding which results in the fracturing of the vertical continuity of vascular tissues, thereby disrupting the host plant's overall metabolism (Florentine *et al.*, 2002). Gall induction usually destroys axillary shoots but the main stem remains unaffected. Under greenhouse conditions in Australia, galling by *C. albocinereus* reduced the shoot biomass by 34%, root biomass by 41%, the number of mature capitula by 21%, and viable seed set by 18%, and the impact was more severe in preflowering plants than in flowering plants (Jean-Marc, 1998). It also appears that more than two larvae per plant are required to have any negative impact on plant vigor and flower production (Jean-Marc, 1998; Fig. 15.9). This insect may have established at a few sites in central Queensland, but is not in sufficient numbers to indicate widespread field establishment.

15.5.6 Platphalonidia mystica *(Razowski and Becker) (Lepidoptera: Totricidae)*

Platphalonidia mystica from Argentina was tested for host specificity in both Argentina and Australia. Larval feeding was evident on members of the subtribe Ambrosiinae, and two members of another subtribe Heliantheae, *Dahlia* and *Helianthus* (Griffiths and McFadyen, 1993). Field releases of this agent commenced in Australia in 1992 (Griffiths and McFadyen, 1993).

Adults are small (6 mm), light gray to white moths, which actively lay eggs in leaf axils during midday. Newly emerged larvae feed on the leaf surface and then enter the main shoot via axillary shoots. Larvae generally feed close to the surface of the phloem tissue of the stem and pupate within the stem. Prior to pupation the larvae cut emergence holes to which the exuvium remains attached following adult emergence. The life cycle from egg to adult takes 8–10 weeks.

Larval feeding results in the death of shoot tips and weakening of the main shoot. Very high galling densities result in the death of the plant (McFadyen, 1990). This agent is not known to be established in the field in Australia.

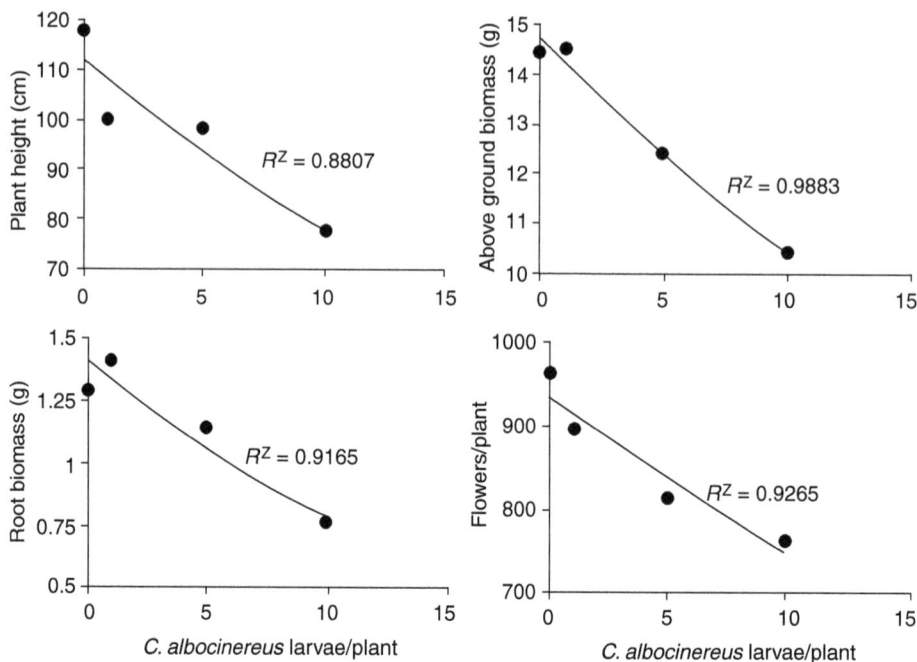

Fig. 15.9 Potential impact of *Conotrachelus albocinereus* on *Parthenium hysterophorus* in relation to larval density in glasshouse trials.

15.5.7 Carmenta *nr.* ithacae *(Beutenmüller) (Lepidoptera: Sesiidae)*

Carmenta ithacae, which has a wide native range distribution in USA (Arizona, Colorado, Kansas, Missouri, North Carolina, northern Florida, New Mexico, New York, Mississippi, Texas, and Wisconsin) and Mexico, has been recorded from *Helenium autumnale* L. and *Heliopsis helianthoides* (L.) Sweet (Asteraceae) in USA and only from *P. hysterophorus* in Mexico (McClay *et al.*, 1995; Withers *et al.*, 1999). The population of *C. ithacae* occurring on *P. hysterophorus* in Mexico may be a different species from the one occurring in USA, and hence is referred to as *C. nr. ithacae* (Withers *et al.*, 1999). *Carmenta* species are mostly host specific and studies in Australia using *C. nr. ithacae* collected from Mexico confirmed that this agent is highly host specific (McFadyen and Withers, 1997; Withers *et al.*, 1999). The clearwing moth was first released in Australia in September 1998. Since then, a total of 12 591 moths have been released at 30 sites in Queensland.

Adults emerge in the mornings (9–11 am), mate in sunshine in the morning, and oviposit from midday to early afternoon. Females oviposit on any part of the plant, and the eggs hatch in 10–14 days. Larvae migrate to the stem base, where they bore into the root and feed on the cortical tissue of the taproot and crown. After 5–6 weeks, the fully developed larvae pupate in the root or stem base in a silk cocoon. Adults emerge after 10–12 days from the clearly visible silk tube protruding from the plant at or above the soil level. Adults live for 3–12 days and females lay from 64 to 235 eggs in their lifetime. Larvae are found on all growth stages of parthenium, and heavily infested plants often die.

This agent has been recovered from the field from only a few sites in central Queensland, but not in sufficient numbers to indicate its widespread field establishment or potential impact.

15.5.8 Stobaera concinna *(Stal) (Homoptera: Delphacidae)*

In its native range in Mexico *S. concinna* was found only on *P. hysterophorus*. At sites in Nuevo Leon, Tamaulipas, and San Luis Potosi in Mexico, the planthopper was generally rare, and dense populations were never found in the field (McClay, 1983a).

Under laboratory conditions eggs are laid singly in stems. There are five nymphal instars and the duration of development from egg to adult lasted 30–56 days. *Stobaera concinna* could reproduce only on *P. hysterophorus* and *Ambrosia* spp. in laboratory host-range testing in Australia.

Nymphs feed on leaves and new shoots, while adults feed on main and axillary shoots. At high population densities, *S. concinna* caused yellowing of the leaves and spindly plant growth. This agent did not establish on parthenium, though the agent has become established on annual ragweed (*Ambrosia artemisiifolia* L.) in southeastern Queensland (McFadyen, 1992).

15.5.9 Bucculatrix parthenica *Bradley (Lepidoptera: Bucculatricidae)*

Bucculatrix parthenica is a highly host-specific agent native to Mexico. It was released in Australia from 1984 and its field establishment was confirmed in 1987 (McClay *et al.*, 1990). The leaf-mining moth became established widely in both central and northern Queensland, but failed to establish in southeastern Queensland.

The leaf-mining moth oviposits on leaves. The first and second instar larvae are leaf miners and later instars feed externally on the leaves. Pupation takes place within the ribbed cocoon adjacent to a midrib, and lasts 7–11 days. Adults can survive up to 14 days. The life cycle is completed in about 25 days under field conditions. By feeding externally, the fourth and fifth instar larvae inflict damage to the plant. Larval feeding is evident on all growth stages of parthenium. In the native range, the moth caused no major damage, due to low population densities and specialist natural enemies. In glasshouse conditions, at high population densities, the agent caused partial defoliation of the plant.

The insect is rare in Mexico, but has become abundant in Queensland at some sites, possibly due to the absence of, or reduced, parasitism. However, further studies are required to confirm this.

15.5.10 Puccinia abrupta *Diet. & Holw. var.* partheniicola *(Jackson)* Parmelee *(Uredinales)*

In the native range of parthenium, the winter rust *P. abrupta* var. *partheniicola* occurs naturally in Argentina, Bolivia, Brazil, Central America, and Brazil (Evans, 1997a). The winter rust collected from the semi-arid, upland regions (1400–1600 m asl) of Mexico (Evans, 1987) was the first pathogen to be released on parthenium in Australia, from 1991 until 1995. It is highly host specific (Parker *et al.*, 1994; Taye *et al.*, 2004a; Tomley, 1990) and its release in Australia began in 1991 and continued until 1995.

The winter rust infects plants when leaves are wet. The highest rate of infection occurs when the temperature is 15 °C and the period of leaf wetness (or dew period) is between 8 and 10 hours. Spore germination is optimal at 15 °C, with a steady decrease in the rate of germination with increasing temperature, until almost no germination occurs at 30 °C (Fauzi *et al.*, 1996, 1999). Frequent drought in Queensland has not helped the spread or effectiveness of this pathogen. In greenhouse studies, winter rust infection hastened leaf senescence, significantly reduced the life span and plant biomass, and reduced the flower production by 90% (Evans, 1987, 1997b; Parker *et al.*, 1994).

The winter rust became established only in a few localized areas in central Queensland (Dhileepan and McFadyen, 1997) with long dew periods and cooler temperatures (Fauzi *et al.*, 1999), but its impact on parthenium in these areas appears to be not significant (Dhileepan, 2003b). This agent did not establish in northern Queensland with warmer and drier conditions (Dhileepan *et al.*, 1996).

Parthenium winter rust has been reported in China (Yun and Xia, 2002), Ethiopia (Taye *et al.*, 2002, 2004a), India (Bagyanarayana and Manoharachary, 1997; Parker *et al.*, 1994), Kenya (Evans, 1997a), Mauritius (Parmelee, 1967), and South Africa (Wood and Scholler, 2002). However, in these countries the winter rust was not intentionally released as a biological control agent, and also the strains known to occur in many of these countries do not appear to be either widespread or aggressive (e.g. Kumar and Evans, 2005). In South Africa, parthenium winter rust was first observed in the town of Brits, in the northwest province (Wood and Scholler, 2002) in 1995, and now also occurs in Mpumalanga and KwaZulu-Natal provinces. In Ethiopia, *P. abrupta* var. *partheniicola* was first reported in 1997, and is now known to occur commonly there in cool and humid areas at high altitudes (1500–2500 m asl) where rainfall varies from 400 to 700 mm (Taye *et al.*, 2002, 2004a). In Ethiopia, the winter rust significantly reduced the plant height, number of leaves, number of branches, and total biomass of parthenium (Taye *et al.*, 2004a). In India, host specificity and suitability as a biological control agent of a highly virulent isolate of *P. abrupta* var. *partheniicola* from Mexico are being explored (Kumar and Evans, 2005).

15.5.11 Puccinia melampodii *Dietel & Holway (Uredinales)*

The summer rust *P. melampodii* occurs naturally in Central America and Mexico (Evans, 1997a). The summer rust collected from low-altitude regions of Mexico and Texas, USA

(Evans, 1997a; Seier *et al.*, 1997; Tomley, 2000), is highly host specific, damaging and adapted to areas with high temperatures and limited periods of humidity (Holden *et al.*, 1995; Seier, 1999; Seier and Tomley, 2000). However, its incidence in Mexico was highest during the wet season. Field release of the summer rust in Queensland commenced in January 2000. Since then, releases have been made at more than 50 sites (Dhileepan *et al.*, 2006a).

The summer rust is a microcyclic rust species which cycles exclusively through the telial stage. This rust species is necessarily autoecious, which means the rust does not require an alternate host but completes its reduced life cycle on one host. The summer rust forms telia predominantly on the lower leaf surface as well as on the stems, initially of distinct sori, which constitute aggregates of individual telia that coalesce over time (Seier, 1999). Sporulation can occur over the entire leaf surface, leading to necrosis and eventual dieback of the affected leaf. Successive infection cycles can cause severe stunting and premature plant death (Seier, 1999).

In glasshouse trials in the UK, summer rust infection reduced parthenium plant height by 53%, number of leaves by 30%, number of side shoots by 46%, and flower production by 100% (Seier, 1999). Rust infection also reduced root, leaf, and stem biomass by 73%, 43%, and 79%, respectively (Seier, 1999).

Field establishment of the summer rust was evident in 88% of the release sites in Australia (Dhileepan *et al.*, 2006a). The summer rust became established immediately, but with higher prevalence and intensity in northern Queensland than in central Queensland. In northern Queensland, in the first year of field release, prevalence of rust infection increased as plants matured. By the end of autumn (May 2000), 71% of the plants were infected with 52% of the leaf area covered with rust. However, the impact of summer rust on seedling establishment, plant height, flower production, plant biomass, and plant density at the end of the first year was not significant (Dhileepan, 2003b). Subsequent dry summers (2004–2006) resulted in low levels of rust incidence (Dhileepan *et al.*, 2006a; Fig. 15.10),

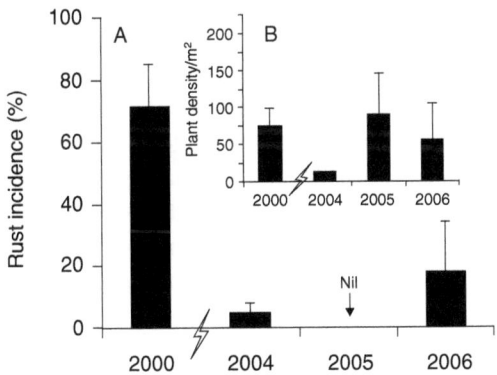

Fig. 15.10 Establishment and prevalence of *Puccinia melampodii* in north Queensland.

with negligible impact on the weed. During dry periods, it was assumed that the pathogen would maintain sufficient residual inoculum on the infected plant (Seier, 1999).

The summer rust was imported into Sri Lanka in 2003 and South Africa in 2004 from Australia, and the pathogenicity and host-specificity tests done indicate that this rust is suitable for release in both Sri Lanka (Jayasuriya, 2005) and South Africa (Ntushelo and Wood, 2008). Although preliminary studies on two highly virulent isolates of *P. melampodii* in quarantine in India suggest that the rust is host specific (Kumar and Evans, 2005), there are no immediate plans to import this rust to India for further studies.

15.6 Impact of biological control

Seven species of insects and two rust fungi have been successfully established as biological control agents in Australia. The impact of these biological control agents on parthenium was evaluated at two properties with contrasting climate (Mt. Panorama and Plain Creek) in Queensland during 1996–2000 based on an exclusion experiment using insecticides and fungicide (Dhileepan, 2001, 2003b, c), and limited mechanical grazing twice a year (spring and autumn). The leaf-feeding beetle *Z. bicolorata* and the stem-galling moth *E. strenuana* were the promising agents at Mt. Panorama. At Plain Creek *E. strenuana* was the predominant agent till the summer of 1999–2000 when the summer rust *P. melampodii* also became prominent. The biological control agents reduced plant height and flower production in 1996–1997, especially at Mt. Panorama. The biological control agents also reduced weed density by 90% at Mt. Panorama in 1996–1997 (Fig. 15.11) and prevented 28% and 18% of plants from producing flowers at Plain Creek in 1996–1997 and 1997–1998 respectively. However, biological control agents had only limited impact on weed population density at Plain Creek (Fig. 15.11). The limited impact of biological control agents in only three out of the four year study was due to below average summer rainfall, which resulted in low *Z. bicolorata* and *E. strenuana* activity at Mt. Panorama and nonsynchrony between parthenium germination and *E. strenuana* emergence at Plain Creek. This resulted in 70% reduction in the soil seed bank at Mt. Panorama over the four-year period, but the reduction at Plain Creek was not significant.

A significant increase in grass biomass production due to biological control was observed at both trial sites, but only in one out of four years at Mt. Panorama and two out of four years at Plain Creek (Dhileepan, 2007). At Mt. Panorama there was a 40% increase in grass biomass in 1997 due to 96% defoliation by the leaf-feeding beetle *Z. bicolorata* and gall induction in 100% of the plants by the moth *E. strenuana*. At Plain Creek, pasture production increased by 52% in 1998 due to reduced parthenium seedling emergence, and by 45% in 2000 (Dhileepan, 2007), due to the combined effects of gall induction by *E. strenuana* and the establishment of the summer rust *P. melampodii* in 72% of the plants. In economic terms, benefits from increased grass production due to biological control are Au$ 1.25/ha/year for buffel grass *Cenchrus ciliaris* L. (Poaceae) in central Queensland, and Au$1.19/ha/year for the Queensland blue grass *Dichanthium sericeus* (R. Br.) (Poaceae) in northern Queensland (Adamson and Bray, 1999). This is in

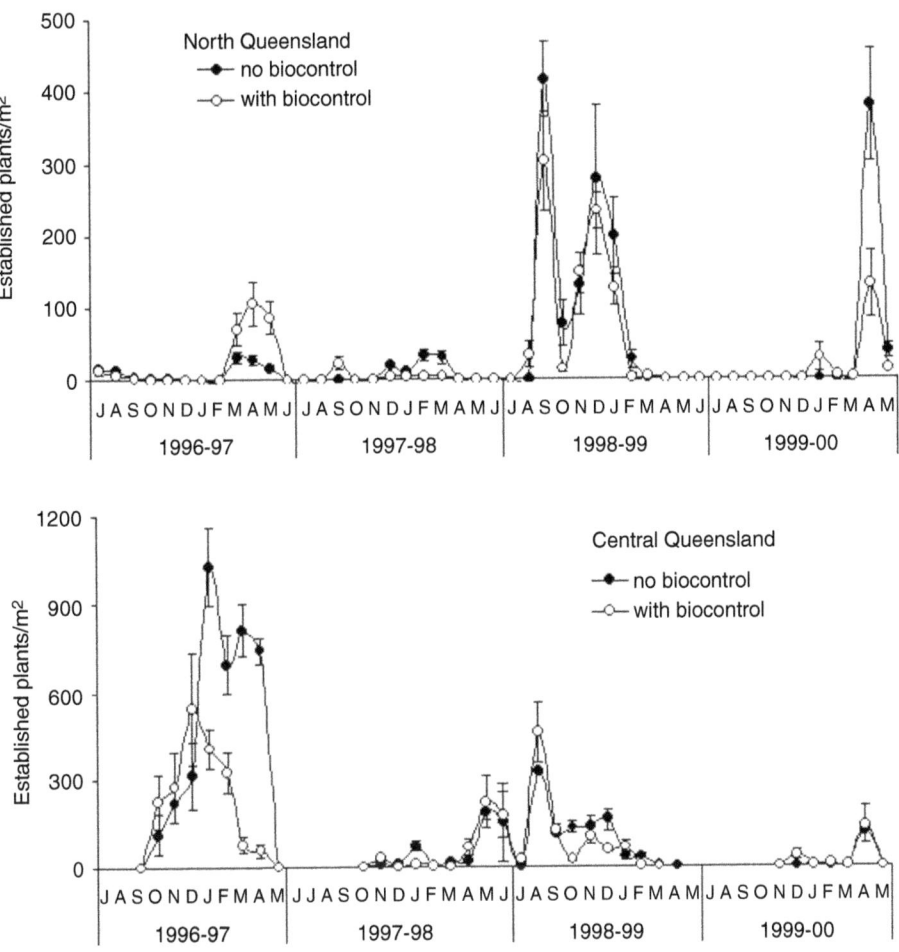

Fig. 15.11 Seasonal variation in the impact of biological control on *Parthenium hysterophorus* populations at two sites in Queensland, Australia.

addition to the saving of Au$8m/year in health costs in treating allergic dermatitis and asthma in workers in infested areas (Page and Lacey, 2006).

In India, defoliation by *Z. bicolorata* resulted in the reestablishment of native vegetation (Jayanth and Visalakshy, 1996; Sridhara *et al.*, 2005). However, information on the long-term impact of defoliation by *Z. bicolorata* in India is lacking.

15.7 Nontarget damage

Helianthus annuus (sunflower), *Helianthus tuberosus* Linn. (Jerusalem artichoke) and *G. abyssinica* (niger) are the only crop plants within Heliantheae. Within Asteraceae, *Carthamus tinctorius* Linn (safflower), *Chrysanthemum cinerariaefolium* Trev. (pyrethrum),

Cichorium intybus Linn. (chicory) and *Lactuca sativa* Linn. (lettuce) are economically important crops. Among various crops in Asteraceae, sunflower and niger are the most widely grown and most economically important crop in many countries where parthenium is currently targeted for biological control. Hence, host-specificity tests should include several varieties of these crops as test plants. The congeneric species *Parthenium argentatum* Gray (Guayule), a rubber-producing shrub, is also often included in the host-specificity tests, even though both plant species share only a limited number of insect species in their native range, due to differences in plant architecture, secondary chemistry and nonoverlapping habitats (McClay *et al.*, 1995).

Sunflower is widely grown in northern Argentina (Wild *et al.*, 1982) and North America (McFadyen, 1982), but none of the agents (*Listronotus setosipennis* and *C. albocinereus* from Argentina and *E. strenuana* from Mexico) collected from these regions has ever been recorded attacking *H. annuus*. In the host-specificity tests of *Z. bicolorata* (McClay, 1985; Jayanth and Nagarkatti, 1987), *L. setosipennis* (Wild *et al.*, 1992), and *P. mystica* (Griffiths and McFadyen, 1993), feeding damage occurred on *H. annuus* and to a lesser extent on ornamental *Dahlia* spp., but the risk of damage to *H. annuus* under field conditions was negligible. Field surveys of *H. annus* crops during 1987–1989 in the central Queensland, Australia, revealed no nontarget incidence on sunflower crops (K. Dhileepan, unpublished data). However, in India, seven years after the introduction and establishment of *Z. bicolorata*, nontarget feeding on sunflower crop was reported (Kumar, 1992; Chakravarthy and Bhat, 1994, 1997; Chakravarthy *et al.*, 1994, 1996). It was even suggested that the beetle observed in *H. annus* was *Zygogramma conjuncta* Rogers (Kumar, 1992; Chakravarthy and Bhat, 1994, 1997; Chakravarthy *et al.*, 1994, 1996) and not *Z. bicolorata*. However, its taxonomic status was confirmed subsequently as *Z. bicolorata* (Jayanth *et al.*, 1997). Further laboratory and field studies in India indicated that nontarget feeding is possibly due to deposition of parthenium pollen on nearby sunflower leaves (Jayanth *et al.*, 1993). On the basis of feeding behavior (Swamiappan *et al.*, 1997a; Jayanth *et al.*, 1998; Withers, 1998, 1999), life-table (Bhumannavar *et al.*, 1998; Viraktamath *et al.*, 2004) and field studies (Jayanth and Visalakshy, 1994b; Jayanth *et al.*, 1997; Swamiappan *et al.*, 1997b; Patel and Viraktamath, 2005), it appears that the chances of *Z. bicolorata* becoming a pest of *H. annus* are negligible. No economic loss due to *Z. bicolorata* feeding on sunflower even at higher insect densities was recorded in India (Kulkarani *et al.*, 2000).

Several species of *Xanthium* and *Ambrosia* represented in Australia, India and South Africa are invasive, and are related closely to *Parthenium*, as they belong to Ambrosiinae. Some of the biological control agents released (i.e. *E. strenuana, Z. bicolorata, C. albocinereus, S. concinna, P. mystica*) were either collected in the native range or fed in host-specificity tests on species of *Xanthium* and *Ambrosia* (Rice, 1937; McClay, 1981, 1983b, 1985, 1987; McFadyen, 1982, 1985, 1990; Goeden and Palmer, 1992). Hence, *E. strenuana* was introduced for the biological control of *Ambrosia artemisiifolia* Linn. (annual ragweed) in China (Wan *et al.*, 1995, 2003), and is now widespread in Hunan, Jiangxi, Jiangsu, and Hubei provinces. Gall induction by *E. strenuana* and defoliation by

Z. bicolorata also occurs on *Xanthium occidentale* Bertol. (Noogoora burr) and *A. artemisiifolia* in Australia (McFadyen, 1987, 1992). In India also, *Z. bicolorata* caused extensive defoliation on *X. strumarium* L. (Kumar, 1992), an introduced weed from North America.

15.8 Factors influencing biological control

15.8.1 Establishment and abundance

In general, 10–20% of biological control agents released fail to establish in a weed biological control project (McFadyen, 2003). In Australia, two of the 11 biological control agents on parthenium (*S. concinna* and *P. mystica*) failed to establish, and several of the other agents on parthenium have failed to reach desired population levels (*S. lutulentus, B. parthenica,* and *L. setosipennis*) or have restricted distribution (*Z. bicolorata, C. ithacae, C. albocinereus,* and *P. abrupta* var. *partheniicola*), thereby limiting their impact on the target weed.

The time taken for establishment of biological control agents also varies widely, possibly due to the climatic differences between the countries of origin and introduction. In India, *Z. bicolorata* took three years to establish (Jayanth and Bali, 1994a). In Australia, *Z. bicolorata* became established on *A. artemisiifolia* in southeastern Queensland in two years of introduction (McFadyen, 1992), but took more than 12 years to become abundant on parthenium in central Queensland (Dhileepan *et al.*, 1996). The longer time taken for *Z. bicolorata* to become abundant in central Queensland is possibly due to the prolonged process for adaptation to extreme local climatic conditions (Rice, 1998; Rice *et al.*, 1999). In Australia, as a result, two genetically distinct allopatric populations of *Z. bicolorata* evolved on two different geographically isolated host plants – on *A. artemisiifolia* in southeastern Queensland and on parthenium in central Queensland – appear to exist (Rice, 1998). The seed-feeding weevil *S. lutulentus* also took more than 16 years to become abundant in Australia (Dhileepan *et al.*, 1996), believed to be due to prolonged adaptation to extreme local climatic conditions. In contrast, *E. strenuana, L. setosipennis,* and *P. melampodii* established with in a few years of field releases in Australia.

15.8.2 Natural enemies

Natural enemies prevent the establishment or reduce the overall effectiveness of introduced weed biological control agents (McFadyen and Spafford-Jacob, 2003). Only on 27 of the 470 insect and mite species introduced worldwide as weed biological control agents have natural enemies been reported (Julien and Griffiths, 1998). Information on the prevalence of natural enemies and their impact on the effectiveness of weed biological control agents are often observational and the significance of natural enemies in the failure or reduced effectiveness of weed biological control agents remains unknown.

In its native range, *E. strenuana* is attacked by around 37 species of parasitoids (McClay, 1981; Table 15.2). In Australia, 10 species of parasitoids attack this species (Table 15.2) and no increase in the parasitoid species assemblage over the 16 years since its introduction has occurred (Erbacher, 1986; Dhileepan *et al.*, 2005). The effectiveness of *E. strenuana* as a biological control agent depends on its ability to induce galls on *Parthenium* at the rosette stage (McFadyen, 1992; Dhileepan and McFadyen, 2001; Dhileepan, 2003b). Parasitism was lower in rosette stages than in flowering-stage plants. Lower levels of parasitism early in the season when most plants are in the rosette stage, as well as lower parasitism in rosettes than in flowering plants suggest that the impact of parasitism on the effectiveness of the gall-inducing insect may not be significant. However, the potential negative impact of higher parasitism levels at the end of the season on the overwintering and the subsequently emerging *E. strenuana* population in the following year cannot be underestimated.

Only two species of parasitoids have been recorded on *Z. bicolorata* in its native range (McClay, 1983b; Table 15.2). In its introduced ranges, only one parasitoid species (Withers *et al.*, 1998; Table 15.2), several generalist predators (Pandey *et al.*, 2002, 2005; Gupta *et al.*, 2004) and the entomopathogenic fungus *Metarhizium anisopliae* (Metchnikoff) Sorokin (Hypocreales: Clavicipitaceae) (Jayanth and Bali, 1994b) have been recorded. However, their impact on the biological control agent population and performance are not known. Although several specialist parasitoid species have been recorded on *L. setosipennis, S. lutulentus, C. ithacae*, and *B. parthenica* (Table 15.2), no information is available on the natural enemies of these agents in Australia.

15.8.3 Abiotic factors

The distribution and abundance of biological control agents and their interactions with their host plants are influenced by a range of abiotic factors (e.g. rainfall, dew period, temperature, humidity, light, and wind). Among them, rainfall, either the total summer rainfall in central Queensland or the timing of the onset of summer rainfall in northern Queensland, is the major factor affecting the effectiveness of parthenium biological control agents (Dhileepan, 2003b).

Gall induction by *E. strenuana* in the field varied considerably (Dhileepan and McFadyen, 1997) and such variations were due to lack of synchrony between parthenium germination and *E. strenuana* emergence (Dhileepan and McFadyen, 2001). In northern Queensland, no relationship between summer rainfall and the proportion of plants with *E. strenuana* gall incidence occurred (Dhileepan, 2003b). But the timing of the onset of summer rainfall had a critical impact on the effectiveness of gall induction by *E. strenuana* (Dhileepan, 2003b). In 1996–1997, due to synchronization between the onset of rainfall (resulting in parthenium germination) and increase in temperature and photoperiod (resulting in the emergence of *E. strenuana* from diapause), 61% of plants experienced gall induction before attaining flowering stage. This prevented 32% of the plants from producing any flowers (Dhileepan, 2001).

Table 15.2 *Natural enemies of parthenium biological control insects in their native and introduced ranges.*

Host insects	Natural enemies	Host stage	Country
Epiblema strenuana	**Hymenoptera: Trichogrammatidae**		
	Trichogrammatoidea sp.	Eggs	Mexico
	Hymenoptera: Braconidae		
	Bracon sp.	Larvae	Australia
	Apanteles sp.	-do-	Australia
	Apanteles sp. *ater* group	-do-	Mexico, Florida
	Chelonus curvimaculatus Cameron	-do-	Australia
	Macrocentrus sp.	-do-	Mexico, Florida
	Microplitis sp.	-do-	-do-
	Hymenoptera: Chalcididae		
	Brachymeria sp.	Larvae/pupae	Australia
	Antrocephalus sp.	Pupae	-do-
	Spilochalcis flavopicta (Cress.)	Pupae	Florida, USA
	Hymenoptera: Elasmidae		
	Elasmus sp.	Larvae	Australia
	Hymenoptera: Eupelmidae		
	Euplemus sp. (*urozonus* group)	Lavae	Mexico, Florida
	Euplemus sp.	Larvae/pupae	Australia
	Hymenoptera: Bethylidae		
	Goniozus sp.	Larvae	Australia
	Goniozus (Prosierola) sp.	Larvae	Mexico, Florida
	Hymenoptera: Eulophidae		
	Pediobium sexdentatus (Girault)	Pupae	Mexico
	Hymenoptera: Eurytomidae		
	Eurytoma sp.	Larvae	Mexico
	Hymenoptera: Perilampidae		
	Perilampus sp. near *tristis* Mayr	Hyperparasitoids?	Mexico
	Hymenoptera: Ichneumonidae		
	Clydonium spp.	Pupae	Mexico
	Temelucha sp.	Pupae	-do-
	Trathala sp.	Pupae	-do-
	Phaeogenes walshi australis (Cush.)	Pupae	Florida, USA
	Glabridorsum sp.	Pupae	Australia
	Xanthopimpla sp.	Pupae	-do-
	Diptera: Tachinidae		
	Lixophaga parva Tnsd.	Pupae	Mexico and Florida
Zygogramma bicolorata	**Hymenoptera: Pteromalidae**		
	Erixestus zygogrammae Cave & Grissell	Eggs	Australia
	Erixestus sp.	Eggs	Mexico

Table 15.2 (cont.)

Host insects	Natural enemies	Host stage	Country
	Diptera: Tachinidae		
	Doryphorophagha hyalinipennis (Wulp)	Pupae?	Mexico
	Hemiptera: Pentatomidae		
	Andrallus spinidens Fab.	Predators	India
	Cantheoconidea furcellata Wolf	-do-	-do-
	Hemiptera: Reduviidae		
	Sycanus pyrrhomelas Walker	Predators	India
Listronotus	**Hymenoptera: Braconidae**		
setosipennis	*Triaspis* sp.	Larvae	Brazil and Argentina
	Apanteles sp. (*laevigatus* group)	-do-	-do-
Smicronyx	**Hymenoptera: Pteromalidae**	Larvae	Mexico
lutulentus	*Zatropis* sp. near *incertus* Ashmead		
	Eutrichosoma mirabile Ashmead	?	Texas, USA
	Hymenoptera: Eulophidae		
	Paracrias sp.	Larvae	
Carmenta	**Hymenoptera: Braconidae**		
ithacae	*Apanteles* sp. *ater* group	Larvae	Mexico
	Hymenoptera: Chalcididae		
	Invreia sp. near *usta* Grissell & Schauff	Larvae/pupae	Mexico
Bucculatrix	**Hymenoptera: Braconidae**		
parthenica	*Apanteles* sp.	Larvae	Mexico

Sources: Stegmaier, 1971; McClay, 1983b, 1985; McFadyen, 1985; McClay *et al.*, 1990; Wild *et al.*, 1992; Withers *et al.*, 1998; Gupta *et al.*, 2004; Dhileepan *et al.*, 2005

In central Queensland, higher levels of defoliation by *Z. bicolorata* and gall induction by *E. strenuana* in 1996–1997 coincided with above-average summer rainfall, but in the following three years with below-average summer rainfall, the defoliation and gall induction levels were much lower (Dhileepan, 2003b). At Mt. Panorama in 1996–1997, higher summer rainfall favored the increase of *Z. bicolorata* populations, resulting in continued defoliation pressure throughout the life of the weed. In three out of four years, due to below-average summer rainfall, *Z. bicolorata* adults, emerging after diapause, fed on parthenium but did not oviposit. Due to low larval population, only low levels of defoliation were achieved and were not sustained throughout the life of the weed. In central Queensland, since 1992, outbreaks of *Z. bicolorata* resulting in complete defoliation have occurred only in three years, when summer rainfall was more than 315 mm (Dhileepan, 2003b). There were no major *Z. bicolorata* outbreaks in years when summer rainfall was below average (<280 mm) (Dhileepan, 2003b). Summer rainfall also affected the incidence of *E. strenuana* gall damage at Mt. Panorama, and as a result, the proportion of plants with galls was higher in 1996–1997 than in the following three years. Higher rust infection in the summer of 2000 and its reappearance in the autumn of 2006 were possibly due to higher rainfall, coupled with prolonged dew periods at the study sites (Dhileepan *et al.*, 2006a).

In the Bangalore region in India, the incidence and abundance of *Z. bicolorata* also coincided with rainfall patterns (Jayanth, 1987a).

15.8.4 Plant vigor preference

The plant vigor hypothesis proposes that insect herbivores, particularly the gall-inducing insects, prefer large, vigorously growing plants or plant modules compared with smaller, less vigorous plants or plant modules (Price, 1991). Such a preference for more vigorous and larger plants by biological control agents can have significant impact on the target weed populations (Dhileepan, 2004). Preference for more vigorous and larger plants could maximize the fitness of the biological control agent, thereby increasing population levels of the agent. However, avoidance of less vigorous and smaller plants leaves a certain proportion of plants undamaged (Dhileepan, 2004), which could potentially limit the ability of the biological control agent to regulate the target weed population.

Selective preference by *E. strenuana* for vigorous plants in Australia resulted in more than a quarter of the parthenium plants, all less vigorous, remaining without galls (Dhileepan, 2003b, 2004). These less vigorous, ungalled plants produced flowers (Dhileepan and McFadyen, 2001), thereby limiting the effects of the gall-inducing insect in reducing the plant population and soil seed bank.

For effective weed biological control, it is important that the insect population regulates the target plant population, rather than the target plant population determining the insect population. As parthenium is an annual herb with a short life cycle, selective preference by *E. strenuana* for more vigorous plants has resulted in the gall-inducing moth population being limited by plant vigor. Preference for plant vigor therefore does not appear to be a beneficial trait for parthenium biological control agents. Moreover this may explain the paucity of successes of biological control against annual weeds using gall-inducing insects.

15.8.5 Genetic impediments

Genetic diversity and fitness of biological control agents are the factors that can affect the establishment and effectiveness of the agents. Potential biological control agents are usually maintained in quarantine for several years before being approved for field release. This can result in a deleterious effect on insect colonies due to inbreeding (i.e. Wardill *et al.*, 2004).

15.9 Future research

Despite ongoing biological control efforts on parthenium over the last three decades in Australia and India, parthenium continues to be a major weed in several parts of the world, including Australia. In many parts of the world, in nonagricultural areas (e.g. forests, marginal vacant land), classical biological control appears to be the only option to manage this weed. In some countries, the high incidence of parthenium may also be attributed to either lack of resources or lack of recognition of the severity of the problem with no ensuing action. Prospects of finding any further host-specific biological control agents in North America, due to extensive surveys already undertaken, are unlikely. It is worth considering South America for further exploration for host-specific biological control agents, as many countries have parthenium populations that originated from these regions. In Australia, currently there are no plans to conduct any more native range exploration studies for parthenium biological control agents. For other countries, research efforts should be initiated or continued, with an emphasis on ecologically based agent prioritization.

15.9.1 Plant response to herbivory

Detailed demographic data on *P. hysterophorus* as an aid to the biological control agent prioritization process are not available from either its native or introduced ranges. In their absence, weaknesses of the target weed identified from simulated herbivory trials in the introduced ranges can be exploited to focus the search for effective agents, thereby enhancing the success rate of biological control efforts. Studying plant response to simulated herbivory to different plant parts/modules can yield significant ecological insights into tolerance of and compensation to herbivory (Raghu and Dhileepan, 2005). Such an approach could help to identify the guilds of herbivores most likely to have a negative impact on plant vigor and reproductive output. This will have major implications for how we can prioritize agents to target the most susceptible stage of parthenium in countries that have young, or not yet established, biological control programs (e.g. Ethiopia, India, Pakistan, Papua New Guinea, Sri Lanka, and South Africa).

15.9.2 Plant genotypic studies

Plant genotype can be used as a filter to identify areas for future agent exploration in the native range (e.g. Dhileepan *et al.*, 2006b). Plant genotypic studies carried out so far are preliminary in nature with samples only from Mexico, USA, and Venezuela. No samples were included from other countries including Bolivia, Chile, Paraguay, Peru, and Uruguay in the native range. A more detailed study with modern techniques and including samples from Central and South America is required to identify and understand the genetic variability in the native range. Further genetic studies are also required in Africa and India, to identify the origin of these populations (Graham and Lang, 1998). At present the sesquiterpene lactones of selected parthenium populations from Argentina, Brazil, South Africa, and Swaziland are being investigated by the University of Hohenheim, Germany.

15.9.3 Plant vigor preference

As parthenium is an annual herb with a short life cycle, selective preference for more vigorous plants resulted in the gall-inducing insect population being limited by plant vigor. The implications of plant vigor preference on the effectiveness of gall insects as weed biological control agents in other annual weeds have not been fully understood. This information is vital to understand the gall-inducing insect–host plant relationships at population levels and for the selection of insects of promise in future weed biological control programs.

15.9.4 Climate matching

Parthenium hysterophorus has a wide geographic distribution in its native and introduced ranges. Natural enemies associated with *P. hysterophorus* in its native range may either be widespread (e.g. *E. strenuana*) or limited (e.g. *P. abrupta* var. *partheniicola*) within the host plant's geographic distribution. The distribution and abundance of natural enemies and their interactions with their host plants are influenced by a range of abiotic factors (e.g. temperature, humidity, rainfall, dew-period). As a result, comparability of abiotic factors between the native and introduced range of an individual biological control agent is critical to its establishment and efficacy in regulating the target weed population. Selection of classical weed biological control agents that are adapted to the climates in areas of intended release demands a thorough analysis of the climates of the source and release sites. Hence, climate matching between the countries of origin and introduction of biological control agents using climate matching softwares (i.e. CLIMEX®, BIOCLIM®, CLIMATE®) appear important for any future biological control agent introduction. The thermal tolerances of *Z. bicolorata* are being investigated in South Africa, with climate matching modeling of its potential geographic range in South Africa.

15.9.5 Field (ecological) host range

Host-specificity tests are usually conducted under quarantine conditions (physiological host range) in the introduced range, and the host range of potential agents under natural conditions (ecological host range) in the native distribution often is not fully known. This is essential because *P. hysterophorus* is closely related to sunflower, and in spite of the fact that many of the biological control agents released on parthenium so far have fed on sunflower in host-specificity trials in quarantine, only *Z. bicolorata* introduced into India has been recorded feeding on sunflower (e.g. Kumar, 1992; Chakravarthy and Bhat, 1994, 1997; Chakravarthy *et al.*, 1994, 1996; Jayanth and Visalakshy, 1994b; Jayanth *et al.*,

1997; Swamiappan *et al.*, 1997b; Patel and Viraktamath, 2005) and not the population introduced into Australia. One possible reason for the differences in the behavior of *Z. bicolorata* populations in India and Australia is that they were collected from two different regions in Mexico. Studies under open field conditions in the native range would greatly enhance the knowledge on ecological host range of candidate agents, thereby minimizing the risk of nontarget damage.

15.9.6 Simulation and system models

A detailed demographic model can help to identify the weak links in plant population dynamics, thereby enhancing the selection of more effective agents (Raghu *et al.*, 2006) and the ability to predict nontarget risk (i.e. Louda *et al.*, 2005). Such models have been developed for many weeds targeted for biological control (i.e. Shea and Kelly, 1998; Kriticos *et al.*, 2003). A model to predict the effectiveness of introduced biological control agents on parthenium is being developed using a systems modeling approach (i.e. STELLA® software).

15.10 Conclusion

That a biological control agent effective in one country will be effective in other countries is a myth. But this need not be the case, due to the factors discussed earlier. Hence agents need to be prioritized specifically for each country, and more specifically for different regions in large countries. For example, most of the biological control agents introduced into Australia are effective only in central Queensland, with limited impact in northern Queensland, possibly due to climatic differences. Hence, future agent prioritization in other countries should focus on agents that are suitable to local climatic conditions, using climate modeling tools. Future research should also focus on the use of plant-based approaches (plant response to herbivory and identifying weak links in *P. hysterophorus* demography) as "predictive" filters for agent prioritization.

Acknowledgments

We thank Wilmot Senaratne, Deanna Bayliss, and Carvert Moya for the parthenium distribution maps; and Bill Palmer and Dane Panetta for comments on the earlier versions of the manuscript.

References

Adamson, D. (1996). Introducing dynamic considerations when economically evaluating weeds. Masters thesis, University of Queensland, Brisbane, Australia.

Adamson, D. C. and Bray, S. (1999). The economic benefit from investing in insect biological control of parthenium weed (*Parthenium hysterophorus*). School of Natural and Rural System Management, University of Queensland, Brisbane, Australia, 44 pp.

Adkins, S. W., Navie, S. C. and Dhileepan, K. (2005). Parthenium weed in Australia: research progress and prospects. In *Proceedings of the Second International Conference on Parthenium Management*, ed. T. V. Ramachandra Prasad, H. V. Nanjappa, R. Devendra, *et al.* Bangalore, India: University of Agricultural Sciences, pp. 11–27.

Anderson, D. N. (1962). A revision of the species *Smicronyx* Schoenherr (Coleoptera: Curculionidae) occurring in America north of Mexico. *Proceedings of the United States Natural Museum*, **113**, 1–185.

Annadurai, R. S. (1989). Reproductive potential in terms of quantitative food utilisation of *Zygogramma bicolorata* (Coleoptera: Chrysomelidae) on *Parthenium hysterophorus* (Asteraceae). In *Proceeding of the VII International Symposium on the Biological Control of Weeds*, held in Rome, ed. E. S. Delfosse. Melbourne, Australia: CSIRO, pp. 385–394.

Auld, B. A., Hosking, J. and McFadyen, R. E. (1983). Analysis of the spread of tiger pear and parthenium weed in Australia. *Australian Weeds*, **2**, 56–60.

Bagyanarayana, G. and Manoharachary, C. (1997). Studies on *Puccinia abrupta* var. *partheniicola* a potential mycoherbicide. In *Proceedings of the First International Conference on Parthenium Management*, Vol. II, ed. M. Mahadevappa and V. C. Patil. Dharwad, India: University of Agricultural Sciences, pp. 95–96.

Basappa, H. (1997). Incidence of biocontrol agent *Zygogramma bicolorata* Pallister on *Parthenium hysterophorus* L. In *Proceeding of the First International Conference on Parthenium Management*, Vol. II, ed. M. Mahadevappa and V. C. Patil. Dharwad, India: University of Agricultural Sciences, pp. 81–84.

Basappa, H. (2005). Parthenium: an alternate host of sunflower necrosis disease and thrips. In *Second International Conference on Parthenium Management*, ed. T. V. Ramachandra Prasad, H. V. Nanjappa, R. Devendra, *et al.*, Bangalore, India: University of Agricultural Sciences, pp. 83–86.

Bennett, F. D. (1976). A preliminary survey of the insects and diseases attacking *Parthenium hysterophorus* L. (Compositae) in Mexico and the USA to evaluate the possibilities of its biological control in Australia. Mimeographical Report. Slough, UK: Commonwealth Institute of Biological Control, 18 pp.

Bhatia, S., Choudhary, R. and Singh, M. (2005). Current status of the invasive weed *Parthenium hysterophorus* (Asteraceae) and impact of defoliation by the biocontrol agent *Zygogramma bicolorata* (Coleoptera: Chrysomelidae) in Jammu (J&K), India. *Eighth International Conference on the Ecology and Management of Alien Plant Invasions*, held 5–12 September 2005 in Katowice, Poland. Leider, Netherland: Backhuys Publ.

Bhumannavar, B. S., Balasubramanian, C. and Ramani, S. (1998). Life table of the Mexican beetle *Zygogramma bicolorata* Pallister on parthenium and sunflower. *Journal of Biological Control*, **12**, 101–106.

Brooks, S. J., Vitelli, J. S., and Rainbow, A. G. (2004). Developing best practice roadside *Parthenium hysterophorus* L. control. In *Proceedings of the 14th Australian Weeds Conference*, ed. B. M. Sindel and S. B. Johnson. Sydney, Australia, Weed *Society of New South Wales*, pp. 195–198.

Butler, D. W. and Fairfax, R. J. (2003). Buffel grass and fire in a Gidgee and Brigalow woodland: a case study from Central Queensland. *Ecological Management and Restoration*, **4**, 120–125.

CABI (2004). Parthenium fact sheet. In *Crop Protection Compendium*. CD-ROM. Wallingford, UK: CAB International.

Chakravarthy, A. K. and Bhat, N. S. (1994). The beetle (*Zygogramma conjuncta* Rogers), an agent for the biological control of the weed *Parthenium hysterophorus* L. in India feeds on sunflower (*Helianthus annus* L.). *Journal of Oilseeds Research*, **11**, 122–125.

Chakravarthy, A. K. and Bhat, N. S. (1997). Ecology of the beetle *Zygogramma conjuncta* (Rogers) on *Parthenium hysterophorus* Linn. In *First International Conference on Parthenium Management*, Vol. II, ed. M. Mahadevappa and V. C. Patil. Dharwad, India: University of Agricultural Sciences, pp. 74–77.

Chakravarthy, A. K., Bhat, N. S. and Sridhar, S. (1994). The beetle *Zygogramma conjuncta* (Rogers), a bioagent for the control of the weed, *Parthenium hysterophorus* L. is oligophagous. *Science and Culture*, **60**, 61–62.

Chakravarthy, A. K., Cox, M. L., Bhat, N. S., Sridhar, S. and Thyagaraj, N. E. (1996). Identification, host specificity and infestation of *Zygogramma conjuncta* Rogers on *Helianthus annus* L. In *Biological and Cultural Control of Insect Pests: An Indian Scenario*, ed. D. P. Ambrose. Thirunelveli, Tamil Nadu, India: Adeline Publishers, pp. 243–250.

Cheney, M. (1998). Determination of the prevalence of sensitivity to *Parthenium* in areas of Queensland affected by the weed. Master of Public Health thesis, Queensland University of Technology, Brisbane, Australia, 118 pp.

Chippendale, J. F. and Panetta, F. D. (1994). The cost of parthenium weed to the Queensland cattle industry. *Plant Protection Quarterly*, **9**, 73–76.

Da Silva, M. C., Izidine, S. and Amuda, A. B. (2004). *A Preliminary Checklist of the Vascular Plants of Mozambique*. South African Botanical Diversity Network Report 30. Pretoria, South Africa: SABONET, 185 pp.

Dale, I. J. (1981). Parthenium weed in the Americas: a report on the ecology of *Parthenium hysterophorus* in South, Central and North America. *Australian Weeds*, **1**, 8–14.

Dhileepan, K. (2001). Effectiveness of introduced biocontrol insects on the weed *Parthenium hysterophorus* (Asteraceae) in Australia. *Bulletin of Entomological Research*, **91**, 167–176.

Dhileepan, K. (2003a). Current status of the stem-boring weevil *Listronotus setosipennis* (Coleoptera: Curculionidae) introduced against the weed *Parthenium hysterophorus* (Asteraceae) in Australia. *Biocontrol Science and Technology*, **13**, 3–12.

Dhileepan, K. (2003b). Seasonal variation in the effectiveness of leaf-feeding beetle *Zygogramma bicolorata* (Coleoptera: Curculionidae) and stem-galling moth *Epiblema strenuana* (Lepidoptera: Totricidae) as biocontrol agents on the weed *Parthenium hysterophorus* (Asteraceae). *Bulletin of Entomological Research*, **93**, 393–401.

Dhileepan, K. (2003c). Evaluating the effectiveness of weed biocontrol at the local scale. In *Improving the Selection, Testing and Evaluation of Weed Biological Control Agents*, ed. H. Spafford Jacob and D. T. Briese. Technical Series 7. Adelaide, Australia: CRC for Australian Weed Management, pp. 51–60.

Dhileepan, K. (2004). The applicability of the plant vigor and resource regulation hypotheses in explaining *Epiblema* gall moth–*Parthenium* weed interactions. *Entomologia Experimentalis et Applicata*, **113**, 63–70.

Dhileepan, K. (2007). Biological control of parthenium (*Parthenium hysterophorus*) in Australian rangeland translates to improved grass production. *Weed Science*, **55**, 497–501.

Dhileepan, K. and McFadyen, R. E. (1997). Biological control of parthenium in Australia: progress and prospects. In *First International Conference on Parthenium*

Management, Vol I, ed. M. Mahadevappa and V. C. Patil. Dharwad, India: University of Agricultural Sciences, pp. 40–44.

Dhileepan, K. and McFadyen, R. E. (2001). Effects of gall damage by the introduced biocontrol agent *Epiblema strenuana* (Lepidoptera: Tortricidae) on the weed *Parthenium hysterophorus* (Asteraceae). *Journal of Applied Entomology*, **125**, 1–8.

Dhileepan, K., Florentine, S. K. and Lockett, C. J. (2006a). Establishment, initial impact and persistence of parthenium summer rust *Puccinia melampodii* in north Queensland. In *Proceedings of the 15th Australian Weeds Conference*, ed. C. Preston, J. H. Watts and N. D. Crossman. Adelaide, Australia: Weed Management Society of South Australia, pp. 577–580.

Dhileepan, K., Lockett, C. J. and McFadyen, R. E. (2005). Larval parasitism by native insects on the introduced stem-galling moth *Epiblema strenuana* Walker (Lepidoptera: Tortricidae) and its implications for biological control of *Parthenium hysterophorus* (Asteraceae). *Australian Journal of Entomology*, **44**, 83–88.

Dhileepan, K., Madigan, B., Vitelli, M., *et al.* (1996). A new initiative in the biological control of parthenium. In *Proceedings of the 11th Australian Weeds Conference*, ed. R. C. H. Shepherd. Melbourne, Australia: Weed Society of Victoria, pp. 309–312.

Dhileepan, K., Setter, S. and McFadyen, R. E. (2000a) Impact of defoliation by the introduced biocontrol agent *Zygogramma bicolorata* (Coleoptera: Chrysomelidae) on parthenium weed in Australia. *BioControl*, **45**, 501–512.

Dhileepan, K., Setter, S. and McFadyen, R. E. (2000b) Response of the weed *Parthenium hysterophorus* (Asteraceae) to defoliation by the introduced biocontrol agent *Zygogramma bicolorata* (Coleoptera: Chrysomelidae). *Biological Control*, **19**, 9–16.

Dhileepan, K., Wilmot, K. A. D. W. and Raghu, S. (2006b). A systematic approach to biological control agent exploration and selection for prickly acacia (*Acacia nilotica* ssp. *indica*). *Australian Journal of Entomology*, **45**, 302–306.

Dhiman, S. C. and Bhargava, M. L. (2005). Seasonal occurrence and bio-control efficacy of *Zygogramma bicolorata* Ballister (Coleoptera: Chrysomelidae) on *Parthenium hysterophorus*. *Annals of Plant Protection Sciences*, **13**, 81–84.

Erbacher, J. (1986). Parasitism by hymenopterous species upon the stem-galling moth, *Epiblema strenuana* Walker (Tortricidae) in southeast Queensland. BSc honours thesis, Department of Biology, Queensland Institute of Technology, Brisbane, Australia, 50 pp.

Evans, H. C. (1983). Parthenium project: report on a visit to Mexico to survey fungal pathogens of *Parthenium hysterophorus*. L. (Compositae). March-May 1983. Kew, Surrey, UK: Commonwealth Mycological Institute, 19 pp.

Evans, H. C. (1987). Life cycle of *Puccinia abrupta* var. *partheniicola*, a potential biological control agent of *Parthenium hysterophorus*. *Transactions of the British Mycological Society*, **88**, 105–111.

Evans, H. C. (1997a). *Parthenium hysterophorus*: a review of its weed status, and the possibilities for biological control. *Biocontrol News and Information*, **18**, 89–98.

Evans, H. C. (1997b). The potential of neotropical fungal pathogens as classical biological control agents for management of *Parthenium hysterophorus* L. In *First International Conference on Parthenium Management*, ed. M. Mahadeveppa and V. C. Patil. Dharwad, India: University of Agricultural Sciences, pp. 55–62.

Farkya, S., Pandey, A. K. and Rajak, R. C. (1994). Evaluation of *Fusarium* sp. as mycoherbicides for *Parthenium hysterophorus*. Host-specificity. *National Academy Science Letters–India*, **17**, 81–82.

Fauzi, M. T., Tomley, A. J., Dart, P. J., Ogle, H. J. and Adkins, S. W. (1996). Effect of temperature and leaf wetness on the infectivity of *Puccinia abrupta* var. *partheniicola*, a potential biocontrol agent of parthenium weed. In *Proceedings of the 11th Australian Weeds Conference*, ed. R. C. H. Shepherd. Melborune, Australia: Weed Science Society of Victoria, pp. 298–300.

Fauzi, M. T., Tomley, A. J., Dart, P. J., Ogle, H. J. and Adkins, S. W. (1999). The rust *Puccinia abrupta* var. *partheniicola*, a potential biocontrol agent of *Parthenium* weed: environmental requirements for disease progress. *Biological Control*, **14**, 141–145.

Fessehaie, R., Chichayibelu, M. and Giorgis, M. H. (2005). Spread and ecological consequences of *Parthenium hysterophorus* in Ethiopia. *Arem*, **6**, 11–21.

Firehun, Y. and Tamado, T. (2006). Weed flora in the Rift Valley sugarcane plantations of Ethiopia as influenced by soil types and agronomic practices. *Weed Biology and Management*, **6**, 139–150.

Florentine, S. K., Raman, A. and Dhileepan, K. (2001). Gall-inducing insects and biological control of *Parthenium hysterophorus* L (Asteraceae). *Plant Protection Quarterly*, **16**, 1–6.

Florentine, S. K., Raman, A. and Dhileepan, K. (2002). Response of the weed *Parthenium hysterophorus* (Asteraceae) to the stem gall-inducing weevil *Conotrachelus albocinereus* (Coleoptera: Curculionidae). *Entomologia Generalis*, **26**, 195–206.

Florentine, S. K., Raman, A. and Dhileepan, K. (2005). Effects of gall induction by *Epiblema strenuana* on gas exchange, nutrients, and energetics in *Parthenium hysterophorus*. *BioControl*, **50**, 787–801.

Frew, M., Solomon, K. and Mashilla, D. (1996). Prevalence and distribution of *Parthenium hysterophorus* L in eastern Ethiopia. In *Proceedings of the First Annual Conference* of the *Ethiopian Weed Science Society*, held 24–25 November 1993, Addis Ababa, Ethiopia, ed. R. Fessehaie. *Arem*, **1**, 19–26.

Gautam, R. D., Khan, M. A. and Garg, A. K. (2006). Ecological adaptability and variations among population of Mexican beetle, *Zygogramma bicolorata* Pallister (Chrysomelidae: Coleoptera). *Journal of Entomological Research*, **30**, 21–23.

Gautam, R. D., Khan, M. A. and Samyal, A. (2005). Release, recovery and establishment of Mexican beetle, *Zygogramma bicolorata* Pallister (Chrysomelidae: Coleoptera) on *Parthenium hysterophorus* Linnaeus proliferating in and around Delhi. *Journal of Entomological Research*, **29**, 167–172.

Goeden, R. D. and Palmer, W. A. (1992). Lessons learned from studies on the insects associated with Ambrosiinae in North America in relation to biological control of weedy members of this group. In *Proceedings of the VIII International Symposium on Biological Control of Weeds*, ed. E. S. Delfosse and R. R. Scott. Melbourne, Australia: DSIR/CSIRO, pp. 565–573.

Govindappa, M. R., Chowda Reddy, R. V., Devaraja, *et al.* (2005). *Parthenium hysterophorus*: a natural reservoir of Tomato Leaf Curl Begomovirus. In *Second International Conference on Parthenium Management*, ed. T. V. Ramachandra Prasad, H. V. Nanjappa, R. Devendra, *et al.* Bangalore, India: University of Agricultural Sciences, pp. 80–82.

Graham, G. C. and Lang, C. L. (1998). Genetic analysis of relationship of *Parthenium* occurrences in Australia and indications of its origins. Internal Report. Cooperative Research Centre for Tropical Pest Management, University of Queensland, Brisbane, Australia.

Griffiths, M. W. and McFadyen, R. E. (1993). Biology and host-specificity of *Platphalonidia mystica* (Lep. Cochylidae) introduced into Queensland to biologically control *Parthenium hysterophorus* (Asteraceae). *Entomophaga*, **38**, 131–137.

Gupta, P. R. and Sood, A. (2002). Spread of *Zygogramma bicolorata* Pallister in Himachal Pradesh. *Insect Environment*, **8**, 101–102.

Gupta, P. R. and Sood, A. (2005). Biological observations on *Zygogramma bicolorata* Pallister on congress grass (*Parthenium hysterophorus* L.), and its activity in mid-hills of Himachal Pradesh. *Pest Management and Economic Zoology*, **13**, 21–27.

Gupta, R. K., Khan, M. S., Bali, K., Monobrullah, M. and Bhagat, R. M. (2004). Predatory bugs of *Zygogramma bicolorata* Pallister: an exotic beetle for biological suppression of *Parthenium hysterophorus* L. *Current Science*, **87**, 1005–1010.

Haseler, W. H. (1976). *Parthenium hysterophorus* L. in Australia. *PANS*, **22**, 515–80.

Hilliard, O. M. (1977). *Compositae in Natal*. Pietermaritzburg, South Africa: University of Natal Press, 659 pp.

Holden, A. N. G., Parker, A. and Tomley, A. J. (1995). Host range screening of *Puccinia abrupta* var. *partheniicola* for the biological control of *Parthenium hysterophorus* in Queensland. In *Proceedings of the VIII International Symposium on Biological Control of Weeds*, ed. E. S. Delfosse and R. R Scott. Melbourne, Australia: DSIR/CSIRO, pp. 555–560.

Holman, D. J. (1981). Parthenium weed threatens Bowen Shire. *Queensland Agricultural Journal*, **107**, 57–60.

Jadhav, R. B. and Varma, A. (2001). Results of field introduction and establishment of *Zygogramma bicolorata* Pallister against *Parthenium hysterophorus* L., weed around sugarcane growing areas of Pravaranagar (Maharashtra). *Indian Journal of Sugarcane Technology*, **16**, 66–69.

Jaipal, S. (2008). Potential of Mexican beetle, *Zygogramma bicolorata* in suppression of *Parthenium hysterophorus* in the Indian subtropics (abstract). *Proceedings of the XII International Symposium on Biological Control of Weeds*, held in La Grande Motte, France, 22–27 April 2007. ed. M. H. Julien *et al*. Wallingfora, UK: CAB International.

Javaid, A. and Anjum, T. (2005). *Parthenium hysterophorus* L. – a noxious alien weed. *Pakistan Journal of Weed Science Research*, **11**, 1–6.

Javaid, A. and Shabbir, A. (2007). First report of biological control of *Parthenium hysterophorus* by *Zyogramma bicolorata* in Pakistan. *Pakistan Journal of Phytopathology*, **18**, 199–200.

Javaid, A., Shafique, S., Bajwa, R. and Shafique, S. (2006a). Effects of aqueous extracts of allelopathic crops on germination and growth of *Parthenium hysterophorus* L. *South African Journal of Botany*, **72**, 609–612.

Javaid, A., Shafique, S., Bajwa, R. and Shafique, S. (2006b). Biological control of noxious alien weed *Parthenium hysterophorus* L. in Pakistan. *International Journal of Biology and Biotechnology*, **3**, 721–724.

Jayachandra (1971). *Parthenium* weed in Mysore State and its control. *Current Science*, **40**, 568–569.

Jayanth, K. P. (1987a). Introduction and establishment of *Zygogramma bicolorata* on *Parthenium hysterophorus* in Bangalore, India. *Current Science*, **56**, 310–311.

Jayanth, K. P. (1987b). Investigations on the host-specificity of *Epiblema strenuana* (Walker) (Lepidoptera: Tortricidae), introduced for biological control trials against *Parthenium hysterophorus* in India. *Journal of Biological Control*, **1**, 133–137.

Jayanth, K. P. and Bali, G. (1993a). Temperature tolerance of *Zygogramma bicolorata* (Coleoptera: Chrysomelidae) introduced for biological control of *Parthenium hysterophorus* (Asteraceae) in India. *Journal of Entomological Research*, **17**, 27–34.

Jayanth, K. P. and Bali, G. (1993b). Diapause behaviour of *Zygogramma bicolorata* (Coleoptera: Chrysomelidae), a biocontrol agent of *Parthenium hysterophorus* in Bangalore, India. *Bulletin of Entomological Research*, **83**, 383–388.

Jayanth, K. P. and Bali, G. (1993c). Biological studies on *Zygogramma bicolorata* Pallister (Coleoptera: Chrysomelidae), a biocontrol agent of *Parthenium hysterophorus* L. (Asteraceae). *Journal of Biological Control*, **7**, 93–98.

Jayanth, K. P. and Bali, G. (1994a). Biological control of parthenium by the beetle *Zygogramma bicolorata* in India. *FAO Plant Protection Bulletin*, **42**, 207–213.

Jayanth, K. P. and Bali, G. (1994b). Life table of the parthenium *beetle Zygogramma bicolorata* Pallister (Coleoptera: Chrysomelidae) in Bangalore, India. *Insect Science and Application*, **15**, 19–23.

Jayanth, K. P. and Nagarkatti, S. (1987). Investigations on the host-specificity and damage potential of *Zygogramma bicolorata* Pallister (Coleoptera: Chrysomelidae) introduced into India for biological control of *Parthenium hysterophorus*. *Entomon*, **12**, 141–145.

Jayanth, K. P. and Visalakshy, P. N. G. (1994a). Dispersal of the parthenium beetle *Zygogramma bicolorata* (Chrysomelidae) in India. *Biocontrol Science and Technology*, **4**, 363–365.

Jayanth, K. P. and Visalakshy, P. N. G. (1994b). Field evaluation of sunflower varieties for susceptibility to the parthenium beetle *Zygogramma bicolorata*. *Journal of Biological Control*, **8**, 48–52.

Jayanth, K. P. and Visalakshy, P. N. G. (1996). Succession of vegetation after suppression of parthenium weed by *Zygogramma bicolorata* in Bangalore, India. *Biological Agriculture and Horticulture*, **12**, 303–309.

Jayanth, K. P., Mohandas, S., Ashokan, R. and Visalakshy, P. N. G. (1993). Parthenium pollen induced feeding by *Zygogramma bicolorata* (Coleoptera: Chrysomelidae) on sunflower. *Bulletin of Entomological Research*, **83**, 595–598.

Jayanth, K. P., Visalakshy, P. N. G., Chaudhary, M. and Ghosh, S. K. (1998). Age-related feeding by the Parthenium beetle *Zygogramma bicolorata* on sunflower and its effect on survival and reproduction. *Biocontrol Science and Technology*, **8**, 117–123.

Jayanth, K. P., Visalakshy, P. N. G., Ghosh, S. K. and Chaudary, M. (1997). Feasibility of biological control of *Parthenium hysterophorus* by *Zygogramma bicolorata* in the light of the controversy due to its feeding on sunflower. In *First International Conference on Parthenium Management*, ed. M. Mahadevappa and V. C. Patil, Dharwad, India: University of Agricultural Sciences, pp. 45–51.

Jayasuriya, A. H. M. (2005). Parthenium weed – status and management in Sri Lanka. In *Second International Conference on Parthenium Management*, ed. T. V. Ramachandra Prasad, H. V. Nanjappa, R. Devendra, *et al*. University of Agricultural Sciences, Bangalore, India, pp. 36–43.

Jean-Marc, D. (1998). Efficacy of the stem galling weevil *Conotrechlus albocinereus* as a biological control agent for parthenium weed (*Parthenium hysterophorus* L.). Report of the Department of Natural Resources, Brisbane, Australia: Alan Fletcher Research Station.

Joel, D. M. and Liston, A. (1986). New adventive weeds in Israel. *Israel Journal of Botany*, **35**, 215–223.

Joshi, S. (1991). Biological control of *Parthenium hysterophorus* L. (Asteraceae) by *Cassia uniflora* Mill (Leguminosae) in Bangalore, India. *Tropical Pest Management*, **37**, 182–184.

Julien, M. H. and Griffiths, M. W. (1998). *Biological Control of Weeds: A World Catalogue of Agents and their Target Weeds*. Fourth edn. Wallingford, UK: CAB International, 223 pp.

Kadhane, D. L., Jangde, C. R., Sadekar, R. D. and Joshirao, M.K. (1992). *Parthenium* toxicity in buffalo calves. *Journal of Soils and Crops*, **21**, 69–71.

Kandasamy, O. S. and Sankaran, S. (1997). Management of parthenium using competitive crops and plants. In *First International Conference on Parthenium Management*, ed. M. Mahadevappa and V. C. Patil. Dharwad, India: University of Agricultural Sciences, pp. 33–36.

Khosla, S. N. and Sobti, S. N. (1979). Parthenium: a national health hazard, its control and utility – a review. *Pesticides*, **13**, 121–127.

Kololgi, P. D., Kololgi, S. D. and Kololgi, N. P. (1997). Dermatologic hazards of parthenium in human beings. In *First International Conference on Parthenium Management*, ed. M. Mahadevappa and V. C. Patil. Dharwad, India: University of Agricultural Sciences, pp. 18–21.

Kriticos, D., Brown, J., Maywald, G. F., *et al.* (2003). SPAnDX: a process-based population dynamics model to explore management and climatic change impacts on an invasive alien plant, *Acacia nilotica*. *Ecological Modelling*, **163**, 187–208.

Kulkarani, K. A., Kulkarani, N. S. and Santoshkumar, G. H. (2000). Loss estimation in sunflower due to *Zygogramma bicolorata* Pallister. *Insect Environment*, **6**, 10–11.

Kumar, A. R. V. (1992). Is the Mexican beetle *Zygogramma bicolorata* (Coleoptera: Chrysomelidae) expanding its host range? *Current Science*, **63**, 729–730.

Kumar, P. S. and Evans, H. C. (2005). The mycobiota of *Parthenium hysterophorus* in its native and exotic ranges: opportunities for biological control in India. In *Proceedings of the Second International Conference on Parthenium Management*, ed. T. V. R. Prasad, H. V. Nanjappa, R. Devendra, *et al.* Bangalore, India: University of Agricultural Sciences, pp. 107–113.

Lewis, W. H., Dixit, A. B. and Wedner, J. H. (1988). Reproductive biology of *Parthenium hysterophorus* (Asteraceae). *Journal of Palynology*, **23–24**, 72–82.

Louda, S. M., Rand, T. A., Arnett, A. E., *et al.* (2005). Evaluation of ecological risk to a population of a threatened plant from an invasive biocontrol insect. *Ecological Applications*, **15**, 234–249.

Macdonald, I. A. W., Reaser, J. K., Bright, C., *et al.* (eds.) (2003). *Invasive Alien Species in Southern Africa*: National Reports and Directory of Resources. Cape Town, South Africa: Global Invasive Species Programme.

Mahadevappa, M. (1997). Ecology, distribution, menace and management of parthenium. In *First International Conference on Parthenium Management*, ed. M. Mahadevappa and V. C. Patil. Dharwad, India: University of Agricultural Sciences, pp. 1–12.

Maninder, S., Singh, J., Brar, K. S. and Bakhetia, D. R. C. (1998). Spread of *Zygogramma bicolorata* Pallister on *Parthenium* in Punjab and adjoining states. *Insect Environment*, **4**, 83.

McClay, A. S. (1979). Preliminary report on the biology and host-specificity of *Smicronyx lutulentus* Dietz (Col.: Curculionidae), a potential biocontrol agent for *Parthenium hysterophorus* L. Unpublished report. Monterrey, Mexico: Commonwealth Institute of Biological Control, Mexican sub-station, 10 pp.

McClay, A. S. (1980). Studies of some potential biocontrol agents for *Parthenium hysterophorus* in Mexico. In *Proceedings of the V International Symposium on Biocontrol of Weeds*, ed. E. S. Delfosse. Melbourne, Australia: CSIRO, pp. 471–482.

McClay, A. S. (1981). Preliminary report on the biology and host-specificity of *Epiblema strenuana* (Walker) (Lepidoptera: Tortricidae), a potential biocontrol agent for *Parthenium hysterophorus* L. (Compositae). Unpublished report. Monterrey, Mexico: Commonwealth Institute of Biological Control, Mexican sub-station, 7 pp.

McClay, A. S. (1983a). Biology and host-specificity of *Stobaera concinna* (Stal) (Homoptera: Delphacidae), a potential biocontrol agent for *Parthenium hysterophorus* L. (Compositae). *Folia Entomologica Mexicana*, **56**, 21–30.

McClay, A. S. (1983b). *Natural Enemies of* Parthenium hysterophorus *L. (Compositae) in Mexico: Final Report* 1978–1983. Wallingford, UK: Commonwealth Institute of Biological Control.

McClay, A. S. (1985). Biocontrol agents for *Parthenium hysterophorus* from Mexico. In *Proceedings of the VI International Symposium on Biological Control of Weeds*, ed. E. S. Delfosse, Vancouver, Canada: Agriculture Canada, pp. 771–778.

McClay, A. S. (1987). Observations on the biology and host specificity of *Epiblema strenuana* (Lepidoptera, Tortricidae), a potential biocontrol agent for *Parthenium hysterophorus* (Compositae). *Entomophaga*, **32**, 23–34.

McClay, A. S. and Anderson, D. M. (1985). Biology and immature stages of *Thecesternus hirsutus* Pierce (Coleoptera: Curculionidae) in north-eastern Mexico. *Proceedings of the Entomological Society of Washington*, **87**, 207–215.

McClay, A. S., McFadyen, R. E. and Bradley, J. D. (1990) Biology of *Bucculatrix parthenica* Bradley sp. n. (Lepidoptera: Bucculatricidae) and its establishment in Australia as a biological control agent for *Parthenium hysterophorus* (Asteraceae). *Bulletin of Entomological Research*, **80**, 427–432.

McClay, A. S., Palmer, W. A., Bennett, F. D. and Pullen, K. R. (1995). Phytophagous arthropods associated with *Parthenium hysterophorus* (Asteraceae) in North America. *Environmental Entomology*, **24**, 796–809.

McFadyen, P. J. (1976). A survey of insects attacking *Parthenium hysterophorus* L. (F. Compositae) in South America. Internal report. Department of Lands, State Government of Queensland, Australia, 8 pp.

McFadyen, P. J. (1979). A survey of insects attacking *Parthenium hysterophorus* L. (Compositae) in Argentina and Brazil. *Dusenia*, **11**, 42–45.

McFadyen, R. E. (1982). Host specificity of parthenium stem galling moth *Epiblema strenuana* Walker from Mexico, a potential biocontrol agent against *Parthenium hysterophorus* in Queensland. Internal report. Sherwood, Brisbane, Australia: Alan Fletcher Research Station, 10 pp.

McFadyen, R. E. (1985). The biological control programme against *Parthenium hysterophorus* in Queensland. In *Proceedings of the VI International Symposium on the Biological Control of Weeds*, ed. E. S. Delfosse. Vancouver, Canada: Agriculture Canada, pp. 789–796.

McFadyen, R. E. (1987). The effect of climate on the stem-galling moth *Epiblema strenuana*. In *Proceedings of the Seventh Australian Weeds Conference*, ed. D. Lemerle and A. R. Leys. Sydney, Australia: Weed Society of New South Wales, pp. 97–99.

McFadyen, R. E. (1990). The stem-boring moth *Platphalonidia mystica* for the biological control of parthenium weed (release application). Alan Fletcher Research Station, Queensland Department of Lands, Brisbane, Australia, 16 pp.

McFadyen, R. E. (1992). Biological control against parthenium weed in Australia. *Crop Protection*, **24**, 400–407.

McFadyen, R. E. (1995). Parthenium weed and human health in Queensland. *Australian Family Physician*, **24**, 1455–1459.

McFadyen, R. E. (2000). Biology and host specificity of the stem galling weevil *Conotrachelus albocinereus* Fielder (Col: Curculionidae), a potential biocontrol agent for parthenium weed *Parthenium hysterophorus* L. (Asteraceae) in Queensland, Australia. *Biocontrol Science and Technology*, **10**, 195–200.

McFadyen, R. E. (2003). Does ecology help in the selection of biocontrol agents? In *Improving the Selection, Testing and Evaluation of Weed Biological Control Agents*, ed. H. Spafford-Jacob and D. T. Briese. Technical Series 7. Adelaide, Australia: CRC for Australian Weed Management, pp. 5–9.

McFadyen, R. E. and McClay, A. R. (1981). Two new insects for the biological control of parthenium weed in Queensland. In *Proceedings of the Sixth Australian Weeds Conference,* ed. B. J. Wilson and J. D. Swarbrick. Brisbane, Australia: Weed Science Society of Queensland, pp. 145–149.

McFadyen, R. E. and Spafford-Jacob, H. (2003). Insects for the biocontrol of weeds: predicting parasitism levels in the new country. In *Proceedings of the XI International Symposium of Biological Control of Weeds*, ed. J. M. Cullen, D. T. Briese, D. J. Kritocos, *et al.* Canberra, Australia: CSIRO, pp. 135–140.

McFadyen, R. E. and Withers, T. M. (1997). Report on biology and host-specificity of *Carmenta ithacae* (Lep: Sesiidae) for the biological control of parthenium weed (*Parthenium hysterophorus*). Internal Report. Queensland Department of Natural Resources, Brisbane, Australia.

Mishra, J., Pandey, A. K., Rajak, R. C. and Hasija, S. K. (1994). Microbial management of *Parthenium hysterophorus*. Host-specificity of *Sclerotium rolfsii* Sacc. *National Academy of Science Letters–India*, **17**, 169–173.

Narasimhan, T. R., Ananth, M., Narayana Swamy, M., *et al.* (1980). Toxicity of *Parthenium hysterophorus* L. in cattle and buffaloes. *Indian Journal of Animal Science*, **50**, 173–178.

Nath, R. (1981). Note on the effect of *Parthenium* extract on seed germination and seedling growth in crops. *Indian Journal of Agricultural Science* **51**, 601–603.

Nath, R. (1988). *Parthenium hysterophorus* L. – a general account. *Agricultural Review*, **9**, 171–179.

Navie, S. C., McFadyen, R. E., Panetta, F. D. and Adkins, S. W. (1996). The biology of Australian weeds. 27. *Parthenium hysterophorus* L. *Plant Protection Quarterly*, **11**, 76–88.

Navie, S. C., Panetta, F. D., McFadyen, R. E. and Adkins, S. W. (1998a). Behaviour of buried and surface-lying seeds of parthenium weed (*Parthenium hysterophorus* L.). *Weed Research*, **38**, 338–341.

Navie, S. C., Panetta, F. D., McFadyen, R. E and Adkins, S. W. (2004). Germinable soil seedbanks of Central Queensland rangelands invaded by the exotic weed *Parthenium hysterophorus* L. *Weed Biology and Management*, **4**, 154–167.

Navie, S. C., Priest, T. E., McFadyen, R. E. and Adkins, S. W. (1998b). Efficacy of the stem-galling moth *Epiblema strenuana* Walk. (Lepidoptera: Tortricidae) as a biological control agent for the ragweed parthenium (*Parthenium hysterophorus* L.). *Biological Control*, **13**, 1–8.

Njoroge, J. M. (1986). New weeds in Kenya coffee. A short communication. *Kenya Coffee*, **51**, 333–335.

Njoroge, J. M. (1989). Glyphosate (Round-up 36% a.i.) low rate on annual weeds in Kenya coffee. *Kenya Coffee*, **54**, 713–716.

Njoroge, J. M. (1991). Tolerance of *Bidens pilosa* and *Parthenium hysterophorus* L. to paraquat (Gramaxone) in Kenya coffee. *Kenya Coffee* **56**, 999–1001.

Ntushelo, K. and Wood, A. R. (2008). Supplementary host specificity testing of *Puccinia melampodii*, a biocontrol agent of *Parthenium hysterophorus* (abstract). In *XII International Symposium on Biological Control of Weeds*, held 22–27 April 2007, La Grande Motte, France, ed. M. H. Julien *et al*. Wallingford, UK: CAB International.

O'Donnell, C. and Adkins, S. W. (2005). Management of parthenium weed through competitive displacement with beneficial plants. *Weed Biology and Management*, **5**, 77–79.

Page, A. R. and Lacey, K. L. (2006). Economic impact assessment of Australian weed biological control. Technical Series 10. Adelaide, Australia: CRC for Australian Weed Management, 150 pp.

Pandey, H. N. and Dubey, S. K. (1989). Growth and population dynamics of an exotic weed *Parthenium hysterophorus* Linn. *Proceedings of the Indian Academy of Sciences (Plant Science)*, **99**, 51–58.

Pandey, S., Joshi, B. D. and Tiwari, L. D. (2001). The incidence of Mexican beetle *Zygogramma bicolorata* Pallister (Coleoptera: Chrysomelidae) on *Parthenium hysterophorus* L. (Asteraceae) from Haridwar and surrounding areas. *Journal of Entomological Research*, **25**, 145–149.

Pandey, S., Joshi, B. D. and Tiwari, L. D. (2002). First report of a new predator, *Andrallus spinidens* (Fabr.) on Mexican beetle *Zygogramma bicolorata* (Pallister) from Uttaranchal, India. *Indian Journal of Entomology*, **64**, 113–116.

Pandey, S., Joshi, B. D. and Tiwari, L. D. (2005). New record of *Eocanthecona furcellata* (Wolff.) (Heteroptera: Pentatomidae) as a predator of the introduced parthenium bio-control agent, *Zygogramma bicolorata* (Pallister) and its cannibalistic feeding behaviour. *Entomon*, **30**, 347–349.

Parker, A., Holden, A. N. G. and Tomley, A. J. (1994). Host specificity testing and assessment of the pathogenicity of the rust, *Puccinia abrupta* var. *partheniicola*, as a biological control agent of parthenium weed (*Parthenium hysterophorus*). *Plant Pathology*, **43**, 1–16.

Parmelee, J. A. (1967). The autoecious species of *Puccinia* on Heliantheae in North America. *Canadian Journal of Botany*, **45**, 2267-2328.

Patel, V. N. and Viraktamath, C. A. (2005). Dispersal of parthenium biocontrol agent *Zygogramma bicolorata* Pallister (Coleoptera: Chrysomelidae) in sunflower and *Parthenium hysterophorus* L. In *Second International Conference on Parthenium Management*, ed. T. V. Ramachandra Prasad, H. V. Nanjappa, R. Devendra, *et al*. Bangalore, India: University of Agricultural Sciences, pp. 120–122.

Picman, A. K. and Towers, G. H. N. (1982). Sesquiterpene lactones in various populations of *Parthenium hysterophorus*. *Biochemical Systematics and Ecology*, **10**, 145–153.

Prasad Rao, R. D., Govindappa, V. J., Devaraja, M. R. and Muniyappa, V. (2005). Role of parthenium in perpetuation and spread of plant pathogens. In *Second International Conference on Parthenium Management*, ed. T. V. Ramachandra Prasad, H. V. Nanjappa, R. Devendra, *et al*. Bangalore, India: University of Agricultural Sciences, pp. 65–72.

Price, P. W. (1991). Plant vigor hypothesis and herbivore attack. *Oikos*, **62**, 244–251.

Raghu, S. and Dhileepan, K. (2005). The value of simulating herbivory in selecting effective weed biological control agents. *Biological Control*, **34**, 265–273.

Raghu, S., Wilson, J. R. and Dhileepan, K. (2006). Refining the process of agent selection through understanding plant demography and plant response to herbivory. *Australian Journal of Entomology*, **45**, 307–315.

Raman, A. and Dhileepan, K. (1999). Qualitative evaluation of damage by *Epiblema strenuana* (Lepidoptera: Tortricidae) to the weed *Parthenium hysterophorus* (Asteraceae). *Annals of Entomological Society of America*, **92**, 717–723.

Raman, A., Madhavan, S., Florentine, S. K. and Dhileepan, K. (2006). Metabolic mobilization into the stem galls of *Parthenium hysterophorus* (Asteraceae) induced by *Epiblema strenuana* (Lepidoptera: Tortricidae) inferred from signatures of isotopic carbon and nitrogen, and total nonstructural carbohydrates. *Entomologia Experimentalis et Applicata*, **119**, 101–107.

Rao, R. S. (1956). Parthenium – a new record for India. *Journal of Bombay Natural History Society*, **54**, 218–220.

Rice, A. D. (1998). Divergent evolution of *Zygogramma bicolorata* during its establishment and proliferation in Queensland, Australia. B.Sc. honours thesis, Department of Zoology, James Cook University of North Queensland, Townsville, Australia, 85 pp.

Rice, A. D., Rowe, R. and Dhileepan, K. (1999). Evidence of local adaptation in the Australian population of *Zygogramma bicolorata* Pallister (Coleoptera: Chrysomelidae): An introduced biocontrol agent for the noxious weed *Parthenium hysterophorus* L. (Asteraceae) (abstract). In *30th Australian Entomological Society Conference*, held in Canberra. Orange, Australia: Australian Entomological Society.

Rice, P. L. (1937). Effect of moisture on emergence of the ragweed borer *Epiblema strenuana* Walker, and its parasites. *Journal of Economic Entomology*, **30**, 108–115.

Robertson, L. N. and Kettle, B. A. (1994). Biology of *Pseudoheteronyx* sp. (Coleoptera, Scarabaeidae) on the central highlands at Queensland. *Journal of the Australian Entomological Society*, **33**, 181–184.

Sarkate, M. B. and Pawar, V. M. (2006). Establishment of Mexican beetle (*Zygogramma bicolorata*) against *Parthenium hysterophorus* in Marathwada. *Indian Journal of Agricultural Sciences*, **76**, 270–271.

Seier, M. K. (1999). Studies on the rust *Puccinia melampodii* Diet. & Holw.: a potential biological control agent for parthenium weed (*Parthenium hysterophorus* L.) in Australia. Report submitted to the Queensland Department of Natural Resources and Mines, February 1999. Silwood Park, UK: CABI Bioscience UK Centre.

Seier, M. K., Harvey, J. L., Romero, A. and Kinnersley, R. P. (1997). Safety testing of the rust *Puccinia melampodii* as a potential biocontrol agent of *Parthenium hysterophorus* L. In *Proceedings of the First International Conference on Parthenium Management*, Vol. I, ed. M. Mahadeveppa and V. C. Patil. Dharwad, India: University of Agricultural Sciences, pp. 63–69.

Seier, M. K. and Tomley A. J. (2000). Host range of *Puccinia melampodii*: implications for its use as a biocontrol agent of parthenium weed in Australia (abstract). In *Proceedings of the X International Symposium on Biological Control of Weeds*, ed. N. R. Spencer. Sidney, MT: USDA-ARS, 686 pp.

Shabbir, A. and Bajwa, R. (2006). Distribution of parthenium weed (*Parthenium hysterophorus* L.), an alien invasive weed species threatening the biodiversity of Islamabad. *Weed Biology and Management*, **6**, 89–95.

Sharma, T. K. and Shujauddin (2006). Incidence and biology of Mexican beetle, *Zygogramma bicolorata* on *Parthenium hysterophorous*. *Bionotes*, **8**, 51.

Shea, K. and Kelly, D. (1998). Estimating biocontrol agent impact with matrix models: *Carduus nutans* in New Zealand. *Ecological Applications*, **8**, 824–832.

Singh, S. P. (1997). Perspectives in biological control of parthenium in India. In *Proceedings of the First International Conference on Parthenium Management*, ed. M. Mahadevappa and V. C. Patil. Dharwad, India: University of Agricultural Sciences, pp. 22–32.

Sridhara, S., Basavaraja, B. K. and Ganeshaiah, K. N. (2005). Temporal variation in relative dominance of *Parthenium hysterophorus* and its effect on native biodiversity. In *Proceedings of the Second International Conference on Parthenium Management*, ed. T. V. Ramachandra Prasad, H. V. Nanjappa, R. Devendra, *et al*. Bangalore, India: University of Agricultural Sciences, pp. 240–242.

Srikanth, J. and Pushpalatha, N. A. (1991). Status of biological control of *Parthenium hysterophorus* L. in India: a review. *Insect Science and Application*, **12**, 347–359.

Stegmaier, C. E. (1971). Lepidoptera, Diptera and Hymenoptera associated with *Ambrosia artemisiifolia* (Compositae) in Florida. *Florida Entomologist*, **54**, 259–272.

Strathie, L. W., Wood, A. R., van Rooi, C. and McConnachie, A. J. (2005). *Parthenium hysterophorus* (Asteraceae) in southern Africa, and initiation of biological control against it in South Africa. In *Proceedings of the Second International Conference on Parthenium Management*, ed. T. V. Ramachandra Prasad, H. V. Nanjappa, R. Devendra, *et al*. Bangalore, India: University of Agricultural Sciences, pp. 127–133.

Subba Rao, P. V., Mangala, A., Subba Rao, B. S. and Prakash, K. M. (1977). Clinical and immunological studies on persons exposed to *Parthenium hysterophorus* L. *Experientia*, **33**, 1387–1388.

Susilkumar (2000). Impact of the introduced Mexican beetle *Zygogramma bicolorata* on the suppression of *Parthenium hysterophorus*: a case study (abstract). *Proceedings of the Third International Weed Science Congress*, held in Foz do Iguassu, Brazil. Oxford, MS: International Weed Science Congress.

Susilkumar and Bhan, V. M. (1998). Establishment and dispersal of introduced exotic parthenium controlling bioagent *Zygogramma bicolorata* in relation to ecological factors at Vindhyanagar. *Indian Journal of Ecology*, **25**, 8–13.

Swamiappan, M., Sethupitchai, U. and Geetha, B. (1997a). Feeding potential of freshly emerged *Z. bicolorata* adults on sunflower and parthenium. In *Prceedings of the First International Conference on Parthenium Management*, ed. M. Mahadevappa and V. C. Patil, Vol. I. Dharwad, India: University of Agricultural Sciences, pp. 52–54.

Swamiappan, M., Sethupitchai, U. and Geetha, B. (1997b). Evaluation of the susceptibility of sunflower cultivars to the Mexican beetle, *Zygogramma bicolorata* Pallister. In *Proceedings of the First International Conference on Parthenium Management*, Vol. II, ed. M. Mahadevappa and V. C. Patil. Dharwad, India: University of Agricultural Sciences, pp. 177–179.

Tamado, T. and Milberg, P. (2000). Weed flora in arable fields of eastern Ethiopia with emphasis on the occurrence of *Parthenium hysterophorus*. *Weed Research*, **40**, 507–521.

Tamado, T., Ohlander, L. and Milberg, P. (2002a). Interference by the weed *Parthenium hysterophorus* L. with grain sorghum: influence of weed density and duration of competition. *International Journal of Pest Management*, **48**, 183–188.

Tamado, T., Schütz, W. and Milberg, P. (2002b). Germination ecology of the weed *Parthenium hysterophorus* in eastern Ethiopia. *Annals of Applied Biology*, **140**, 263–270.

Taye, T. (2005). Investigation of pathogens for biological control of parthenium (*Parthenium hysterophorus* L.) in Ethiopia. In *Programme and Abstracts of the Seventh Annual Conference of the Ethiopian Weed Science Society*, held 24–25 November 2005. Addis Ababa, Ethiopia: Ethiopian Weed Science Committee, pp. 21–22.

Taye, T., Einhorn, G., Gossmann, M., Büttner, C. and Metz, R. (2004a). The potential of parthenium rust as biological control of parthenium weed in Ethiopia. *Pest Management Journal of Ethiopia*, **8**, 39–50.

Taye, T., Einhorn, G. and Metz, R. (2004b). *Parthenium hysterophorus*, an invasive species in Ethiopia: investigations on the occurrence and on its pathogens. *Journal of Plant Diseases and Protection*, **19**, 271–278.

Taye, T., Gossmann, M., Einhorn, G., Büttner, C., Metz, R. and Abate, D. (2002). The potential of pathogens as biological control of parthenium weed (*Parthenium hysterophorus* L.) in Ethiopia. *Mededelingen – Faculteit Landbouwkundige en Toegepaste Biologische Wetenschappen, Universiteit Gent*, **67**, 409–420.

Taye, T., Obermeier, C., Einhorn, G., Seemüller, E. and Büttner, C. (2004c). Phyllody disease of parthenium weed in Ethiopia. *Pest Management Journal of Ethiopia*, **8**, 83–95.

Tomley, A. J. (1990). Parthenium weed rust, *Puccinia abrupta* var. *partheniicola*. In *Proceedings of the Ninth Australian Weeds Conference*, ed. J. W. Heap. Adelaide, Australia: Crop Science Society of South Australia, pp. 511–512.

Tomley, A. J. (2000). *Puccinia melampodii* (summer rust) a new biocontrol agent for parthenium weed. In *Proceedings of the Sixth Queensland Weed Symposium*, ed. J. T. Swarbrick. Brisbane, Australia: Weed Society of Queensland, pp. 126–129.

Towers, G. H. N. (1981). Allergic exzematous contact dermatitis from parthenium weed (*Parthenium hysterophorus*). In *Proceedings of the Sixth Australian Weeds Conference*, eds. B. J. Wilson and J. T. Swarbrick. Brisbane, Australia: Wees Society of Queensland, pp 565–469.

Tudor, G. D., Ford, A. L., Armstrong, T. R. and Bromage, E. K. (1982). Taints in meat from sheep grazing *Parthenium hysterophorus*. *Australian Journal of Experimental Agriculture and Animal Husbandry,* **22**, 43–46.

Uniyal, V. P., Mukherjee, S. K., Goyal, C. P. and Mathur, P. K. (2001). Defoliation of parthenium by Mexican beetle (*Zygogramma bicolorata*) in Rajaji National Park. *Annals of Forestry*, **9**, 327–330.

van der Laan, M. (2006). Allelopathic interference potential of the alien invader plant *Parthenium hysterophorus*. M.Sc. thesis, University of Pretoria, South Africa, 95 pp.

Viraktamath, C. A., Bhumannavar, B. S. and Patel, V. N. (2004). Biology and ecology of *Zygogramma bicolorata* Pallister, 1953. In *New Developments in the Biology of Chrysomelidae*, ed. P. Joliver, J. A. Santiago-Blay and M. Schmitt. The Hague, The Netherlands: SPB Academic Publishing, pp. 767–777.

Vogler, W., Navie, S., Adkins, S. and Setter, C. (2002). Use of fire to control parthenium weed. A report for the Rural Industries Research and Development Corporation, Australia, 41 pp.

Wan, F., Wang, R. and Ding, J. (1995). Biological control of *Ambrosia artemisiifolia* with introduced insect agents, *Zygogramma suturalis* and *Epiblema strenuana*, in China.

In *Proceedings of the Eighth International Symposium on Biological Control of Weeds*, ed. E. S. Delfosse and R. R. Scott. Melbourne, Australia: DSIR/CSIRO, pp. 193–200.

Wan, F., Ma, J., Guo, J. and You, L. (2003). Integrated control effects of *Epiblema strenuana* (Lepidoptera: Tortricidae) and *Ostrinia orientalis* (Lepidoptera: Pyralidae) against ragweed, *Ambrosia artemisiifolia* (Compositae). *Acta Entomologica Sinica*, **46**, 473–478.

Wardill, T. J., Graham, G. C., Playford, J., *et al.* (2004). The importance of species identity in the biocontrol process. In *Proceedings of the Fourteenth Australian Weeds Conference*, ed. B. M. Sindel and S. B. Johnson. Sydney, Australia: Weed Society of New South Wales, pp. 364–367.

Wedner, H. J., Zenger, V. and Lewis, W. (1986). Identification of American feverfew (*Parthenium hysterophorus*) as an allergen in the United-States Gulf-Coast. *Journal of Allergy and Clinical Immunology*, **77**, 198.

Wild, C. H. (1980). Preliminary report on the biology and host-specificity of *Hyperodes* sp. (Coleoptera: Curculionidae), a potential biological control agent for *Parthenium hysterophorus* L. (Compositae). Unpublished report, Alan Fletcher Research Station, South American Field Office, Londrina, Brazil. 31 pp.

Wild, C. H., Wilson, B. W. and Tomley, A. J. (1982). A further report on the host specificity of *Listronotus setosipennis* (Hustache) (formerly *Hyperodes* sp.) (Col: Curculionidae): a potential biocontrol agent for *Parthenium hysterophorus* (Compositae). Internal report. Alan Fletcher Research Station, Land Administration Commission, Government of Queensland, Australia, 19 pp.

Wild, C. H., McFadyen, R. E., Tomley, A. J. and Wilson, B. W. (1992). The biology and host specificity of the stem-boring weevil *Listronotus setosipennis* (Coleoptera: Curculionidae), a potential biocontrol agent for *Parthenium hysterophorus* (Asteraceae). *Entomophaga*, **37**, 591–598.

Withers, T. M. (1998). Influence of plant species on host acceptance behavior of the biocontrol agent *Zygogramma bicolorata* (Col.: Chrysomelidae). *Biological Control*, **13**: 55–62.

Withers, T. M. (1999). Examining the hierarchy threshold model in a no-choice feeding assay. *Entomologia Experimentalis et Applicata*, **91**, 89–95.

Withers, T. M., Christensen, J. A. and Burwell, C. J. (1998). *Erixestus zygogrammae* Cave and Grissell (Hymenoptera: Pteromalidae): an exotic egg parasitoid of *Zygogramma bicolorata* Pallister (Coleoptera: Chrysomelidae) in Australia. *Australian Journal of Entomology*, **37**, 83–84.

Withers, T. M., McFadyen, R. E. and Marohasy, J. (1999). Importation protocols and risk assessment of weed biological control agents in Australia: the example of *Carmenta* nr. *ithacae*, In *Nontarget Effects of Biological Control*, ed. P. A. Follett and J. J. Duan. Boston, MD: Kluwer, pp. 195–214.

Wood, A. R. and Scholler, M. (2002). *Puccinia abrupta* var. *partheniicola* on *Parthenium hysterophorus* in Southern Africa. *Plant Disease*, **86**, 327.

Wood, J. M. (1897). *Report on Natal Botanic Gardens for the Year 1896*. Durban, South Africa: Durban Botanic Society, 31 pp.

Yun, Z. and Xia, W. (2002). Notes on some *Puccinia* species from tropical China. *Mycosystema*, **21**, 448–451.

16

Passiflora mollissima (HBK) Bailey (Passifloraceae)

George P. Markin

16.1 Introduction

Originally from the Andes Mountains of South America, the woody vine *Passiflora mollissima* (HBK) Bailey (Passifloraceae) (banana passion fruit) has been widely disseminated through the world as an ornamental for its large, showy trumpet-shaped flower and large elongate yellow fruit (Fig. 16.1) (Vanderplank, 1991). It is referred to as curubra or tumbo in South America, and as banana poka in Hawaii; *poka* being a Hawaiian word for twine or twisting. Disseminated by fruit-feeding birds, it has become feral in many semitropical areas of the world, including the island of Madeira in the North Atlantic, in southern Africa (MacDonald, 1987; Henderson, 1995), Kenya (De Wilde, 1976), New Zealand (Young, 1970), and the Hawaiian Islands in the central Pacific.

Introduced as an ornamental vine to the island of Kauai in Hawaii in the 1890s, it was moved to the Island of Hawaii in 1932, and subsequently to the island of Maui. On all three islands it found the mid-elevation (500–2200 m) mountain rain forest (Cuddihy, 1989) an ideal habitat and the introduced birds and feral pigs effective means for dissemination of its seeds. By the mid 1970s, its ability to invade the Hawaiian rain forests and form dense mats of vines that smothered understory plants and broke down mature trees led it to be recognized as a threat to the continued survival of Hawaiian rain forests (Beardsley and Smith, 1978; Waage *et al.*, 1981; Warshauer *et al.*, 1983; LaRosa, 1984).

A consortium of land managers on the Island of Hawaii, recognizing conventional control was ineffective due to inaccessible terrane and damage to desirable native species, considered biological control as an alternative. In 1982, a preliminary survey of *P. mollissima*'s natural enemies in the Andes of South America and a literature review (Pemberton, 1982, 1989) documented a large number of insects and pathogens attacking *P. mollissima*, and closely related species of *Passiflora*. A decision was made to proceed with a major biological control effort, which was in operation from 1985 to 1995.

Biological Control of Tropical Weeds using Arthropods, ed. R. Muniappan, G. V. P. Reddy, and A. Raman. Published by Cambridge University Press. © Cambridge University Press, 2009.

Fig. 16.1 *Passiflora mollissima*, called banana poka in Hawaii, is widely grown as a domestic plant in mid-elevations of the Andes in South America. Because of its showy flower and edible fruit, it has been moved through the tropics and semitropics of the world where it has often escaped cultivation. (Photo by G. P. Markin.)

16.2 *Passiflora mollissima*, the plant

16.2.1 Range in South America

The plant order Passifloraceae includes more than 400 species, most within *Passiflora* (Killip, 1938). *Passiflora mollissima* is the most cold tolerant and occurs throughout the Altiplano on the western side of the Andes from western Venezuela to Bolivia and northern Chile. In the Altiplano, *P. mollissima* is always associated with humans as an ornamental or in small plots for sale in local markets. Because its juice was found comparable to that of the common passion fruit (*P. edulis*), numerous 5–10 ha size commercial plantings were made in Colombia and Venezuela in the 1970s and 1980s.

16.2.2 Origin

The inability to study the plant growing in any native ecosystem, where it should have its highest concentration of coevolved natural enemies, has left the question of its location of origin unanswered. The Altiplano, located at 2800–3500 m, is an area of high human occupancy primarily along the western, drier side of the Andes Mountains (Weberbauer, 1936), and has been under intense cultivation for thousands of years. It is possible that during this time the original ecosystem in which *P. mollissima* evolved was destroyed. A second explanation is that in Hawaii *P. mollissima* is capable of reaching the tops of trees forming the upper canopy (*c.* 30 m height) of rain forests, and it exists in areas with over 300 cm of rainfall; therefore *P. mollissima*'s origin may not be in the Altiplano but in a high-elevation temperate rain forest on the wetter, eastern side of the Andes, possibly the "Ceja de la Montana" (brow of the forest) in Peru (Weberbauer, 1936). Unfortunately, due to rebel activity, we were never able to visit this area.

16.2.3 Taxonomy

All Passifloraceae are either tropical or subtropical, and a vast majority are indigenous to the New World (Hutchinson, 1967). *Passiflora* contains approximately 400 species, most of which are true tropical lianas (woody vines) of the low-elevation rain forests of Central and South America (Killip, 1938). Only a few closely related species in the subgenus *Tacsonia*, to which *P. mollissima* belongs, occur in the cooler mid-elevations of the Andes. The complex of cultivated plants to which *P. mollissima* belongs was under domestication by the Incas in pre-Colonial time (Cook, 1925; Latcham, 1936; Sauer, 1952; Popenoe *et al.*, 1989). The flowers of *Tacsonia* have an elongate corolla tube, large fleshy fruit, and comprise a complex of closely related forms that readily interbreed and hybridize (Killip, 1938; Martin and Nakasone, 1970; Escobar, 1981; Vanderplank, 1991).

   The highly invasive form found in Hawaii and the target of the biological control program is the same form that has turned invasive in New Zealand and South Africa. It is commonly referred to as *P. mollissima*, but has also been determined as *P. triparteda*, *P. mixta*, and *P. terminata*, or a combination of these as varieties in South America. Plant taxonomists in Hawaii have concluded that, as a cultivated plant, it should be described as cultivar *Passiflora banana poka* (Green, 1990, 1994) or *Passiflora mollissima sensu* (Escobar, 1980) or *P. mollissima (sensu lato)* (LaRosa, 1985) according to the rules for domestic plant taxonomy (Brickell, 1980). In light of the taxonomic confusion due to the South American taxonomists treating each form as a naturally occurring species, most recently Coppens *et al.* (2001) and the Hawaiian taxonomists who recognized it as a human-derived cultivated plant, I will continue to refer to it by the most commonly used name *P. mollissima*, the name used during the biological control program, and still used in the current Hawaiian botanical literature (Wagner *et al.*, 1990; Starr *et al.*, 2003).

16.3 Economics of the biological control effort

Estimates of the total cost for the ten years the program was in full operation indicate that between US$2.5 and US$3 million were invested jointly by the US Forest Service, the Hawaii Department of Agriculture, and the University of Hawaii. However, while the program was considered to have been relatively successful (primarily due to the release of an introduced pathogen and not the introduced insects) determining the value of the resource protected is difficult.

Analysis using conventional cost-benefit for this program is impossible since in only a few very limited areas has direct control been attempted. An unreported amount of money was spent previous to 1982 to physically remove or spray with herbicides the plant in a few limited "special ecological areas" within Volcanoes National Park, but due to the ineffectiveness of the treatments and particularly the impact on the native ecosystems being protected, the program was discontinued in 1982 in favor of biological control. In the late 1980s, a relatively new infestation of *P. mollissima* in the Kula area on the island of Maui expanded into a rural housing development. Under pressure from the local landowners, the state of Hawaii appropriated US$500 000 for its eradication from approximately 100 hectares. Crews of five to seven operators between 1989 and 1992 cut and sprayed the plant in and around the area, but it was soon realized the control effort was ineffective and quietly discontinued.

Due to the nature of the problem caused by *P. mollissima* in Hawaii where the infestations are primarily restricted to stands of native mountain rain forests, the justification for the biological control program was therefore not made on conventional economic values but as an effort to protect these endangered ecosystems due to their historic and scientific values.

16.4 Selection of biological control agents

As it was a widely grown domestic plant in South America, a preliminary survey and a literature review (Pemberton, 1982, 1989) identified more than 80 arthropods and several pathogens attacking *P. mollissima* in South America. Of the 20 or so other species of *Passiflora* established in Hawaii (Wagner *et al.*, 1990), all were minor, low-elevation ornamentals, or weeds, on which attack by an introduced biocontrol agent would probably be acceptable. The exception was *P. edulis*, which at the time of this program was being grown in several hundreds of hectares as a commercial crop in Hawaii for its juice (Anonymous, 1956; Seale and Sherman, 1960). All potential natural enemies known to attack cultivated *P. edulis* were, therefore, eliminated from consideration. Thirty-five possible candidates (Causton *et al.*, 1999) were eventually selected as potential biological control agents but narrowed down (primarily based on finding suitable populations accessible to work with) to a final 12.

16.5 Agents released

16.5.1 Cyanotrica necryai *Felder and Ragenhofer (Lepidoptera: Notodontidae = Dioptidae)*

The metallic blue moth *Cyanotrica necryai* (3–4 cm wingspan) was abundant in western Colombia and is the most damaging pest capable of totally defoliating large commercial plantings. Studies in Colombia (Casanas-Arango *et al.*, 1991) and host testing in quarantine in Hawaii (Markin and Nagata, 1989; Markin *et al.*, 1989) resulted in *C. necryai* being approved for release in 1986. In Hawaii it could be propagated readily in small cages under ambient outdoor conditions in the rain forests at Hawaii Volcanoes National Park but, despite releases of 10 000 individuals as eggs, larvae, or adults over three years, none established and all subsequent surveys have not detected the moth. The colony used for release was disease free and eggs, larvae, and pupae exposed in the field were not attacked by local parasites. This left by default the possibility that failure to establish might be due to either predation or a nectar resource missing for the adults in the introduced areas (Campbell *et al.*, 1993).

16.5.2 Pyrausta perelegans *Hampson (Lepidoptera: Pyralidae)*

The white and pink moth *Pyrausta perelegans* (3–3.5 cm wingspan) occurs from Peru to western Venezuela (Pemberton, 1989). The larva mines the growing tip of the vine first and then enters the covering bracts of the developing flower bud. Single larva can attack several developing buds causing them to abort. It is considered a major pest of commercial plantings of *P. mollissima* in Colombia (Rojas de Hernandez and Ulloa, 1982). The population tested in Hawaii and eventually released, however, came from western Venezuela (Markin and Nagata, 2000). Approval was obtained and releases began in Hawaii in 1991 and within six months, establishment was confirmed. Subsequent releases were successfully made on the Island of Hawaii and island of Maui (Campbell *et al.*, 1993). Initially, the population increased at all sites and within three years had dispersed throughout the *P. mollissima* infestation. Between 1992 and 1995, the population stabilized with only 1–2% of the buds being attacked and at no location were more than 10% of the buds destroyed. Monitoring was discontinued in 1996 when the program was terminated, but incidental surveys since then have shown no change in the population.

The reason for the failure to establish populations capable of causing a significant impact on *P. mollissima* in Hawaii was investigated by Campbell *et al.* (1993), who found that eggs suffered 50% attack by trichogrammatid wasps, and hymenopteran parasitoids attacked 10–50% of the larvae. It was not proven, but is suspected, that in Hawaii parasitism is the main biotic factor preventing *P. perelegans* developing population levels capable of noticeable decrease in fruit production.

16.5.3 Josia fluonia *(Lepidoptera: Notodentidae = Dioptidae)*

With the failure to establish the moth *C. necryai*, efforts were made to test a new defoliator, *Josia fluonia* was rare in Peru and Colombia and never found in Venezuela, but usable populations were located in Ecuador where its biology was initially studied. Colonies were sent to quarantine in Hawaii for host testing in 1992 (Friesen *et al.*, 1994) and a permit for its release was issued in 1996. *Josia fluona*, despite being approved, was not released before the program was terminated.

16.5.4 Septoria passiflorae, *SID (Coelomycete: Sphaeropsidales)*

As part of the biological control program, plant pathogen *Septoria passiflorae* was released at several locations in Hawaii in 1995 (Trujillo *et al.*, 1994; Norman and Trujillo, 1995). The disease readily established and, by causing early loss of mature leaves, appeared to cause a reduction in biomass; after four years, it was claimed *P. mollissima* was completely controlled (Trujillo *et al.*, 2001). However, foresters familiar with this plant agree that while a noticeable reduction in the *P. mollissima* biomass had occurred, no mortality had been observed and the vigorous regrowth and the continual spread of *P. mollissima* appeared to indicate the pathogen had lost much of its initial virulence.

16.6 Other insects tested or considered

At the termination of the *P. mollissima* program in 1996, several insects had been tested and discarded as unsuitable or were at uncompleted stages of evaluation.

16.6.1 Zapriothrica *sp., near* salebrosa *Wheeler* *(Diptera: Drosophilidae)*

Because of the spread of *P. mollissima* by seeds, finding a fruit- or flower-destroying natural enemy was given high priority and *Zapriothrica salebrosa*, described in the literature as a common pest of *P. mollissima*, was selected as a potential candidate (Pemberton, 1982, 1989). However, specimens from Colombia and Venezuela sent for confirmation of identification (C. Vilela, personal communication) came back as a new species, which for the duration of the studies, therefore, was referred to as *Zapriothica near* salebrosa (Casanas-Arango *et al.*, 1996).

Initial studies in Colombia and quarantine in Hawaii were unsuccessful when the fly failed to mate in captivity. However, by using an environmental chamber, adding mist as the temperature was lowered, and providing open flowers in which the fly could aggregate, mating was achieved and a colony was established in Hawaii. However, it was eventually realized that an introduced drosophilid fly would likely be attacked by endemic parasitoids of the very numerous native Hawaiian drosophilid flies (Hardy, 1965), which shifted emphasis to the flower-feeding moth *P. perelegans*. By 1994 when it appeared that *P. perelegans* would not be an effective biological control agent, a new colony was established in quarantine in Hawaii, but studies ended with the termination of the program in 1996.

16.6.2 Dasiops caustonae *Norrbom and McAlpine (Diptera: Lochaedae)*

In Venezuela, *Zapriothica* sp. near *salebrosa* was replaced by another flower-attacking fly, *D. caustonae*, which has a biology similar to that of *Zapriothica* sp. near *salebrosa*. Adults of both drosoplhilids laid eggs on the flower bud, and the larvae fed within developing flowers. Studies in Venezuela were completed in 1994 and indicated the fly had potential as a biological control agent (Causton and Rangel, 2002), but before this fly could be introduced into quarantine in Hawaii, the program was terminated.

16.6.3 Josia lingata *Walker (Lepidoptera: Notodontidae = Dioptidae)*

Josia fluonia was being studied in Ecuador when a second, similar-appearing species was found on *P. mollissima*. The two species of *Josia* were brought into quarantine in Hawaii and jointly tested, but *J. lingata* was found to be a generalist that attacked most Hawaiian *Passiflora* including *P. edulis* (commercial passion fruit). *Josia lingata* was dropped as a potential biocontrol agent, but there were concerns that the two could be confused and future shipments of *J. fluonia* to Hawaii, or elsewhere, could accidentally be contaminated with this similar-appearing generalist.

16.6.4 Acrocercops *sp. near* pylonias *Meyrick (Lepidotera: Gracillariidae)*

In Ecuador, Colombia, and Peru, but not Venezuela, *P. mollissima* leaves were attacked by a gracillariid larva that created hollow blisters in the mesophyll. In low-elevation locations in Colombia where *P. mollissima* grew adjacent to the common passion fruit, it did not appear to attack *P. edulis* and initially was studied as a potential biological control agent. However, gracillariids belong to one of the few families of Lepidoptera that are endemic to Hawaii (30 species native to Hawaii; Zimmerman, 1978), and all were attacked by both endemic and introduced hymenopteran parasites. Previously, an attempt to release another gracillariid moth as a biological control agent for *Myriea faya* Aiton (Myricaceae) (Markin,

2001) was unsuccessful because of these parasitoids and *Acrocercops* sp. near *pylonais* was dropped from further considerations.

16.6.5 Odonna passiflorae *Clarke (Lepidoptera: Oecophoridae)*

The gregarious larvae of this moth mine the crowns and base of mature plants (Chacon and de Hernandez, 1981). In Colombia near the border of Ecuador, *Odonna passiflorae* were occasionally found and its impact was observed to be severely damaging. Working with this insect was a high priority, but unfortunately this area soon became inaccessible because of drug-plant growing activity and work shifted to Venezuela and Ecuador, where *O. passiflorae* was not found.

16.7 Other stem-mining insects

Defoliators and flower-attacking insects were common, but the third trophic group, stem-mining insects, was relatively rare. In Venezuela (Causton *et al.*, 1999), a number of cerambycids were recovered, but they attacked only already dead stems. In Colombia, several stem-mining insects had been described in the Cauca Valley, including the moth *Aepytus* (*Pseudolaca*) *serta* (Schaus) (Lepidoptera: Hepislidae) (Rojas and Ulloa, 1980) and the cerambycids, *Heterachles* sp., *Nyssodrys* sp., *Ibidion* sp., and *Eurysthen oblicus* Serville (Chacon and Rojas, 1984). Studying this complex was prevented when this area became unsafe due to drug-plant growing activities.

16.7.1 Heliconiine *butterflies (Lepidoptera)*

Early in the *P. mollissima* program, heliconiid butterflies (Lepidoptera: Heliconiidae) were suggested based on their well-known coevolution with *Passiflora* sp. (Waage *et al.*, 1981). However, surveys made it evident that the heliconiids were exclusively restricted to tropical lowlands and none had adapted to the higher altitude *Passiflora* in the Andes.

16.7.2 Agraulis vanilla *(L.) (Lepidoptera: Nymphalidae)*

Agraulis vanilla occurs commonly throughout the range of *P. mollissima* in the Altiplano of the Andes and initially there was an interest in introducing the high-altitude Andean strain of this butterfly into Hawaii. Unfortunately, early in the program it was discovered that this butterfly was already established at lower altitudes in Hawaii. It appears to have been a deliberate but unapproved introduction either attempting to enlarge the small butterfly complex of Hawaii or an amateur attempt at biological control. Unfortunately, the population introduction contained a highly virulent nuclear polyhedrosis virus which would have attacked the high-altitude Andes strain, if it had been introduced, thereby eliminating a biological control agent of promise.

16.8 Discussion

In Hawaii the *P. mollissima* program terminated more than 10 years ago, but recently there has been interest in its resurrection. Interest has also developed in implementing a new biological control program in New Zealand (Hugh Goarlay, personal communication). Knowing the wide distribution of this plant in tropical and semitropical parts of the world and its ability to escape from cultivation into native ecosystems, we predict that at some time in the future other countries will also be interested in their own biological control program. Accordingly, it will be only a matter of time before other entomologists follow up on the work done in the Andes. We hope that this description of our program and the insects tested, found, or considered will provide the basic foundation on which to build future programs.

Future workers, however, should first address the problems of *P. mollissima* being a domesticated plant, which has caused taxonomic confusion, and prevented our locating its center of origin. In-depth study of genetics of the *P. mollissima* complex, as well as detailed ecological surveys in the Andes will be needed to determine if the center of origin can be found.

While initially impressed by the number of natural enemies reported attacking *P. mollissima* in the Andes, we found that, as the study progressed, the small number that appeared to be monophagous and had potential as biocontrol agents was disappointing. For Hawaii, and possibly other countries, future programs will have to address the problem that many promising potential biocontrol agents will feed on many species of *Passiflora*. This means resolving ahead of time the risk-benefit conundrum of secondary attack on ornamental species of *Passiflora* and locally cultivated commercial passion fruit, *P. edulis*.

The third problem that will face future workers is the political instability of the area. Initially, access to Peru and Colombia was possible, but within a few years both were closed due to drug-plant growing or rebel activities. This caused a shift of our work to Venezuela and Ecuador: two countries at the time friendly with the USA. However, at the time of this writing (2006) both countries have undergone major changes in government and visiting scientists are reporting that it is becoming difficult to travel, obtain collecting permits, and particularly to find local scientists willing to cooperate. Thus if a part of the Andes is accessible, work should be done as rapidly as possible since there is no guarantee that even within a few years the country will still be accessible.

This study has shown that there are enough natural enemies in the Andes that a complex of biological control agents probably exists, which should give suitable control of this plant in Hawaii or other locations through the world where *P. mollissima* is or will become a pest. Future programs, however, will not be quick, simple, or cheap. Future scientists should plan for a program of long duration and convince their funding sources of the need for a long-term commitment. This will mean future biocontrol programs must first develop a strong justification based on scientific and economic data.

Acknowledgments

Over the ten-year program, a very large number of students, postdocs, and technicians participated or supported the work in quarantine or in the field in Hawaii who as a group I would like to acknowledge for their invaluable support. However, the group that deserves special recognition are the scientists who under difficult and often dangerous conditions worked in South America. These include Drs. Rex Friesen and Robert Pemberton, for the extensive visits they made to the Andes to supervise, set up, and coordinate the work, and our key local cooperators Marta Rojas de Hernandez in Colombia, Charlotte Causton in Venezuela, and Giovanna Orone in Ecuador. I would also like to thank the Hawaii Volcanoes National Park for providing the quarantine facilities and the Hawaii Department of Lands and Natural Resources, Division of Forestry for much of the funding to support the work in South America.

References

Anonymous (1956). *Passion Fruit Culture in Hawaii.* Circular 345. Honolulu, HI: Cooperative Extension Service. University of Hawaii, 22 pp.

Beardsley, J. W. and Smith, C. W. (1978). Biological control of forest weed pests in Hawaii – is it a feasible solution? In *Proceedings of the Second Conference in Natural Sciences, Hawaii Volcanoes National Park*, ed. C. W. Smith. Honolulu, HI: University of Hawaii at Manoa.

Brickell, C. D. (1980). *International Code of Nomenclature for Cultivated Plants – 1980.* Utrecht, Netherlands: Bohn, Scheltama & Holkema, 104 pp.

Campbell, C. L., Markin, G. P. and Johnson, M. W. (1993). Fate of *Cyanotricha necyria* (Lepidoptera: Notodontidae) and *Pyrausta perelegans* (Lepidoptera: Pyralidae) released for biological control of banana poka (*Passiflora mollissima*) on the Island of Hawaii. *Proceedings of the Hawaiian Entomological Society*, **32**, 123–130.

Casanas-Arango, A. D., Trujillo, E. E., Friesen, R. D. and Rojas de Hernandez, A. M. (1996). Field biology of *Zapriothrica* sp. Wheeler (Diptera, Drosophilidae), a pest of *Passiflora* spp. of high elevation possessing long tubular flowers. *Journal of Applied Science*, **120**, 111–114.

Casanas-Arango, A., Trujillo, E. E., de Hernandez, A. M. and Taniguchi, G. (1991). Field biology of *Cyanothrica necryai* Felder (Lep., Dioptidae), a pest of *Passiflora* spp., in southern Colombia's and Ecuador's Andean region. *Journal of Applied Entomology*, **109**, 93–97.

Causton, C. E. and Rangel, A. P. (2002). Field observations on the biology and behavior of *Dasiops caustonae* Norrbom and McAlpine (Dipt., Lonchaeidae), as a candidate biocontrol agent of *Passiflora mollissima* in Hawaii. *Journal of Applied Entomology*, **126**, 169–174.

Causton, C. E., Markin, G. P. and Friesen, R. (1999). Exploratory survey in Venezuela for biological control agents of *Passiflora mollissima* in Hawaii. *Biological Conservation*, **18**, 110–119.

Chacon, P. and Rojas, M. (1984). Entomofauna asociada a *Passflora mollissima*, *P. edulis* f. *flavicarpa* y *P. quadrangularis* en el Departamento del Valle del Cauca. *Turrialba*, **34**, 297–311.

Chacon, P. and de Hernandez, M. (1981). Immature stages of *Odonna passiflorae* Clarke (Lepidoptera: Óecophoridae): biology and morphology. *Journal of Research Lepidoptera*, **20**, 43–45.

Cook, O. F. (1925). Peru as a center of domestication. *Journal of Heredity*, **16**, 33–46, 95–110.

Coppens, G. E., Barney, V. E., Jorgensen, P. M. and MacDougal, J. M. (2001). *Passiflora terminiana*, a new cultivated species of *Passiflora* subgenus *Tacsonia* (Passifloraceae). *Novon*, **11**, 8–15.

Cuddihy, L. W. (1989). Vegetation zones of the Hawaiian Islands. In *Conservation Biology in Hawaii*, ed. C. P. Stone and D. B. Stone. Honolulu, HI: University of Hawaii Cooperative National Park Resources Studies Unit, pp. 27–37.

De Wilde, W. J. J. O. (1976). *Passiflora mollissima*. In *Flora of Tropical East Africa: Passifloraceae*. London: Crown Agents for Oversea Governments and Administrations, pp. 14–15.

Escobar, L. K. (1980). Interrelationships of the edible species of *Passiflora* centering around *Passiflora mollissima* (H.B.K.) Bailey subgenus *Tacsonia*. Ph.D. dissertation, University of Texas at Austin, 824 pp.

Escobar, L. K. (1981) Experimentos preliminares en la hibridación de especies comestibles de Passiflora. *Actualidades Biologicas*, **10**, 103–111.

Friesen, R., Markin, G. P. and Nagata, R. F. (1994). Host suitability studies of the moth *Josia fluonia* (Lepidoptera: Dioptidae) as a biological control agent for the weed *Passiflora mollissima* in Hawaii forests: a petition for the approval to release an insect biological control agent in Hawaii. Unpublished report. Hawaii Department of Agriculture, Plant Quarantine Branch, Honolulu, Hawaii.

Green, P. (1990). On the name of the weedy *Passiflora* known as banana poka. *Newsletter of Hawaiian Botanical Society*, **29**, 43–45.

Green, P. (1994). *Passiflora mollissima* Passifloraceae. *Kew Magazine*, **11**, 183–187.

Hardy, E. (1965). *Insects of Hawaii: Diptera Cyclorrhapha*. Vol. **12**. Honolulu, HI: University of Hawaii Press, 814 pp.

Henderson, L. (1995). *Passiflora mollissima*. Plant invaders of Southern Africa. Pretoria, South Africa: Plant Protection Research Institute, 107 pp.

Hutchinson, J. (1967). *The Genera of Flowering Plants (Angiospermae). Dicotyledons*. Vol. II, 107. Passifloraceae. Oxford, UK: Clarendon Press, pp. 364–374.

Killip, E. P. (1938). *The American Species of Passifloraceae*. Chicago, IL: Field Museum of Natural History, 331 pp.

LaRosa, A. M. (1984). *The biology and Ecology of* Passiflora mollissima *in Hawaii*. Technical Report. Manoa, HI: Cooperative National Park Resources Studies, University of Hawaii. 50 pp.

LaRosa, A. M. (1985). Note on the identity of the introduced passionflower vine "banana poka" in Hawaii. *Pacific Science*, **39**, 369–371.

Latcham, R. E. (1936). *La Agricultura precolombiana en Chile y los Paises Vecinos*. Santiago, Chile: Ediciones de la Universidad de Chile, 336 pp.

MacDonald, I. A.W. (1987). Banana poka in the Knysna Forest. *Veld and Flora*, **73**, 133–134.

Markin, G. P. (2001). Notes on the biology and release of *Caloptilia* sp. nr. *schinella* (Walsingham) (Lepidoptera: Gracilariidae), a biological control moth for the control of the weed firetree (*Myrica faya* Aiton) in Hawaii. *Proceedings of the Hawaiian Entomological Society*, **35**, 67–76.

Markin, G. P. and Nagata, R. F. (1989). *Host Preference and Potential Climatic Range of* Cyanotricha necryai *(Lepidoptera: Dioptidae), Potential Biological Control Agent of the weed* Passiflora mollissima *in Hawaiian Forests.* Technical Report 67. Honolulu, HI: Cooperative National Park Resources Study Unit, University of Hawaii at Manoa, 35 pp.

Markin, G. P. and Nagata, R. F. (2000). Host suitability of the moth, *Pyrausta perelegans* Hampson (Lepidoptera: Pyralidae), as a control agent of the forest weed banana poka *Passiflora mollissima* (HBK) Bailey, in Hawaii. *Proceedings of the Hawaiian Entomological Society,* **34**, 169–179.

Markin, G. P., Nagata, R. F., and Taniguchi, G. (1989). Biology and behavior of the South American moth, *Cyanothrica necryai* (Felder and Rogenhofer) (Lepidoptera: Notodontidae), a potential biocontrol agent in Hawaii of the forest weed, *Passiflora mollissima* (HBK) Bailey. *Proceedings of the Hawaiian Entomological Society,* **29**, 115–123.

Martin, F. W. and Nakasone, H. Y. (1970). The edible species of *Passiflora. Economic Botany,* **24**, 333–343.

Norman, D. J. and Trujillo, E. E. (1995). Development of *Colletotrichum gloeosporioides* f.sp. *clidemiae* and *Septoria passiflorae* into two mycoherbicides with extended viability. *Plant Diseases,* **79**, 1029–1032.

Pemberton, R. W. (1982). Exploration for natural enemies of *Passiflora mollissima* in the Andes. Unpublished report on file Hawaii Department of Lands and Natural Resource, Division of Forestry and University of Hawaii in Manoa Library, 125 pp.

Pemberton, R. W. (1989). Insects attacking *Passiflora mollissima* and other *Passiflora* spp: field survey in the Andes. *Proceedings of the Hawaiian Entomological Society,* **29**, 71–84.

Popenoe, H., King, S. R., Leon, J. and Kalinowski, L. S. (1989). *Lost Crops of the Incas.* Washington, DC: National Academy Press, 415 pp.

Rojas, M. A. and Ulloa, C. P. (1980). *Aepytus* (Pseudolaca) *serta* (Schaus) Barronador del Tallo de la Curuba. *Revista Colombiana de Entomología,* **6**, 66–67.

Rojas de Hernandez, M. and Ulloa, P. C. (1982). Contribución a la biología de *Pyrausta perelegans* Hampson (Lepidoptera: Pyralidae). *Brenesia,* **19**, 325–331.

Sauer, C. O. (1952). Cultivated plants of South and Central America. In *Handbook of South American Indians.* Vol. 6. *Agricultural Origins and Dispersals.* Bulletin 143. New York: Bureau of American Ethnology, pp. 487–543.

Seale, P. E. and Sherman, G. D. (1960). *Commercial Passion Fruit Processing in Hawaii.* Circular 58. Honolulu, HI: Hawaii Agricultural Experiment Station; University of Hawaii, 18 pp.

Starr, F., Starr, K. and Loope, L. (2003). *Passiflora mollissima*: banana poka, Passifloraceae. Plants of Hawaii: Reports. http://www.hear.org/star/hiplants/reports/html/passiflora_mollissima.htm.

Trujillo, E. E., Kadooka, C., Tanimoto, V., *et al.* (2001). Effective biomass reduction of the invasive weed species banana poka by *Septoria* leaf spot. *Plant Disease,* **85**, 357–361.

Trujillo, E. E., Norman, D. J. and Killgore, E. M. (1994). *Septoria* leaf spot, a potential biological control for banana poka vine in forests of Hawaii. *Plant Disease,* **78**, 883–885.

Vanderplank, J. (1991). *Passion Flowers and Passion Fruit:* P. mollissima *(HBK) Bailey.* Cambridge, MA: MIT Press, pp. 114–116.

Waage, J. K., Smiley, J. T. and Gilbert, L. E. (1981). The *Passiflora* problem in Hawaii: prospects and problems of controlling the forest weed *P. mollissima* [Passifloraceae] with *Heliconiine* butterflies. *Entomophaga*, **26**, 275–284.

Wagner, W. L., Herbst, D. R. and Sohmer, S. H. (1990). *Passiflora mollissima* (Kunth). In *Manual of the Flowering Plants of Hawaii*, ed. L. H. Bailey. Special Publication 83. Honolulu, HI: Bishop Museum, pp. 1012–1013.

Warshauer, F. R., Jacobi, J. D., LaRosa, A. M., Scott, M. J. and Smith, C. W. (1983). *The Distribution, Impact, or Potential Management of the Introduced Vine* Passiflora mollissima *(Passifloraceae) in Hawaii*. Technical Report 48. Manoa, HI: Cooperative National Park Resource Study Unit, University of Hawaii.

Weberbauer, A. (1936). Phytogeography of the Peruvian Andes. In *Flora of Peru*, Part I, ed. F. McBride. Botany Series 13, Pub. 351. Chicago, IL: Field Museum of Natural History, pp. 13–81.

Young, B. R. (1970). Identification of passionflowers in New Zealand (Dicotyledones: Passifloraceae). *Records of the Auckland Institute and Museum*, **7**, 147–169.

Zimmerman, E. C. (1978). *Insects of Hawaii*. Vol. 9. *Microlepidoptera*. Honolulu, HI: University of Hawaii Press, 1903 pp.

17

Pistia stratiotes L. (Araceae)

Peter Neuenschwander, Mic H. Julien, Ted D. Center,
and Martin P. Hill

17.1 Taxonomy

Earliest descriptions of *Pistia stratiotes* L. (*Araceae*) were by the ancient Egyptians and by the Greek philosophers Dioscorides and Theophrastus.This plant has also been mentioned by Plinius (Stoddard, 1989). According to Bogner and Nicolson (1991) *P. stratiotes* is the solitary member of the subfamily Pistioidea in *Araceae*. However, USDA (2008) places it in the subfamily Aroideae along with the numerous other genera. The many synonyms and obsolete subspecific names (*Plantatlas*, 2006) attest to the variability of this taxonomically isolated species, which is the only free-floating aroid. The plant is known as water lettuce; other common names are available in Randall (2002).

17.2 Description

Pistia consists of a rosette of obovate to spatulate, velvety, light-green leaves (up to 40 cm long in African and American clones) (Fig. 17.1a, b), covered by short hairs, which trap air bubbles and thus enable buoyancy. The underside of leaves is densely hairy and almost white, with longitudinal ribs with embedded veins. The long feathery roots hang freely in the water. A clonal plant forms small colonies through stolons. Inflorescences are inconspicuous (7–12 × 5 mm) with short peduncles in the center of the rosette, growing on a stem. The spadix, enclosed in a whitish spathe, is pale green, hairy outside and glabrous inside. The spathe generally shows a constriction between the groups of male and the female flowers. The spathe below the constriction opens first in the morning hours to expose the wet stigma, whereas the male flowers remain enclosed. Some hours later, the spathe opens completely and exposes the part bearing male flowers. After fertilization, the peduncle bends and pulls the developing fruit (2 mm long) underwater where the seeds are released (Buzgó, 2006).

Biological Control of Tropical Weeds using Arthropods, ed. R. Muniappan, G. V. P. Reddy, and A. Raman. Published by Cambridge University Press. © Cambridge University Press, 2009.

(a)

(b)

Fig. 17.1 (a) *Pistia stratiotes* (photo by Ted Center), (b) flower (photo by Ken Harley).

17.3 Biology

It was thought that *P. stratiotes* would not reproduce sexually in many parts of its adventive range. Depending on the conditions and clones, seed production (4–6 seeds/ fruit) has, however, been observed. Seeds sink to the bottom of the water body, where they form a persistent seed bank (densities of up to 4000 seeds/m^2) (Dray and Center, 1989). Seeds germinate readily in warm (>20 °C), shallow water under high-light intensities (Pieterse *et al.*, 1981). They remain dormant for long periods in dry sediments when water levels recede in dry seasons, and readily germinate when rehydrated during rains. *Pistia stratiotes* develops within the range of 15 and 35 °C, with optimum growth achieved between 22 and 30 °C. Seeds do not germinate at temperatures less than 20 °C; but survive at least two months in cold water at 4 °C and several weeks in ice at –5 °C. The floating plant is frost susceptible and – as there are no persisting vegetative organs – dies completely in areas where a cold season occurs, reemerging from buried seeds in the following rainy season. In some tropical areas, an annual dieback linked to an aphid-transmitted virus has been reported (Pettet and Pettet, 1970). Dispersal is by vegetative means and also potentially by seed. Daughter plants detach from parent plants or colonies and get dispersed either by water currents or by animals. Growth of *P. stratiotes* is inhibited at pH 4 and optimal at pH 7 (Pieterse *et al.*, 1981).

17.4 Ecology

In Laguna Grande (Monagas State, Venezuela) a reduction in forest cover and water surface, the transformation from graminaceous to nongraminaceous herbaceous swamp land, and eutrophication led to a marked decrease of emergent plants and an increase in the floating aquatic weeds water hyacinth, *Eichhornia crassipes* (Mart.) Solms-Laubach (Pontederiaceae), and *P. stratiotes* in the course of 45 years (Gordon, 2001). The results of a four-year study on the Minjiang River (Fujian Province, China) concur with the Venezuelan study: growth of *P. stratiotes* (together with *E. crassipes*) was linked to increased concentrations of nitrogen, phosphorus, and microorganisms, lower pH, and reduced total mass and species diversity of planktonic forms (Cai, 2006). *Pistia stratiotes* therefore appears to thrive particularly well in ecologically disturbed habitats and in impoundments that usually do not dry. This could explain why references to its occurrence in several countries are of rather recent origin.

17.5 Distribution and pest status

Pistia stratiotes was first reported from Florida by J. and W. Bartram in 1765 (Stuckey and Les, 1984), which led to the belief that this plant could be a native to North America. The presence of co-evolved herbivorous insects in South America (Dray *et al.*, 1993; Cordo and Sosa, 2000) suggested, however, that *Pistia* originated from South America. Today, the earlier descriptions from antiquity (Stoddard, 1989) and recent studies based on chloroplast- and mitochondrial-DNA sequences, including those from other aroids, together with fossil evidence, point to a palearctic origin (Renner and Zhang, 2004). According to the Renner and Zhang study, which used Bayesian divergence time inference, the *Pistia* lineage branched off 90 to 76 million years ago, with the oldest closely related fossils of 45 million years ago found in Germany, suggesting that *Pistia* originated in the Tethys region. *Pistia stratiotes* might have been distributed widely in the northern hemisphere during the Eocene (Stoddard, 1989).

The fact that post-Pleistocene strata, in what are now temperate regions, lack *P. stratiotes* fossils suggests that dramatic climatic changes associated with the ice ages may have caused *P. stratiotes* to become extinct locally in these regions. According to this theory, extant populations of this species in the United States would have derived from *P. stratiotes* that arrived in Florida with the earliest European settlers (Stuckey and Les, 1984), most likely by way of Spanish ships out of Central America.

Today, the species has a worldwide distribution in the tropics and subtropics, and is absent only in Antarctica (Parsons and Cuthbertson, 2001). In the last ten years, it has been newly recorded from several European countries (Austria, the Netherlands, Portugal, Russia, Slovenia, and Spain). These new records may be attributable to people simply paying better attention to exotic invaders, even to changing conditions caused by global warming, but most importantly to increased inadvertent as well as commercial transport. *Pistia stratiotes* is regularly imported for commercial trade into Europe, although its invasive potential is well known (EPPO, 2007).

Pistia stratiotes was or is considered a serious pest in some parts of the tropics, but generally it is of less importance than other exotic floating water weeds like *E. crassipes* or *Salvinia molesta* D. S. Mitchell (water fern, giant salvinia) (Salviniaceae). In Africa, it has received most attention since the early 1960s, for instance in Nigeria (Pettet and Pettet, 1970). In the Republic of South Africa (RSA), *P. stratiotes* is of minor importance, but is a declared weed mainly to prevent the sale and distribution of the plant (Cilliers *et al.*, 2003). In Asia it is known from most countries on the mainland, with a greater level of reporting beginning in the 1970s (Waterhouse, 1994). In China, *P. stratiotes* is now present in 14 southern provinces and is ranked in the top 22 of 87 major weeds (Li-ying *et al.*, 1997). It is also reported from most countries in the Pacific region, where *P. stratiotes* is still grown as an ornamental in countries such as Fiji Islands (Waterhouse, 1994). The plant is widely distributed in South America and the Caribbean and recent studies show that it forms dense mats in areas such as the southern Pantanal of Brazil (Coelho *et al.*, 2005). The species is also reported as a potential nuisance species from the Itaipu Reservoir in Brazil (Thomaz *et al.*, 1999), and it has been increasing in density (e.g. Venezuelan study; Gordon, 2001). It seems that *P. stratiotes* in South America is increasingly becoming conspicuous as the numbers of impoundments which are suffering eutrophication increase.

17.6 Possible utilization and impact of *Pistia stratiotes*

Like other floating water plants, *P. stratiotes* has a remarkable capacity to build up biomass rapidly (Reddy and De Busk, 1984). This high productivity and its high nutritive value have attracted interest to use *P. stratiotes* for methane production, as feed (Henry-Silva and Camargo, 2000) or green leaf manure (Raju and Gangwar, 2004).

Use of *P. stratiotes* as cattle feed has, however, limitations because the plant bioaccumulates considerable quantities of heavy metals (Odjegba and Fasidi, 2004). Dead *P. stratiotes* was therefore also used as a biosorbent material to remove metals derived from industrial activities (Miretzky *et al.*, 2006) and the living plant is sought after to clear biological waste of water treatment plants (Koné *et al.*, 2002; Zimmels *et al.*, 2006) or polluted ponds (Vardanyan and Ingole, 2006). *Pistia stratiotes* is also used to reconstruct wetlands (Chen *et al.*, 2006) or to monitor water quality in rivers (Klumpp *et al.*, 2002).

In most natural situations, however, *P. stratiotes* is a weed because it can quickly overgrow and cover still-water surfaces. Like other floating water weeds, it impedes fishing, boat traffic, and flood control, and even relatively small mats can threaten hydropower generation dams. By covering water surfaces, it affects habitats and biodiversity. It blocks out sunlight affecting native water plants; it reduces oxygen movement and either displaces or kills native organisms, such as fish. Its root system contributes to increased siltation making the substrate unsuitable for nesting sites for fish. Evidence for this is, however, only anecdotal.

Pistia stratiotes may also harbor disease-carrying mosquitoes such as species of the malaria vector *Anopheles* and *Mansonia*. The larvae of *Mansonia* perforate leaves and roots of *P. stratiotes* to reach air chambers (Lounibos and Dewald, 1989).

In small-scale experiments, water loss through evapotranspiration by *P. stratiotes* has been found to exceed evaporation of open water considerably; but in large areas covered by floating aquatic vegetation the ratio is expected not to exceed 1.0 (Allen *et al.*, 1997).

Pistia stratiotes is easily spread to new areas. Plants may attach to boats or fishing equipment and become transferred to distant locations. Silt including *P. stratiotes* seeds could move by flow or other mechanical means, and even potentially attached to animals. Moreover, as an ornamental for outdoor ponds and open aquaria (it does not grow well in covered aquaria) it is a favored organism in international trade, where it is sold by aquarium supply dealers and even over the Internet.

17.7 Management

To the best of our knowledge, no public debate among various stakeholders, especially those who gain and those who suffer from high *P. stratiotes* populations, has ever taken place, as has happened for *E. crassipes*. There seems to be, however, a general agreement that *P. stratiotes* in open natural water systems is a nuisance. Although it is not one of the world's most important weeds, it was included in the list of the World's Worst Weeds (Holm *et al.*, 1977), and is officially listed as an exotic invader and noxious plant in Australia, USA, Puerto Rico, and South Africa. In many other countries, where it might be a weed, legislation for its official declaration is lacking. Although *P. stratiotes* is recommended for use in waste treatment plants or as mulch or feed, care must be taken not to violate quarantine regulations and transport it to uninfested watersheds.

Mechanical removal of floating water weeds (occasionally using harvesters) and the installation of booms on slow flowing rivers are often practiced. Among the herbicides the same compounds (diquat, glyphosate, 2,4-D) evaluated for *E. crassipes* have been investigated for the management of *P. stratiotes* (in South Africa: Cilliers *et al.*, 1996). Recently, new compounds (Koschnik *et al.*, 2004) and a now patented mycoherbicide (de Jong and de Voogd, 2003) have been developed for use against floating water weeds. Mechanical clearing combined with herbicide applications is, however, costly. An estimated US$4 million/year was spent in Florida for the control of *P. stratiotes* up to the early 1990s (Habeck and Thompson, 1994). No other published document relative to damage assessment is available.

Reinfestation usually occurs because of either missed plants during harvest or plants that grow in inaccessible locations and because of invasion from other areas via seeds. In Côte d'Ivoire, a freshwater coastal lagoon was mechanically opened to the sea as an outlet for weeds such as *E. crassipes* (Sankaré *et al.*, 1991). However, this effort provided a temporary relief as the created channel closed and weeds invaded the lagoon.

Despite this plant being regarded as a weed worldwide, its biological control has often been undertaken only as a minor component of biological control projects concerned with *E. crassipes*. The first agent that was studied was considered successful in most situations where it was later released. Subsequently very few specifically targeted surveys have been undertaken and the few quantitative ecological data that are available come mostly from release sites.

17.8 Phytophagous species associated with *Pistia stratiotes*

In a study on insects associated with *P. stratiotes* in Florida (Dray *et al.*, 1993), six species were recovered feeding on it: *Petrophila drumalis* (Dryar), *Synclita obliteralis* (Walker), *Samea multiplicalis* Guenée (Lepidoptera: Pyralidae), *Rhopalosiphum nymphaeae* L. (Homoptera: Aphididae), *Draeculacephala inscripta* Van Duzee (Homoptera: Cicadellidae), and *Tanysphyrus lemnae* (F.) (Coleoptera: Curculionidae). Most of these insects were either polyphagous or stenophagous. *Samea multiplicalis* is by far the most common and damaging of these herbivores.

In comparison, Chaco Province in Argentina has 17 species feeding on *P. stratiotes* with 10 species being oligophagous and seven weevils feeding almost exclusively on this plant (Dray *et al.*, 1993).

In the Old World, the only weevil recorded from *P. stratiotes* is *Bagous pistiae* Marshall (Coleoptera: Curculionidae). In addition, in Africa, *Angionychus lividus* Klug (Coleoptera: Carabidae) and *Lepidocyrtus pistiae* Paulin and Delmare-Deboutteville (Collembola: Entomobryidae) were observed to damage *P. stratiotes* severely. Caterpillars reported on *P. stratiotes* from Africa include *Nymphula* sp. and *Spodoptera* sp. (Dray *et al.*, 1993).

In mainland Asia and Indonesia, larvae of seven moth species damaged *P. stratiotes* heavily (Dray *et al.*, 1993), with *Spodoptera pectinicornis* (Hampson) (Lepidoptera: Noctuidae) being totally restricted to this weed (Habeck and Thompson, 1994).

In Australia, two native species, *Nymphula tenebralis* Lower and *N. turbata* Butler (Lepidoptera: Pyralidae), were reported (Gillett *et al.*, 1988) together with *Samea multiplicalis* that had been introduced for biological control of *S. molesta* (Sands and Kassulke, 1984).

To date, 46 species of phytophagous insects have been recorded on *P. stratiotes* (South America: 25 species, Asia: 13 species, Africa: 8 species) (Cordo and Sosa, 2000). Most of these are oligophagous or polyphagous, but 11 species, weevils belonging to *Neohydronomus*, *Pistiacola*, and *Argentinorhynchus*, are assumed to be monophagous. Although damaging, the *Argentinorhynchus* species have not been considered for release as the pupae require the water body to dry before adults emerge. Overall, only two species, a weevil and a noctuid moth, have been given special attention as possible biological control agents.

17.8.1 Neohydronomus affinis *Hustache (Coleoptera: Curculionidae)*

In Argentina, *P. stratiotes* is attacked by several insects, including a weevil that was referred to as *N. pulchellus* Hustache by DeLoach *et al.* (1976), but was subsequently identified as *N. affinis* Hustache by O'Brien and Wibner (1989).

DeLoach *et al.* (1976) collected this weevil in Buenos Aires, Chaco, and Formosa in Argentina and at Piracicaba in Brazil and studied it (Fig. 17.2). The weevils (1.7–2.3 mm long) were brown to bluish gray with dense plumose scales on the elytra, with a nearly straight rostrum that was strongly constricted ventrally at the base. The sex ratio in the field was about 1:1. During feeding, the adults made circular holes and occasionally burrowed in the spongy mesophyll. Eggs were oval (0.33 × 0.40 mm) and were deposited singly below the epidermis, usually on the leaf's upper surface. Oviposition punctures were sealed with a black material. The rate of oviposition was one egg/day/female. Eggs hatched in 3–4 days and the three larval instars developed in 11–14 days. When the larvae

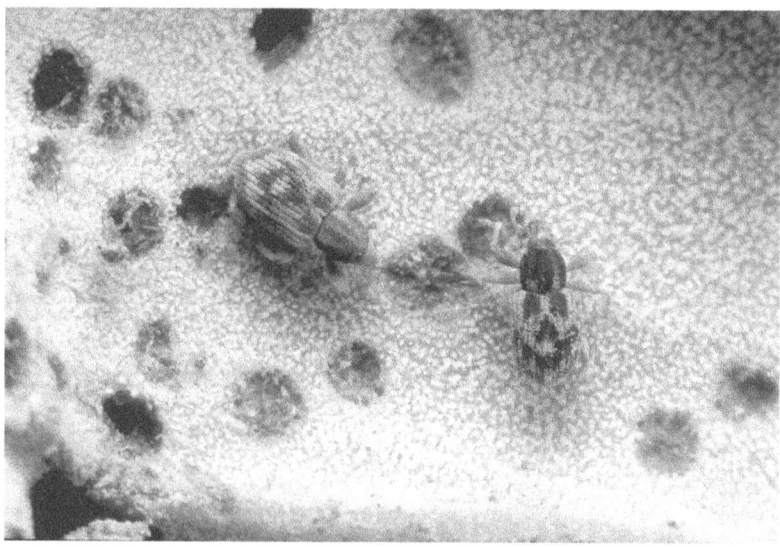

Fig. 17.2. *Neohydronomus affinis* (photo by Ted Center)

were 2.5–3 mm in length, they mined the spongy leaf tissue and transformed to naked pupae within leaf-tissue pockets. Generation time, depending on temperature, was 4–6 weeks. The insect had three generations/year, overwintering mostly as adults, but sometimes as pupae.

Depending on weevil density, adult feeding and burrowing along with larval mining can kill plants (128 weevils/plant killed the plants in 3 days). At peak field densities of about 8 weevils/plant (520/m^2) plants showed little damage for several days before collapsing totally (DeLoach *et al.*, 1976).

In host-specificity tests, adults fed slightly (1–5%) on three plant species other than *P. stratiotes*, namely the floating *Lemna* sp. and *Spirodela* sp. (both *Lemnaceae*) and *Limnobium* (*Hydrocharis*) sp. (*Hydrocharitaceae*) and nibbled on six other tested species. Occasionally, eggs were deposited on four other plants (1% of the number found on *P. stratiotes*), but on 19 other test-plant species no eggs were laid. In the field, however, *N. pulchellus* was never observed attacking any other plant than *P. stratiotes* (DeLoach *et al.*, 1976). In a similar study, no feeding was observed on 30 test plant species (Harley *et al.*, 1990). During starvation tests conducted in Florida, the adult weevils fed and oviposited also on three species of Lemnaceae, *Limnobium spongia* (Bosc.) Steud. (Hydrocharitaceae), *Azolla caroliniana* Willd. (Azollaceae) and *Salvinia minima* Baker (Salviniaceae) (apart from nibbling or a single egg deposition on two nonaquatic plants) (Thompson and Habeck, 1989). However, when the test-plant species were retested in choice tests, the weevils fed and oviposited only on *P. stratiotes*, indicating that this weevil is a biological control agent of promise.

No insect parasitoids were observed on adult weevils, but high numbers of diverse nonidentified generalist predators existed among *P. stratiotes* mats, suggesting that *N. pulchellus* was possibly exposed to predation (DeLoach *et al.*, 1976).

17.8.2 Spodoptera (Epipsammea) pectinicornis *(Hampson)*
(Lepidoptera: Noctuidae)

This species has been listed in the most recent *World Catalogue of Agents and their Target Weeds* (Julien and Griffiths, 1998). It had previously been placed in the genus *Epipsammea* and has been variously cited as belonging to the genera *Athetis, Caradrina, Namangana, Proxenus,* or *Xanthoptera* (and was also misspelled as *Episammia* or *Episamea*) (see also Julien and Griffiths, 1998).

This small (wing span: 19 mm in ♂, 23 mm in ♀), grayish-speckled moth (Fig. 17.3) has bipectinate antennae, which are conspicuous in the male, and gender is best distinguished by observation of genitalia. Eggs are subspherical, greenish when freshly deposited, then turning yellow. They are laid in masses of up to 150 eggs each, which are covered by scales from the female's abdomen. The ovipositional period lasts 2–6 days and each female lays up to 990 eggs (with an average of 666 eggs). The incubation period ranges from 3 to 6 days. The creamy-white first-instar larvae feed within the spongy mesophyll and progress through seven instars in 17–20 days. They attain 25 mm length

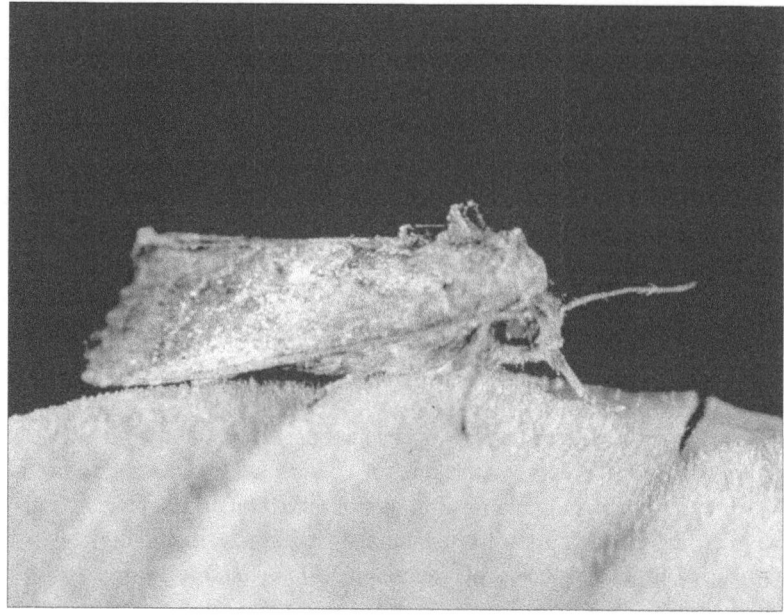

Fig. 17.3. *Spodoptera pectinicornis* (photo by Ted Center)

and pupate in the leaf base between the leaves or between the thick veins on the underside of the leaf. The prepupal period lasts 1–2 days and the pupal period lasts 3.5–5.5 days. The total generation time is about 30 days (Center *et al.*, 2002).

Larvae feed on leaves and destroy the apical meristem, thus preventing leaf replacement. Moth populations fluctuate, but reproduce continuously in overlapping generations (Center *et al.*, 2002).

Spodoptera pectinicornis is native to India, Sri Lanka, Thailand, Indonesia, and Papua New Guinea (Waterhouse, 1994). The species is associated with slow-moving waterways and other bodies of fresh water, where the larvae feed on *P. stratiotes*. Larvae severely damage leaves and meristems (Suasa-ard and Napompeth, 1982). In an experimental set-up, high fertilizer treatment of *P. stratiotes* produced significantly heavier adults, which lived about a day less than those reared on unfertilized plants, but laid high egg numbers. Overall, most eggs were laid in egg masses, the percentage being higher on well-fertilized host plants. During the first days of their adult lives, females laid proportionally more eggs on the youngest leaves (Wheeler *et al.*, 1998).

Spodoptera pectinicornis was introduced from Thailand into quarantine in Florida in 1986, where it was studied in detail (Habeck and Thompson, 1994). In cages, adult females would occasionally lay eggs on plants other than *P. stratiotes*. In host-specificity tests on 61 plant species belonging to 32 families, however, less than 1% of all neonate larvae survived on plants other than *P. stratiotes* for more than 3 days and all died within 6 days. Even third-instar larvae did not survive more than 6 days on other hosts. The exception concerned *Impatiens* spp. (Balsaminaceae), on which larvae survived 25 days, but could not complete development. Similar studies in Thailand (Suasa-ard and Napompeth, 1982) support the conclusion that this insect is specific to *P. stratiotes*.

17.9 Implementing biological control

The two natural enemies, *N. affinis* and *S. pectinicornis*, have been used in biological control projects of different types.

17.9.1 Neohydronomus affinis

The first attempt at classical biological control of *P. stratiotes* was by CSIRO Australia. *Neohydronomus affinis* was collected at Pelotas, Brazil, and brought to Australia, host-specificity tested, and released near Brisbane in 1982 (Harley *et al.*, 1984). Within five months, weevil populations increased to about $1000/m^2$ and in one year the biomass of *P. stratiotes* dropped by 80%. In the next few years, *P. stratiotes* populations showed strong declines each year from midwinter to early summer and an overall decline in plant biomass, which was closely followed by the population density of the weevil. Finally, the weed disappeared while the weevil dispersed 75 km (Harley *et al.*, 1984, 1990).

   Weevils were also supplied to RSA, where they were released in the Pafuri area of the Kruger National Park in December 1985. Weevil populations and plant damage initially fluctuated with seasonal conditions, but by October 1986 *P. stratiotes* could no more be found in this water pan (Cilliers, 1987). *Neohydronomus affinis* was released into the Sabie River in 1987. After a relatively slow start, weevil populations increased and in two years 80% of plants were damaged. However, this declined to 57% after a few months when infested plants were washed out of the study area. In 1988 the weevil was released into a large dam in Mmabatho, where *P. stratiotes* had been introduced three years earlier. Despite severe winter frosts in the area and a temporary increase in the plant population, the weevil brought the weed under control by 1990. Overall, the weevil was most successful in lentic water bodies (less so in fast-flowing streams). Though eutrophication or higher levels of nitrates and phosphates may influence the success of biological control programs on aquatic weeds (Hill and Olckers 2001), biological control of *P. stratiotes* has also been achieved in two highly eutrophic systems, Sunset Dam (Kruger National Park) and the Port Elizabeth waste water treatment plant (Cape Recife). *Neohydronomus affinis* was introduced to Sunset Dam in the early 1990s and initially reduced the *P. stratiotes* infestation from 100% to less than 20% coverage in the first summer season. However the dam became completely covered again during winter (Cilliers *et al.*, 1996). This cycle of summer clearing and winter return continued until the late 1990s, after which the dam cleared during the summer and has remained at 20% cover. In the Cape Recife waste water treatment ponds, the cover had been reduced from 100% to less than 5% 12 months after the release (Moore, 2005). Another rise to around 60% cover occurred in 2004 winter (May to July), which was reduced to less than 5% in the following spring, and has remained at this level since then. These two examples illustrate that biological control of *P. stratiotes* is possible, even under highly eutrophic conditions, but that it might take up to five years (as in the Sunset Dam example) to achieve the preferred outcomes.

First release of *N. affinis* in Papua New Guinea was made in June 1985. The weevil established in Wewak and on the Sepik River in the East Sepik Province, and at Bulolo in the Morobe Province. *Neohydronomus affinis* spread to other locations via humans or via natural means. For example, it was deliberately released near Kimbe on the island of West New Britain, and from its release near Angoram on the lower Sepik River it spread throughout the Sepik River lagoons. In many localities, *P. stratiotes* is now either scarce or present as few, small isolated clumps hosting the weevil. However, in some locations, although *P. stratiotes* infestations have been reduced, the high level of control seen elsewhere has not occurred. Overall, seasonal flooding and the indigenous moth, *S. pectinicornis*, may have reduced *P. stratiotes* populations (Laup, 1987), but it is now clear that many lagoons, ponds, and creeks remained heavily infested until after *N. affinis* was introduced. In December 2006, *N. affinis* collected from Papua New Guinea was released in Vanuatu, where it is being monitored (W. Orapa, personal communication, 2007).

In the USA, *P. stratiotes* had become widespread by the mid 1700s, but was reduced when *E. crassipes* invaded North America. Following the reduction of *E. crassipes* due to biological control, it was feared that *P. stratiotes* might regain its past weed status. *Neohydronomas affinis* was therefore introduced from Australia into quarantine in 1981 and releases were made into several habitats in southern Florida in 1987 and 1988. Weevil populations persisted at low levels in winter, but eventually increased rapidly. The weevil was therefore considered established (Dray *et al.*, 1990) and good control was achieved in most, although not in all locations (Dray and Center, 1992). The weevil has subsequently spread on its own as far as Louisiana and has been introduced into southern Texas (Grodowitz *et al.*, 1992).

In Zimbabwe, *N. affinis* from Australia was released during April 1988 in the Manyame River, upstream of Lake McIllwaine, the main water source for Harare. By July 1988, the weevil was well established, *P. stratiotes* coverage was declining and *E. crassipes* began to invade the area. By October, the weevil was found on a 14-km stretch of the river. By February 1989, *P. stratiotes* was no longer considered a problem (Chikwenhere and Forno, 1991; Chikwenhere, 1994). In 1998, infestations estimated at 56 ha appeared in the eastern part of the country and *N. affinis* was released, immediately resulting in control (G. Chikwenhere and M. Hill, personal communication, 2007).

In the Congo, biological control had been implemented against all three major floating water weeds since 1999. *Pistia stratiotes* had a rather patchy distribution, mainly in the central Cuvette and in the southern Kouilou Province. Control was felt immediately following the releases of *N. affinis*. By 2003, no *P. stratiotes* could be found in the release area in the Cuvette and coverage on lakes in the south had diminished considerably. Because of the mobility of this weevil, the releases in the Congo (Brazzaville) may spread to infestations in the Democratic Republic of Congo (RDC), which at present are of difficult to access (Mbati and Neuenschwander, 2005).

Neohydronomus affinis was also released in Senegal in 1994, Ghana in 1996 (Julien and Griffiths, 1998), Côte d'Ivoire in 1997, as well as in Kenya (Cilliers *et al.*, 2003).

In Senegal and Mauritania, *P. stratiotes* threatened the World Heritage Site of the Djoudj National Bird Sanctuary and the adjacent Diawling National Park (Diop, 2007). Eight months after the first releases on Lake Guiers, *P. stratiotes* mats were destroyed and within 18 months *P. stratiotes* populations dropped to acceptable levels at Djoudj Park, 150 km northwest of Lake Guiers. In 2005, *P. stratiotes* reappeared in new sites (following the introduction of aquaculture projects), but was quickly subdued by new releases of *N. affinis* (Diop, 2007).

In Côte d'Ivoire, *N. affinis* was received from the International Institute of Tropical Agriculture in Benin in December 1997. It was released at numerous sites throughout the southern portion of the country beginning in early 1998. By November 1998 it had reduced the weed by over 95% at the Barrage de Ayame II (Fig. 17.4 a, b), and was causing noticeable damage at most other sites. By November 1999 (less than two years post release), *N. affinis* had controlled the weed at the six major infestations that were inspected (Z. Mesmer and M. Julien, unpublished observations).

In Benin, *N. affinis* was imported from Zimbabwe in 1993 and released in 1995 at six sites in the southern Mono Province and in 1996 at one site in the northern Borgou Province, which drains towards the Niger River. Two years later, the weevil had spread 90 km and by 2000 it was found 250 km to the northwest within a different watershed. In two monitored sites in the south, water levels, total plant biomass, and weed cover fluctuated considerably in response to rains. After 3 to 4 years, *P. stratiotes* had disappeared almost completely (Fig. 17.5) (Ajuonu and Neuenschwander, 2003).

It seems that the weevil has not yet been released in Asia. Certainly, no biological control agents have been released against *P. stratiotes* in China (D. Jianqing, personal communication, 2007).

In conclusion, it seems that at subtropical latitudes seasonal reductions in the populations of *N. affinis* allow *P. stratiotes* to grow unchecked for a portion of the year and levels of biological control therefore fluctuate strongly. In tropical regions, however, control is approaching 100% in 1–2 years. In this context, *N. affinis* seems to be more efficient in tropical areas, where it has been introduced, than in subtropical Rio Grande do Sul of southeastern Brazil, where this weevil is native and where the recorded generalist predators might hamper its impact (DeLoach *et al.*, 1976).

17.9.2 Spodoptera (Epipsammea) pectinicornis

This species was, and still is, used for "inundative" biological control in Thailand. Augmentation of natural *S. pectinicornis* populations with cultured insects prevents destruction of rice seedlings by invading *P. stratiotes* and reduces the threat of water lettuce affecting hydroelectric power plants (Napompeth, 1990). However, no detailed quantitative evaluation of the impact of *S. pectinicornis* in Asia seems to have been made.

(a)

(b)

Fig. 17.4. View of Ayame Dam II in Côte d'Ivoire (a) in February 1998, before release of *Neohydronomus affinis* (photo by Mic Julien), and (b) in November 1998, after release (photo by Mic Julien).

Because *N. affinis* caused *P. stratiotes* declines on several waterways in Florida, but was not effective at every release site and did not establish at some sites (Dray and Center, 1992), an additional agent was sought. Following additional host range studies, *S. pectinicornis* was approved for use in Florida and releases began in December 1990 (Grodowitz *et al.*, 1992). Between 1990 and 1997 great numbers (332 000 larvae, pupae and adults) were released in a total of 22 sites (Dray *et al.*, 2001). Initially, relatively small numbers (a few thousand per site) were released infrequently in many localities; but

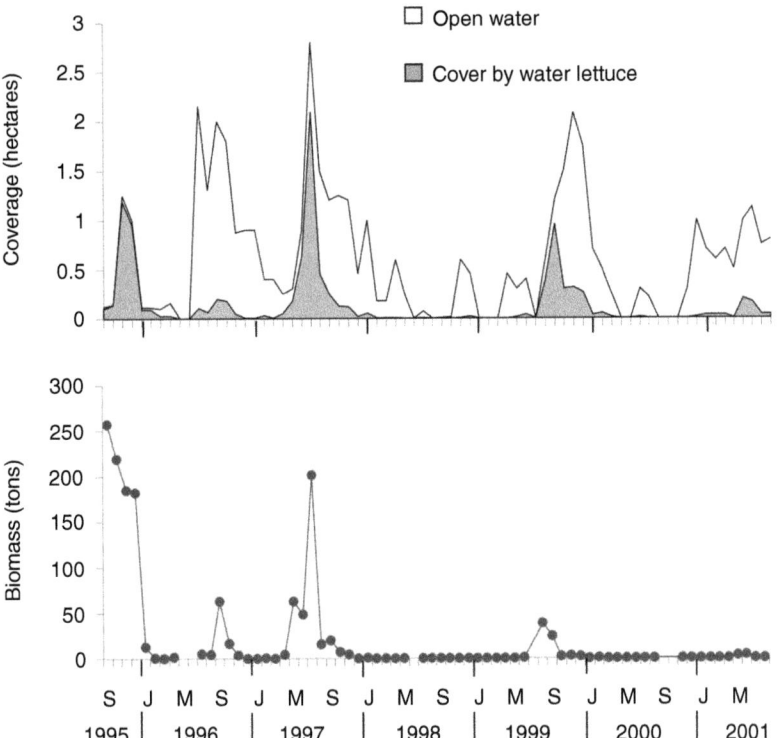

Fig. 17.5. Impact of *Neohydronomus affinis* on coverage (ha) and biomass (tons per ha) of water lettuce, *Pistia stratiotes*, on a pond at Sé, Benin (modified from Ajuonu and Neuenschwander, 2003).

dispersal of adults and predation by birds, ants, and spiders prevented populations from persisting. In later attempts, cages and plant-free zones around these cages restricted adult dispersal and excluded flying and crawling predators. They also allowed resupplying *S. pectinicornis* repeatedly (1200 to 2000 larvae, pupae and adults every 5 to 10 days) and providing fresh plant material on a regular basis. By this technique up to seven postrelease generations were obtained, with up to >20 individuals/m². Unfortunately, all these incipient populations declined until *S. pectinicornis* was undetectable. Although that specific biological control project was an apparent failure, the authors caution that small, undetected populations of this species might still exist in Florida and that further attempts with other species are still warranted.

17.10 Competition with other floating water weeds

In many areas such as the USA, Benin, and others, *P. stratiotes* was observed to be abundant long before the introduction of *E. crassipes*. It was often observed that wherever

E. crassipes was spreading, *P. stratiotes* was displaced. Based on this general pattern, it was considered early on that biological control of the three major floating water weeds, *E. crassipes*, *S. molesta*, and *P. stratiotes*, should be developed concurrently. It was assumed that if this approach was not adopted, reduction in one of these weeds would simply create space for another to increase and the problem of floating aquatic weeds would not be reduced, since all three plants were thought to have similar requirements.

This strategy has been implemented and biological control of *P. stratiotes* was usually attached as a minor component to a larger biological control project against the other weeds (e.g. *N. affinis* was collected from Brazil for Australia when the main target was *Salvinia molesta* and at the same time that biological control of *E. crassipes* and *Althernanthera philoxeroides* (Mart.) Griseb. (Amaranthaceae) was underway). However, biological control of *E. crassipes* is generally rather slow, taking 2 to 8 years before a new lower equilibrium is achieved (Center *et al.*, 1989; Julien *et al.*, 1999; Ajuonu *et al.*, 2003), and often not completely successful as to permanently leave open water surfaces for reinvasion by *P. stratiotes*.

Since *Neochetina* spp. were often already present where *N. affinis* were being released, the possible antagonism between the two floating water weeds often cannot be subjected to a rigorous analysis. It can now only be deduced from analysis of historical data or experimental studies. In competition experiments in confined spaces, *E. crassipes* usually wins out over *P. stratiotes*. If *E. crassipes* is, however, attacked by phytophagous insects, even seemingly small damage already weakens and reduces the competitive capacity of *E. crassipes*, and *P. stratiotes* takes advantage (Center *et al.*, 2005; Coetzee *et al.*, 2005).

The suggestion that the three floating water weeds have similar requirements is true only in a generic context. *Eichhornia crassipes* and *P. stratiotes* share the same narrow pH range, but *P. stratiotes* is slightly less tolerant of saltwater intrusions (Haller *et al.*, 1974). This seems to be borne out in the field, but has never been subjected to a rigorous analysis. Remote-sensing techniques are now available for distinguishing the floating water weeds over large areas. Field reflectance measurements showed that *P. stratiotes* had higher visible green reflectance than associated plant species and could be detected in both aerial color-infrared (CIR) photography and videography, where it showed up in light pink color (Everitt *et al.*, 2003).

Nontarget, indirect impacts of biological control of *P. stratiotes* have barely been investigated. When dense mats of these exotic weeds eventually succumb to the action of biological control agents, they sink to the bottom; some may move away by flow. Especially in confined water bodies, their gradual decomposition leads to a surge in nutrients, which is, however, never as dramatic as when herbicide-treated mats sink to the bottom. Sometimes this leads to the growth of a floating plant mass (sudd) involving *Echinochloa* spp., *Phragmites* sp. (Poaceae), *Typha* spp. (Typhaceae), and *Scirpus* spp. (Cyperaceae), as observed after the control of *E. crassipes*. If eutrophication does not continue due to other causes, any problems including excessive growth of grasses should gradually dissipate.

17.11 Conclusion

Pistia stratiotes occurs today in all continents (except Antarctica). In open waters, *P. stratiotes* is a weed and might be indicative of an environment disturbed by eutrophication, deforestation, and impounding aquatic ecosystems, which slow flow rates. In most cases, *P. stratiotes* is not considered a weed of high importance, especially because *E. crassipes* and *S. molesta* have a more devastating impact.

Biological control has been the management method of choice and two insects have been used. Though the evolutionary origin of *Pistia* seems to be in the Tethys region in southern Europe, the observed, often spectacularly successful, control by the neotropical weevil points to a more recent coevolution of the plant and its weevil in South America. The Asian moth, which does not seem to control *P. stratiotes* well throughout Asia, failed to establish following releases in the USA. The result of potential future releases of the neotropical weevil into Asia and comparative molecular studies of *P. stratiotes* across its full range could eventually clarify the associative relationships between *P. stratiotes* and its two phytophagous insects in this unusual biological control program.

Pistia stratiotes has recently been investigated extensively for use as feed and as an agent for clearing polluted waters. These schemes are still experimental and a potentially negative impact by established biological control agents has not yet been observed nor discussed.

Judging from the available studies, it seems that in most tropical areas, where biological control by means of *N. affinis* has been implemented, no future actions are needed. The weevil has the unusual tendency to exterminate its host locally, which might disassociate the natural enemy from its host and eventually lead to resurgences. Because of the good dispersal capacity of the weevil, these new foci of weed infestation are, however, expected to be detected in good time. If not, repeated local releases might be the answer. In subtropical regions, control might not be as complete.

Benefits relating to biological control of all Australian targeted weeds have been assessed (Page and Lacey, 2006). It was, however, not possible to separate the benefits concerning biological control of *E. crassipes*, *S. molesta*, and *P. stratiotes*, as all three plants occur in similar water bodies and in similar locations. The combined costs of the biological control projects were AU$5.1 million (1974 to 1993). The *P. stratiotes* project cost Au$306 900 over five years (1978 to 1982). The combined benefit:cost ratio was 27.5. This figure documents the high cost efficiency of biological control projects and corresponds to the benefits documented for other floating water weed species that had been studied individually.

References

Ajuonu, O. and Neuenschwander, P. (2003). Release, establishment, spread and impact of the weevil *Neohydronomus affinis* (Coleoptera: Curculionidae) on water lettuce (*Pistia stratiotes*) in Benin, West Africa. *African Entomology*, **11**, 205–211.

Ajuonu, O., Schade, V., Veltman, B., Sedjro, K. and Neuenschwander, P. (2003). Impact of the weevils *Neochetina eichhorniae* and *N. bruchi* (Coleoptera: Curculionidae) on water hyacinth, *Eichhornia crassipes* (Pontederiaceae), in Benin, West Africa. African Entomology, **11**, 153–161.

Allen, L. H., Jr., Sinclair, T. R. and Bennett, J. M. (1997). Evapotranspiration of vegetation of Florida: perpetuated misconceptions versus mechanistic processes. *Proceedings of the Soil and Crop Science Society of Florida*, **56**, 1–10.

Bogner, J. and Nicolson, D. H. (1991). A revised classification of Araceae with dichotomous keys. *Wildenowia*, **21**, 35–50.

Buzgó, M. (2006). *The Genus Pistia Benth & Hook*. International Aroid Society. www.aroid.org/genera/pistia

Cai, L. (2006). Impact of floating vegetation in Shuikou impoundment, Minjiang River, Fujian Province. *Hupo Kexue*, **18**, 250–254.

Center, T. D., Cofrancesco, A. F. and Balciunas, J. K. (1989). Biological control of aquatic and wetland weeds in the Southeastern United States. *Proceedings of the VII International Symposium on Biological Control of Weeds*, held in Rome, 6–11 March, 1988, ed. E. S. Delfosse. Melbourne, Australia: CSIRO, pp. 239–262.

Center, T. D., Dray, F. A., Jubinsky, G. P. and Grodowitz, J. (2002). *Insects and Other Arthropods that Feed on Aquatic and Wetland Plants*. Technical Bulletin 1870. Washington DC: USDA, Agricultured Research Service.

Center, T. D., Van, T. K., Dray, F. A., *et al.* (2005). Herbivory alters competitive interactions between two invasive aquatic plants. *Biological Control*, **33**, 173–185.

Chen, T. Y., Kao, C. M., Yeh, T. Y., Chien, H. Y. and Chao, C. (2006). Application of a constructed wetland for industrial wastewater treatment: a pilot-scale study. *Chemosphere*, **64**, 497–502.

Chikwenhere, G. P. (1994). Biological control of water lettuce in various impoundments of Zimbabwe. *Journal of Aquatic Plant Management*, **32**, 27–29.

Chikwenhere, G. P. and Forno, I. W. (1991). Introduction of *Neohydronomus affinis* for biological control of *Pistia stratiotes* in Zimbabwe. *Journal of Aquatic Plant Management*, **29**, 53–55.

Cilliers, C. J. (1987). First attempts at and early results on the biological control of *Pistia stratiotes* L. in South Africa. *Koedoe*, **30**, 35–40.

Cilliers, C. J., Hill, M. P., Ogwang, J. A. and Ajuonu, O. (2003). Aquatic weeds in Africa and their control. In *Biological Control in IPM Systems in Africa*, ed. P. Neuenschwander, C. Borgemeister and J. Langewald. Wallingford, UK: CABI Publishing, pp. 161–178.

Cilliers, C. J., Zeller, D. and Strydom, G. (1996). Short- and long-term control of water lettuce (*Pistia stratiotes*) on seasonal water bodies and on a river system in the Kruger National Park, South Africa. *Hydrobiology*, **340**, 173–179.

Coelho, F. F., Deboni, L. and Lopes, F. S. (2005). Density-dependent reproductive and vegetative allocation in the aquatic plant *Pistia stratiotes* (Araceae). *Revista de Biologia Tropical*, **53**, 369–376.

Coetzee, J. A., Center, T. D., Byrne, M. J. and Hill, M. P. (2005). Impact of the biocontrol agent *Eccritotarsus catarinensis* on the competitive performance of waterhyacinth, *Eichhornia crassipes*. *Biological Control*, **32**, 90–96.

Cordo, H. A. and Sosa, A. (2000). The weevils *Argentinorhynchus breyeri*, *A. bruchi* and *A. squamosus* (Coleoptera: Curculionidae), candidates for biological control of waterlettuce (*Pistia stratiotes*). In *Proceedings of the X International Symposium on*

Biological Control of Weeds, held at Montana State University, Bozeman, 4–14 July 1999, ed. N. Spencer, Sidney, MT: USDA-ARS, pp. 325–335.

de Jong, M. D. and de Voogd, W. B. (2003). *Novel Mycoherbicide for Biological Control of Aquatic Weeds Such as Water Hyacinth and Water Lettuce*. Patent disclosure, application: EPO-DG1 (102).

DeLoach, C. J., DeLoach, A. D. and Cordo, H. A. (1976). *Neohydronomus pulchellus*, a weevil attacking *Pistia stratiotes* in South America: biology and host specificity. *Annals of the Entomological Society of America*, **69**, 830–834.

Diop, O. (2007). Management of invasive aquatic weeds with emphasis on biological control in Senegal. Ph.D. thesis, Rhodes University, Grahamstown, South Africa.

Dray, F. A. and Center, T. D. (1989). Seed production by *Pistia stratiotes* L. (water lettuce) in the United States. *Aquatic Botany*, **33**, 155–160.

Dray, F. A. and Center, T. D. (1992). *Biological control of* Pistia stratiotes *L. (Waterlettuce) using* Neohydronomus affinis *Hustache (Coleoptera: Curculionidae)*. Miscellaneous Paper A-92-1. Vicksburg, MS: U.S. Army Corps of Engineers, Waterways Experiment Station.

Dray, F. A., Center, T. D. and Habeck, D. H. (1993). Phytophagous insects associated with *Pistia stratiotes* in Florida. *Environmental Entomology*, **22**, 1146–1154.

Dray, F. A., Center, T. D., Habeck, D. H., *et al.* (1990). Release and establishment in the southeastern United States of *Neohydronomus affinis* (Coleoptera: Curculionidae), an herbivore of waterlettuce. *Environmental Entomology*, **19**, 799–802.

Dray, F. A., Center, T. D. and Wheeler, G. S. (2001). Lessons from unsuccessful attempts to establish *Spodoptera pectinicoris* (Lepidoptera: Noctuidae), a biological control agent of waterlettuce. *Biocontrol Science and Technology*, **11**, 301–316.

EPPO Reporting Service. (2007). Number 1, Paris. Available from: http://archives.eppo.org/EPPOReporting/Reporting_Archives.htm (accessed 28 July 2008).

Everitt, J. H., Yang, C. and Flores, D. (2003). Light reflectance characteristics and remote sensing of waterlettuce. *Journal of Aquatic Plant Management*, **41**, 39–44.

Gillett, J. D., Dunlop, C. R. and Miller, I. L. (1988). Occurrence, origin, weed status, and control of water lettuce (*Pistia stratiotes* L.) in the Northern Territory. *Plant Protection Quarterly*, **3**, 144–148.

Gordon, E. (2001). Cambios de la vegetación durante 45 años (1947–1992) en Laguna Grande (Estado Monagas, Venezuela). *Acta Botánica Venezuelica*, **24**, 37–57.

Grodowitz, M. J., Johnson, W. and Nelson, L. D. (1992). *Status of Biological Control of Waterlettuce in Louisiana and Texas using Insects*. Miscellaneous Paper A-92–3. Vicksburg, MS: U.S. Army Corps of Engineers, Waterways Experiment Station.

Habeck, D. H. and Thompson, C. R. (1994). Host specificity and biology of *Spodoptera pectinicornis* (Lepidoptera: Noctuidae), a biological control agent of waterlettuce (*Pistia stratiotes*). *Biological Control*, **4**, 263–268.

Haller, W. T., Sutton, D. L. and Barlowe, W. C. (1974). Effects of salinity on growth of several aquatic macrophytes. *Ecology*, **55**, 891–894.

Harley, K. L. S., Forno, I. W., Kassulke, R. C. and Sands, D. P. A. (1984). Biological control of water lettuce. *Journal of Aquatic Pest Management*, **22**, 101–102.

Harley, K. L. S., Kassulke, R. C., Sands, D. P. A. and Day, M. D. (1990). Biological control of water lettuce *Pistia stratiotes* [Araceae] by *Neohydronomus affinis* [Coleoptera: Curculionidae]. *Entomophaga*, **35**, 363–374.

Henry-Silva, G. G. and Camargo, A. F. M. (2000). Composicão química de quatro espécies de macrófitas aquáticas e possibilidade de uso de suas biomassas. *Naturalia São Paulo*, **25**, 111–125.

Hill, M. P. and Olckers, T. (2001). Biological control initiatives against water hyacinth in South Africa: constraining factors, success and new courses of action. In *Biological and Integrated Control of Water Hyacinth*, Eichhornia crassipes: *Proceedings of the Second Global Working Group Meeting for the Biological and Integrated Control of Water Hyacinth*, held in Beijing, China, 9–12 October 2000, ed. M. H. Julien, M. P. Hill, T. D. Center and Ding Jianqing. Proceedings 102. Canberra, Australia: ACIAR, pp. 33–38.

Holm, L. G., Plucknett, D. L., Pancho, J. V. and Herberger, J. P. (1977). *The World's Worst Weeds: Distribution and Biology*. Honolulu, HI: University Press of Hawaii.

Julien, M. H. and Griffiths, M. W. (1998). *Biological Control of Weeds. A World Catalogue of Agents and Their Target Weeds*. Fourth edn. Wallingford, UK: CABI Publishing.

Julien, M. H., Griffiths, M. W. and Wright, A. D. (1999). *Biological Control of Water Hyacinth. The Weevils* Neochetina bruchi *and* N. eichhorniae: *Biologies, Host Ranges, and Rearing, Releasing and Monitoring Techniques for Biological Control of* Eichhornia crassipes. Monograph 60. Canberra, Australia: ACIAR.

Koschnick, T. J., Haller, W. T. and Chen, A. W. (2004). Carfentrazone-ethyl pond dissipation and efficacy on floating plants. *Journal of Aquatic Plant Management*, **42**, 103–108.

Klumpp, A., Bauer, K., Franz-Gerstein, C. and de Menezes, M. (2002). Variation of nutrient and metal concentrations in aquatic macrophytes along the Rio Cachoeira in Bahia (Brazil). *Environment International*, **28**, 165–171.

Koné, D., Cissé, G., Seignez, C. and Holliger, C. (2002). Le lagunage à laitue d'eau (*Pistia stratiotes*) à Ouagadougou: une alternative pour l'épuration des eaux usées destinées à l'irrigation. *Cahiers Agricultures*, **11**, 39–43.

Laup, S. (1987). Biological control of water lettuce: early observations. *Harvest*, **12**, 41–43.

Li-ying, L., Wang, Ren and Waterhouse, D. F. (1997). *The Distribution and Importance of Arthropod Pests and Weeds of Agriculture and Forestry Plantations in Southern China*. Canberra, Australia: ACIAR.

Lounibos, L. P. and Dewald, L. B. (1989). Oviposition site selection by *Mansonia* mosquitoes on water lettuce. *Ecological Entomology*, **14**, 413–422.

Mbati, G. and Neuenschwander, P. (2005). Biological control of three floating water weeds, *Eichhornia crassipes*, *Pistia stratiotes*, and *Salvinia molesta* in the Republic of Congo. *BioControl*, **50**, 635–645.

Miretzky, P., Saralégui, A. and Cirelli, A. F. (2006). Simultaneous heavy metal removal mechanism by dead macrophytes. *Chemosphere*, **62**, 247–254.

Moore, G. R. (2005). The role of nutrients in the biological control of water lettuce, by the leaf-feeding weevil, with particular reference to eutrophic conditions. Masters thesis, Rhodes University, Grahamstown, South Africa.

Napompeth, B. (1990). Biological control of weeds in Thailand – a country report. In *BIOTROP Special Publication 38*, ed. B. A. Auld, R. C. Umaly and S. S. Tjitrosomo. Bogor, Indonesia: SEAMEO-BIOTROP, pp. 23–36.

O'Brien, C. W. and Wibner, G. J. (1989). Revision of the neotropical genus *Neohydronomus* Hustache (Coleoptera: Curculionidae). *The Coleopterists Bulletin*, **43**, 291–304.

Odjegba, V. J. and Fasidi, I. O. (2004). Accumulation of trace elements by *Pistia stratiotes*: implications for phytoremediation. *Ecotoxicology*, **13**, 637–646.

Page, A. R. and Lacey, K. L. (2006). Economic impact assessment of Australian weed biological control. *CRC Technical Series for Australian Weed Management*, **10**, pp. 150.

Parsons, W. T. and Cuthbertson E. G. (2001). *Noxious Weeds of Australia*. Canberra, Australia: CSIRO Publishing.

Pettet, A. and Pettet, J. (1970). Biological control of *Pistia stratiotes* L. in Western State, Nigeria. *Nature*, **226**, 282.

Pieterse, A. H., DeLange, L. and Verhagen, L. (1981). A study on certain aspects of seed germination and growth of *Pistia stratiotes* L. *Acta Botanica Neerlandica*, **30**, 47–57.

Plantatlas (2006). *Pistia stratiotes* (http://www.plantatlas.usf.edu/synonyms.asp).

Raju, R. A. and Gangwar, B. (2004). Utilization of potassium-rich green-leaf manures for rice (*Oryza sativa*) nursery and their effect on crop productivity. *Indian Journal of Agronomy*, **49**, 244–247.

Randall, R. P. (2002). *A Global Compendium of Weeds*. Meredith, Australia: R. G. and F. J. Richardson.

Reddy, K. R. and De Busk, W. F. (1984). Growth characteristics of aquatic macrophytes cultured in nutrient-enriched water. I. Water hyacinth, water lettuce, and pennyworth. *Economic Botany*, **38**, 229–239.

Renner, S. S. and Zhang, L.-B. (2004). Biogeography of the *Pistia* clade (Araceae): based on chloroplast and mitochondrial DNA sequences and Bayesian divergence time inference. *Systematic Biology*, **53**, 422–432.

Sands, D. P. A. and Kassulke, R. C. (1984). *Samea multiplicalis* (Lepidoptera: Pyralidae) for biological control of two water weeds, *Salvinia molesta* and *Pistia stratiotes*, in Australia. *Entomophaga*, **29**, 267–273.

Sankaré, Y., Amon Kothias, J. B. and Konan, A. A. (1991). Les effets de la reouverture de l'embouchure du fleuve Comoé sur la végétation littorale lagunaire (Lagune Ebrié – Côte d'Ivoire). *Journal Ivoirien d'Océanologie et de Limnologie, Abidjan*, **1**, 71–79.

Stoddard, A. A. (1989). The phytogeography and paleofloristics of *Pistia stratiotes* L. *Aquatics*, **11**, 21–24.

Stuckey, R. L. and Les, D. H. (1984). *Pistia stratiotes* (water lettuce) recorded from Florida in Bartram's travels, 1765–74. *Aquaphyte*, **4**, 6.

Suasa-ard, W. and Napompeth, B. (1982). *Investigations on* Episammia pectinicornis *(Hampson) (Lepidoptera: Noctuidae) for Biological Control of the Waterlettuce in Thailand*. Technical Bulletin 3, Bangkok, Thailand: National Biological Control Research Center.

Thomaz, S. M., Bini, L. M., de Souza, M. C., Kita, K. K. and Camargo, A. F. M. (1999). Aquatic macrophytes of Itaipu Reservoir, Brazil: survey of species and ecological considerations. *Brazilian Archives of Biology and Technology*, **42**, 15–22.

Thompson, C. R. and Habeck, D. H. (1989). Host specificity and biology of the weevil *Neohydronomus affinis* [Coleoptera: Curculionidae] a biological control agent of *Pistia stratiotes*. *BioControl*, **34**, 299–306.

USDA (2008). *Germplasm Resources Information Network – (GRIN)* [Online Database]. National Germplasm Resources Laboratory, Beltsville, Maryland. (http://www.ars-grin.gov/cgi-bin/npgs/html/taxon.pl?102491).

Vardanyan, L. G. and Ingole, B. S. (2006). Studies on heavy metal accumulation in aquatic macrophytes from Sevan (Armenia) and Carambolim (India) lake systems. *Environment International*, **32**, 208–218.

Waterhouse, D. F. (1994). *Biological Control of Weeds: Southeast Asian Prospects.* Monograph 26. Canberra, Australia: ACIAR.

Wheeler, G. S., Van, T. K. and Center, T. D. (1998). Fecundity and egg distribution of the herbivore *Spodoptera pectinicornis* as influenced by quality of the floating aquatic plant *Pistia stratiotes. Entomologia Experimentalis et Applicata*, **86**, 295–304.

Zimmels, Y., Kirzhner, F. and Malkovskaja, A. (2006). Application of *Eichhornia crassipes* and *Pistia stratiotes* for treatment of urban sewage in Israel. *Journal of Environmental Management*, **81**, 420–428.

18

Prosopis species (Leguminosae)

Rieks D. van Klinken, John H. Hoffmann,
Helmuth G. Zimmermann, and Anthony P. Roberts

18.1 Introduction

The taxon *Prosopis* (Leguminosae, mesquite) includes some of the most common tree species in the dry tropics (Pasiecznik *et al.*, 2004). Several species, all of American origin, have been intentionally distributed throughout the tropical world, because of their acclaimed roles as fast-growing, drought-tolerant, multipurpose trees. They are valued as a rehabilitation tool for degraded rangelands, shade, fodder (the pods are palatable for livestock and humans), honey, charcoal, timber, fuel, and several other resources (Fagg and Stewart, 1994; Felker and Moss, 1996; Pasiecznik *et al.*, 2001). Large-scale systematic plantings have been made in parts of Africa, several oceanic islands, the Middle East and the Indian subcontinent since the mid 1900s, but *ad hoc* introductions have been common since the early nineteenth century (Harding, 1988; Fagg and Stewart, 1994; Felker and Moss, 1996; Tewari *et al.*, 1998; Pasiecznik *et al.*, 2001; van Klinken and Campbell, 2001; Mauremootoo, 2006; Ogutu and Mauremootoo, 2006).

Introduced mesquite species have become invasive in many countries, while some species are a nuisance to humans and livestock within their native ranges (DeLoach, 1985; Dussart *et al.*, 1998). As a result, mesquite is now seen to be causing substantial negative economic, environmental, and social impacts over large parts of the world (van Klinken and Campbell, 2001; Mauremootoo, 2006; Ogutu and Mauremootoo, 2006; Zimmermann *et al.*, 2006). Direct economic impacts in the USA alone have been estimated at US$200–500m annually (DeLoach, 1985). However, indications are that the per capita impacts of invasive populations could be greater in their exotic ranges. For example, mesquite invasions in Australia contrast with invasive populations in their native range in being up to 10-fold more dense (average 4860 adult trees/ha), excluding the herbaceous layer, and failing to act as nursery plants for native shrubs and trees (van Klinken *et al.*, 2006). In Australia mesquite is one of 20 weeds of national significance (Thorpe and Lynch, 2000). In parts of the Republic of South Africa (RSA), its area has been doubling every five years with open areas becoming smothered in 10–24 years (Harding and Bate, 1991). Dense stands are estimated to use the equivalent of 1100 mm

Biological Control of Tropical Weeds using Arthropods, ed. R. Muniappan, G. V. P. Reddy, and A. Raman. Published by Cambridge University Press. © Cambridge University Press, 2009.

Fig. 18.1 Mesquite invasion in Gewane (Ethiopia) that resulted from intentional introductions made in the 1980s (Photo by Andy Kenyon, Oxfam, Australia, September 2003).

of rainfall/year (194 million m³), the equivalent to almost four times the average rainfall in the North Cape Province (Versveld *et al.*, 1998; Zimmermann *et al.*, 2006). In Kenya and Ethiopia, plantings made in the 1970s and 1980s have led to invasions that are causing substantial social impacts, threatening the livelihood of many local communities (Choge *et al.*, 2002; Mauremootoo, 2006; Ogutu and Mauremootoo, 2006) (Fig. 18.1). Following "small" introductions to the Lake Chad region in 1977, mesquite now extends over more than 300 000 ha, causing enormous problems to farmers and fishing communities (Geesing *et al.*, 2004).

Managing mesquite is expensive, and currently no techniques for controlling large-scale infestations are available, particularly the fire-tolerant species (van Klinken *et al.*, 2006). Most studies into mesquite management have been done in the USA and Australia (Schuster, 1969; Jacoby and Ansley, 1991; van Klinken and Campbell, 2001; Osmond *et al.*, 2003). Herbicides, when applied individually to trees (through basal barking or cut stumps), can cause high mortality rates, but this technique is prohibitively expensive in dense and/or extensive stands (Osmond *et al.*, 2003). Foliar herbicides are effective only against young plants (<1.5 m tall). Mesquite can be killed mechanically, provided the plants are severed below the bud zone, or approximately 30 cm below ground. This is a relatively cost-effective control technique in dense infestations, provided suitable machinery is available. Fire can be an effective management tool for fire-sensitive species such as *P. pallida* (Campbell *et al.*, 1996). However, most invasive mesquite species are fire tolerant, and high kill rates can only be achieved by intense fire regimens, which are often not feasible due to either lack of fuel or associated risks of fire to property and people. The conversion of weedy stands to agroforestry systems has been widely advocated as a management strategy, and has the advantage of generating income to the grower (Patch and Felker, 1997; Pasiecznik *et al.*, 2004; Pasiecznik, 2006). However, agroforestry is often not attractive socioeconomically (e.g. because of high labour demands), or may not always be feasible if the problem is already great (Geesing *et al.*, 2004; Zimmermann *et al.*, 2006).

Biological control, either as a stand-alone control option or, more realistically, as an integral part of a strategic management program, may offer a long-term chance for solving large-scale mesquite invasions. Although biological control has been contemplated in the USA (DeLoach, 1981) and has been attempted on Ascension Island (Cheesman, 2006), only the RSA and Australia have embarked on intensive biological control programs so far, resulting in the release of three and four agents respectively. In this chapter we provide an overview of mesquite and the current biological control efforts, and characterize the potential for further biological control.

18.2 Taxonomy and distribution of invasive mesquite species

Forty-four species within the taxon *Prosopis* are known today; 40 being native to the Americas (from western North America to Patagonia), and four to southwest Asia and Africa (Burkart, 1976; Pasiecznik *et al.*, 2001). Definitive determination of mesquite is often difficult (van Klinken and Campbell, 2001; Pasiecznik *et al.*, 2004), and therefore, resolving taxonomic issues is a high priority. At least 12 species have been introduced to other parts of the world. Of these, *P. pallida*, *P. velutina*, *P. glandulosa* (both var. *glandulosa* and var. *torreyana*), and *P. juliflora* have naturalized and become highly invasive in dry regions of Africa, Asia, and Australia (Pasiecznik *et al.*, 2004). They all belong to section Algarobia within the genus *Prosopis*, can interbreed, and hereafter are referred collectively as "mesquite." Most of these interbreeding species co-occur following human introductions, and as a consequence, introgression is common, and a wide range of invasive hybrids have resulted (Harding, 1988; van Klinken and Campbell, 2001; Pasiecznik *et al.*, 2004). Two species are invasive only in their native ranges: *P. ruscifolia* in South America and *P. farcta* in the Middle East, although these species have not been considered further in this chapter.

18.3 Biology of the mesquite

Invasive mesquite species are well adapted to the dry tropics. We will discuss their biology briefly here, with a particular focus on features that enable them to be invasive and on aspects that challenge management strategies.

Mesquite reproduces sexually (Fig. 18.2), although vegetative propagation is possible in artificial circumstances. Most reproduction occurs in summer in subtropical regions and year-round in the tropics. Each inflorescence bears 200–400 flowers, and can produce from one or two pods to several dozens. Pods are indehiscent and usually contain 15–30 seeds. Adults usually begin producing seeds when they are between two and five years

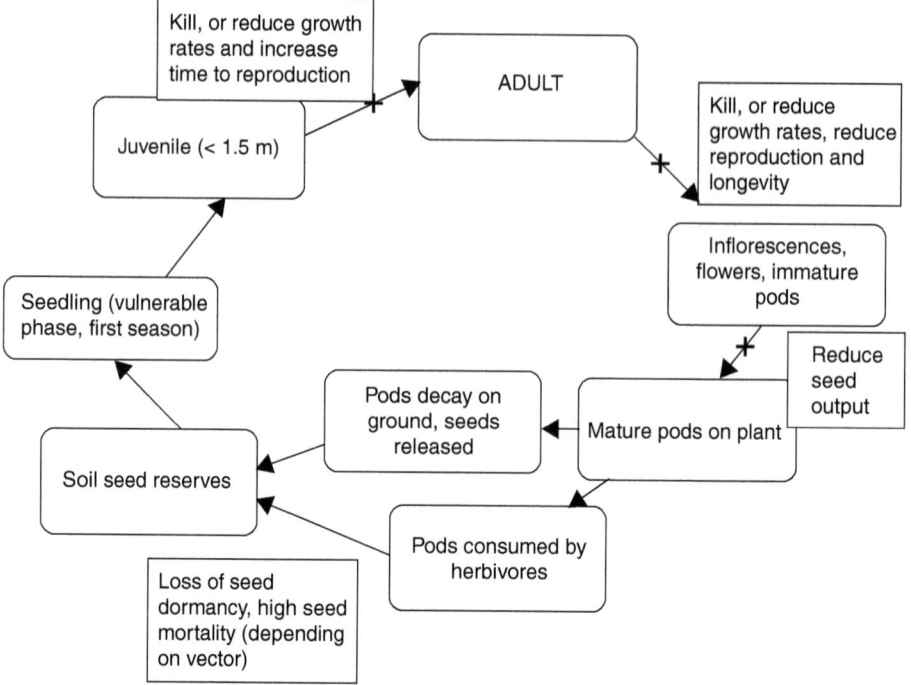

Fig. 18.2 Lifecycle of mesquite, with a cross indicating transitions where herbivory would cause the greatest population impacts.

old. Seed production increases with plant size, with large trees able to produce many millions of seeds, although this can vary considerably between years and site (Pasiecznik *et al.*, 2001), and can be very low on plants that are stressed by either prolonged drought or herbivory (R. D. van Klinken, unpublished data).

Mature pods are highly nutritious and are favored food for a range of herbivores, including livestock (e.g. cattle, horses, sheep, goats, camels), feral animals (e.g. pigs), and wildlife (e.g. macropods and emus in Australia; porcupines, antelope and warthogs in Africa) (Harding, 1991; Sawal *et al.*, 2004; Shiferaw *et al,.* 2004; Stein *et al.*, 2005). Seed survival following passage through the animal gut depends on the herbivore species and possibly the tree species, but is usually <60% (Harding, 1991; Shiferaw *et al.*, 2004), being the lowest for species that masticate their food, such as sheep and macropods. Unconsumed pods may remain on the soil surface for several months, but eventually decay and release seeds. Pods float and so can be dispersed by appropriately timed flood waters washing around the trees.

The dispersal vector has important consequences for how mesquite invades, be it water, vertebrate herbivores, or humans (Robinson *et al.*, 2008). Most seeds pass through the gut of an animal before entering the seed bank, with the pattern of deposition depending on the herbivore species. For example, pigs deposit most seeds under mesquite canopy, whereas cattle disperse them more uniformly throughout the surrounding landscape (Harding, 1988).

Mesquite seeds have hard-seeded (physical) dormancy and can be very long lived (van Klinken and Campbell, 2001), especially in dry conditions, such as during prolonged droughts. Dormancy levels depend on the history of the seed. For example, approximately 95% of seeds in fresh pods are dormant, but seeds that have been passed intact through sheep have no dormancy and germinate as soon as moisture is available (Roberts, 2006).

Optimal conditions for recruitment require high summer rainfall to germinate the seeds and good follow-up rain within several weeks to ensure high seedling survival (Fravolini *et al.*, 2005). However, significant establishment can also occur after single rainfall events, and in below-average rainfall years. Seedlings and juveniles can be very resilient to mechanical damage. Two-week-old plants survive top removal, as long as the removal is above the cotyledonary node, and one-year-old plants survive intense fires. Juveniles can remain quiescent and stunted for many years in unfavorable conditions, such as under dense canopies and during prolonged drought (van Klinken *et al.*, 2006; Roberts, 2006).

Adults are long-lived, averaging between 33 and 44 years of age in one native-range study of *P. velutina* in the USA (Archer, 1989), and trees over 100 years are common (Archer, 1989; van Klinken and Campbell, 2001). Some *P. pallida* trees are reportedly over 1000 years old in their native range in Peru (Pasiecznik *et al.*, 2004). Invasive populations are therefore still very young, with senescing adults rare in Australia (van Klinken *et al.*, 2006).

Adults are well adapted to xeric conditions (Harding and Bate, 1991; van Klinken and Campbell, 2001). Roots grow to depths of 80 m and tap groundwater (Le Maitre, 1999) while surface roots extend laterally for up to 30 m exploiting the limited moisture that accumulates during light rainfall periods (Harding, 1988; Gile *et al.*, 1997). The roots store large volumes of carbohydrate reserves, allowing plants to survive extreme environmental stress periods, repeated pruning, and many years of continuous defoliation. Other important adaptations include an ability to alter foliage properties to minimize water loss, defoliate during stressful drought conditions, utilize brackish water, photosynthesize actively even in conditions of low soil moisture, and readily coppice when damaged (van Klinken and Campbell, 2001). Foliage is almost entirely unpalatable to livestock, giving mesquite another competitive advantage over other, more palatable species.

Only a few comparative studies have been conducted on the ecology and management of invasive mesquite species and their hybrids. However, key differences between some of the species exist (Pasiecznik *et al.*, 2001). For example, the tree-like *P. pallida* is intolerant to fire and to the use of heavy knock-down chains (Osmond *et al.*, 2003). This contrasts with species such as *P. velutina* and *P. glandulosa*, which are fire tolerant and generate root and stem suckers readily when felled. Mesquite species also require varying climate and edaphic conditions, as reflected in their native-range distributions (Simpson, 1977).

The causes of mesquite invasions both in its native and introduced range continue to be debated, but are likely to be multiple (Archer *et al.*, 1988; Geesing *et al.*, 2004; Shiferaw *et al.*, 2004; van Klinken *et al.*, 2006). Hypotheses include: more effective seed dispersal following the introduction of livestock; overgrazing resulting in less competition from grasses and/or less combustible organic material for fires; increased atmospheric CO_2 favoring C4 species; and fixation of nitrogen by mesquite giving it a competitive advantage over grasses in low-nitrogen soils. Release from natural enemies is another explanation for the much higher seedling and juvenile densities observed under parent plants in Australia (van Klinken *et al.*, 2006), but studies have not been conducted to confirm this or other potential population-level effects from natural enemies.

18.4 Challenges for biological control

18.4.1 Criteria for a successful mesquite biological control program

The criteria against which a successful biological control program can be judged depend on the ecological and management context in which the weed occurs, and the negative impacts that biological control is seeking to address (van Klinken and Raghu, 2006). For mesquite, the key issues are: the social, economic, and environmental impacts caused by high densities of plants, an ability to invade new environments, rapid rates of increase, and the cost and/or practicality of existing management options (chemical, mechanical, fire, or utilization), especially when managing extensive, dense infestations. Consequently we propose here two possible levels of success that a mesquite biological control program could achieve.

Complete success: Canopy cover is reduced to, or prevented from reaching, 10% (i.e. about 100 medium-sized trees/ha) by biological control agents, and the long-term removal of all adult mesquite plants is feasible using other management strategies.
Partial success: Canopy cover target of 10% can be realistically met through a combination of biological control and other control techniques.

A 10% cover target was set because higher densities are expected to cause negative impacts, such as competition with fodder, erosion, and impeding access, but would still provide a substantial standing biomass for harvesting plant products. Also, management options are available for removing standing trees at low densities (i.e. <10% cover), but biological control could assist by restricting the volume and rate of regrowth and recruitment, and thereby reducing the need for expensive follow-up treatments.

Successful or partly successful biological control, as determined here, will be more difficult to achieve in regions where communities, such as nomadic and subsistence farmers in Africa, do not have easy access to heavy machinery and herbicides to implement complementary control methods against invasive populations.

18.4.2 Predicting efficacy

Although natural enemies attack all mesquite parts (Fig. 18.2), the successful outcome of a biological control program will depend on researchers being able to select agents that inflict the kind of damage that is needed to give a desired outcome (Raghu *et al.*, 2006).

Population models can assist in addressing this question (Buckley *et al.*, 2004), but suitable models have not yet been developed for mesquite (Golubov *et al.*, 1999). We therefore develop predictions based on the known biology of mesquite, and lessons learned from other woody weeds.

We predict that damage to three life stages will achieve the greatest impact: adults, reproductive parts prior to seed maturation, and established juvenile plants (Fig. 18.2). Biological control agents that attack mature seeds are unlikely to be effective on their own, given the short period that mature pods are available prior to consumption and dispersal by vertebrate vectors (Moran *et al.*, 1993; Impson *et al.*, 1999).

A reduction in mesquite canopy cover to 10% within a reasonable time frame will require agents that kill mature plants, given their natural longevity and ability to survive repeated top-killing from a young age and high levels of defoliation over several years (R. van Klinken, unpublished data). Candidate biological control agents might include either stem-gall inducers that act as resource sinks (Dennill, 1988; Dorchin *et al.*, 2006) or microbial pathogens that could attack the root system (Diplock *et al.*, 2006). However, to be effective, these agents would almost certainly need to inflict high levels of sustained damage over many years to be effective.

Preventing canopy cover from increasing above 10% will require juvenile mortality and/or recruitment being either slowed down or stopped completely. The mesquite life stage most vulnerable to herbivory is likely to be newly established juveniles (0.3–1.5 m height), based on population modeling of similar woody shrubs (Kriticos, 2003; Buckley *et al.*, 2004). However, juvenile mesquite plants are robust, and like adults will probably require sustained damage to cause mortality. Effective candidate biological control agents that could cause substantial mortalities include stem borers, especially if they tunnel into the root crown.

A dramatic reduction in viable seed production is most likely to be achieved indirectly by natural enemies that stress adult plants so that either they flower less or they are induced to abscise immature pods. A wide range of arthropod herbivores such as gall inducers, defoliators, stem borers, girdlers, and microbial pathogens can inflict prolonged and extensive stress. Alternatively, herbivores that can feed and damage green pods extensively could be effective, provided their life cycles are well synchronized with plant reproduction.

The population impacts outlined in preceding paragraphs, that is, adult and juvenile mortality and reduction in seed production, will make other management options more cost effective, especially by reducing spread rates, rates of increase, and the need for follow-up control work. Additional benefits would be increased time to reproduction from germination and following top-kill (such as from mechanical control or fire) and increased susceptibility to other control techniques (such as fire and greater competition from the herbaceous layer). However, biological control can also make other control options less effective. For example, trees stressed by a defoliating moth (*Evippe* sp. #1, Lepidoptera: Gelechiidae) in Western Australia are more difficult to kill with herbicides (R. Climas, personal communication), although the net impact of the moth is over-whelmingly positive (Anderson *et al.*, 2006).

Candidate natural enemies must be adapted to the target environment if they are to become established, and reach sufficient densities at the right time and duration to cause impacts (Zalucki and van Klinken, 2006). Predicting which candidate agent will do the best in a particular environment is critical, particularly for a target such as mesquite, which occurs in diverse environments, in both its native and introduced contexts (van Klinken *et al.*, 2003; Roberts, 2006). In Australia, for example, the leaf-tying moth (*Evippe* sp. #1) has only reached damaging densities in some climatic regions, and therefore other agents are required that are better adapted to the remaining invaded regions.

18.4.3 Host-specificity requirements

Biological control agents for mesquite must ideally be sufficiently general to perform well on all invasive mesquite species, including their hybrids. Available evidence indicates that natural enemies generally cannot distinguish between species within the section Algarobia, which includes all highly invasive species. For example, the five biological control agents that have already been released against mesquite perform similarly on all tested *Prosopis* species (section Algarobia) (Zimmermann, 1991; van Klinken and Heard, 2000; van Klinken, 1999) and *Evippe* sp. #1 released in Australia is causing substantial impacts on mesquite species other than its native hosts.

The nontarget plants most at risk are *Prosopis* species native to tropical Africa and Asia (Pasiecznik and Felker, 2006) and other genera within the Tribe Mimoseae. Four *Prosopis* species are native to tropical Africa and Asia. They belong to sections Prosopis (*P. cineraria, P. farcta,* and *P. koelziana*) and Anonychium (*P. africana*) (Burkart, 1976), although *P. africana* may more properly belong in a separate genus (Pasiecznik *et al.*, 2004). Both *P. africana* and *P. cineraria* are economically important resources and under some threat from overexploitation and land-use changes (Pasiecznik and Felker, 2006). Genetic analyses suggest that these native Afro-Asian species are distinct and are distantly related to the exotic invasive species in section Algarobia (Pasiecznik *et al.*, 2006). Other genera in the tribe Mimoseae are well represented in Africa, but there are only nine species native to Australia (Lewis and Elias, 1981).

Available evidence suggests that natural enemies may commonly differentiate between Algarobia (including the invasive species) and mesquite species that are native to Africa and Asia, and potential agents that are sufficiently host specific for release in Africa and Asia may therefore be readily available. For example, in Yemen, where the seed feeder *Algarobius prosopis* (Le Conte) (Coleoptera: Bruchidae) was first recorded in 1987, probably after being inadvertently introduced with mesquite seeds (Delobel and Fédière, 2002), no crossover in seed-feeding insects between the native and exotic mesquite species has been reported. *Algarobius prosopis* is inflicting extensive damage to exotic mesquite, but has not been recorded on its native congener, *P. cinerea,* growing in close proximity, while two native species, *Bruchidius andrewesi* Pic. and *Caryedon serratus* (Olivier) (Coleoptera: Bruchidae) (Grobbelear E., ID 2005–213, National Collection of

Insects, Pretoria, South Africa), have been recorded only from *P. cineraria*. A surprisingly small number of the many herbivorous insect species associated with Mimoseae in Africa are able to utilize introduced *Prosopis* species as host plants (van Tonder, 1985; F. A. C. Impson, personal communication). For example, in South Africa, where substantial observations have been made, most native insects that are encountered on the introduced mesquite species are transient and scarce.

The dispersal ability of potential biological control agents needs to be considered when identifying plant species to be included in host-specificity tests. At least two of the existing mesquite biological control agents, *A. prosopis* and *Evippe* sp. #1, are excellent dispersers. Host-test lists should therefore be based on biogeography rather than country; that is, for Australia, southern Africa, and North and East Africa through to Asia.

18.4.4 *Conflict between benefits and costs*

The use of biological control to manage mesquite needs to be placed within the context of a broader long-term strategy for its sustainable management. This has already been done in Australia where mesquite is not utilized and hence no constraints on biological control exist (ARMCANZ, 2001). The situation is complex in Africa and Asia where mesquite simultaneously has negative impacts on the environment and agriculture, while providing multiple benefits to people (Pasieznik *et al.*, 2001; Pasiecznik, 2006). Except South Africa, a general sense of reticence in using biological control in Africa and Asia exists for the fear that the useful properties of the plant will be diminished. Even in South Africa, the biological control program is limited to the use of agents that can destroy seeds, while not affecting either the nutritional value of mature pods or any other positive attributes of the plant (Zimmermann, 1991).

The challenge is to identify and capitalize on the roles that all available management tools can play in the sustainable management of mesquite, including utilization and biological control. In addition, the potential substitution of mesquite with other non-invasive agroforestry species, or industries, also needs to be considered. Deciding how to manage invasive populations becomes especially problematic when the needs of local communities are superimposed on the needs of societies at a national or even regional scale (e.g. southern Africa) and when the requirements of the communities change with time. For example, benefits are likely to decrease in situations where mesquite becomes increasingly invasive or where the genetic integrity of exploitable, noninvasive species such as *P. pallida* (Pasiecznik, 2006) is altered through hybridization (van Klinken and Campbell, 2001; van Klinken *et al.*, 2006). Biological control using seed feeders is least likely to be effective in these instances, where mesquite is already in advanced stages of invasion.

In spite of their apparent incompatibility, effective utilization and biological control are the two available options for managing invasive mesquite populations in Africa and Asia (Pasiecznik *et al.*, 2004, Zimmermann and Pasiecznik, 2005; Pasiecznik, 2006; Zimmermann *et al.*, 2006). Several projects have recently been initiated (e.g. in Chad,

Yemen and Kenya) by the FAO and HDRA to apply improved utilization as a management tool (N. Pasiecznik, personal communication), and has resulted in the imminent release of the seed-feeding biological control agent, *A. prosopis*, in Kenya to be placed on hold until at least May 2009. The initial focus of the utilization program is pods for livestock feed and human dietary foods (Pasiecznik *et al.* 2006), which is expected to have the added benefit of reducing the risk of future recruitment (one tonne of pods is worth almost US$100 to the collectors and contains approximately 2 million seeds). Management through utilization is most likely to be successful at a local scale (Zimmermann *et al.*, 2006; Mauremootoo, 2006), with the possibilities of a knock-on effect regionally, should appropriate markets develop. However, many invasive plant populations are either difficult or uneconomical to manage, being impenetrable thickets of thorny, multistemmed shrubs in which harvestable material is limited and can only be obtained by processing large volumes of waste. Also, infestations frequently extend into remote, inhospitable areas where labor shortages, transport difficulties, and limited markets prevail.

If mesquite is to be managed sustainably, an unrestricted biological control program relevant to the invasive mesquite species is necessary wherever it cannot be managed adequately through utilization. Reluctance to adopt biological control in Africa and Asia appears to be based on a cautious attitude. Even if a suite of the most damaging agents is introduced, the weed is likely to remain common and widespread, and, at worst, slightly less valuable as a resource. The invasive shrub *Mimosa pigra* L., a comparable perennial weedy tree with several damaging agents, is a good example of where this has happened (Buckley *et al.*, 2004). Another example is the biological control of *Opuntia ficus-indica* (L.) Miller (Cactaceae) in South Africa where the plant remains sufficiently abundant to be exploited by local communities as a source of fruit and emergency fodder (Zimmermann and Moran, 1991) while being under excellent biological control (Moran and Zimmermann, 1984; Shackleton *et al.*, 2007), although the latter study also found that most local people would now prefer more opuntia in their neighborhoods, and some are propagating it, even when plants have to be protected against the introduced biological control agents. Biological control agents that feed on vegetative parts could therefore be considered, even where mesquite is used as a resource (perhaps other than for pods and honey). However, consensus would be required at a regional level (e.g. southern, northern and eastern Africa and Asia) if this were to happen, given the likely dispersal ability of biological control agents.

A move in the direction of promoting biological control was initiated recently in the RSA when the guidelines for biological control of mesquite were widened to include consideration of potential agents that attack flower buds, flowers, and immature pods, which was taken unanimously by 50 stakeholders including conservationists, agriculturalists, legislators, and local landowners in November 2001. This meeting decided that the potential benefits of biological control outweighed the potential damage to seed pods in South Africa (Zimmermann *et al.*, 2006). The biological control program subsequently expanded, albeit with restrictions on agents that might damage the vegetative parts of mesquite. However, a region-wide consultation process has not yet been conducted.

18.5 Natural enemies

18.5.1 Introduced range

A wide range of generalist natural enemies are known to occur on mesquite in its introduced range, and some of them inflict substantial damage (Tewari *et al.*, 1998; van Klinken and Campbell, 2001). For example, *Oxyrachis tarandus* Fabricius (Hemiptera: Membracidae) has been reported to kill trees; *Taragama siva* Lefevre (Lepidoptera: Lasiocampidae) reported to defoliate plants completely; and *Poekilocerus pictus* Fabricius (Orthoptera: Acrididae) to skeletonize plants (Yousuf and Gaur, 1998). In Australia, an unnamed cerambycid borer kills *P. pallida* seedlings in north Queensland (S. D. Campbell, unpublished data, 2000), stem-boring moths readily kill juvenile plants in pots, and the crusader bug *Mictis profana* (F.) (Hemiptera: Coreidae) attacks immature foliage and immature reproductive organs. The crusader bug also attacks a range of other plant species, and probably reduces growth rate and seed production on *M. pigra* in Northern Territory, Australia (Flanagan, 1994).

The management benefits provided by these native, natural enemies are probably limited, as they are generally uncommon and patchily distributed in time and space. Furthermore, most or all are probably not sufficiently host specific for introduction into other regions. Also, we are not aware of any examples where the impacts of native natural enemies have successfully been enhanced, such as through augmentative release or habitat manipulation.

Surveys of diseased mesquite material in the RSA and Namibia reported 151 taxa of fungi and these were subsequently screened for potential use as mycoherbicides (Zimmermann *et al.*, 2006). Although several of these taxa showed promise, they failed extensive efficacy tests. Commercially existing mycoherbicides were also tested, but none showed potential for mesquite control (Zimmermann *et al.*, 2006).

Herbivorous vertebrates can help in the management of some invasive woody shrubs, mainly through browsing, but mesquite foliage is generally unpalatable to all livestock.

18.5.2 Native range

Biological control candidates that evolved on mesquite species (section Algarobia) that are not being targeted for use as biological control agents have been shown to be effective on target species within this section. Natural enemies of mesquite species native to the Americas are relatively well documented. Over 945 phytophagous insect species which attack all parts of mesquite are recorded (Swenson, 1969; Ward *et al.*, 1977; Johnson, 1983; Cordo and DeLoach, 1987; Silva, 1988; Pasiecznik *et al.*, 2001). In South America, systematic surveys have been conducted in Argentina and Paraguay (Cordo and DeLoach, 1987). Surveys in North America have been rather ad hoc, but results have been summarized (Ward *et al.*, 1977). Many of the reported natural enemies are unsuitable as biological control agents as they are known to have wide field host ranges, are transients on mesquite, or their feeding damage is unlikely to have any significant impact. Several of the most damaging insects in North America have been studied, but most of these are unlikely to be sufficiently host specific (De Loach, 1982, 1983 a,b, 1994; Cuda *et al.*, 1990; DeLoach and Cuda, 1994; Cuda and DeLoach, 1998).

Twenty-seven fungal isolates were isolated from green pods in North and South America, of which nine showed promise for classical biological control (Zimmermann *et al.*, 2006). However, none was found suitable following extensive testing in 2000–2002, although one isolate warranted further study as a potential mycoherbicide.

Although extensive, the cataloged list of natural enemies from the Americas is unlikely to be comprehensive. Much of the distribution of mesquite remains poorly explored, particularly in northern South America (Peru, Ecuador, Colombia, and Venezuela), Central America, and the Caribbean. Also, to our knowledge, either no or extremely limited systematic searches for pathogens, stem-borers, root-feeders, and natural enemies that preferentially attack immature mesquite plants exist.

18.6 Biological control practiced currently

Interest in mesquite biological control first began in the USA where native mesquite species are highly invasive (DeLoach, 1988). It resulted in surveys conducted by the United States Department of Agriculture in North America and later in South America (Cordo and DeLoach, 1987). However, no subsequent work was done because of concerns regarding the use of classical biological control to target native species. Biological control programs were then initiated in South Africa in 1985 (Zimmermann, 1991; Moran *et al.*, 1993; Impson *et al.*, 1999) and in Australia in the early 1990s (van Klinken and Campbell, 2001). More recently Kenya has conducted host-specificity testing on the seed-feeding beetle *A. prosopis*. The results confirm that the beetles are sufficiently specific and permission is being sought for their release in Kenya (W. Ogutu, personal communication).

18.6.1 Biological control agents that have been considered

A wide range of potential biological control agents have been tested for host specificity (Table 18.1). Testing conducted by South Africa was restricted to natural enemies that impacted on flowers or pods. Selection of new agents in Australia was not constrained by social conflict, and they were short-listed on the basis of abundance and type of damage caused in their native range (T. A. Heard, personal communication).

The sap-sucking bug *Mozena obtusa* Uhler (Hemiptera: Coreidae), did not appear to be sufficiently host specific for release in Australia. Moreover, results suggested that inter-actions between host suitability and plant nitrogen could influence the degree of nontarget attack (van Klinken, 1999). *Oncideres rhodosticta* Bates (Coleoptera: Cerambycidae), a stem borer and girdler, could not be cultured successfully. Field-collected adults fed on nontarget plants, but their performance was relatively poor (R. van Klinken, unpublished data). *Heteropsylla flava* Crawford (Hemiptera: Pysllidae) did oviposit and complete development on a small number of nontarget hosts, but performance was relatively poor

Table 18.1 *Natural enemies that have been imported into quarantine for host-specificity testing, and the outcome of the work*

Natural enemy	Guild (origin)	Testing: year (agency)	Outcome[a]
Coleoptera: Bruchidae			
Algarobius prosopis (Le Conte)	Mature seeds (USA)	1985–87 (PPRI), 1994–96 (QNRM)	Released and established (1987: South Africa; 1996: Australia; 1997: Ascension Island); Permission to release being sought in Kenya (1,3,8)
Algarobius bottimeri Kingsolver	Mature seeds (USA)	1985–90(PPRI), 1994–97 (QNRM)	Released and initially established (1990: South Africa; 1997: Australia), but not recovered for several years (South Africa) (3,8)
Neltumius arizonensis (Schaeffer)	Mature seeds (USA)	1992–94 (PPRI)	Released and established (1993: South Africa). Released, probably not established (1997: Ascension Island) (1,3)
Coleoptera: Cerambycidae			
Onciders rhodosticta Bates	Stem borer (USA)	1996–98 (CSIRO)	Testing terminated, adults may not be sufficiently host-specific (8)
Coleoptera: Curculionidae			
Coelocephalapion gandolfoi Kissinger	Green pods (Argentina)	2001–06 (ARS Argentina/ PPRI)	Host-specific to Algarobia section. Release application submitted and being assessed (9)
Hemiptera: Coreidae			
Mozena obtusa Uhler	Sap sucker (North America)	1996 (CSIRO)	Not sufficiently host-specific for Australia (4)
Hemiptera: Psyllidae			
Heteropsylla texana Crawford	Sap sucker (North America)	1991 (QNRM)	May be sufficiently host-specific for release in Australia, but permit to release never applied for (2)
Prosopidopsylla flava Burckhardt	Sap sucker (Argentina)	1996–98 (CSIRO)	Released, tenuously established (1998: Australia) (5,6)
Lepidoptera: Gelechiidae			
Evippe sp. #1	Leaf tier (Argentina)	1996–98 (CSIRO)	Released and established (1998: Australia) (6,7)
Evippe sp. #1	Leaf tier (Argentina)	1996–98 (CSIRO)	Cultured, but not tested (8)

Table 18.1 (cont.)

Natural enemy	Guild (origin)	Testing: year (agency)	Outcome[a]
Diptera: Cecidomyiidae			
Asphondylia prosopidis Cockerell	Flower bud galler (USA)	2005–(PPRI)	Host-testing commenced (see text)
Fungal pathogens			
27 Fungal isolates	Green pods	1994–2003 (PPRI)	None considered suitable following extensive testing (see text)

[a] References: 1 Cheesman, 2006; 2 Donnelly, 2002; 3 Roberts, 2006; 4 van Klinken, 1999; 5 van Klinken, 2000; 6 van Klinken *et al.*, 2003; 7 van Klinken and Heard, 2000; 8 van Klinken, unpublished data; 9 A. Witt, pers. comm. 2007.

(Donnelly, 2002). Further testing is required to determine whether it was safe for release in Australia, and elsewhere.

Coelocephalapion gandolfi Kissinger (Coleoptera: Curculionidae) and *Asphondylia prosopidis* Cockerell (Diptera: Cecidoymiidae) are under consideration by South Africa as part of an effort to introduce insects that destroy the reproductive structures that are still attached to the trees and are not yet utilized by vertebrate herbivores. *Coelocephalapion gandolfoi* damages immature seeds in green pods that are still attached to plants. Although *C. gandolfoi* originates from South America, it looks promising because it only develops on pods of species within the Algarobia section, including the two main invasive species in South Africa, *P. velutina* and *P. glandulosa* var. *torreyana* (A. Witt, personal communication). Submission of a request for clearance for release of this agent in South Africa is pending. The gall-inducing cecidomyiid *Asphondylia prosopidis* is currently being investigated in the USA. *Asphondylia prosopidis* induces galls on the inflorescences of *Prosopis* species, completing five generations through summer and overwintering as first-instar larvae in galls (Beuhler, 2006). This work is at an early stage, focusing on developing techniques for manipulating the midges so that host-specificity tests can be done.

18.6.2 Ecology and impacts of biological control agents currently in use

Five biological control agents have been released so far, three in the RSA and four in Australia, including two species in both countries. Although all established, the current status of two of these, *Prosopidopsylla flava* Burckhardt (Hemiptera: Psyllidae) (in Australia) and *Algarobius bottimeri* Kingsolver (Coloptera: Bruchidae) (in South Africa and Australia) is questionable (Table 18.1). All of the five species are multivoltine, and all performed equally well in laboratory tests on *Prosopis* species within the section Algarobia (at least in the laboratory). Postrelease evaluation of abundance and impacts has been done on each species.

Seed feeders (Bruchidae)

Three seed-feeding bruchids from North America have been released in South Africa, Australia and/or Ascension Island (British overseas territory) (Table 18.1). All three bruchid species are easily mass reared and distributed. In addition, *A. prosopis* is currently present in Egypt (first recorded in 2001), Saudi Arabia (1980), Dubai (1983), and Yemen (1987) (Delobel and Fédière, 2002). It was probably accidentally introduced into the region with mesquite seeds introduced for reforestation (Delobel and Fédière, 2002). It could also be present in Chad as the pods there are heavily infested by an as yet unidentified bruchid beetle (Geesing *et al.*, 2004).

All three species are specific to mature seeds, but they differ in their oviposition habits and in the behavior of the neonate larvae. The two *Algarobius* species (*A. prosopis* and *A. bottimeri*) have similar behavior and deposit clusters of eggs in cracks, holes, or damaged ends of broken mature seeds pods. *Algarobius prosopis* also lays eggs on, and subsequently emerge from, dung pellets of sheep that have been feeding on mesquite and adult beetles have been recorded emerging from donkey dung (Roberts, 2006), cattle dung, and emu scats (R. van Klinken, unpublished data, 2006). The newly hatched larvae are mobile, burrowing through pods or dung in search of seeds and moving considerable distances over the substrate when suitable seeds are unavailable in the immediate vicinity. In contrast, *Neltumius arizonensis* (Schaeffer) (Coleoptera: Bruchidae) females only lay eggs on pristine portions of mature seed pods (Strathie, 1995). Each egg is glued in place and the hatching larva tunnels directly into the adjacent seed within the pod. Once inside seeds, the larvae of all three species take approximately a month to complete their development and adults emerge following pupation in the hollowed-out seed.

Algarobius prosopis appears to be the only bruchid species that has become abundant. In South Africa, *A. bottimeri* was recovered in low numbers at only two release sites for three years. Although widely released in Australia, it has not been recorded during recent surveys (R. van Klinken, unpublished data, 2005–7). Its failure to survive in South Africa and Australia remains unexplained, but is unlikely to be the result of parasitism (Hoffmann *et al.*, 1993). *Neltumius arizonensis* (Schaeffer) (Coleoptera: Bruchidae) is widely established in South Africa, but is responsible for only about 1% of the seed damage caused by bruchids (Roberts, 2006). High levels of parasitism suppress *N. arizonensis* numbers (Coetzer and Hoffmann, 1997; Roberts, 2006). Moreover, in most instances the abscised mature seed pods are eaten almost immediately by domestic and wild ungulates and most of the undigested seeds remain inaccessible to *N. arizonensis* within dung pellets. In contrast, *A. prosopis* is able to locate and oviposit in seed-containing dung at great distances (>170 m) from mesquite trees. This behavior enables *A. prosopis* to exploit more of the common resource and may account for the numerical dominance of this species within the system.

In South Africa, seed destruction by *A. prosopis* increases with time both in pods and in sheep dung, frequently reaching up to >90% levels within 10–12 months after pod maturation, provided either the pods are not consumed or the seeds in dung do not germinate (Roberts, 2006). Late-season rains in more arid regions, where pods are mostly consumed by sheep, enable *A. prosopis* to utilize and regularly destroy almost all dung-borne seeds (Roberts, 2006). In Australia, mesquite occurs in regions now occupied by cattle. Predation of seeds in pods is low, and emergence from seed-containing cattle dung and emu scats is rare (R. van Klinken, unpublished data, 2005–7). On Ascension Island, *A. prosopis* is well established, but considered unimportant (Cheesman, 2006) and mechanical removal of plants has been prioritized to protect turtle breeding areas (Broderick et al., 2002).

Potential for further use: Algarobius prosopis has the potential to achieve high levels of seed predation, especially of postdispersal seeds in areas of low rainfall. Where conditions are right, seed predation by *A. prosopis* may constrain the rates at which the weed spreads and at which dense thickets form, providing time for other management procedures. *Neltumius arizonensis* is a less promising agent but other habitats and climatic zones probably exist where this species would thrive, although it is likely to be highly susceptible to egg parasitoids in Australia because it does not conceal its eggs (van Klinken and Flack, 2008).

Sap-sucker (Prosopidopsylla flava)

Prosopidopsylla flava is a highly host-specific psyllid native to Argentina that is intimately linked with its host throughout its life cycle (van Klinken, 2000). Eggs are inserted into plant tissue. Nymphs are free living but relatively sedentary, and do not generate a lerp. Adults are relatively short lived (average 20 days), highly fecund (average 232 eggs/female) and will die within several days in the absence of their host (mesquite species). Generation time in the laboratory is approximately 46 days (van Klinken, 2000). There is no evidence of diapause, at least not in subtropical Brisbane with monthly winter temperatures of 9.5–20.6 °C, although developmental times are slowed considerably in cool conditions. Nymphs and adults suck sap of mature foliage, and nymphs also feed between overlapping pinnules of immature leaves causing characteristic distortion of growing tips.

A large-scale release program was conducted across Australia, with almost 183 000 adults released over a 21-month period from October 1998 (van Klinken *et al.*, 2003). However, *P. flava* only tenuously established in one region (northwestern New South Wales) by 2001, but it has not been surveyed since. The most likely explanation for its failure to establish were climatic factors (establishment was successful in summer-rainfall, semi-arid regions with cool winters and warm–hot summers), predation by ants, or laboratory inbreeding.

Potential for further use: The psyllid (*P. flava*) has the potential to inflict considerable damage, and it has been reported to do so in its native range. Unfortunately the reasons why it failed to establish in Australia are unknown. It may warrant rerelease in Australia, and new releases elsewhere in the world, with further attention on predictive climate modeling, use of fresh genetic material, and exclusion of potential predators during the establishment phase.

Leaf feeders (Evippe *sp. #1*)

Evippe sp. #1 is an undescribed species from Argentina. Evippe sp. #1 oviposits in cracks and fissures on the plant (van Klinken and Heard, 2000). It has four larval instars. The first instar mines the leaf before exiting and constructing a leaf tie involving 2–10 leaflets. Each successive instar constructs a new, but larger leaf tie (involving up to 20 leaflets), before pupating in the final leaf tie. They prefer fully formed leaflets, although will use immature foliage, if required. Adults live up to 20 days in the laboratory, but most eggs are oviposited in the first week. Development takes between 34 and 48 days in the laboratory. It enters diapause within the leaf tie as a fourth instar when days shorten and exit when days lengthen (*c.* April–July in the southern hemisphere). Limited larval development occurs on *Leucaena leucocephala* (Lamk) de Wit (also in tribe Mimoseae) in laboratory tests, with only one-fifth the number of leaf mines when compared with mesquite, small leaf mines and leaf ties, and no adult emergence (van Klinken and Heard, 2000).

This species established easily in most regions across Australia. Rates of increase were greatest at sites with warm winters and hot summers, possibly allowing a greater number of generations per year. *Evippe* sp. # 1 is an excellent disperser. It spread 1.3–3.6 km/year following release (almost certainly an underestimate), *c.* 115 km from one release within three years (van Klinken *et al.*, 2003), and over 1300 km between isolated mesquite populations, presumably by wind (R. van Klinken, unpublished observations). It has attracted a diverse range of larval and pupal parasitoids in Australia; however, parasitism rates remain low, averaging 1.8% (van Klinken and Burwell, 2005). Although it performs equally well on all tested mesquite species in the laboratory, field observations in Australia suggest that it may not perform as well on *P. pallida* (R. van Klinken, unpublished observations).

Extensive postrelease evaluations have been conducted in the Pilbara region of Western Australia since its release there in 1998 (van Klinken *et al.*, 2003; Anderson *et al.*, 2006; R. van Klinken unpublished data). Over 90% of leaves were tied within approximately 12 months following release. High levels of defoliation have continued since (Fig. 18.3), generally lagging a few months after reflush of foliage (Fig. 18.4). Limited seed set occurred throughout the period, although this could be partly due to a prolonged drought between 2000 and 2004; reflushing following rainfall events was limited, and adult mortality began to occur after approximately eight years of prolonged defoliation.

Potential for further use: Of the existing biological control agents, the leaf-tying moth (*Evippe* sp. #1) offers the greatest potential. It has been introduced throughout Australia, but performs best where conditions are warm to hot all year round (mean average monthly temperature 19.8–30.1 °C) such as the Pilbara region of Western Australia. Although parasitized by a diverse range of parasitoids in Australia, parasitism rates remain low (van Klinken and Burwell, 2005), and the same could be true elsewhere. It maintains exceptionally high densities in the Pilbara region, resulting in greatly reduced growth rates and seed production. It does not appear to perform as well on *P. pallida*, although this requires

Fig. 18.3 Defoliation caused by the leaf-tying moth (Gelechiidae: *Evippe* sp. #1) at the largest mesquite infestation in Australia (the Pilbara Region of Western Australia) (Photo by R. D. van Klinken, November 2004).

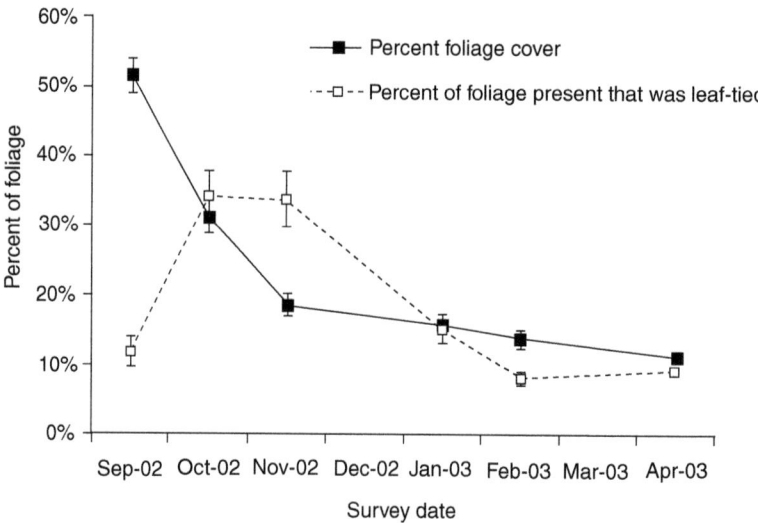

Fig. 18.4 Average ± SE foliage cover (%), and the proportion of that foliage cover that was leaf-tied by *Evippe* sp. #1, along a single 2-km transect ($n = 60$ adult trees) in the Pilbara Region of Western Australia (R. D. van Klinken, unpublished data). Data is shown for a full growing season, four years after the release of the biological control agent. Trees are relatively well foliated in September, prior to postdiapause increase in leaf-tie activity.

confirmation, and tree mortality remains limited even after prolonged defoliation. It would, however, compromise the use of mesquite for human and animal food, most notably pod products and honey.

18.7 The potential for new biological control measures

Mesquite is causing dramatic negative economic, social, and environmental impacts despite still being in early phases of invasion in most parts of the world. Effective biological control agents may be an important element of a management strategy to address the mesquite problem in the long term, especially given the geographical scale at which impacts are and will be occurring. However, the social conflicts in Africa and Asia between those who value it as a resource and those who view its impacts negatively must be addressed if the most effective agents are to be made available there. The social conflict needs to be reassessed to determine under what circumstances existing management approaches will remain adequate, and what the long-term effects will be in terms of socioeconomic changes (e.g. changes that might result in increased or decreased reliance on mesquite as a fuel), continued invasions, and possible hybridization between useful (especially *P. pallida*) and problem mesquite species. In areas where trade-offs between benefits and costs is successfully met through utilization, the phasing in of noninvasive shrubs or trees may also need to be considered, so that it does not prevent effective management in other areas where biological control is the primary option. However, even in areas where mesquite is controlled through utilization, natural regeneration may still exceed utilization, and a broader suite of biological control agents may be acceptable.

The future of the biological control program in South Africa will be guided by the outcome of the cost-benefit studies that are being planned. Recent evidence that mesquite is causing death of a keystone species in the Kalahari desert, *Acacia erioloba*, is pushing sentiments towards more drastic control actions against mesquite (S. Woodborne *et al.*, CSIR unpublished data).

Postrelease evaluation of the leaf-tying moth provides confidence that there are other potential biological control agents available that can achieve successful control. However, to be effective, biological control agents will need to reach high densities for prolonged periods, probably more so than is generally reported in their native range.

Considerable work is still required to understand the ecology of mesquite better across its introduced range, identify what type of damage mesquite is likely to be most vulnerable to, better categorize the natural enemies of mesquite occurring throughout the native range, especially guilds that are most likely to be effective, and conduct further studies on efficacy and host specificity of the prioritized shortlist. This will include collating and synthesizing available data on the natural enemies of mesquite across its native range. Further, targeted surveys of natural enemies are probably also required, focusing on climatically suitable regions, and taxa such as pathogens and natural enemies on juvenile plants that have been largely overlooked. Greater international collaborative efforts for biological control, and management generally, will result in substantial benefits, as will further discussion with all stakeholders concerned in the impacts of mesquite invasion throughout the world.

Acknowledgments

We thank Nick Pasiecznik and Celine Clech-Goods for valuable comments on an earlier draft of this manuscript.

References

Anderson, L. J., van Klinken, R. D., Parr, R. J., Climas, R. and Barton, D. (2006). Integrated management of hybrid mesquite: a collaborative fight against one of Australia's worst woody weeds. In *Fifteenth Australian Weeds Conference, Papers and Proceedings*, ed. C. Preston, J. H. Watts and N. D. Crossman. Adelaide, South Australia: Weed Management Society of South Australia, pp. 239–242.

Archer, S. (1989). Have southern Texas savannas been converted to woodlands in recent history? *American Naturalist*, **134**, 545–561.

Archer, S., Scifres, C. and Bassham, C. R. (1988). Autogenic succession in a subtropical savanna: conversion of grassland to thorn woodland. *Ecological Monographs*, **58**, 111–127.

ARMCANZ (2001). *Agriculture and Resource Management Council of Australia and New Zealand Environment and Conservation Council and Forestry Ministers Weeds of National Significance Mesquite (Prosopis species) Strategic Plan.* Launceston, Tasmania: National Weeds Strategy Executive Committee, 25 pp.

Beuhler, H. (2006). *Asphondylia prosopidis: Summary Report for 2006.* Report on file. Pretoria, South Africa: Plant Protection Research Institute.

Broderick, A. C., Glen, F., Godley, B. J. and Hays, G. C. (2002). *A Management Plan for the Marine Turtles of Ascension Island.* University of Wales Swansea, UK: Marine Turtle Research Group, 33 pp.

Buckley, Y. M., Rees, M. and Paynter, Q. (2004). Modelling integrated weed management of an Invasive shrub in tropical Australia. *Journal of Applied Ecology*, **41**, 547–560.

Burkart, A. (1976). A monograph of the genus *Prosopis* (Leguminosae subf. Mimosoideae). *Journal of the Arnold Arboretum*, **57**, 219–249, 450–525.

Campbell, S. D., Setter, C. L., Jeffrey, P. L. and Vitelli, J. (1996). Controlling dense infestations of *Prosopis pallida*. *Proceedings of Eleventh Australian Weeds Conference*, ed. R. C. H. Shepherd. Melbourne, Australia: Weed Science Society of Victoria, pp. 231–232.

Cheesman, O. D. (2006). Biological control of *Prosopis* on Ascension Island. *Biocontrol News and Information*, **27**, 34–35.

Choge, S. K., Ngujiri, F. D., Kuria, M. N., Busaka, E. A. and Muthondeki, J. K. (2002). The status and impact of *Prosopis* spp. in Kenya. Nairobi, Kenya: KEFRI, 59 pp.

Coetzer, W. and Hoffmann, J. H. (1997). Establishment of *Neltumius arizonensis* (Coleoptera: Bruchidae) on mesquite (*Prosopis* species: Mimosaceae) in South Africa. *Biological Control*, **10**, 187–192.

Cordo, H. A. and DeLoach, C. J. (1987). *Insects that Attack Mesquite (*Prosopis *spp.) in Argentina and Paraguay: Their Possible Use for Biological Control in the United States*. ARS-62; United States Department of Agriculture, 36 pp.

Cuda, J. P. and DeLoach, C. J. (1998). Biology of *Mozena obtusa* (Hemiptera: Coreidae), a candidate for biological control of mesquite, *Prosopis* spp. *Biological Control*, **13**, 101–110.

Cuda, J. P., DeLoach, C. J. and Robbins, T. O. (1990). Population dynamics of *Melipotis indomita* (Lepidoptera: Noctuidae), an indigenous natural enemy of mesquite, *Prosopis* spp. *Environmental Entomology*, **19**, 415–422.

DeLoach, C. J. (1981). Prognosis for biological control of weeds of southwestern U.S. rangelands. In *Proceedings of the Fifth International Symposium on the Biological Control of Weeds, 22–27 July 1980, Brisbane, Australia*, ed. E. S. Delfosse. Melbourne, Australia: CSIRO Publishing, pp. 175–199.

DeLoach, C. J. (1982). Biological studies on a mesquite leaf tier, *Tetralopha euphemella*, in central Texas. *Environmental Entomology*, **11**, 261–267.

DeLoach, C. J. (1983a). Life history of a mesquite looper, *Semiothisa cyda* (Lepidoptera: Geometridae), in central Texas. *Annals of the Entomological Society of America*, **76**, 83–86.

DeLoach, C. J. (1983b). Field biology and host range of a mesquite looper, *Semiothisa cyda* (Lepidoptera: Geometridae), in central Texas. *Annals of the Entomological Society of America*, **76**, 87–93.

DeLoach, C. J. (1985). Conflicts of interest over beneficial and undesirable aspects of mesquite (*Prosopis* spp.) in the United States as related to biological control. In *Proceedings of the Sixth Symposium of Biological Control of Weeds*, ed. E. S. Delfosse. Vancouver, Canada: Agriculture Canada, pp. 301–340.

DeLoach, C. J. (1988). Mesquite: a weed of southwestern rangelands: a review of its taxonomy, biology, harmful and beneficial values, and its potential for biological control. A petition to the technical advisory group on introduction of biological control agents of weeds, USDA, Animal and Plant Health Inspection Service.

DeLoach, C. J. (1994). Feeding behaviour of *Melipotis indomita* (Lepidoptera: Noctuidae), a herbivore of mesquite (*Prosopis* spp.). *Environmental Entomology*, **23**, 161–166.

DeLoach, C. J. and Cuda, J. P. (1994). Host range of the mesquite cutworm, *Melipotis indomita* (Lepidoptera: Noctuidae), a potential biological control agent for mesquite (*Prosopis* spp.). *Biological Control*, **4**, 38–44.

Delobel, A. and Fédière, G. (2002). First report in Egypt of two seed-beetles (Coleoptera: Bruchidae) on *Prosopis* spp. *Bulletin of Faculty of Agriculture, Cairo University*, **53**, 129–139.

Dennill, G. B. (1988). Why a gall former can be a good biocontrol agent: the gall wasp *Trichilogaster acaciaelongifoliae* and the weed *Acacia longifolia*. *Ecological Entomology*, **13**, 1–9.

Diplock, N., Galea, V., van Klinken, R. D. and Wearing, A. (2006). A preliminary investigation of dieback on *Parkinsonia aculeata*. In *Fifteenth Australian Weeds Conference, Papers and Proceedings*, ed. C. Preston, J. H. Watts and N. D. Crossman. Adelaide, Australia: Weed Management Society of South Australia, pp. 585–587.

Donnelly, G. P. (2002). The host range and biology of the mesquite psyllid *Heteropsylla texana*. *Biocontrol*, **47**, 363–371.

Dorchin, N., Cramer, M. D. and Hoffmann, J. H. (2006). The carbon costs of wasp-induced galls on *Acacia pycnantha*. *Ecology*, **87**, 1781–1791.

Dussart, E., Lerner, P. and Peinetti, R. (1998). Long term dynamics of two populations of *Prosopis caldenia* Burkart. *Journal of Range Management*, **51**, 685–691.

Fagg, C. W. and Stewart, J. L. (1994). The value of *Acacia* and *Prosopis* in arid and semi-arid environments. *Journal of Arid Environments*, **27**, 3–25.

Felker, P. and Moss, J., eds. (1996). Prosopis: *Semi-arid Fuelwood and Forage Tree Building Consensus for the Disenfranchised*. Texas: Centre for Semi-Arid Forest Resources, pp vii–xiii.

Flanagan, G. J. (1994). The Australian distribution of *Mictis profana* (F.) (Hemiptera: Coreidae) and its lifecycle on *Mimosa pigra* L. *Journal of the Australian Entomological Society*, **33**, 111–114.

Fravolini, A., Hultine, K. R., Brugnoli, E., *et al.* (2005). Precipitation pulse use by an invasive woody legume: the role of soil texture and pulse size. *Oecologia*, **144**, 618–627.

Geesing, D., Al-Khawlani, M. and Abba, M. L. (2004). Management of introduced *Prosopis* species: can economic exploitation control an invasive species? *Unasylva*, **55**(217): 36–44.

Gile, L. H., Gibbens, R. P. and Lenz, J. M. (1997). The near-ubiquitous pedogenic world of mesquite roots in an arid basin floor. *Journal of Arid Environments*, **35**, 39–58.

Golubov, J., del Carmen Mandujano, M., Franco, M., *et al.* (1999). Demography of the invasive woody perennial *Prosoips glandulosa* (honey mesquite). *Journal of Ecology*, **87**, 955–962.

Harding, G. B. (1988). The genus *Prosopis* as an invasive alien in South Africa. Unpublished Ph. D. thesis, University of Port Elizabeth, South Africa, 195 pp.

Harding, G. B. (1991). Sheep can reduce recruitment of invasive *Prosopis* species. *Applied Plant Science*, **5**, 25–28.

Harding, G. B. and Bate, G. C. (1991). The occurrence of invasive *Prosopis* species in the north-western Cape, South Africa. *South African Journal of Science*, **87**, 188–192.

Hoffmann, J. H., Impson, F. A. C. and Moran, V. C. (1993). Biological control of mesquite weeds in South Africa using a seed-feeding bruchid, *Algarobius prosopis*: initial levels of interference by native parasitoids. *Biological Control*, **3**, 17–21.

Impson, F. A. C., Moran, V. C. and Hoffmann, J. H. (1999). A review of the effectiveness of seed-feeding bruchid beetles in the biological control of mesquite, *Prosopis* species (Fabaceae), in South Africa. In *Biological Control of Weeds in South Africa (1990–1998)*, ed. T. Olckers and M. P. Hill. African Entomology Memoir 1. Johannesburg, South Africa: Entomological Society of Southern Africa, pp. 81–88.

Jacoby, P. and Ansley, R. J. (1991). Mesquite: classification, distribution, ecology and control. In *Noxious Range Weeds*, ed. L. F. James. J. O. Evans, M. H. Ralphs and R. D. Child. Boulder, CO: Westview Press, 466 pp.

Johnson, C. D. (1983). *Handbook on Seed Insects of Prosopis*. Rome, Italy: Food and Agriculture Organisation of the United Nations, 55 pp.

Kriticos, D. J. (2003). SPAnDX: a process-based population dynamics model to explore management and climate change impacts on an invasive alive plant, *Acacia nilotica*. *Ecological Modelling*, **163**, 187–208.

Le Maitre, D. (1999). *Prosopis and Groundwater: A Literature Review and Bibliography*. Report ENV-S-C 99077, Stellenbosch, South Africa: CSIR, 36 pp.

Lewis, G. P. and Elias, T. S. (1981). Mimoseae Bronn (1822). In *Advances in Legume Systematics*, Part 1, ed. R. M. Polhill and P. H. Raven. Kew, UK: Royal Botanic Gardens, pp. 155–168.

Mauremootoo, J. R. (2006). Current status and future prospects for *Prosopis juliflora* in Ethiopia. *Biocontrol News and Information*, **27**, 37–40.

Moran, V. C., Hoffmann, J. H. and Zimmermann, H.G. (1993). Objectives, constraints and tactics in the biological control of mesquite weeds (*Prosopis*) in South Africa. *Biological Control*, **3**, 80–83.

Moran, V. C. and Zimmermann, H. G. (1984). The biological control of cactus weeds: achievements and prospects. *Biocontrol News and Information*, **5**, 297–320.

Ogutu, W. O. and Mauremootoo, J. R. (2006). *Prosopis* in Kenya: acquiring the knowledge for informed management. *Biocontrol News and Information*, **27**, 35–37.

Osmond, R., van Klinken, R. D., March, N., Cobon, R. and Campbell, S. (2003). The mesquite control toolbox. In *Mesquite Best Practice Manual: Control and Management Options for Mesquite* (Prosopis *spp.*) *in Australia*, ed. R. Osmond. Brisbane, Australia: Queensland Department of Natural Resources and Mines, pp. 37–60.

Pasiecznik, N. M., Harris, P. J. C., Harsh, L. N., *et al.* (2001). *The* Prosopis juliflora–Prosopis pallida *Complex: a Monograph*. Coventry, UK: HDRA Publishing, 162 pp.

Pasiecznik, N. M., Harris, P. J. C. and Smith, S. J. (2004). *Identifying Tropical Prosopis Species: a Field Guide*. Coventry, UK: HDRA Publishing, 29 pp.

Pasiecznik, N. M. (2006). Limits to *Prosopis* biocontrol: utilisation and traditional knowledge could fill the gap. *Biocontrol News and Information*, **27**, 5N–6N.

Pasiecznik N. M. and Felker, P. (2006). Safeguarding valued Old World native *Prosopis* species from biocontrol introductions. *Biocontrol News and Information*, **27**, 3N–4N.

Pasiecznik, N. M., Choge, S. K., Muthike, G. M., *et al.* (2006). *Putting Knowledge on Prosopis into use in Kenya*. Pioneering Advances in 2006. Coventry, UK: HDRA Publishing, 17 pp. (www.gardenorganic.org.uk/pdfs/international_programme/ProsopisKenyaSummaryReport.pdf).

Patch, N. L. and Felker, P. (1997). Influence of silvicultural treatments on growth of mature mesquite (*Prosopis glandulosa* var. *glandulosa*) nine years after initiation. *Forest Ecology and Management*, **94**, 37–46.

Raghu, S., Dhileepan, K. and Wilson, J. (2006). Refining the process of agent selection through understanding plant demography and plant response to herbivory. *Australian Journal of Entomology*, **45**, 308–316.

Roberts, A. (2006). Biological control of alien invasive mesquite species (*Prosopis*) in South Africa: the role of introduced seed-feeding bruchids. Unpublished Ph.D. thesis, University of Cape Town, South Africa, 233 pp.

Robinson, T. P., van Klinken, R. D. and Metternicht G. (2008). Spatial and temporal rates and patterns of mesquite (*Prosopis* species) invasion in Western Australia. *Journal of Arid Environments*, **72**, 175–188.

Sawal, R. K., Ratan, R. and Yadav, S. B. S. (2004). Mesquite (*Prosopis juliflora*) pods as a feed source for livestock: a review. *Asian–Australian Journal of Animal Sciences*, **17**, 719–25.

Schuster, J. L., ed. (1969). *Literature on the Mesquite* (Prosopis *L.*) *of North America: An Annotated Bibliography*. Special Report 26. Lubbock, TX: Texas Tech University, 83 pp.

Shackleton, C., McGarry, D., Fourie, S., *et al.* (2007). Assessing the effects of invasive alien species on rural livelihoods: case examples and a framework for South Africa. *Human Ecology*, **35**, 113–27.

Shiferaw, H., Teketasy, D., Nemomissa, S. and Assefa, F. (2004). Some biological characteristics that foster the invasion of *Prosopis juliflora* (Sw.) DC at middle Awash rift valley area, north-eastern Ethiopia. *Journal of Arid Environments*, **58**, 135–154.

Silva, S. (1988). *Prosopis juliflora* (Sw) DC. in Brazil. In *The Current State of Knowledge on* Prosopis juliflora, ed. M. A. Habit and J. C. Saavedra. Recife, Brazil: Food and Agriculture Organisation of the United Nations, pp. 29–55.

Simpson, B. B., ed. (1977). *Mesquite, Its Biology in Two Desert Scrub Ecosystems.* Stroudsburg, PA: Dowden, Hutchinson and Ross Inc., 250 pp.

Stein, R. B. D., de Toledo, L. R. A., de Almeida, F. Q., *et al.* (2005). *Brazilian Journal of Animal Science*, **34**, 1240–1247.

Strathie, L. W. (1995). Ovipositional behaviour and efficacy of *Neltumius arizonensis*, a potential biological control agent of *Prosopis* weeds in South Africa. Unpublished M. Sc. thesis, University of Cape Town, South Africa, 122 pp.

Swenson, W. H. (1969). Natural enemies of mesquite. In *Literature on the Mesquite (*Prosopis *L.) of North America, an Annotated Bibliography*, ed. J. L. Schuster. Lubbock, TX: Texas Tech University, pp. 23–24.

Tewari, J. C., Pasiecznik, N. M., Harsh, L. N. and Harris, P. J. C., eds (1998). Prosopis *Species in the Arid and Semi-Arid Zones of India.* Jodhpur, India: The *Prosopis* society of India and the Henry Doubleday Research Association, 127 pp.

Thorpe, J. R. and Lynch, R. (2000). *The Determination of Weeds of National Significance.* Launceston, Tasmania: National Weeds Strategy Executive Committee, 234 pp.

van Klinken, R. D. (1999). Developmental host-specificity of *Mozena obtusa* (Heteroptera: Coreidae), a potential biocontrol agent for mesquite (*Prosopis* species). *Biological Control*, **16**, 283–290.

van Klinken, R. D. (2000). Host-specificity constrains evolutionary host change in the psyllid *Prosopidopsylla flava*. *Ecological Entomology*, **25**, 413–422.

van Klinken, R. D. and Burwell, C. (2005). Evidence from a gelechiid leaf-tier on mesquite (Mimosaceae: *Prosopis*) that semi-concealed Lepidopteran biological control agents may not be at risk from parasitism in Australian rangelands. *Biological Control*, **32**, 121–129.

van Klinken, R. D. and Campbell, S. (2001). Australian weeds series: *Prosopis* species. *Plant Protection Quarterly*, **16**, 1–20.

van Klinken, R. D. and Flack, L. (2008). What limits predation rates by the specialist seed-feeder *Penthobruchus germaini* on an invasive shrub? *Journal of Applied Ecology*, in press.

van Klinken, R. D. and Heard, T. A. (2000). Estimating fundamental host range: a host-specificity study of a biocontrol agent for *Prosopis* species (Leguminosae). *Biocontrol Science and Technology*, **10**, 331–342.

van Klinken, R. D. and Raghu, S. (2006). A scientific approach to agent selection. *Australian Journal of Entomology*, **45**, 253–258.

van Klinken, R. D., Fichera, G., and Cordo, H. (2003). Targeting biological control across diverse landscapes: the release, establishment and early success of two insects on mesquite (*Prosopis*) in rangeland Australia. *Biological Control*, **26**, 8–20.

van Klinken, R. D., Graham, J. and Flack, L. (2006). The demography of exotic, invasive fire-tolerant mesquite (*Prosopis*) in Australia and the implications for ecosystem impacts and management. *Biological Invasions*, **8**, 727–741.

van Tonder, S. J. (1985). Annotated records of southern African Bruchidae (Coleoptera) associated with Acacias, with a description of a new species. *Phytophylactica,* **17**, 143–148.

Versveld, D. B., Le Maitre, D. C. and Chapman, R. A. (1998). *Alien Invading Plants and Water Resources in South Africa: A Preliminary Assessment.* Tt99/98. Pretoria, South Africa: Water Research Commission.

Ward, C. R., O'Brien, C. W., O'Brien, L. B., Foster, D. E., and Huddleston, E. W. (1977). Annotated checklist of New World insects associated with *Prosopis* (mesquite). *United States Department of Agriculture Research Service Bulletin,* **1557**, 115 pp.

Yousuf, M. and Gaur, M. (1998). Some noteworthy insect pests of *Prosopis juliflora* from Rajasthan. In Prosopis *Species in the Arid and Semi-Arid Zones of India,* ed. J. C. Tewari, N. M. Pasiecznik, L. N. Harsh and P. J. C. Harris. Jodhpur, India: The *Prosopis* Society of India and the Henry Doubleday Research Association, pp. 91–94.

Zalucki, M. and van Klinken, R. D. (2006). Predicting population dynamics and abundance of introduced biological agents: science or gazing into crystal balls. *Australian Journal of Entomology,* **45**, 331–344.

Zimmermann, H. G. (1991). Biological control of mesquite, *Prosopis* spp. (Fabaceae), in South Africa. *Agriculture, Ecosystems and Environment,* **37**, 175–186.

Zimmermann, H. G. and Moran, V. C. (1991). Biological control of prickly pear, *Opuntia ficus-indica* (Cactaceae), in South Africa. *Agriculture, Ecosystems and Environment,* **37**, 29–35.

Zimmermann, H. and Pasiecznik, N. M. (2005). *Realistic Approaches to the management of* Prosopis *Species in South Africa.* Policy brief pamphlet. Coventry, UK: HDRA Publishing, 4 pp.

Zimmermann, H. G., Hoffmann, J. H. and Witt, A. B. R. (2006). A South African perspective on *Prosopis. Biocontrol News and Information,* **27**, 6–10.

19

Salvinia molesta D. S. Mitchell (Salviniaceae)

Mic H. Julien, Martin P. Hill, and Philip W. Tipping

19.1 Introduction

Salvinia molesta D. S. Mitchell (Salviniaceae) (salvinia) is a floating water fern of tropical and subtropical distribution worldwide. Its center of origin is southeastern Brazil. It is an extremely important invasive species and its biological control is an extraordinary, contemporary, success story.

Salvinia molesta is named after Antonio Maria Salvini (1633–1729), University of Florence. The specific epithet *molesta* originates from the Latin *molestus* meaning "troublesome," "annoying," referring to its weediness (Parsons and Cuthbertson, 2001).

19.2 Taxonomy

Salviniaceae in Hydropteridales comprises the monotypic taxon *Salvinia* with 10–12 species (Hassler and Swale, 2002): *S. minima* Baker, *S. oblongifolia* Martius, and four species in the *S. auriculata* complex originating in the tropical Americas. The *S. auriculata* complex comprises species in which the upper section of each leaf hair forms an "egg-beater" or "cage" shape by splitting apart below the tip and joining at the tip (Fig. 19.1) (Forno, 1983) and includes *S. auriculata* Aublet, *S. biloba* Raddi, *S. herzogii* de la Sota, and *Salvinia molesta* D. S. Mitchell. *Salvinia molesta* was separated from *S. auriculata* by Mitchell (1972). Most literature that refers to *S. auriculata* as a pest species outside South America and Trinidad actually refers to *S. molesta*.

Herzog (1935) recognized *S. auriculata* Aublet, and this name was applied to the invasive species that occurred outside South America. De la Sota (1962) recognized that *S. auriculata* comprised a number of species and described *S. herzogii*. Mitchell and Thomas (1972) described four species in the *auriculata* complex and their distributions: three were from South America and the fourth, from other countries, was named *S. molesta* (Mitchell 1972). The native range of *S. molesta* in South America has been described in Forno and Harley (1979) and Forno (1983).

Biological Control of Tropical Weeds using Arthropods, ed. R. Muniappan, G. V. P. Reddy, and A. Raman. Published by Cambridge University Press. © Cambridge University Press, 2009.

Fig. 19.1 *Cyrtobagous salviniae* on the characteristic "egg-beater" shaped leaf hairs of salvinia (photo by S. Bauer, USDA-ARS, Bugwood.org).

The most widely used common name is salvinia; however, other names include water fern, Kariba weed, African payal, African pyle (in Africa); giant salvinia, watervaring, koi kandy, water spangles, floating fern (in USA); giant azolla, and Australian azolla (in the Philippines) (Room and Julien, 1995; Randall, 2002).

19.3 Description

Salvinia molesta is a free-floating aquatic fern. It has a horizontal stem that lies at or just below the water surface. Buoyancy is facilitated by the formation of aerenchyma tissue in the stems and leaves (Barrett, 1989). A pair of leaves and a submerged "root," which is actually a modified leaf (Croxdale, 1978, 1979, 1981), occurs at each node and together (the node and its leaves and "root") they form a ramet. A single ramet or, usually, a cluster of joined ramets forms a plant. The apical and axillary buds give rise to the next new nodes and every node bears lateral buds that develop into branches under the right conditions (Room, 1988). *Salvinia molesta* growth is apically dominant and under optimal growth conditions the plant assumes a deltoid form (Fig. 19.2) with a bud at the terminal and successively older branches towards the base. Under poor growing conditions just a single stem with no branches may occur. Stems tend to grow in a zig-zag manner. Leaves are oval with a crease along the long axis. The edges tend to fold upwards along this crease as they age and with crowding. The upper side of the leaves is covered with rows of hairs, the tips of which are split into four and which join at the tip to form the egg-beater shape that is characteristic of *S. molesta* (Fig. 19.1) and its close relatives in

Fig. 19.2 *Salvinia molesta*, tertiary form (photo by M. Julien, CSIRO).

the *S. auriculata* complex. Other surfaces are covered in sharply pointed trichomes, including on "roots." The "root" has a short stalk between the node and the numerous fine "root" filaments. A subsessile to sessile raceme of papillate sporocarps occurs on a stalk among the filaments of the "root."

The features that distinguish *S. molesta* from other species within the *S. auriculata* complex include the shape of sporocarps, the arrangement of sporangia, and the leaf venation pattern (Mitchell and Thomas, 1972; Mitchell, 1972; Forno, 1983).

This plant shows considerable phenotypic plasticity, mainly because of environmental conditions such as crowding and nutrient availability. Mitchell and Tur (1975) described three basic forms associated with degrees of crowding: (1) a small-leafed, primary form (Fig. 19.3) typical of plants invading open water, (2) a slightly larger-leaved secondary form with leaves slightly folded (Fig. 19.4), and (3) a tertiary form (Fig. 19.5), typical of mature stands, with larger, deeply folded and densely packed leaves. These growth forms are dynamic and often comprise a continuum of growth forms within a population that change temporally and spatially in response to the local environment. The differences in appearance among plant growth forms may confuse the observer into believing that they represent more than one species.

19.4 Biology

Salvinia molesta is pentaploid ($x = 45$) and is sterile (Loyal and Grewal, 1966). Reproduction is entirely asexual: colony increase is through vegetative growth and the spread occurs only via movement of viable fragments (nodes or apical buds) and, therefore, *S. molesta* represents one massive clone worldwide. Fragmentation usually occurs at the

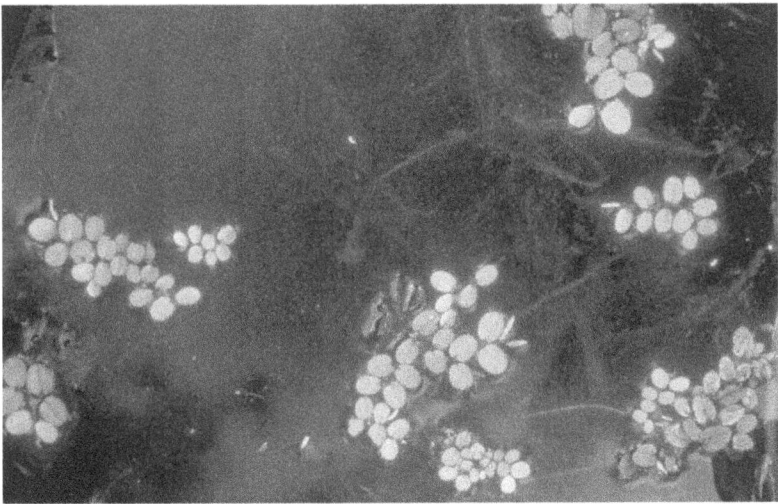

Fig. 19.3 Salvinia primary growth form (photo by M. Julien, CSIRO, Bugwood.org).

Fig. 19.4 Salvinia secondary growth form (photo by M. Julien, CSIRO, Bugwood.org).

internodes because of either aging or mechanical forces such as the movement of floating mats, animals, or boats. Each node is normally capable of producing three axillary buds (Room, 1988) but under stress conditions up to six develop (Julien and Bourne, 1986) leading to the formation of new plants.

Fig. 19.5 Salvinia tertiary growth form (photo by M. Julien, CSIRO).

The key factors affecting growth are temperature and nutrients. Optimal temperature for growth is 30°C while no growth occurs at either below 10°C or above 40°C (Room, 1986). *Salvinia molesta* was killed when exposed to temperatures below –3°C and above 43°C for more than two hours (Whiteman and Room, 1991). However, in tropical conditions high leaf temperatures above 40°C did not kill the plants due to the buffering effect of the water, which retained a temperature less than 40°C (Storrs and Julien, 1996). Frosts kill exposed plants, but plants survive when protected by other *S. molesta* (or other vegetation) or by the temperature-buffering effect of water. The buffering effect of water bodies may act to extend the range of *S. molesta* into cooler climates.

In most parts of the world where *S. molesta* grows, temperatures remain in the optimal range for most of the year and so the availability of nutrients is a key growth factor. The content of nitrogen in *S. molesta* ranged from 0.6 to 4.0% dry mass (Room and Thomas, 1986). At low levels of nitrogen "roots" are usually longer, leaves larger, sporocarps occur more frequently, and there are fewer branches than under high nitrogen levels. The proportion of buds that develop are independent of temperature but usually correlate with the nitrogen content of the plant (Room, 1983; Julien and Bourne, 1986).

Room (1986) calculated that at maximum rates of growth *S. molesta* could take up 6000 kg nitrogen/ha/year, whereas Finlayson *et al.* (1982) measured uptake from a sewage lagoon at 1580 kg nitrogen/ha/year. *Salvinia molesta* grows in water with conductivities from 100 µS/cm to 1400 µS/cm (Mitchell *et al.* 1980; Room and Gill, 1985). Divakaran *et al.* (1980) reported a 25% reduction in *S. molesta* growth when grown in 10% seawater (4800 µS/cm) while Room and Julien (1995) found slow growth in 20% seawater and complete mortality after 30 minutes exposure to seawater itself. Although the optimum pH for growth is 6.5 (Cary and Weerts, 1984), *S. molesta* is found growing in the field where pH ranges from 5.2 to 9.5 (Holm *et al.*, 1977; Mitchell *et al.*, 1980).

Salvinia molesta compensates for destruction of buds (Julien and Bourne, 1986), and the level of compensation increases with availability of nitrogen (Julien *et al.*, 1987). However, the plant does not compensate for damage to leaves and stems (Julien and Bourne, 1988).

19.5 Ecology

When conditions are optimal for growth, *S. molesta* can express its full growth potential, resulting in rapid spread over the surface of water. Once coverage is complete the mats then thicken vertically, by layering, and may reach 0.5–1 m, sufficient to support a human being. In high-nutrient and artificial conditions, biomass and numbers of ramets doubled in 2–4 days (Mitchell and Tur, 1975; Cary and Weerts, 1983). Dry weight doubled in 5–30 days in Kakadu National Park (Storrs and Julien, 1996), while on Lake Kariba ramet numbers doubled in 9–17 days (Mitchell and Tur, 1975; Oliver, 1993 citing Mitchell, 1979).

At high densities (2500 tertiary form ramets/m^2 in nutrient-poor waters or 30 000 tertiary form ramets/m^2 in nutrient-rich waters), natality equals mortality. Live biomass was 670–1620 g/m^2 dry weight (Mitchell, 1979). Biomass of living and attached dead shoots and "roots" can exceed 1600 g/m^2 or 400 t/ha of fresh weight, and *S. molesta* is about 95% water by weight (Room and Julien, 1995).

Stable mats of *S. molesta* are often invaded by other plants, initially grasses and sedges, but small trees and shrubs may also colonize mats and form floating communities (or sudds) that may float freely or remain attached to the bank or the substrate beneath the water body. In this context *S. molesta* can be a transformer species by providing the substrate for other plant species that eventually cover and fill shallow water bodies.

In natural systems that seasonally flood, mats that develop in the "dry season" are flushed downstream, often to sea where salinity kills the plants, or they are deposited on floodplains where they desiccate and die. However, individual plants may survive through a dry season under a mulch of dead *S. molesta* plants, especially in lower, moist areas (Storrs and Julien, 1996). Flushing events can be variable since they are dependent on precipitation and may be mediated by surrounding vegetation. For example, in Kakadu National Park, lower than normal precipitation during a wet season may provide insufficient energy in floodwaters to flush mats of *S. molesta* from the remnant floodplain lakes during the dry season, particularly when these were surrounded by trees. Thus the mats increase in biomass over the following "dry season" and greater flow energy is required to remove the increased biomass during following flood events (Julien and Storrs, 1996; Storrs and Julien, 1996).

In southern Africa, *Eichhornia crassipes* (Mart.) Solms (water hyacinth) Pontederiaceae has been the most dominant of the exotic floating aquatics. However, there is a replacement process with other floating aquatic plant species. When water hyacinth is under successful management, *Pistia stratiotes* (water lettuce) takes over. If that is well managed, *S. molesta* takes over; manage *S. molesta* and azolla takes over, manage azolla and duckweeds (*Lemna* sp., *Spirodela* sp. and *Wolfia* sp.) take over. This succession of invading plant species reinforces the need to consider management of all floating exotics at the same time. Although water hyacinth is considered the dominant floating plant invader, Kammathy (1968) found that *S. molesta* successfully competed with and replaced *E. crassipes* and *P. stratiotes* in canals, rice paddies, and backwaters in Kerala, India.

19.6 Invasion pathways

In natural situations, spread of *S. molesta* is by water flow or wind and by animals that use waterways; birds (everywhere), capybara (South America), hippopotamus (Africa), and water buffalo (Australasia) (Room and Julien, 1995; Forno and Smith, 1999). *Salvinia molesta* and other floating aquatic weeds, notably, *E. crassipes* and *Pistia stratiotes*, have attracted attention as aquaria and water garden plants and have been spread around the world. This demand and supply continues today and humans are the key contemporary vector for *S. molesta* and other aquatics. For example, on average in 2006, 3131 legal consignments of aquatic plants/month, including *E. crassipes*, *P. stratiotes*, and *Salvinia* spp., were processed through Charles de Gaulle Airport, Paris (EPPO, 2007). *Salvinia molesta* was one of a number of prohibited taxa that contaminated commercial plant orders entering Minnesota (Maki and Galatowitsch, 2004). It was spread throughout Senegal as a chicken feed. Inevitably, propagules of some of these plants, or their progeny, end up in natural water systems or constructed impoundments where, if temperatures are favorable, they often encounter eutrophic conditions that contribute to rapid growth. *Salvinia molesta* has been used as a biological weapon. During the period around 1979, people living on the Sepik River in Papua New Guinea deliberately spread *S. molesta* to destroy fishing grounds during disputes (Gewertz, 1983).

19.7 Distribution and pest status

The native range of *S. molesta* encompasses a relatively small area in southeastern Brazil in the States of São Paulo, Paraná, Santa Catarina, and Rio Grande do Sul, between 24° 05′ S and 32° 05′ S; up to 200 km inland and up to 500 m altitude (Forno and Harley, 1979). It grows in wetland communities – natural lagoons and swamps, along the margins of streams and rivers, and in artificial dams and drainage channels (Forno and Harley, 1979) – rarely forming single-species mats, but mostly mixed with other vegetation and often supporting a variety of herbivores (Forno and Bourne, 1984). Its exotic range includes cosmopolitan tropical, subtropical in every continent, and some warm temperate regions such as parts of southern Africa, the United States, and southern Australia.

Table 19.1 *Regions and countries where* Salvinia molesta *is invasive*

Regions	Countries
Africa	Botswana, Cameroon, Central African Republic, Côte d'Ivoire, Democratic Republic of Congo, Ghana, Kenya, Lesotho, Malawi, Mozambique, Mauritius, Madagascar, Mauritania, Namibia, Senegal, Swaziland, South Africa, Tanzania, Uganda, Zambia, Zimbabwe
Asia	India, Indonesia, Malaysia, Thailand, Singapore, Philippines, Sri Lanka
Central America and the Caribbean	Cuba, Mexico, Trinidad and Tobago
North America	USA
South America	Argentina, Brazil, Colombia, Guyana
Oceania	Australia, Fiji, French Polynesia, Papua New Guinea, New Caledonia, New Zealand

[*] Information was derived from GISP database; CABI Compendium; Waterhouse and Norris (1987) and W. Orapa, personal communication, for French Polynesia; Mbati and Neuenschwander (2005) for Republic of Congo; O. Ajuonu and P. Neuenschwander, personal communication, for West African countries, and authors' observations.

Salvinia molesta was first reported in Sri Lanka in 1939 (Williams, 1956). It was first found in Australia in 1952 and grows at latitudes of about 35° S (Room and Julien, 1995). First recorded naturalized in USA in 1995, it is now established in 12 states including Hawaii (Jacono *et al.*, 2001), as far 34° N. It is widespread in southern and Southeast Asia (at least seven countries, although it is not recorded in China), and Africa (at least 20 countries). It also occurs in Papua New Guinea, New Caledonia, Fiji, and New Zealand in the South Pacific (Waterhouse and Norris, 1987), and in French Polynesia (W. Orapa, personal communication).

Salvinia molesta has been found in Europe. It is growing in a botanical garden pond at Ponta Delgarda, Azores Islands, Portugal (M. Julien, personal observation) and is growing in Pozzo del Merro, Latium, in Italy (Giardini, 2004; M. Giardini, personal communication). In Europe, low winter temperature is likely to be the only limiting factor since most systems are eutrophic and the commercial pathway via the aquaria and water-garden trades is wide open. In Flora Europaea (http://rbg-web2.rbge.org.uk/FE/fe.html accessed 22 October 2007) it is recorded from Hs (Spain) citing reference Brit. Fern Gaz. 10: 251 (1972). This reference is Mitchell (1972) and there is no information in it, or elsewhere, of *S. molesta* having been found growing in Spain.

Salvinia molesta is a very serious invader worldwide (Holm *et al.*, 1977). It is included in the Australian top 20 Weeds of National Significance. In many countries it is listed as a noxious weed and its growth and movement is prohibited, for example in Australia (van Oosterhout, 2006). In the USA, *S. molesta* and other *Salvinia* species are on the Federal Noxious Weed List as directed by the Noxious Weed Act of 1974, which prevents the movement of listed species into or through the US without a permit.

However, it does not regulate intrastate movement. In South Africa it is listed as a category 1 weed that has to be controlled under the Conservation of Agricultural Resources Act. Although no formal legislation occurs in most African countries, it is regarded as an aquatic invader throughout the continent and attempts to control it have been initiated in many countries.

19.8 Possible utilization of *Salvinia molesta*

Rapid growth response under favorable conditions permits *S. molesta* to extract large quantities of nutrients from water bodies. Consequently, these plants have been used to remove nutrients from water bodies. Conversely, they lock up nutrients that would otherwise be available to other organisms. *Salvinia molesta* can also take up heavy metal contaminants and, therefore, should not be utilized in situations where further contamination to land or produce could occur unless they are known to be free of contaminants.

In potassium-deficient soils in an Indian delta, potassium-rich plants such as *P. stratiotes* and *S. molesta* increased seedling performance when used as green manure in seedling nurseries. Grain yield was best with *P. stratiotes* and then with either *S. molesta* or a complete fertilizer (Raju and Gangwar, 2004).

Salvinia molesta has been used as compost, mulch, and as livestock feed or to supplement livestock fodder (Thomas and Room, 1986b). For example, dried *S. molesta* has also been used in Senegal as chicken feed. The chemical composition of *S. molesta* was studied using plants from India (Moozhiyil and Pallauf, 1986). The concentration of crude protein (12.4%) was similar to conventional forage. However, the high levels of crude ash (17.3% in dry matter) and of lignin (13.7%) and the presence of tannins (0.93%) may reduce acceptance and digestibility and therefore reduce the use and value of *S. molesta* as a feed source for ruminants. High levels of trace elements such as iron and manganese could also cause problems in animals. They also concluded that the high moisture content and therefore the bulk of material needed to supply forage make sun-drying impractical, particularly in tropical countries. Others have investigated using *S. molesta* for paper production and generation of biofuel. None of the uses for *S. molesta* has led to large-scale development because of the difficulty and cost of collecting the large bulk of material to provide dry matter when fresh *S. molesta* is about 95% water (Thomas and Room, 1986b).

19.9 Impact of *Salvinia molesta*

The rapid growth response by *S. molesta* ensures its weedy status in practically all natural situations from the most pristine to the most eutrophic water bodies. Just a few *S. molesta* plants are all that is necessary to start a major infestation over vast areas. Such blankets of *S. molesta* (Fig. 19.6) prevent light penetration, reduce oxygenation, increase carbon

Fig. 19.6 Salvinia covering a pond (photo by T. Center, USDA-ARS, Bugwood.org).

dioxide and hydrogen sulphide, smother aquatic flora, and displace aquatic fauna by altering habitats, destroying niches, and reducing or eliminating food sources (food plants, benthic biota, and other fauna). As mentioned previously, persistent mats often support other vegetation resulting in the formation of sudds which increase sedimentation and block water flow.

Salvinia molesta invades flood canals, rice paddies, and artificial water bodies used for irrigation, hydroelectricity, and town water supply reservoirs (Barrett, 1989). It is a pest of rice paddies in many countries where it competes for water, nutrients, and space, thereby reducing yields. Doeleman (1989) estimated that *S. molesta* reduced rice yields by 3% and fish catch by 20–40% in Sri Lanka. It replaces native flora, disrupts or prevents services to communities by clogging pipe intakes (water supply) and turbines (electricity supply). For communities largely dependent on waterway transport it prevents access to markets, schools, hospitals, and public administration. People's livelihoods and cash economies may be seriously impacted in subsistent and semisubsistent communities by reducing fisheries, preventing fishing and access to gardens, hunting grounds, and markets, and sometimes causing displacement of populations and the abandonment of villages (Gewertz, 1983; Thomas and Room, 1986a; Mbati and Neuenschwander, 2005; O. Diop, personal communication).

Salvinia molesta harbors snails that are the intermediate host for schistosomiasis (Thomas and Room, 1986a) leading to greater incidence of disease (Bennett, 1966). It also harbors *Mansonia* mosquitoes, a vector of rural encephalitis (Pancho and Soerjani, 1978), and other mosquitoes that are responsible for transmitting encephalitis, malaria (e.g. *Anopheles*), and dengue fever (Creagh, 1991/1992).

19.10 Management

Physical removal using booms to accumulate or control the location of the mats and machines to collect and remove the weed have been used in many instances, rarely with great success and always at great expense, for example, on the Hawkesbury River, Australia (Coventry, 2006). However, in most instances the costs of such operations exceed the benefits because weed growth can exceed removal rates, or the lack of follow-up management allows recolonization by remaining plants (Room and Thomas, 1986). Intensive monitoring of treated sites is essential to deal with reinvasion, or more usually, rapid recolonization from missed plants. However, because post-treatment monitoring is expensive, it is rarely conducted over the long term and, therefore, mechanical methods usually fail to provide acceptable and sustainable levels of weed control. Similarly, the use of herbicides may quickly reduce biomass but they do not provide long-term cost-effectiveness (Thomas, 1985) or environmentally friendly solutions in most situations. A comparison of the costs and benefits for different attempted control options in Zimbabwe showed that physical and chemical controls were expensive and ineffective, while biological control was effective and inexpensive (Chikwenhere and Keswani, 1997). However, it should be noted that the scale of the infestation can influence the prospects of controlling *S. molesta* successfully. Both mechanical and chemical control methods, either alone or in combination, can be very useful strategies for small defined water bodies where access is relatively easy, especially as the plant is sterile and so no recruitment from spores occurs. However, they generally are not practical or affordable in large natural systems or in inaccessible areas.

Herbicides utilized for *S. molesta* control in Australia include diquat, glyphosate, calcium dodecyl benzene sulphonate, and orange oil (van Oosterhout, 2006). In South Africa, diquat and 2,4-D are no longer permitted and only glyphosate is currently registered. Herbicides permitted in the United States include diquat dibromide, fluridone, glyphosate, and several chelated copper compounds. Herbicides are usually applied using hand guns or booms from boats, including airboats and sometimes aircraft. The application of herbicides has limitations because they impact nontarget plants, some *S. molesta* infestations are not easily located or accessible, materials can be expensive, and the application of treatments can be very time consuming require repeated applications. Costs for the chemicals alone can range from US\$210 to \$900 per ha. In some countries there are prohibitions or strict requirements for their use on or near water and in the US postspray intervals must be observed before cattle can be allowed access to the water body, except when the copper compounds are used.

Biological control has proven to be very effective against *S. molesta* and is now the preferred strategy, sometimes in combination with other strategies, in most situations. This success was not achieved until the native range *of S. molesta* in southern Brazil was identified and natural enemies were selected from that area. Earlier attempts to use insects collected from other species in the *S. auriculata* complex were not successful.

19.11 Phytophagous species associated with *Salvinia molesta*

Before 1978 *S. molesta* was considered a chance hybrid species derived from plants within the *S. auriculata* complex and, therefore, with parental origins in South America but no actual native range (Mitchell, 1978). The first surveys for biological control agents for *S. molesta* were conducted during 1961 to 1963 in parts of the native range of *S. auriculata*, in Trinidad, British Guyana, and Brazil, (Bennett, 1966, 1975). About 25 phytophagous insect species were found associated with the *S. auriculata* complex (Bennett, 1975) and three species, a weevil *Cyrtobagous singularis* Hustache (Coleoptera: Curculionidae), a moth *Samea multiplicalis* (Guenée) (Lepidoptera: Pyralidae) and a grasshopper *Paulinia acuminata* (De Geer) (Orthopetra: Pauliniidae), were selected for further study. These host-specificity studies were conducted at Belem, Brazil, and Curepe, Trinidad, in 1964–1965 (Bennett, 1966).

Later, with knowledge of the native range of *S. molesta* (Forno, 1983), surveys were conducted on that species as well as on the other members of the *S. auriculata* complex. All insect and mite species found are listed in Forno and Bourne (1984) including those collected by Bennett (1966). Forno and Bourne (1984) also indicated that the three main phytophages on *S. molesta* were those selected by Bennett (1975) with one important difference. The *Cyrtobagous* weevil, although appearing similar to *C. singularis*, was a new species, later named *Cyrtobagous salviniae* (Calder and Sands, 1985) and this weevil is an effective biological control agent. Some literature published before *C. salviniae* was recognized as a new species refers to *C. singularis* when the data actually refer to *C. salviniae* (e.g. Forno, 1981; Room *et al*, 1981; Forno *et al*., 1983). Forno and Bourne (1984) found 31 insect and mite species associated with *S. molesta*, two of which had been recorded from other *Salvinia* species by Bennett (1975). Several other moths, a grasshopper, various fungi, and a snail have been reported attacking *S. molesta* in its exotic range (Bennett, 1975; Waterhouse and Norris, 1987). None has been reported to be providing significant control.

19.11.1 **Cyrtobagous singularis** *Hustache (Coleoptera: Curculionidae)*

Cyrtobagous singularis is a small dark weevil (about 2 mm long), similar to *C. salviniae*. Features to distinguish it from *C. salviniae* are available in May and Sands (1986) for larvae and in Calder and Sands (1985) for adults. The type specimens were collected from Corumba, Matto Grosso, Brazil, in 1949. This species has been collected from *S. auriculata* at Trinidad, Georgetown and Ogle Estate in Guyana, Belem, Obidos, Manaus, and Recife in Brazil, Paraguay and northern Argentina, and from *S. oblongata* in Brazil (Bennett, 1966, 1975). It has established following releases in Botswana (Procter, 1984) and Namibia (Sands and Schotz, 1985). *Cyrtobagous* weevils collected from *S. minima* in Florida, USA, in 1962 and 1964, were considered to be *C. singularis* (Kissinger, 1966) but later identified as *C. salviniae* (Calder and Sands, 1985). Female adults lay their eggs singly in feeding holes made in the leaf veins and stems below the surface of the water and hatch in 7–8 days. Larvae complete development after 25 days. Late larvae construct a cocoon among the fine "root" hairs within which pupation occurs. Pupation takes 6–8 days. Oviposition begins several days after the new adult emerges from the cocoon. Adults may remain submerged for several hours and they live for 55 days or more (Bennett, 1966). Sands and Schotz (1985) compared feeding behavior of *C. singularis* with *C. salviniae*. Soon after hatch larvae of *C. singularis* began to feed on "roots" and buds and thereafter fed mostly externally on submerged parts of the plant including buds, rhizome aerenchyma, and leaf petioles. They destroyed buds and damaged rhizomes and petioles, causing discoloration, but failed to prevent growth of the plants. Adults fed on young apical leaves, not the apical bud, and on leaves of the next one or two nodes and associated internodes. This insect is restricted to *Salvinia* (Bennett, 1966) but it did not cause widespread damage to *Salvinia* in its native range (Bennett, 1975).

19.11.2 Cyrtobagous salviniae *Calder and Sands (Coleoptera: Curculionidae)*

This weevil (Fig. 19.1) is now commonly known as the "salvinia weevil." Detailed descriptions of *C. salviniae* are given for larvae by May and Sands (1986) and for adults by Calder and Sands (1985). Adults are small dark weevils, about 2 to 3 mm long. On first emergence they are brown and over the first five days they darken to black (Forno *et al.*, 1983). These sub-aquatic adults can be found on or under young leaves, in the leaf buds or among the "roots." They feed and oviposit on *S. molesta* underwater by respiring from air bubbles held between their legs and lower body. By spending much of their life underwater they avoid extreme temperatures and thus populations can survive when temperatures are high (40°C or more under shade) at northern Australian sites (Room *et al.*, 1984; Julien and Storrs, 1993) and low (e.g. –9 °C, Tipping and Center, 2003; and 40 days of frost, Room *et al.*, 1984).

In studies conducted at 25.5 °C (Forno *et al.*, 1983), adults mated more than once, beginning five days after emergence; preoviposition activity spread over 6–14 days; eggs hatched after about 10 days; and oviposition rate was one egg every two to five days for 60 days. Eggs were laid in feeding holes in young unopened leaves or in the developing root mass below the young unopened leaves, and were 0.5×0.24 mm. Young larvae are white, 1 mm long, crescent shaped, and feed initially in the young terminal buds and later in the rhizomes, petioles and "roots." There are three instars and larval development was complete after 23 days at 25.5°C, when older larvae are 2.6 mm long. Larval development is dependent on temperature and nitrogen levels in the host between 21°C and 31°C, and larvae failed to develop at 17°C (Sands *et al.*, 1983). A cocoon (about 2–2.6 mm diam.) of root hairs is attached to "roots." Development in the cocoon is about 12.6 days and a generation takes about 55 days at 25.5°C (Forno *et al.*, 1983). Negligible oviposition occurred at or below 21°C, though, with acclimation to cool temperatures, eggs may be laid at 19°C (Hennecke and Postle, 2006) and eggs did not hatch at or below 19°C or at 37°C. Adult fed at temperatures between 13 and 33°C (Forno *et al.*, 1983).

Adults preferentially feed on the tender apical buds and leaves and on roots. First-instar larvae tunnel and feed inside the young buds while second- and third-instar larvae tunnel and feed inside the rhizome and roots. Feeding increases with increased temperature but not with increased nitrogen in the host plant (Forno *et al.*, 1983). Increased nitrogen stimulates an increase in oviposition (Sands *et al.*, 1986) and both feeding and oviposition occur mostly at night (Schotz and Sands, 1988). The combined effect of feeding by adults and larvae prevents growth by destroying meristematic tissues and destroying rhizomes and ramets, severing the root–shoot link. As a result the plant (or mat of plants) rots, becomes waterlogged and sinks (Julien *et al.*, 1987). The salvinia weevil is a relatively slow disperser until high weevil populations cause a decrease in availability of food source (young buds of *S. molesta*), then an apparent behavioral switch to flight dispersal occurs (Thomas and Room, 1986a).

The type specimens of *C. salviniae* were collected in March 1979 at Joinville, Brazil. It has been collected from each of the species in the *S. auriculata* complex: *S. molesta*, *S. auriculata*, *S. herzogii*, and *S. biloba* in Brazil (Forno, 1983). Madeira *et al.* (2006) also collected it from *S. mimima* in Argentina and Parauay. It has also been collected in Paraguay and northern Argentina (Calder and Sands, 1985). Host-specificity testing demonstrated that *C. salviniae* is restricted to plants in the taxon *Salvinia* (Bennett, 1966; Forno *et al.*, 1983). Following releases for biological control, *C. salviniae* is established in many countries.

19.11.3 Samea multiplicalis *(Guenée) (Lepidoptera: Pyralidae)*

This pyralid was described from Brazil by Guenée (1854). It occurs from southeastern USA to Argentina and has been collected in Trinidad, Argentina, Guyana, Brazil, Uruguay, Florida, and Panama from species in the *Salvinia auriculata* complex (Bennett, 1966; Forno, 1981).

Adults are tan with dark markings on the wings (Fig. 19.7) and females tend to be lighter colored than the males (forewing length: 6.5–10.5 mm). Detailed wing coloration and markings, and sex differentiation in adults are available in Sands and Kassulke (1984). Larvae (Fig. 19.8) construct a canopy of silk and host-plant hairs and feed on the leaves under the canopy. As larvae grow they tend to eat and move through a number of vertical leaves in the tertiary growth form of *S. molesta* causing a "shot-hole" effect. When larvae are numerous, much of the leaf material is eaten, the plant is ragged and messy with frass and the remaining pieces of leaf material decompose. For pupation, a silken cocoon is attached to the upper surface of a large leaf. Adults emerge usually at night, mate after 24 hours and begin to lay eggs on the third night. Eggs are deposited singly among hairs on the upper surface of leaves (Sands and Kassulke, 1984). On an average, laboratory-reared females laid 73 eggs, while wild-collected females laid 94 eggs, with the maximum laid by one female being 291 (Knopf and Habeck, 1976). Bennett (1966) recorded 140 eggs/female. This moth appears to be restricted to aquatic habitats as it uses only aquatic plants for food and development (Bennett, 1975).

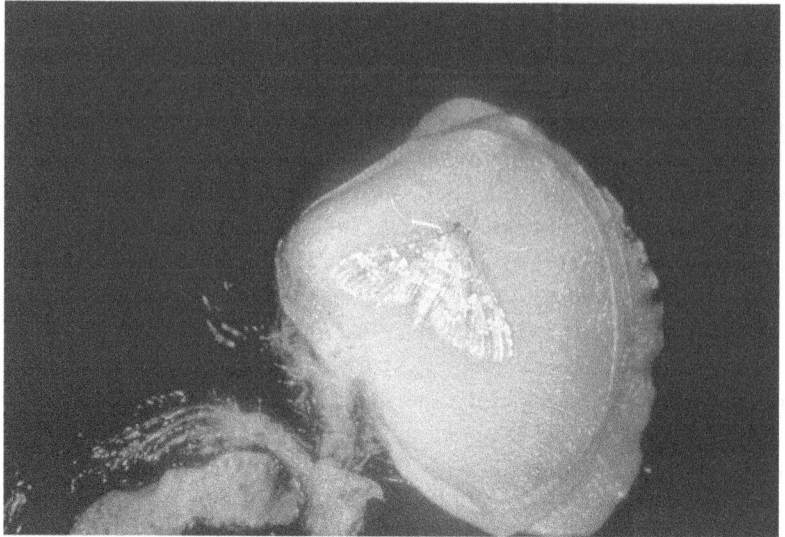

Fig. 19.7 *Samea multiplicalis* adult (photo by P. Room, CSIRO, Bugwood.org).

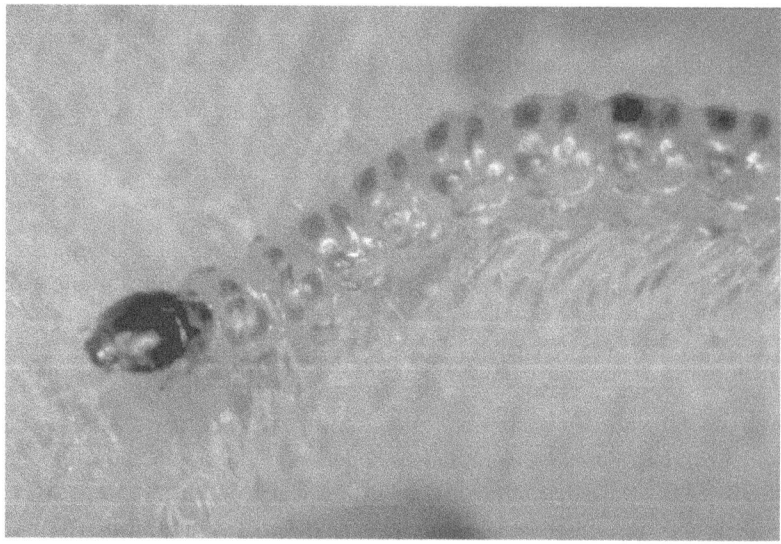

Fig. 19.8 *Samea multiplicalis* larva (photo by P. Room, CSIRO, Bugwood.org).

Under laboratory conditions (28°C; 14 h photoperiod), 64% of larvae completed five instars and 36% completed six instars in 15.6 days, while feeding on *S. minima*. Egg deposition to hatch averaged four days and the pupal stage took just over five days (Knopf and Habeck, 1976). In studies by Bennett (1966), where temperature was not indicated, egg hatch required either 5 or 6 days and larval development 16–25 days. Adult females lived an average of 4.2 days and males lived for 3.6 days. Sands and Kassulke (1984) found that females fed on honey and water laid 145 eggs and those without laid half that number. Larvae had 5–7 instars (usually six) and preferred younger plant material. Mean development times were 4 days for eggs, 29 days for larvae, 18 days for pupae, and 5 days for adults with a one-day preoviposition period (Sands and Kassulke, 1984). Larval nutrition had a major influence on fecundity (Taylor and Sands, 1986) and larval development was faster with fewer instars when young larvae were provided nitrogen-rich *S. molesta* (Taylor, 1984). The biology of this insect when reared on *P. stratiotes* was also studied by DeLoach *et al.* (1979).

The three main hosts are *Salvinia minima* (previously called *S. rotundifolia*), *Azolla caroliniana*, and *P. stratiotes*. In choice oviposition tests, adults preferred *P. stratiotes* over *S. minima* or *A. caroliniana*. Limited feeding occurred on *Lemna* sp. and *E. crassipes*, although larvae could not complete development on these plants (Bennett, 1966). Older larvae that were transferred from natural hosts were able to complete their development on *Lemna* sp., *Pontederia rotundifolia*, *E. crassipes* and *Brassica oleracea* (DeLoach *et al.*, 1979). *Samea* occasionally attacked *E. crassipes* (Knopf and Habeck, 1976) and could be reared from it (Bennett, 1966) though it appeared not a preferred host.

Bennett (1966) noted that *Samea* is heavily parasitized in Trinidad and Argentina and is heavily predated on by semi-aquatic syrphids, dyticids, and hydrophilids. A Braconidae and two Ichneumonidae parasites caused 52% parasitism in the field in USA (Knopf and Habeck, 1976). Semple and Forno (1987) found five parasitoids and three pathogens attacking *Samea* four years after its release in Australia.

19.11.4 *Paulinia acuminata* (De Geer) (Orthoptera: Pauliniidae)

This semi-aquatic grasshopper requires high humidity for population survival. The adults are green and brown and 3–4 cm long (Fig. 19.9). They have five or six nymphal stages, the later more common when temperatures are low (Thomas, 1980), although nutrition level experienced by the nymphs could be another reason (Taylor, 1984). Adults and nymphs eat leaves, preferring the tertiary growth form but will eat secondary form plants. In laboratory experiments, egg sacs are attached to the abaxial surface of leaves, while remaining in contact with the surface of the water. An average of one egg sac/day was produced by females ($n = 20$) with a mean of 3.9 eggs per egg sac on *S. molesta* with fewer produced on *Azolla* and *P. stratiotes* (Sands and Kassulke, 1986). Whereas Vieira and Adis (2000) observed an average 7.3 ± 2.8 eggs per egg sac ($n = 100$) on *Salvinia* spp., *Azolla* cf. *microphylla* and *P. stratiotes*. Egg development took 17–25 days, larval development of the five or six instars took 47 days, and preoviposition lasted 8–10 days at 25°C (Sands and Kassulke, 1986; Vieira and Adis, 2000). The life cycle was completed in 67 days and 93.5 days at 25°C and 29°C, respectively (Sands and Kassulke, 1986; Vieira and Adis, 2000). Thomas (1980) estimated the life cycle to vary between 85 and 33 days

Fig. 19.9 *Paulinia acuminata* (photo by P. Room, CSIRO, Bugwood.org).

at temperatures 25°C to 36°C and found that nymphal mortality was about 50% regardless of temperature.

Paulinia acuminata feeds on a wide range of plants with some differences in feeding host acceptance between nymphs and adults (Sands and Kassulke, 1986). Bennett (1966) found complete development only on *Salvinia* spp., *P. stratiotes*, *Azolla* spp., and *Hydromystria* sp., whereas Sands and Kassulke (1986) recorded complete development on *Azolla pinnata* and Vieira and Adis (2000) found complete development on *Azolla* cf. *microphylla*, *S. auriculata*, *S. minima*, and *S. sprucei*. Adults oviposited on *E. crassipes*, after feeding on that plant, but the eggs failed to hatch (Sands and Kassulke, 1986).

This grasshopper has been recorded from Argentina, Brazil, Paraguay, Trinidad, Uruguay, and is known to occur widely in South America (Bennett, 1966; Forno, 1981), The holotype is from Surinam and was cited by De Geer in 1773 (http://osf2.orthoptera. org/HomePage.aspx, accessed 11 July 2007). It is present in Botswana, Fiji, India, Sri Lanka, Zambia, and Zimbabwe following releases for biological control, and in Mozambique as a result of downstream movement from Zambia and Zimbabwe (Julien and Griffiths, 1998).

19.12 Implementing biological control

Biological control has had two distinct stages. The first was from the mid 1960s to mid 1970s when three insect species were collected from the *S. auriculata* complex and released in various countries but they did not provide control. The second stage began in the late 1970s. Two insect species were used, but one, a small weevil, proved to be the most effective and has been introduced to new locations as recently as 2003 into USA (Table 19.2).

Table 19.2 *Biological control of* Salvinia molesta: *releases of the salvinia weevil,* Cyrtobagous salviniae; *country and year of first release and the documented time to achieving control of the weed*

Country	Year	Level of control achieved (time between release and control)	Reference
Australia	1980	Excellent (14 mths[b]; 13 mths to 4 yrs[c])	[b] Thomas and Room, (1986b); [c] Forno, (1987)
Botswana	[a]	Excellent (1–5 yrs)	Procter (1984)
Congo, Democratic Republic	2000	Excellent (<2 yrs)	Mbati and Neuenschwander (2005)
Fiji	1991	Good	Julien and Griffiths (1998)
Ghana	1996	Good	Julien and Griffiths (1998)
India	1983	Excellent (<3 yrs)	Jayanth (1987)
Indonesia	1997	Unknown	
Kenya	1991	Good	Julien and Griffiths (1998)
Ivory Coast	1998	Unknown	
Papua New Guinea	1982	Excellent (11 mths for thin mats; 17 mths for thick mats; >2 yrs for sudd)	Thomas and Room, (1986a)
South Africa	1985	Excellent (1–3 yrs)	Cilliers (1991)
Malaysia	1989	Good (14 mths)	Julien and Griffiths (1998)
Mauritania	2000	Excellent (18 mths)	Dieme (2002) in Pieterse *et al.* (2003)
Namibia	1984	Excellent	Schlettwein and Hamman (1984)
Philippines	1991	Unknown	
Senegal	2001	Excellent (18 mths)	Pieterse *et al.* (2003)
Sri Lanka	1986	Excellent (usually 12–24 mths; 3 mths at one pond; >30 mths at another)	Room and Fernando (1992)
USA	2003	Excellent	
Zambia	1981	Excellent	Julien and Griffiths (1998)
Zimbabwe	1992	Excellent (2 yrs)	Chikwenhere and Keswani (1997)

[a] Spread into Botswana from Namibia.

19.12.1 Cyrtobagous singularis *Hustache (Coleoptera: Curculionidae)*

The first release of *Cyrtobagous singularis* was made in 1971 by CIBC (now CABI) on Lake Kariba, Zambia. Recovery was not possible as the release area became part of a military zone but by 1984 it was recovered in neighboring Zimbabwe. During 1971 to 1976 releases were made on the Chobe River, Botswana (Bennett, 1975; Julien and

Griffiths, 1998), where it is established but failed to provide control of the weed (Procter, 1984). Between 1972 and 1974 the weevil was released on the Linyanti River in the Caprivi Strip, Namibia, and in 1981 it was recovered on Lake Liambezi. Although it did cause some damage, reduction in the weed population was not significant (Schlettwein, 1985). From there it spread into the rest of Namibia with similar results. Finally it was released in Fiji in 1979, but not recovered until 1991 (Julien and Griffiths, 1998).

This species has not contributed significantly to the control of *S. molesta* despite establishing widespread populations in some areas. Its feeding behavior damages leaves and superficially damages stems and petioles but none is critical and the plant continues to grow (Sands and Schotz, 1985). In addition, its lower fecundity, lower larval and pupal survivorship, and greater dependence on nitrogen in its host plant, compared with *C. salviniae*, limit its ability to develop large and damaging populations (Sands *et al.*, 1986).

19.12.2 Cyrtobagous salviniae *Calder and Sands (Coleoptera: Curculionidae)*

The first release of this weevil was made on Lake Moondarra near Mt. Isa, Australia, in 1980 and this resulted in a spectacular reduction in weed abundance (Room *et al.*, 1981) (Figs. 19.10 and 19.11). Subsequently successful control was repeated at a number of large and small infestations such as the Sepik River in Papua New Guinea (Thomas and Room, 1986a, b) and the many ponds throughout Sri Lanka (Room and Fernando, 1992) as well as numerous smaller infestations in Australia and elsewhere (Table 19.2). The success of these projects led to a number of awards for the scientists involved including being awarded the UNESCO Science Prize in 1984.

To date at least 18 countries have benefited from releases of this insect (Table 19.2) with enormous implications for the welfare of human communities and wetland eco-logical systems. The Foreign Affairs Minister for Australia commented in parliament that this little insect was one of the best foreign aids Australia has developed.

This weevil has also been successful in some warm temperate areas in Australia providing control in ponds in the Sydney region (latitude 34°S) in under three years. However, in one small, exposed pond, populations persisted for five years without reducing the population of the weed, at which time it was removed using herbicide. In other areas at the southern extent of the weed, control has not been successful, possibly because releases of the weevil were made late in the growing season which reduced overwintering survival. Earlier releases in the following season resulted in establishment and control. There is great opportunity to use this insect in integrated management strategies in areas that are marginal for the weed and less than optimal for the weevil.

Nitrogen has been implicated in the successful establishment of the salvinia weevil in some locations (Room and Thomas, 1985). This insect will establish on nitrogen-poor *S. molesta* (Forno, 1987) but population development by the weevil takes longer as a low nitrogen food source leads to lower development rates and lower fecundity (Sands *et al.*, 1986). Under optimal conditions salvinia weevil populations have been known to increase

Fig. 19.10 Lake Moondarra, Australia, before biological control (photo by P. Room, CSIRO, Bugwood.org).

Fig. 19.11 Lake Moondarra, Australia, after biological control (photo by P. Room, CSIRO, Bugwood.org).

and destroy mats in as little as three months (Room and Fernando, 1992) but usually it has taken 1–2 years, even in large systems (Table 19.2).

The salvinia weevil has been released at a number of sites throughout Africa and has successfully brought the weed under control throughout the continent. This insect is not a rapid disperser and, where new *S. molesta* infestations appear, control is achieved more quickly if insects are released at these sites rather than relying on natural dispersal. While the control of *S. molesta* under tropical conditions usually occurs within 18 months, in cooler, high elevation areas and under eutrophic conditions, for example in South Africa, the control requires longer, sometimes up to three years (Cilliers, 1991). The weevil has not been reported to have recruited any parasitoids in the regions of introduction. However, White *et al.* (2007) recorded a parasitic pathogenic green alga, *Helicosporidum* sp. (Chlorophta: Trebouxiophyceae) from a culture of the weevil imported from South Africa to the USA. Although the impact of this alga on weevil populations is unknown, it could explain why control sometimes takes longer in South Africa under cooler conditions where the weevils are stressed.

In June 1999, weevils thought to be *C. salviniae* (Kissinger, 1966) were collected from *S. minima* in Florida and released in Texas to control *S. molesta*. This was done to minimize the risk of introducing new pathogens or parasites from foreign locations. Weevil recoveries were made but lasting control was not achieved because most of the release sites were destroyed by drought or herbicides. Molecular studies suggested that the weevil collected from Florida may be a new species (Goolsby *et al.*, 2000). This led to a temporary cessation of further releases and refocused efforts on release of the Brazil population of *C. salviniae* that was acquired from Australia (Tipping and Center, 2005). A subsequent study found that the Florida and Brazil populations were both *C. salviniae* (Madeira *et al.*, 2006).

19.12.3 Samea multiplicalis *(Guenée) (Lepidoptera: Pyralidae)*

This moth was first released on Lake Kariba, Zambia, in 1970 (Bennett, 1975, 1984; Mitchell and Rose, 1979) and then in Botswana in 1972 (Julien and Griffiths, 1998), but in both cases it failed to establish. In 1976 and 1981 it was released and became established in Fiji (Kamath, 1979; Julien and Griffiths, 1998) and in Australia (Room *et al.*, 1984), respectively. This species is commonly found attacking *S. molesta* in the US but has not contributed to control.

In Australia *Samea* quickly dispersed (170 km in 20 months (Room *et al.*, 1984)) and is now widespread. It causes patchy and sometimes very severe defoliation to the weed but does not provide control, even though compensatory growth is not stimulated (Julien and Bourne, 1988). A single generation of larvae may cause considerable defoliation and may reduce growth but larvae do not harm the growing points or affect the root–shoot connection (Julien and Bourne, 1988). Once that generation of larvae pupate, the plant is able to outgrow the damage. Because adults seek undamaged plants for oviposition (Taylor and Forno, 1987) there is little or no follow-up defoliation until the *S. molesta* has regenerated. As a consequence mats of *S. molesta* under heavy attack by *Samea* may have patches in different stages of defoliation and regrowth.

19.12.4 Paulinia acuminata *(De Geer) (Orthoptera: Pauliniidae)*

The first release of *P. acuminata* as a biological control agent was of a biotype (color morph) from Trinidad liberated into cages, from which they escaped, on Lake Kariba on the Zimbabwe side in 1969. Open field releases were made in 1971. Releases of the Trinidad type were made on the Zambia side in 1970. Material from Uruguay (a different color morph) was released on the Zimbabwe side in 1971. Both biotypes became established but after about a year the Trinidad strain was predominant (Julien and Griffiths, 1998).

Four years after its release on Lake Kariba, *P. acuminata* developed large populations up to $27/m^2$ and was spreading away from release sites. During 1973 the amount of *S. molesta* in the south (Zimbabwe) side of the lake (Mitchell and Rose, 1979) declined suddenly from around 400 km^2 to about 77 km^2 and even further to 39 km^2 over the next few years. This reflected the decline on all of the lake but access to the north was not possible. Mitchell and Rose (1979) considered that other changes that were taking place in this relatively young lake may have contributed to the decline in *S. molesta* but believed that the introduction of *P. acuminata* contributed markedly to the reduction of the weed. Marshall and Junor (1981) in their assessment of factors that were changing in the lake concluded that the decrease in *S. molesta* was associated mainly with competition for resources amongst the lake biota. They suggested that the introduction of the grasshopper accelerated a process that would have occurred anyway.

In 1984, fifteen years after release on Lake Kariba, field observations on the Zambezi River between Lake Kariba and Mana Pools (90 km downstream) indicated that populations were less than $3/m^2$. Observers noted that they grazed the uppermost leaves but did not damage the buds, hence the plants kept growing (Sands and Kassulke, 1986). At Hippo Creek, 5 km upstream of Victoria Falls on the Zambezi River, *P. acuminata* was observed feeding on the secondary growth form of *S. molesta* with less than one individual per m^2. The highest field densities recorded were 45–54 *P. acuminata* per m^2 at Mana Pools, 90 km below Lake Kariba in 1984, when 87% of leaves were damaged. Again, there was no damage to apical tips, and the damage did not destroy the plants (Sands and Kassulke, 1986).

Other releases of this grasshopper were made in Kenya in 1970 and Botswana in 1971 where it did not become established. This insect is present in Botswana following a second release in 1975. It is also present in Sri Lanka, India, and Fiji from releases made in 1973, 1974, and 1975, respectively (Julien and Griffiths, 1998). Although imported and studied in a quarantine facility in Australia (Sands and Kassulke, 1986) this insect was never released there. With the possible exception of the decline of *S. molesta* on Lake Kariba, *P. acuminata* has not been an effective biological control agent at any location where it has become established.

19.13 Integrated management of *Salvinia molesta* with biological control

Sometimes additional weed control techniques have been used in conjunction with releases of the salvinia weevil, *C. salviniae*. Thick mats of *S. molesta* and *S. molesta* supporting other plant species (sudds), which are not preferred by the weevil, have been reduced by herbicide application (Room and Thomas, 1986). After the thick mat loses buoyancy and sinks, a new rapidly growing layer of *S. molesta* forms over the open water which is much more effectively controlled by the weevil. Once populations buildup they are much better able to deal with the thicker sections of the mats. In the US it is common on larger drainages for managers to spray around boat ramps to minimize the chances of *S. molesta* hitching a ride while leaving the larger areas for the weevil.

Monitoring of weevil populations and weed biomass and relating them to seasonal conditions to make decisions about herbicide application to assist biological control was recommended for a tropical wetland area in Australia (Julien and Storrs, 1996). Booms, "harvesting" machines and herbicides were used to reduce the biomass of a large infestation of *S. molesta* near Sydney, Australia (Coventry, 2006). The salvinia weevil was released and is providing control of remaining *S. molesta* on the major waterways. It remains to be seen if the weevil will be able to maintain populations in this area where winters can be cool or if augmentative releases will be required early in the growing season (P. Sullivan and L. Postel, personal communication). Besides cool temperatures, the biggest threat to the weevils is follow-up management when herbicides are used to "mop up" small areas of the weed.

Many water bodies do not have native plants to fill the "gap" once *S. molesta* is brought under control. Consequently, they are susceptible to further invasion by other floating weeds such as *E. crassipes*, *P. stratiotes* and *Azolla* spp. For this reason it is sensible to develop management strategies for all of the weeds that threaten a water system. A sound strategy will preserve biological control agents already released and, where possible, will reduce the flow of nutrients entering waterways as these contribute to rapid rates of growth and greater biomass to be managed.

19.14 Benefit:cost of biological control of *Salvinia molesta*

Chikwenhere and Keswani (1997) detailed the costs of various methods of control for *S. molesta* covering commercial ponds in Zimbabwe. Physical and chemical controls were more expensive and less effective whereas biological control provided benefit to cost ratio of 10.6:1. Benefits relating to biological control have been assessed for Australia (Page and Lacey, 2006). However, it was not possible to separate the benefits concerning biological control of *E. crassipes*, *P. stratiotes* and *S. molesta*, as all three plants occur in similar water bodies and in similar locations. The combined costs of the biological control projects was AU$5.1 million (1974 to 1993). The *S. molesta* project cost AU$4.2 million over 11 years during the period 1978 and 1993. The combined benefit:cost ratio was 27.5:1. In a study on the biological control of *S. molesta* in Sri Lanka, Doeleman (1989) estimated a benefit to cost ratio of 53:1 in terms of money and 1671:1 in terms of labor for the complete control of the weed in the early 1980s. Such studies serve to highlight the huge economic benefits to be gained by using biological control.

19.15 Conclusion

In some situations, for example, at the edge of the cool range of *S. molesta*, integrated strategies may be required to manage this weed. However, they should always include biological control. Even in less than optimal climates, once the salvinia weevil is established it is probable that it will reduce *S. molesta* growth rates and biomass accumulation. This will assist other forms of control if they are required; there will be less weed to treat or treatment will be required less often. In cases where thick mats have formed or sudds developed, breaking those up with herbicides will assist biological control. These situations are not the norm. In most situations *S. molesta* can be effectively and efficiently managed by release of the salvinia weevil. The biological control of *S. molesta* is an extraordinary success story with a single weevil species. The success of this insect as a biological control agent has been repeated in a range of countries, climates, and habitats. There have been no reported adverse effects on other organisms during or following biological control of *S. molesta*. Therefore the first priority for the management of *S. molesta* should always be the release of the salvinia weevil.

References

Barrett, S. C. H. (1989). Waterweed invasions. *Scientific America*, **261**, 90–97.
Bennett, F. D. (1966). Investigations on the insects attacking the aquatic ferns, *Salvinia* spp. in Trinidad and northern South America. In *Proceedings of the Southern Weed Conference*, **19**, 497–504.
Bennett, F. D. (1975). Insects and plant pathogens for the control of *Salvinia* and *Pistia*. In *Proceedings of the Symposium on Water Quality Management Through Biological Control* , ed. P. L. Brezonik and J. L. Fox. Gainesville, FL: University of Florida, pp. 28–35.
Bennett, F. D. (1984). Biological control of aquatic weeds. In *Proceedings of the International Conference on Water Hyacinth*, held in Hyderabad, India, 7–11 February 1983, ed. G. Thyagarajan. Nairobi, Kenya: UNEP, pp. 14–40.
Calder, A. A. and Sands, D. P. A. (1985). A new Brazilian *Cyrtobagous* Hustache (Coleoptera: Curculionidae) introduced into Australia to control salvinia. *Journal of the Australian Entomological Society*, **24**, 57–64.
Cary, P. R. and Weerts, P. G. J. (1983). Growth of *Salvinia molesta* as affected by water and temperature and nutrition. I. Effects of nitrogen level and nitrogen compounds. *Aquatic Botany*, **16**, 163–172.

Cary, P. R. and Weerts, P. G. J. (1984). Growth of *Salvinia molesta* as affected by water and temperature and nutrition. III. Nitrogen-phosphorus interactions and effect of pH. *Aquatic Botany*, **19**, 171–182.

Chikwenhere, G. P. and Keswani, C. L. (1997). Economics of biological control of Kariba weed (*Salvinia molesta* Mitchell) at Tengwe in north-western Zimbabwe – a case study. *International Journal of Pest Management*, **43**, 109–112.

Cilliers, C. J. (1991). Biological control of water fern, *Salvinia molesta* (Salviniaceae), in South Africa. *Agriculture, Ecosystems and Environment*, **37**, 219–224.

Coventry, R. (2006). Hawkesbury River: managing salvinia on the Hawkesbury – a $1.8 million cooperative effort. In *Salvina Control Manual: Management and Control for Salvinia (*Salvinia molesta*) in Australia*, ed. E. van Oosterhout. Orange, Australia: NSW Department of Primary Industries, pp. 64–69.

Creagh, C. (1991/1992). A marauding weed in check. *Ecos*, **70**, 26–29.

Croxdale, J. G. (1978). Salvinia leaves. I. Origin and early differentiation of floating and submerged leaves. *Canadian Journal of Botany*, **56**, 1982–1991.

Croxdale, J. G. (1979). Salvinia leaves. II. Morphogenesis of the floating leaf. *Canadian Journal of Botany*, **57**, 1951–1959.

Croxdale, J. G. (1981). Salvinia leaves. III. Morphogenesis of the submerged leaf. *Canadian Journal of Botany*, **59**, 2065–2072.

DeLoach, C. J., DeLoach, D. J. and Cordo, H. A. (1979). Observations on the biology of the moth, *Samea multiplicalis*, on waterlettuce in Argentina. *Journal of Aquatic Plant Management*, **17**, 42–44.

De la Sota, E. R. (1962). Contribución al conocimiento de las *Salviniaceae* neotropicales. III. *Salvinia herzogii* nov. sp. *Darwiniana*, **12**, 514–520.

Divakaran, O., Arunachalam, M. and Nair, N. B. (1980). Growth rates of *Salvinia molesta* Mitchell with special reference to salinity. *Proceedings of the Indian Academy of Sciences (Plant Science)*, **89**, 161–168.

Doeleman, J. A. (1989). *Biological Control of* Salvinia molesta *in Sri Lanka. An Assessment of Costs and Benefits*. ACIAR Technical Report 12. Canberra, Australia: Union Offset, 14 pp.

EPPO (2007). EPPO Reporting Service no. 1 (Rse-0701.doc) (http://archives.eppo.org/EPPOReporting/2007/Rse-0701.pdf; accessed 17 May 2007).

Finlayson, C. M., Farrell, T. P. and Griffiths, D. J. (1982). *Treatment of Sewage Effluent Using the Water Fern Salvinia*. Technical Report 57. Kingsford, Australia: Water Research Foundation of Australia, 37 pp.

Forno, I. W. (1981). Progress in the exploration for biological control agents for *Salvinia molesta*. In *Proceedings of the V International Symposium on Biological Control of Weeds*, held 22–29 July 1980, Brisbane, ed. E. Delfosse. Melbourne, Australia: CSIRO, pp. 167–173.

Forno, I. W. (1983). Native distribution of the *Salvinia auriculata* complex and keys to species identification. *Aquatic Botany*, **17**, 71–83.

Forno, I. W. (1987). Biological control of the floating fern *Salvinia molesta* in north-eastern Australia: plant-herbivore interactions, *Bulletin of Entomological Research*, **77**, 9–17.

Forno, I. W. and Bourne, A. S. (1984). Studies in South America of arthropods on the *Salvinia auriculata* complex of floating ferns and their effects on *S. molesta*. *Bulletin of Entomological Research*, **74**, 609–621.

Forno, I. W. and Harley, K. L. S. (1979). The occurrence of *Salvinia molesta* in Brazil. *Aquatic Botany*, **6**, 185–187.

Forno, I. W. and Smith, P. A. (1999). Management of the alien weed, *Salvinia molesta*, in the wetlands of the Okavango, Botswana. In *An International Perspective on Wetland Rehabilitation*, ed. W. Streever. Dordrecht, Netherlands: Kluwer Academic Publishers, pp. 159–166.

Forno, I. W., Sands, D. P. A. and Sexton, W. (1983). Distribution, biology and host specificity of *Cyrtobagous singularis* Hustache (Coleoptera: Curculionidae) for the biological control of *Salvinia molesta*. *Bulletin of Entomological Research*, **73**, 85–95.

Gewertz, D. B. (1983). *Salvinia molesta*: the destruction of an ecosystem. In *Sepik River Societies: A Historical Ethnography of the Chambri and Their Neighbours*, ed. D. B. Gewertz. New Haven, CT: Yale University Press, pp. 196–217.

Giardini, M. (2004). *Salvinia molesta* D.S. Mitchell (Salviniaceae): seconda segnalazione per l'Italia (Lazio) e considerazioni sul controllo di questa specie infestante. *Webbia*, **59**, 457–467.

Goolsby, J. A., Tipping, P. W., Center, T. D. and Driver, F. (2000). Evidence of a new *Cyrtobagous* species (Coleoptera: Curculionidae) on *Salvinia minima* Baker in Florida. *Southwestern Entomologist*, **25**, 299–301.

Guenée, A. (1854). *Species général des lépidoptéres*. Vol. 8. *Deltoides et Oyralites*. Paris: Roret (cited in Knopf and Habeck, 1976).

Hassler, M. and Swale, B. (2002). Family Salviniaceae, genus *Salvinia*; world species list. http://homepages.caverock.net.nz/~bj/fern/salvinia.htm (accessed 20 July 2007).

Hennecke, B. and Postle, L. (2006). The key to success: an investigation into oviposition of the salvinia weevil in cool climate regions. In *Proceedings of the 15th Australian Weeds Conference*, held 24 to 28 September 2006, Adelaide, ed. C. Preston, J. H. Watts and N. D. Crossman. Adelaide, Australia: Weed Management Society of South Australia, pp. 780–783.

Hertzog, R. (1935). Einm Beitrag zur Systematik der Gattung *Salvinia*. *Hedwigia*, **74**, 257–284.

Holm, L. G., Plucknett, D. L., Pancho, J. V. and Herberger, J. P. (1977). *The World's Worst Weeds: Distribution and Biology*. Honolulu, HI: University Press of Hawaii, 609 pp.

Jacono, C. C., Davern, T. R. and Center, T. D. (2001). The adventive status of *Salvinia minima* and *S. molesta* in the southern U.S. and the related distribution of the weevil *Cyrtobagous salviniae*. *Castane*, **66**, 214–226.

Jayanth, K. P. (1987). Biological control of the water fern *Salvinia molesta* infesting a lily pond in Bangalore (India) by *Cyrtobagous salviniae*. *Entomophaga*, **32**, 163–165.

Julien, M. H. and Bourne, A. S. (1986). Compensatory branching and changes in nitrogen content in the aquatic weed *Salvinia molesta* in response to disbudding. *Oecologia*, **70**, 250–257.

Julien, M. H. and Bourne, A. S. (1988). Effects of leaf-feeding by the larvae of the moth *Samea multiplicalis* Guen. (Lep., Pyralidae) on the floating weed *Salvinia molesta*. *Journal of Applied Entomology*, **106**, 518–526.

Julien, M. H. and Griffiths, M. W. (1998). *Biological Control of Weeds. A World Catalogue of Agents and Their Target Weeds*. Fourth edn. Wallingford, UK: CABI Publishing, 223 pp.

Julien, M. H. and Storrs, M. J. (1993). *Salvinia molesta* in Kakadu National Park: biological control. In *Proceedings of the 10th Australian Weeds Conference and the*

14th Asian Pacific Weed Science Society Conference, held September 6–10 in Brisbane Australia, ed. J. T. Swarbrick. Brisbane, Australia: Weed Society of Queensland, pp. 220–224.

Julien, M. H. and Storrs, M. J. (1996). Integrating biological and herbicidal controls to manage Salvinia in Kakadu National Park, northern Australia. In *Proceedings of the IX International Symposium on Biological Control of Weeds*, held 19–26 January 1996, ed. V. C. Moran and J.H. Hoffman. Stellenbosch, South Africa: University of Cape Town, pp. 445–449.

Julien, M. H., Chan, R. R. and Bourne, A. S. (1987). Effects of adult and larval *Cyrtobagous salviniae* on the floating weed *Salvinia molesta*. *Journal of Applied Ecology*, **24**, 935–944.

Kamath, M. K. (1979). A review of biological control of insect pests and noxious weeds in Fiji 1969–1978. *Agricultural Journal of Fiji*, **41**, 55–72.

Kammathy, R. V. (1968). *Salvinia auriculata* Aublet – a rapidly spreading exotic weed in Kerala. *Science and Culture*, **34**, 396.

Kissinger, D. G. (1966). *Cyrtobagous* Hustache, a genus of weevils new to the United States fauna (Coleoptera: Curculionidae: Bagoini). *Coleopterists' Bulletin*, **20**, 125–127.

Knopf, K. W. and Habeck, D. H. (1976). Life history and biology of *Samea multiplicalis*. *Environmental Entomology*, **5**, 539–542.

Loyal, D. S. and Grewal, R. K. (1966). Cytological studies on sterility in *Salvinia auriculata* Aublet with a bearing on its reproductive mechanism. *Cytologia*, **31**, 330–338.

Madeira, P. T, Tipping, P. W., Gandolfo, D. E., *et al.* (2006). Molecular and morphological examination of *Cyrtobagous* sp. collected from Argentina, Paraguay, Brazil, Australia, and Florida. *BioControl*, **51**, 679–701.

May, B. M. and Sands, D. P. A. (1986). Descriptions of larvae and biology of *Cyrtobagous* (Coleoptera: Curculionidae): agents for biological control of salvinia. *Proceedings of the Entomology Society of Washington*, **88**, 303–312.

Maki, K. and Galatowitsch, S. (2004). Movement of invasive aquatic plants into Minnesota (USA) through horticultural trade. *Biological Conservation*, **118**, 389–396.

Marshall, B. E. and Junor, F. J. R. (1981). The decline of *Salvinia molesta* on Lake Kariba. *Hydrobiologia*, **83**, 477–484.

Mbati, G. and Neuenschwander, P. (2005). Biological control of three floating water weeds, *Eichhornia crassipes*, *Pistia stratiotes*, and *Salvinia molesta* in the Republic of Congo. *BioControl*, **50**, 635–645.

Mitchell, D. S. (1978). *Aquatic Weeds in Australian Waters*. Canberra, Australia: Australian Government Publishing Service, 189 pp.

Mitchell, D. S. (1979). The incidence and management of *Salvinia molesta* in Papua New Guinea. Draft Report, Office of Environment and Conservation, Papua New Guinea. Cited in Oliver (1993).

Mitchell, D. S. (1972). The Kariba weed: *Salvinia molesta*. *British Fern Gazette*, **10**, 251–252.

Mitchell, D. S. and Rose, D. J. W. (1979). Factors affecting fluctuations in extent of *Salvinia molesta* on Lake Kariba. *PANS*, **25**, 171–177.

Mitchell, D. S. and Thomas, P. A. (1972). *Ecology of Water Weeds in the Neotropics*. Technical Papers in Hydrology 12. Paris: UNESCO, 50 pp.

Mitchell, D. S. and Tur, N. M. (1975). The rate of growth of *Salvinia molesta* (*S. auriculata* Auct.) in laboratory and natural conditions. *Journal of Applied Ecology*, **12**, 213–225.

Mitchell, D. S., Petr, T. and Viner, A. B. (1980). The water fern *Salvinia molesta* in the Sepik River, Papua New Guinea. *Environmental Conservation*, **7**, 115–122.

Moozhiyil, M. and Pallauf, J. (1986). Chemical composition of water fern, *Salvinia molesta*, and its potential as feed source for ruminants. *Economic Botany*, **40**, 375–383.

Oliver, J. D. (1993). A review of the biology of giant salvinia (*Salvinia molesta* Mitchell). *Journal of Aquatic Plant Management*, **31**, 227–231.

Page, A. R. and Lacey, K. L. (2006). Economic impact assessment of Australian weed biological control. *CRC for Australian Weed Management Technical Series*, **10**, 123–127.

Pancho, J. V. and Soerjani, M. (1978). *Aquatic Weeds of Southeast Asia*. Quezon City, Philippines: National Publ. Coop., 129 pp.

Parsons, W. T. and Cuthbertson E. G. (2001). *Noxious Weeds of Australia*. Collingwood, Australia: CSIRO Publishing. pp. 609–612.

Pieterse, A., Kettunen, M., Diouf, S., *et al.* (2003). Effective biological control of *Salvinia molesta* in the Senegal River by means of the weevil *Cyrtobagous salviniae*. *Ambio*, **23**, 458–462.

Procter, D. L. C. (1984). Biological control of the aquatic weed *Salvinia molesta* D. S. Mitchell in Botswana using the weevils *Cyrtobagous singularis* and *Cyrtobagous* sp. nov. *Botswana Notes and Records*, **15**, 99–101.

Raju, R. A. and Gangwar, B. (2004). Utilisation of potassium-rich green-leaf manures for rice (*Oryza sativa*) nursery and their effects on crop productivity. *Indian Journal of Agronomy*, **49**, 244–247.

Randall, R. P. (2002). *A Global Compendium of Weeds*. Meredith, Victoria: R. G. and F. J. Richardson, 944 pp.

Room, P. M. (1983). Falling-apart as a lifestyle – the rhizome architecture and population growth of *Salvinia molesta*. *Journal of Ecology*, **17**, 349–365.

Room, P. M. (1986). Equations relating growth and uptake of nitrogen by *Salvinia molesta* to temperature and the availability of nitrogen. *Aquatic Botany*, **24**, 43–59.

Room, P. M. (1988). Effects of temperature, nutrients and a beetle on branch architecture of the floating weed *Salvinia molesta* and simulations of biological control. *Journal of Ecology*, **76**, 826–848.

Room, P. M. and Fernando, I. V. S. (1992). Weed invasions countered by biological control: *Salvinia molesta* and *Eichhornia crassipes* in Sri Lanka. *Aquatic Botany*, **42**, 99–107.

Room, P. M. and Gill, Y. J. (1985). The chemical environment of *Salvinia molesta* Mitchell: ionic concentrations of infested waters. *Aquatic Botany*, **23**, 127–135.

Room, P. M. and Julien, M. H. (1995). *Salvinia molesta* D. S. Mitchell. In *The Biology of Australian Weeds*, Vol. 1, ed. R. H. Groves, R. C. H Shepherd and R. G. Richardson. Melbourne, Australia: R.G. and F. J. Richardson, pp. 217–230.

Room, P. M. and Thomas, P. A. (1985). Nitrogen and establishment of a beetle for biological control of the floating weed salvinia in Papua New Guinea. *Journal of Applied Ecology*, **22**, 139–156.

Room, P. M. and Thomas, P. A. (1986). Nitrogen, phosphorus and potassium in *Salvinia molesta* Mitchell in the field: effects of weather, insect damage, fertilizers and age. *Aquatic Botany*, **24**, 213–232.

Room, P. M., Forno, I. W. and Taylor, M. F. J. (1984). Establishment in Australia of insects for biological control of the floating weed *Salvinia molesta* Mitchell. *Bulletin of Entomological Research*, **74**, 505–516.

Room, P. M., Harley, K. L. S., Forno, I. W. and Sands, D. P. A. (1981). Successful biological control of the floating weed salvinia. *Nature*, **294**, 78–80.

Sands, D. P. A. and Kassulke, R. C. (1984). *Samea multiplicalis* [*Lep. Pyralidae*], for biological control of two water weeds, *Salvinia molesta* and *Pistia stratiotes* in Australia. *Entomophaga*, **29**, 267–273.

Sands, D. P. A. and Kassulke, R. C. (1986). Assessment of *Paulinia acuminata* [Orthoptera: Acrididae] for the biological control of *Salvinia molesta* in Australia. *Entomophaga*, **31**, 11–17.

Sands, D. P. A. and Schotz, M. (1985). Control or no control: a comparison of the feeding strategies of two salvinia weevils. In *Proceedings of the VI International Symposium on Biological Control of Weeds*, ed. E. S. Delfosse. Vancouver, Canada: Agriculture Canada, pp. 551–556.

Sands, D. P. A., Schotz, M. and Bourne, A. S. (1983). The feeding characteristics and development of larvae of a salvinia weevil *Cyrtobagous* sp. *Entomologia Experimentalis et Applicata*, **34**, 291–296.

Sands, D. P. A., Schotz, M. and Bourne, A. S. (1986). A comparative study of the intrinsic rates of increase of *Cyrtobagous singularis* and *C. salviniae* on the water weed *Salvinia molesta*. *Entomologia Experimentalis et Applicata*, **42**, 231–237.

Schotz, M. and Sands, D. P. A. (1988). Diel pattern of feeding and oviposition by *Cyrtobagous salviniae* Calder and Sands (Coleoptera: Curculionidae). *Australian Entomological Magazine*, **15**, 31–32.

Schlettwein, C. H. G. (1985). Distribution and densities of *Cyrtobagous singularis* Hustache (Coleoptera: Curculionidae) on *Salvinia molesta* Mitchell in the Eastern Caprivi Zipfel. *Madoqua*, **14**, 291–293.

Schlettwein, C. H. G. and Hamman, P. F. (1984). The control of *Salvinia molesta* in the Eastern Caprivi Zipfel, *South West Africa Annual*, **1984**, 49–51.

Semple, J. L. and Forno, I. W. (1987) native parasitoids and pathogens attacking *Samea multiplicalis* Guenée (Lepidoptera: Pyralidae) in Queensland. *Journal of the Australian Entomological Society*, **26**, 365–366.

Storrs, M. J. and Julien, M. H. (1996). Salvinia: *A Handbook for the Integrated Control of* Salvinia molesta *in Kakadu National Park*. Northern Landscapes Occasional Papers No. 1. Darwin, Australia: Australian Nature Conservation Agency, 58 pp.

Taylor, M. F. J. (1984). The dependence of development and fecundity of *Samea multiplicalis* on early larval nitrogen intake. *Journal of Insect Physiology*, **30**, 779–785.

Taylor, M. F. J. and Forno, I. W. (1987). Oviposition preferences of the salvinia moth *Samea multiplicalis* Guenée (Lep., Pyralidae) in relation to hostplant quality and damage. *Zeitschrift für angewandte Entomologie*, **104**, 73–78.

Taylor, M. F. J. and Sands, D. P. A. (1986). Effects of ageing and nutrition on the reproductive system of *Samea multiplicalis* Guenée (Lepidopteran: Pyralidae). *Bulletin of Entomological Research*, **76**, 513–517.

Thomas, P. A. (1980). Life-cycle studies on *Paulinia acuminata* (De Geer) (Orthoptera: Pauliniidae) with particular reference to the effects of constant temperature. *Bulletin of Entomological Research*, **70**, 381–389.

Thomas, P. A. (1985). Management of *Salvinia molesta* in Papua New Guinea. *FAO, Plant Protection Bulletin*, **33**, 50–56.

Thomas, P. A. and Room, P. M. (1986a). Successful control of the floating weed *Salvinia molesta* in Papua New Guinea: a useful biological invasion neutralizes a disastrous one. *Environmental Conservation*, **13**, 242–248.

Thomas, P. A. and Room, P. M. (1986b). Taxonomy and control of *Salvinia molesta*. *Nature*, **320**, 581–584.

Tipping, P. W. and Center, T. D. (2003). *Cyrtobagous salviniae* (Coleoptera: Curculionidae) successfully overwinters in Texas and Louisiana. *Florida Entomologist*, **86**, 92–93.

Tipping, P. W. and Center, T. D. (2005). Influence of plant size and species on preference of *Cyrtobagous salviniae* adults from two populations. *Biological Control*, **3**, 263–268.

van Oosterhout, E. (2006). *Salvinia Control Manual: Management and Control Options for Salvinia* (Salvinia molesta) *in Australia*. Orange, Australia: NSW Department of Primary Industries. 80 pp.

Vieira, M. F. and Adis, J. (2000). Aspectos da biologia e etologia de *Paulinia acuminata* (De Geer), 1773 (Orthoptera: Pauliniidae), um gafanhoto semi-aquático, na Amazônia Central. *Acta Amazonica*, **30**, 333–346.

Waterhouse, D. F. and Norris, K. R. (1987). *Biological Control Pacific Prospects*. Melbourne, Australia: Inkata Press, 454 pp.

White, S. E., Tipping, P. W. and Becnel, J. J. (2007). First isolation of a *Helicosporidum* sp. (Chlorophyta: Trebouxiophyceae) from the biological control agent *Cyrtobagous salviniae* (Coleoptera: Curculionidae). *Biological Control*, **40**, 243–245.

Whiteman, J. B. and Room, P. M. (1991). Temperature lethal to *Salvinia molesta* Mitchell. *Aquatic Botany*, **40**, 27–35.

Williams, R. H. (1956). *Salvinia auriculata* Aubl.: the chemical eradication of a serious aquatic weed in Ceylon. *Tropical Agriculture*, 33 145–157.

20

Solanum mauritianum Scopoli (Solanaceae)

Terry Olckers

20.1 Introduction

The South American *Solanum mauritianum* Scopoli (Solanaceae) comprises an unarmed, branched shrub or small tree, 2–4 m tall, which threatens many commercial activities and natural habitats worldwide, particularly in subtropical and tropical regions (ISSG, 2006). Introductions of the plant have occurred via several routes, notably accidental transfer by seafaring human colonists, deliberate importations for ornamental purposes, and long-distance dispersal by frugivorous birds (Roe, 1972; ISSG, 2006). Known by several common names worldwide, the most important being bugweed, tree tobacco, and woolly nightshade, the plant is emerging as an important environmental weed in many countries (Florentine and Westbrooke, 2003; PIER, 2005; ISSG, 2006).

In South Africa, where the weed has proved to be particularly invasive, infestations affect agricultural lands, forestry plantations, riverine habitats, and conservation areas, particularly in the eastern, higher rainfall regions of the country (Henderson, 2001). The high weed status of *S. mauritianum* in South Africa resulted in its targeting for biological control in 1984 (Olckers and Zimmermann, 1991). South Africa is currently the only country that has fully implemented a biological control program against *S. mauritianum*, including exploration for candidate agents, host-specificity testing and release and establishment of agents (Olckers, 1999). New Zealand has recently funded research on some of the agents that have been used in South Africa and may soon consider the importation of the most promising species (Withers *et al.*, 2002; Borea, 2006).

In this chapter, I review the global distribution and ecology of *S. mauritianum* as well as the biocontrol initiatives that have been undertaken so far, including: the problems that are unique to this program; the nature of the biocontrol agents that were considered; the efficacy and economics of these initiatives; and the prospects for the successful biological control of this weed worldwide.

20.2 Weed distribution and ecology

Several published and electronic sources provide a description of the plant, which is characterized by the densely pubescent foliage (caused by fine, whitish trichomes) and the

Biological Control of Tropical Weeds using Arthropods, ed. R. Muniappan, G. V. P. Reddy, and A. Raman. Published by Cambridge University Press. © Cambridge University Press, 2009.

profusion of lilac blue flowers which produce compact terminal clusters of green berries that ripen to a dull yellowish color (Kissmann and Groth, 1997; Henderson, 2001; PIER, 2005; ISSG, 2006). In South Africa, frugivorous birds extensively use the ripe fruits, which are generally available throughout the year (Oatley, 1984; Pooley, 1993). The excessively high levels of fruit set in countries like South Africa has facilitated the rapid spread and establishment of dense infestations of the weed.

The native range of *S. mauritianum* is considered to cover northern Argentina, southern Brazil, Uruguay, and Paraguay (Roe, 1972). Several very closely related species, including *S. erianthum* D. Don, *S. granuloso-leprosum* Dun., *S. riparium* Pers., and *S. verbascifolium* L., coexist with *S. mauritianum* or occur further north in South America, even into North America (Kissmann and Groth, 1997). These may well comprise geographical forms of a widespread and variable species complex that includes *S. mauritianum*.

Global databases list *S. mauritianum* as an introduced species over a wide range, including several African countries, Australia, India, New Zealand, and many islands in the Atlantic, Indian, and Pacific oceans (PIER, 2005; ISSG, 2006). The plant is considered to be invasive in many countries, notably South Africa, several Pacific Islands (e.g. Fiji, Hawaii, and Tonga), New Zealand, and some Indian Ocean islands (e.g. Madagascar, Mauritius, and La Reunion). Although *S. mauritianum* has been promoted as a nursery plant to facilitate forest regeneration in tropical Australia, recent evidence suggests that it is emerging as an important invader that may well have allelopathic properties (Florentine and Westbrooke, 2003).

The life history and ecology of *S. mauritianum* predispose the plant to rapid invasion (see Olckers and Zimmermann, 1991; ISSG, 2006 and references therein). The plant reproduces primarily by seeds and is capable of rapid regrowth when mechanical destruction is carried out without herbicidal follow-up. Seeds of *S. mauritianum* are dispersed by several species of frugivores, notably birds (Oatley, 1984; Pistorius, 2005) and bats (Parsons *et al.*, 2006). In particular, populations of Rameron pigeon (*Columba arquatrix*) have been linked with high levels of seedling recruitment in disturbed habitats, especially forestry plantations, in South Africa (Oatley, 1984). The plant also displays very rapid growth rates and an ability to produce flowers within the first year after germination. Plants are capable of self-pollination (Rambuda and Johnson, 2004) and also produce flowers and fruits all year round. Although soil seed banks persist for relatively short periods, seed germination may be stimulated by fire, and seedlings that become established in summer are able to flower by autumn. Within 2–3 years, plants reach a height of several meters.

In disturbed habitats, notably pastoral land, native forest margins, forestry plantations, and urban areas, the weed forms dense stands that overcrowd and shade out native plants, thereby inhibiting their growth (Henderson, 2001). In South Africa, the weed is legislated as a "transformer," recognizing the ability of monospecific stands to "dominate or replace any canopy or sub-canopy layer of a natural or semi-natural ecosystem thereby altering its structure, integrity and functioning" (Henderson, 2001). Similar effects occur in commercial pine plantations, where inhibition of growth of young forestry plants has been observed (see references in Olckers and Zimmermann, 1991). The foliage is unpalatable to wild and domestic animals and all parts of the plant are toxic to humans and animals, especially the green berries, which are rich in alkaloids (ISSG, 2006). In addition, the fruit provides alternative hosts for several species of fruit flies (Tephritidae) that are pests of cultivated fruit in Africa, and serve as reservoirs for these pests throughout the year (Olckers and Zimmermann, 1991; Copeland and Wharton, 2006).

20.3 Biological control initiatives

Biological control was initiated in South Africa because conventional control methods are either ineffective or unsustainable over the long term (Olckers and Zimmermann, 1991; Olckers, 1999). Besides hand pulling of seedlings and ring barking of trees, mechanical control is largely ineffective due to rapid regrowth of felled plants. In addition, the fine trichomes that cover the foliage may cause skin and respiratory irritations to humans when they are dislodged during clearing operations (Henderson, 2001; ISSG, 2006). Chemical control is more effective and the plant is easily killed by herbicides, with chemicals such as glyphosate, triclopyr, and imazapyr being registered for the control of *S. mauritianum* in South Africa. However, the density and extent of existing infestations and the rate at which cleared areas are reinvaded by bird-dispersed seedlings renders chemical control unsustainable over the long term (see Section 20.5). Biological control was thus invoked in South Africa to augment conventional methods within an integrated control framework.

Biological control initiatives until 1999 have been reviewed elsewhere (Olckers and Zimmermann, 1991; Olckers, 1999) and are briefly summarized. Surveys for potential agents were initiated in 1984 and continued opportunistically until 1994 when the program was officially inaugurated. Lists of arthropods associated with *S. mauritianum* in South America are provided elsewhere (Neser *et al.*, 1990; Olckers *et al.*, 2002; Pedrosa-Macedo *et al.*, 2003). Subsequent to the political and social changes in South Africa in 1994, official collaboration with South American institutions and scientists has greatly facilitated surveys for candidate agents as well as host-range studies in their native habitats, notably in Argentina (Olckers *et al.*, 2002) and Brazil (Pedrosa-Macedo *et al.*, 2003).

The first agents introduced into quarantine were selected on the basis of their abundance and obvious damage and not in any order of priority. Since 1994, agents that reduce fruiting have been prioritized, given the high levels of fruit set of *S. mauritianum* in South Africa and the need to limit the spread and invasion of the weed through long-range seed dispersal (Olckers, 1999). In addition, agents that reduce growth rates and possibly the density of existing populations have also been considered, given the weed's high rates of vegetative growth. In this regard, defoliating agents with high reproductive and feeding rates as well as stem-boring agents have the potential to debilitate plants by damaging photosynthetic and structural tissues. The ultimate aim is to establish a complex of agents from these feeding guilds to maximize the stress of high herbivore loads on plant populations and augment conventional control methods (Olckers, 1999).

Some 15 insect species have been introduced into quarantine in South Africa for screening as candidate agents (Table 20.1). The prioritization of these for testing was influenced by the

Table 20.1 *South American insects introduced into South Africa for consideration as biological control agents for* Solanum mauritianum

Insect species	Year	Origin	Damage	Outcome
Agents that reduce photosynthetic area				
Corythaica cyathicollis (Costa) (Tingidae)	1984	Argentina	Sap-sucking	Rejected; crop pest
Acrolepia xylophragma (Meyrick) (Acrolepiidae)	1984–1990	Argentina	Leaf-mining	Rejected; suitable for other countries?
Platyphora species (Chrysomelidae)	1994	Argentina, Brazil	Leaf-chewing	Rejected; suitable for other countries?
Acallepitrix sp. nov. (Chrysomelidae)	1994–1998	Argentina, Brazil	Leaf-mining	Rejected; suitable for other countries?
Gargaphia decoris Drake (Tingidae)	1995, 2002	Argentina, Brazil	Sap-sucking	Released; established and causing localized damage
Collabismus notulatus Boheman (Curculionidae)	1995, 1998	Argentina	Shoot-galling	Untested; possibly suitable
Agents that cause structural damage				
Nealcidion bicristatum (Bates) (Cerambycidae)	1984, 1995	Argentina	Stem-boring	Rejected; crop pest
Adesmus hemispilus (Germar) (Cerambycidae)	1995, 1997	Argentina, Brazil	Stem-boring	Untested; possibly suitable
Conotrachelus squalidus Boheman (Curculionidae)	1995, 1998	Argentina, Paraguay	Stem-boring	Untested; possibly suitable
Agents that reduce fruit production				
Anthonomus santacruzi Hustache (Curculionidae)	1995, 1998	Argentina, Paraguay	Flowerbud-feeding	Cleared for release; releases pending
Anthonomus morticinus Clark (Curculionidae)	1998	Argentina, Paraguay	Flowerbud-feeding	Partially tested; probably suitable

ease with which ectophagous leaf-feeding species were collected, cultured, and tested relative to the more challenging endophagous flowerbud-feeding and stem-boring species. In addition, the screening of agents was strongly influenced by difficulties in demonstrating host specificity during routine laboratory testing procedures (Olckers, 1999). Indeed, *S. mauritianum* is a particularly difficult target for biological control given the problem of expanded host ranges during laboratory tests which has been aggravated by the number of

Solanum species that are either native to or cultivated in South Africa. In particular, the ease with which candidate agents accept cultivated eggplant (*Solanum melongena* L.) during cage trials is a constant problem. Of these 15 agents, nine were rejected because of a lack of demonstrable host specificity, three were not tested due to culturing difficulties, one was released and has become established, and two are under consideration for release. Details on these agents are provided in Section 20.4.

Slow progress in South Africa, caused largely by the difficulty in obtaining clearance for the release of agents, could jeopardize the launch of biocontrol programs against *S. mauritianum* in other countries. Approaches that could be used to resolve these problems are discussed later. Since 1998, there have been no introductions of new agents into South Africa and the program against *S. mauritianum* was suspended in 2003, but was resumed in late 2007.

20.4 Biology and ecology of natural enemies

20.4.1 Agents deemed unsuitable

Despite their unsuitability for release in South Africa, the oligophagous nature of some of these agents may permit their use in other countries that have depauperate *Solanum* floras or where cultivated *Solanum* species, notably *S. melongena*, have low economic status.

A stem-boring beetle *Nealcidion bicristatum* (Bates) (Coleoptera: Cerambycidae) and a leaf-sucking lace bug *Corythaica cyathicollis* (Costa) (Hemiptera: Tingidae) were the first two agents to be introduced in 1984. Both were rejected before any host-specificity tests were conducted because of published host records in Brazil that included cultivated plants (Olckers, 1999). Subsequent field surveys in Argentina, Brazil, and Paraguay (Olckers *et al.*, 2002; Pedrosa-Macedo *et al.*, 2003) confirmed that neither of these insects is host specific and that, despite the high levels of damage inflicted on *S. mauritianum*, they are not suitable for release anywhere in the world.

A leaf-mining moth *Acrolepia xylophragma* (Meyrick) (Lepidoptera: Acrolepiidae) was imported on several occasions, starting in 1984, because of impressive damage in which the caterpillars cause blotch mines that destroy the leaves of young plants growing in shady habitats. Although the moths are restricted to the genus *Solanum*, larvae developed on several nontarget species during host-specificity tests, including potato (*S. tuberosum* L.) and eggplant, which have never been recorded as hosts in South America. Research on the moth was suspended in the early 1990s, when the *S. mauritianum* project was temporarily suspended and the moth has not been reconsidered since then (Olckers, 1999). More intensive evaluations, including multichoice trials under less restrictive quarantine conditions and open-field trials in South America, are required before it can be concluded whether or not the moth is suitable for release in South Africa and elsewhere.

Five species of leaf-feeding beetles of the genus *Platyphora* Gistel (Coleoptera: Chrysomelidae) were introduced in 1994. Despite displaying narrow host ranges in the field, their laboratory host ranges were considerably broader and included eggplant, potato, and several native *Solanum* species (Olckers, 1998, 2000a). Although these results contradicted published information on the genus and may well be laboratory artifacts, all five *Platyphora* species were rejected because their sensitivity to food quality and microhabitat made them unlikely to establish widely under South African conditions anyway. These beetles prefer the softer foliage of small plants, seedlings, and coppice of *S. mauritianum* growing in cool, moist, and shaded habitats (e.g. plantations, forest margins, and clearings) (Olckers, 1998, 2000a). Although the beetles did not appear to cause extensive damage to their host, some species may be considered for introduction and evaluation in New Zealand, where the conditions may be more suitable than in South Africa (R. L. Hill, personal communication).

An undescribed flea beetle, *Acallepitrix* sp. nov. (Coleoptera: Chrysomelidae), which occurs mostly on plants growing in shaded or semishaded habitats, was tested after it was introduced and successfully cultured in 1997. Extensive damage by the leaf-chewing adults and leaf-mining larvae causes leaf abscission and thus has the potential to reduce plant growth rates. Although confined to the genus *Solanum*, an unacceptably broad host range during quarantine tests, which included eggplant and some native species, resulted in its rejection (Olckers, 2004). To preclude the possibility of laboratory artifacts, open-field trials in Brazil are required before it can be determined whether or not the beetle is suitable for release in South Africa and elsewhere.

20.4.2 Untested candidate agents

Three candidate agents that were introduced during 1995–1998 were not tested because of difficulties in initiating or sustaining cultures in quarantine. All are endophagous species that include two stem-boring beetles, *Adesmus hemispilus* (Germar) (Coleoptera: Cerambycidae) and *Conotrachelus squalidus* Boheman (Coleoptera: Curculionidae), and a weevil *Collabismus notulatus* Boheman (Coleoptera: Curculionidae) whose larvae were presumed to be stem borers but are now thought to form galls on the shoot tips.

Adults of *A. hemispilus* oviposited in captivity, but proved problematic because the adults are aggressive and attack other conspecifics in cages and because the larvae develop slowly, taking 7–12 months to reach adulthood (Olckers, 1999). Although the two cerambycids, *N. bicristatum* and *A. hemispilus*, laid eggs on stem sections which were able to support larval development, this was not true of the weevil *C. squalidus*. Adults of *C. squalidus* oviposited and produced larvae on potted plants, but no larvae were reared to adulthood, and attempts to transfer field-collected larvae into fresh stem sections failed because the larvae rejected the fresh material. It seems that larvae of *C. squalidus* require succulent stems that are typical of young plants or coppice and are not suited to older woodier stems (Olckers, 1999). Adults of *C. notulatus* that were introduced into quarantine did not oviposit and very little is known about the biology of this weevil.

All three species may be suitable for reintroduction since they were recorded only on *S. mauritianum* during field surveys of native and cultivated *Solanum* species in Argentina and Brazil (Olckers *et al.*, 2002; Pedrosa-Macedo *et al.*, 2003) and may thus be host specific. *Conotrachelus squalidus* is considered to be the most promising of the stem-boring candidates but culturing problems may constrain progress with host-specificity testing.

20.4.3 Agents released

The leaf-sucking lace bug *Gargaphia decoris* Drake (Hemiptera: Tingidae) is the only agent that has been released against *S. mauritianum* anywhere in the world. First imported from a small founder colony collected at a single locality in Argentina in 1995, the insect was tested and later released in South Africa in 1999 (Olckers, 2000b). This occurred despite feeding on eggplant and some native *Solanum* species during host-specificity tests. Further attempts to introduce new genetic stocks from Argentina failed because the insect was not encountered again (Olckers *et al.*, 2002) and releases from 1999–2001 were conducted using the original material. However, in 2002, fresh stocks of *G. decoris*, comprising a much broader genetic base, were introduced from the First Plateau of Paraná in southern Brazil (Pedrosa-Macedo *et al.*, 2003), where the cooler climatic conditions are more similar to the warm temperate regions in South Africa where *S. mauritianum* is particularly invasive. Releases from both provenances have resulted in establishments in the field, although most have arisen from the original Argentinean material. Releases were carried out in five South African provinces, with establishment so far confirmed in KwaZulu-Natal, Limpopo, and Mpumalanga. Surveys at the original release sites have shown that the insects do not colonize the nontarget *Solanum* species that were fed on in the laboratory trials, justifying the decision to release the lace bug (Olckers and Lotter, 2004).

Preliminary laboratory experiments revealed that high levels of feeding damage affected the growth of *S. mauritianum* plants, with damaged plants being stunted and containing 33% less biomass than the undamaged controls (T. Olckers, unpublished data). However, with one notable exception (see below), high levels of damage by *G. decoris* have mostly not been realized in the field. Indeed, the insect has proved inconsistent, with many early releases failing to establish colonies (Lotter, 2004). However, observations during 2006–2007 revealed many established field populations in close proximity to some of the release sites, even ones where releases were deemed to have failed. This suggests that establishment success may well be higher than the 18% reported from KwaZulu-Natal province, where over 148 000 insects were released at 32 sites (Lotter, 2004). Nevertheless, damage to weed populations has mostly been moderate because lace bug populations have remained at low densities and have mostly not reached the outbreak levels that are needed to inflict severe damage (but see below).

Initially, cold winter temperatures were suspected of suppressing populations of *G. decoris* (Lotter, 2004) but temperature tolerance trials revealed that populations from both provenances are cold tolerant, with the Brazilian population able to tolerate lower temperatures (Barker and Byrne, 2005). Instead, predation of the eggs and early nymphal instars by several generalist predators is believed to have limited population increases, but has not yet been quantified.

In April 2007, an outbreak of *G. decoris* was reported in an invaded forestry plantation in Mpumalanga province. Massive numbers of adults and nymphs had inflicted severe damage on the *S. mauritianum* plants, causing extensive, and sometimes total, defoliation, an absence of fruit and flowers, and even mortality of both seedlings and larger trees (A. B. R. Witt, personal communication). Similar outbreaks were reported from other sites in the area. Monitoring of the site is in progress to determine whether this is a sporadic event or whether the damage will be sustained over time as well as to quantify the impact on the weed population. At this stage, it is unclear why similar outbreaks have not been observed elsewhere in South Africa. In any event, interest in *G. decoris* has been renewed by this recent occurrence and the agent may have more potential than previously thought.

Gargaphia decoris is being considered for introduction into New Zealand. Additional host-specificity testing in South Africa, involving other cultivated *Solanum* species and ones that are native to New Zealand, further confirmed that the insect is suitable for release in New Zealand (Withers *et al.*, 2002; Borea, 2006). Field surveys of predators associated with *S. mauritianum* populations in New Zealand are planned to assess the potential threat to *G. decoris* establishment and efficacy (R. L. Hill, personal communication). In addition, it seems prudent that the impact of *G. decoris* on the plant's growth and reproduction should be quantified to provide a more compelling case for the release of the agent in New Zealand.

20.4.4 *Agents proposed for release*

The most promising agents are considered to be flowerbud-feeding weevils of the genus *Anthonomus* Germar (Coleoptera: Curculionidae) which occur throughout the natural range of *S. mauritianum* in South America and appear to be largely responsible for the low levels of fruiting typical of the plants in their natural habitats. Feeding by high adult populations causes abortion and abscission of the flowers and flower buds, while the endophagous larvae develop inside the flower buds and destroy them (Olckers, 2003). Two species, *Anthonomus santacruzi* Hustache and *A. morticinus* Clark, are commonly associated with *S. mauritianum* populations in South America, and often coexist at the same localities (Olckers *et al.*, 2002). Despite their potential and several introductions into quarantine, excessive mortality during importations and inadequate culturing procedures constrained progress for several years.

These problems were resolved in 1998 and biology studies and host-specificity tests on *A. santacruzi* were completed in 2002 (Olckers, 2003). Despite some nontarget feeding on eggplant and some native *Solanum* species during these trials, several lines of evidence, including host records, field surveys in South America, and a risk assessment, suggested that the ambiguous laboratory results were further examples of artificially expanded host ranges (Olckers *et al.*, 2002; Olckers, 2003). An application for permission to release the weevil was submitted in 2003 and after a protracted delay, permission was granted in mid 2007. Since the laboratory cultures died out during the four-year interim period, releases will probably commence in 2008, once stocks of the weevil have been reintroduced.

Anthonomus morticinus was accidentally imported into quarantine in South Africa, together with *A. santacruzi*, in 1998. Following confirmation that two species were involved, all individuals of *A. morticinus* were destroyed to prevent contamination of the *A. santacruzi* culture and possible hybridization. While *A. morticinus* and *A. santacruzi* were collected at the same sites in Argentina and Paraguay (Olckers *et al.*, 2002), *A. santacruzi* was rare at sites on the First Plateau of Paraná in southern Brazil, where *A. morticinus* was very common (Pedrosa-Macedo *et al.*, 2003). Although *A. morticinus* has not been subjected to quarantine trials, field surveys and open-field trials in southern Brazil have revealed that this species also has a very narrow host range (J. H. Pedrosa-Macedo, unpublished) and could thus also be considered for release.

20.5 Economics and efficacy of control

Considerable funds are allocated annually for the clearing of *S. mauritianum* infestations in South Africa. Although various agencies are involved, the only records of costs available are those of the "Working for Water" program, a national initiative focused on the removal of alien vegetation from various lands and water resources, with an annual operating budget of about US$80 million (Zimmermann *et al.*, 2004; Moran *et al.*, 2005). Preliminary figures from the 2002/03 financial year revealed that of the US$18 million that was spent on the mechanical clearing and chemical control of a complex of weed species nationally, US$1 million (5.6%) was spent on *S. mauritianum* alone (Marais *et al.*, 2004). Although the weed is effectively controlled by such actions, follow-up treatments are needed to cope with rapid reinvasion that is facilitated by bird-dispersed seeds and rapid growth rates. Indeed, it was calculated that at least 23 years are required to clear the estimated 89 500 ha that are currently covered by *S. mauritianum* in South Africa (Marais *et al.*, 2004). Ignoring annual cost increases, at least US$24 million is required to control the weed over the long term, underlying the need for biological control to reduce these costs.

So far, the biological control program against *S. mauritianum* in South Africa has not yielded any measurable benefits. Although *G. decoris* has become established and is dispersing, for the most part, damage to *S. mauritianum* populations is trivial (i.e. some damage, but survival, growth, and seed production of the plants appears unaffected) and is currently making no contribution to the management of the weed (Zimmermann *et al.*, 2004). Considerable time and money, including some 14–20 scientist years, were invested into this program until its termination in 2003 but have not delivered the favorable benefit-cost outcomes typical of several other weed biocontrol initiatives in South Africa (van Wilgen *et al.*, 2004). This may change if the recent outbreak of *G. decoris* is sustained and becomes more widespread.

20.6 Prospects for biological control

A major gap in the program is knowledge of whether or not the invasiveness of *S. mauritianum* is a result of its escape from herbivory. Although this has not been demonstrated quantitatively, lack of insect herbivore attack on *S. mauritianum* populations in South Africa (Olckers and Hulley, 1995) compared with high insect herbivore diversity and damage on those in South America (Neser *et al.*, 1990) has been assumed to contribute to the weed's invasiveness. Indeed, the high levels of fruit production in South Africa are not typical of South American populations, where several natural enemies, notably florivorous species (Olckers, 2003), are believed to influence seedling recruitment. Ecological studies on the plant's population dynamics in both native and invaded habitats are needed to determine whether natural enemies regulate natural populations and thus whether biological control is likely to be effective.

This program has been a challenging one because of the difficulties experienced during host-specificity testing and in advocating agents for release. Indeed, many countries might not have considered targeting plants such as *S. mauritianum* for biological control because of their relatedness to not only the native flora (e.g. Pemberton, 2000), but also cultivated crops. However, other countries that are starting to experience problems with the plant could consider the launch of biological control programs, building on the South African experience and mindful of the constraints. Besides quantitative consideration of the role of natural enemies, a country's decision to implement a biological control program against *S. mauritianum* could be influenced by: (i) the availability of agents from climatically suitable regions to facilitate establishment; (ii) the diversity of native species of *Solanum* in the country; (iii) the economic importance and extent of cultivation of crops in the genus *Solanum*; (iv) whether or not the regulatory authorities are prepared to accept an element of risk, given the ambiguity of the results likely to be generated by quarantine host-specificity tests; and (v) the extent to which native predators and parasites are likely to interfere with imported agents. These considerations are further discussed below.

Solanum mauritianum occurs over a very wide geographic and climatic range in South America (Lorenzi, 1991), including warm temperate and tropical regions. Countries where the weed occurs in colder climates, e.g. South Africa and New Zealand, have an opportunity to source agents from similar areas, e.g. the southern States of Paraná and Rio Grande do Sol in Brazil. Alternatively, tropical and subtropical countries, e.g. Australia (Queensland) and various Pacific Islands, could focus their importations on northern Argentina, Paraguay, and the northern limits of *S. mauritianum* in Brazil, e.g. Minas Gerais and São Paulo States. Consequently, the availability of suitably adapted agent populations should not constitute a limitation.

Countries with a high diversity of native *Solanum* species may decide against the implementation of biological control because of fears of nontarget damage. The southern hemisphere supports the highest diversity of *Solanum* species globally, with South America, which supports at least 1000–1100 species, regarded as the main center of speciation (D'Arcy, 1979; Hunziker, 1979; Symon, 1981). Other centers of speciation include Africa with some 110 species (34 of these occurring in South Africa) and Australia with some 94 species (Symon, 1981; Jaeger and Hepper, 1986; Welman, 2003). In contrast, New Zealand supports only three native *Solanum* species (Withers *et al.*, 2002). Countries with a low *Solanum* diversity, or that have no native species, would thus carry a much lower risk of having agents rejected because of expanded host ranges in quarantine or because of fears of host range extension in the field. Countries with a native *Solanum* flora should also consider the initiation of preintroduction surveys of *S. mauritianum* as well as native *Solanum* species (Olckers and Hulley, 1995) to determine whether native solanaceous insects have extended their host ranges to the weed or whether potential agents have already been introduced inadvertently. Such surveys in South Africa have demonstrated that native solanaceous insects do not readily transfer to "new" hosts (Olckers and Hulley, 1995) and have provided some support for the release of imported biocontrol agents on the basis that they are similarly unlikely to do so (Olckers, 2000b, 2003).

Countries that produce solanaceous crops intensively may also be suspicious of introducing agents against *S. mauritianum*, in light of the expanded host ranges experienced during host-range testing. Although potato is the most widely grown of these, the agents tested so far have mostly been unable to survive on this crop (Olckers, 1999). Instead, eggplant has consistently proven to be able to support survival and development of several agents introduced against *S. mauritianum*, even though the crop has never been attacked by them in South America where it has been exposed to the agents for decades. Other cultivated *Solanum* species include *S. muricatum* Aiton (pepino dulce), *S. quitoense* Lam. (naranjilla) and *S. macrocarpon* L. (African eggplant), but these have mostly not been considered since they are not routinely planted in South Africa. Cultivated *Solanum* species can be considered to be more at risk than native congeners, since the former have mostly been artificially selected for human palatability and may well have lost the majority of their chemical and physical defences through domestication. Indeed, eggplant in South Africa is particularly susceptible to generalist pests and is mostly cultivated under intensive pesticide regimes (Olckers and Hulley, 1994, 1995). Countries that grow eggplant and related crops under these conditions could argue that pesticide regimes will preclude damage in the unlikely event that they are colonized by the agents. The argument is obviously weakened when such crops are grown organically, in the absence of pesticides.

Countries that have a long history of weed biocontrol may be more amenable than others to accepting the arguments that were used in South Africa to argue against the ambiguous results of laboratory host-range tests and put forward a case that advocates release of the agents (Olckers, 2000b, 2003). Such arguments will need to be invoked because of the certainty that expanded host ranges will pose a problem during risk analyses and result in the unnecessary rejection of promising agents. Countries where the regulatory authorities are more risk averse and demand clear-cut results with no nontarget damage during no-choice and choice tests would be advised to not consider biocontrol in the case of *S. mauritianum*, unless these countries do not cultivate crops or harbor native species in the genus *Solanum*. The conservative nature of no-choice tests should be recognized and emphasis should be placed on the results of choice tests in quarantine or surveys and open-field trials in the agents' country of origin. In many cases, it may be necessary to resolve any doubts or ambiguous results by resorting to open-field trials in the country of origin or in countries like South Africa where the agents have been released. Countries that are successful in securing releases of agents against *S. mauritianum* should include studies of potential nontarget effects as part of their postrelease evaluations (Olckers and Lotter, 2004) to demonstrate that the predictions of no, or very little, nontarget damage were realized and thus that the decision to release was scientifically sound.

Although it is very difficult to predict whether or not predation will influence bio-control programs, some evidence to support or disprove a suspicion may facilitate a decision on whether or not to proceed. In South Africa, *S. mauritianum* populations seem to support high numbers of generalist predators, notably ants, ladybirds, and mirid bugs, which may be associated with nutrients (e.g. sugars) that are produced by the ever-present flowers and ripe fruit. Several of these generalist predators prey on the eggs and early nymphal stages of *G. decoris* and are believed to be reducing their impact in the field. Consideration of potential predators during preintroduction surveys of herbivores may provide some insight into the possibilities for success.

20.7 Conclusions

It should be acknowledged that *S. mauritianum* is a difficult target for biocontrol, not because of a lack of candidate agents but rather difficulties in demonstrating their safety for release. This may indeed be a deterrent to many countries where there are higher priorities for new biocontrol programs with better prospects for success. However, countries where the risk of nontarget effects, whether real or perceived, is considerably lower could benefit considerably from the efforts undertaken so far in South Africa. Such countries could focus on promising agents that have already been identified (Table 20.1) or undertake surveys in South America, in areas not covered by South African scientists, to source either "new" agents or climatically adapted "biotypes" of prioritized species. In addition, quantitative studies on the impact of individual agents, which are carried out in the laboratory or in the country of origin prior to release, can advance a case for release, since regulators may be more willing to tolerate risk if efficacy is clearly demonstrated.

Guidelines for countries embarking on a new biocontrol program against *S. mauritianum* include: (i) initiating studies to elucidate the demography and impact of the weed so as to establish a baseline for evaluating the role of biological control in weed management as well as its success or failure; (ii) conducting preintroduction surveys of the weed and any native or cultivated *Solanum* species to record any host-range extensions of native herbivores and determine the "pool" of natural enemies that could affect introduced agents; (iii) choosing host-range testing methodologies that are less likely to yield ambiguous results (e.g. multichoice and open-field trials); (iv) evaluating the status of all native *Solanum* species (e.g. rare, endangered, weedy, etc.) that could potentially be subject to nontarget effects; (v) determining the economic value and susceptibility to pests (e.g. extent of pesticide use) of all *Solanum* crops; (vi) quantifying the economic and environmental damage already caused by *S. mauritianum* in relation to any potential risks that might be realized; and (vii) conducting quantitative prerelease studies on the impact of promising candidate agents to demonstrate their efficacy and justify their release. Such considerations may be used to alleviate the risk and implications of nontarget effects. Although the *S. mauritianum* program in South Africa has so far not reaped the desired outcomes, there is considerable potential for better control.

Acknowledgments

I thank M. P. Hill (Rhodes University), J. H. Hoffmann (University of Cape Town), A. B. R. Witt (ARC-Plant Protection Research Institute) and anonymous reviewers for comments on the manuscript. The "Working for Water" program (Department of Water Affairs and Forestry), the ARC-PPRI and, more recently, the University of KwaZulu-Natal and Landcare Research, New Zealand are acknowledged for their financial support of the program.

References

Barker, A. L. and Byrne, M. J. (2005). Biotypes and biocontrol: physiological factors affecting the establishment of a South American lace bug *Gargaphia decoris*, a biocontrol agent of bugweed, *Solanum mauritianum*. *Proceedings of the 15th Entomological Congress*, ed. M. H. Villet. Pretoria, South Africa: Entomological Society of Southern Africa, 8 pp.

Borea, C. K. (2006). Host specificity and risk assessment of the leaf-sucking lace bug *Gargaphia decoris* (Tingidae), a potential biocontrol agent against the invasive weed *Solanum mauritianum* (Solanaceae) in New Zealand. Unpublished B.Sc. (Hons) thesis, University of KwaZulu-Natal, Pietermaritzburg, South Africa.

Copeland, R. S. and Wharton, R. A. (2006). Year-round production of pest *Ceratitis* species (Diptera: Tephritidae) in fruit of the invasive species *Solanum mauritianum* in Kenya. *Annals of the Entomological Society of America*, **99**, 530–535.

D'Arcy, W. G. (1979). The classification of the Solanaceae. In *The Biology and Taxonomy of the Solanaceae*, ed. J. G. Hawkes, R. N. Lester and A. D. Skelding. Linnean Society Symposium Series 7. London: Academic Press, pp. 3–47.

Florentine, S. K. and Westbrooke, M. E. (2003). Allelopathic potential of the newly emerging weed *Solanum mauritianum* Scop. (Solanaceae) in the wet tropics of north-east Queensland. *Plant Protection Quarterly*, **18**, 23–25.

Henderson, L. (2001). *Plant Protection Research Institute Handbook*. Vol. 12. Alien Weeds and Invasive Plants. Cape Town, South Africa: Paarl Printers.

Hunziker, A. T. (1979). South American Solanaceae: a synoptic survey. In *The Biology and Taxonomy of the Solanaceae*, ed. J. G. Hawkes, R. N. Lester and A. D. Skelding. Linnean Society Symposium Series 7. London: Academic Press, pp. 49–85.

ISSG (2006). Ecology of *Solanum mauritianum*. In *Global Invasive Species Database*, ed. Invasive Species Specialist Group (ISSG) (http://www.issg.org).

Jaeger, P. L. and Hepper, F. N. (1986). A review of the genus *Solanum* in Africa. In *Solanaceae Taxonomy and Biology*, ed. W. G. D'Arcy. New York: Columbia University Press, pp. 41–45.

Kissmann, K. G. and Groth, D. (1997). *Plantas Infestantes e Nocivas*, Tomo 3, 2 Edição. São Paulo, Brasil: BASF.

Lorenzi, H. (1991). *Plantas Daninhas do Brasil: Terrestres, Aquáticas, Parasitas, Tóxicas e Medicinais*, 2 Edição. Nova Odessa, Brazil: Editora Plantarum.

Lotter, W. D. (2004). The establishment and ecological impact of the leaf-sucking lace bug *Gargaphia decoris* on *Solanum mauritianum* in KwaZulu-Natal.

Unpublished M.Tech. thesis, Tshwane University of Technology, Pretoria, South Africa.

Marais, C., van Wilgen, B. W. and Stevens, D. (2004). The clearing of invasive alien plants in South Africa: a preliminary assessment of costs and progress. *South African Journal of Science*, **100**, 97–103.

Moran, V. C., Hoffmann, J. H. and Zimmermann, H. G. (2005). Biological control of invasive alien plants in South Africa: necessity, circumspection, and success. *Frontiers in Ecology and the Environment*, **3**, 77–83.

Neser, S., Zimmermann, H. G., Erb, H. E. and Hoffmann, J. H. (1990). Progress and prospects for the biological control of two *Solanum* weeds in South Africa. In *Proceedings of the VII International Symposium on Biological Control of Weeds*, held in Rome, ed. E. S. Delfosse. Melbourne, Australia: CSIRO, pp. 371–381.

Oatley, T. B. (1984). Exploitation of a new niche by the Rameron pigeon *Columba arquatrix* in Natal. *Proceedings of the fifth Pan-African Ornithological Congress*, ed. J. Ledger. Johannesburg, South Africa: Southern African Ornithological Society, pp. 323–330.

Olckers, T. (1998). Biology and host range of *Platyphora semiviridis*, a leaf beetle evaluated as a potential biological control agent for *Solanum mauritianum* in South Africa. *BioControl*, **43**, 225–239.

Olckers, T. (1999). Biological control of *Solanum mauritianum* Scopoli (Solanaceae) in South Africa: a review of candidate agents, progress and future prospects. In *Biological Control of Weeds in South Africa (1990–1998)*, ed. T. Olckers and M. P. Hill. African Entomology Memoir 1. Johannesburg, South Africa: Entomological Society of Southern Africa, pp. 65–73.

Olckers, T. (2000a). Biology and physiological host range of four species of *Platyphora* Gistel (Coleoptera: Chrysomelidae) associated with *Solanum mauritianum* Scop. (Solanaceae) in South America. *The Coleopterists Bulletin*, **54**, 497–510.

Olckers, T. (2000b). Biology, host specificity and risk assessment of *Gargaphia decoris*, the first agent to be released in South Africa for the biological control of the invasive tree *Solanum mauritianum*. *BioControl*, **45**, 373–388.

Olckers, T. (2003). Assessing the risks associated with the release of a flowerbud weevil, *Anthonomus santacruzi*, against the invasive tree *Solanum mauritianum* in South Africa. *Biological Control*, **28**, 302–312.

Olckers, T. (2004). Biology, host specificity and risk assessment of the leaf-mining flea beetle, *Acallepitrix* sp. nov., a candidate agent for the biological control of the invasive tree *Solanum mauritianum* in South Africa. *BioControl*, **49**, 323–339.

Olckers, T. and Hulley, P. E. (1994). Resolving ambiguous results of host-specificity tests: the case of two *Leptinotarsa* species (Coleoptera: Chrysomelidae) for biological control of *Solanum elaeagnifolium* Cavanilles (Solanaceae) in South Africa. *African Entomology*, **2**, 137–144.

Olckers, T. and Hulley, P. E. (1995). Importance of preintroduction surveys in the biological control of *Solanum* weeds in South Africa. *Agriculture, Ecosystems and Environment*, **52**, 179–185.

Olckers, T. and Lotter, W. D. (2004). Possible non-target feeding by the bugweed lace bug, *Gargaphia decoris* (Tingidae), in South Africa: field evaluations support predictions of laboratory host-specificity tests. *African Entomology*, **12**, 283–285.

Olckers, T. and Zimmermann, H. G. (1991). Biological control of silverleaf nightshade, *Solanum elaeagnifolium*, and bugweed, *Solanum mauritianum* (Solanaceae) in South Africa. *Agriculture, Ecosystems and Environment*, **37**, 137–155.

Olckers, T., Medal, J. C. and Gandolfo, D. E. (2002). Insect herbivores associated with species of *Solanum* (Solanaceae) in northeastern Argentina and southeastern Paraguay, with reference to biological control of weeds in South Africa and the United States of America. *Florida Entomologist*, **85**, 254–260.

Parsons, J. G., Cairns, A., Johnson, C. N., *et al.* (2006). Dietary variation in spectacled flying foxes (*Pteropus conspicillatus*) of the Australian Wet Tropics. *Australian Journal of Zoology*, **54**, 417–428.

Pedrosa-Macedo, J. H., Olckers, T., Vitorino, M. D. and Caxambu, M. G. (2003). Phytophagous arthropods associated with *Solanum mauritianum* Scopoli (Solanaceae) in the First Plateau of Paraná, Brazil: a cooperative project on biological control of weeds between Brazil and South Africa. *Neotropical Entomology*, **32**, 519–522.

Pemberton, R. W. (2000). Predictable risk to native plants in weed biological control. *Oecologia*, **125**, 489–494.

PIER (2005). *Solanum mauritianum*. Pacific Island Ecosystems at Risk (PIER) (http://www.hear.org/pier).

Pistorius, P. (2005). Potential dispersal of invasive alien South African plant fruits by dark-capped bulbuls (*Pycnonotus tricolor*): a dietary perspective. Unpublished B.Sc.(Hons.) thesis, University of KwaZulu-Natal, Pietermaritzburg, South Africa.

Pooley, E. (1993). *The Complete Field Guide to Trees of Natal, Zululand and Transkei*. Durban, South Africa: Natal Flora Publications Trust.

Rambuda, T. D. and Johnson, S. D. (2004). Breeding systems of invasive alien plants in South Africa: does Baker's rule apply? *Diversity and Distributions*, **10**, 409–416.

Roe, K. E. (1972). A revision of *Solanum* sect. *Brevantherum* (Solanaceae). *Brittonia*, **24**, 239–278.

Symon, D. E. (1981). A revision of the genus *Solanum* in Australia. *Journal of the Adelaide Botanical Gardens*, **4**, 1–367.

van Wilgen, B. W., De Wit, M. P., Anderson, H. J., *et al* (2004). Costs and benefits of biological control of invasive alien plants: case studies from South Africa. *South African Journal of Science*, **100**, 113–122.

Welman, W. G. (2003). Solanaceae. In *Plants of Southern Africa: an Annotated Checklist*, ed. G. Germishuizen and N. L. Meyer. Strelitzia 14. Pretoria, South Africa: National Botanical Institute, pp. 913–918.

Withers, T. M., Olckers, T. and Fowler, S. V. (2002). The risk to *Solanum* spp. in New Zealand from *Gargaphia decoris* (Hem.: Tingidae), a potential biocontrol agent against woolly nightshade, *S. mauritianum*. *New Zealand Plant Protection*, **55**, 90–94.

Zimmermann, H. G., Moran, V. C. and Hoffmann, J. H. (2004). Biological control in the management of invasive alien plants in South Africa, and the role of the Working for Water program. *South African Journal of Science*, **100**, 34–40.

21

Application of natural antagonists including arthropods to resist weedy *Striga* (Oranbanchaceae) in tropical agroecosystems

Joachim Sauerborn and Dorette Müller-Stöver

21.1 Introduction

Parasitic flowering plants are defined by the production of specialized nutrition-deriving structures, the haustoria, that form a functional link to their hosts. Species of *Striga* (witchweeds) are obligate hemiparasites, and connection to a host plant is fundamental for them to survive. Seeds of *Striga* cannot germinate until a "chemical" such as strigol and sorgolactone exuded by the host root indicates the vicinity of a host. Host-recognition factors that can activate development programs in *Striga* spp. are termed xenognosins (Lynn *et al.*, 1981). Atsatt (1977) proposed that parasitic plants probably use host defence chemicals as cues to stimulate the germination and growth of the haustorium, and which have originally evolved in the host to deter harmful organisms. Akiyama *et al.* (2005) suggest that plants release chemicals (sesquiterpene lactones) from their roots as signals fostering their symbiosis with arbuscular mycorrhizal fungi, and that these signals are used by *Striga* to detect host roots. Several strigolactones found in root exudates of various plant species (Yasuda *et al.*, 2003) stimulated germination in seeds of *Striga* species under laboratory conditions.

Striga is an r-strategist; that is, it allocates lots of energy to produce large numbers of minute seeds to reduce the risk associated with host finding. Producing many minute seeds increases the chance that at least a few seeds will get close enough to the roots of a suitable host plant. Numbers of seeds per plant average 58 000 in *S. asiatica*, and numbers over 200 000 almost certainly occur in well-grown *S. hermonthica* (Parker and Riches, 1993). In agroecosystems, the frequent and dense occurrence of host plants results in improved conditions for the reproduction of *Striga*, supporting the build-up of an abundant soil seed bank that can contain up to several million seeds per square meter. The resulting large, persistent seed bank – seeds may remain viable in soil for 7–14 years – causes a high infection probability in host plants (Bebawi *et al.*, 1984; Sand, 1990). *Striga* species exhibit two life phases: autotrophic and heterotrophic (Fig. 21.1). The autotrophic

Biological Control of Tropical Weeds using Arthropods, ed. R. Muniappan, G. V. P. Reddy, and A. Raman. Published by Cambridge University Press. © Cambridge University Press, 2009.

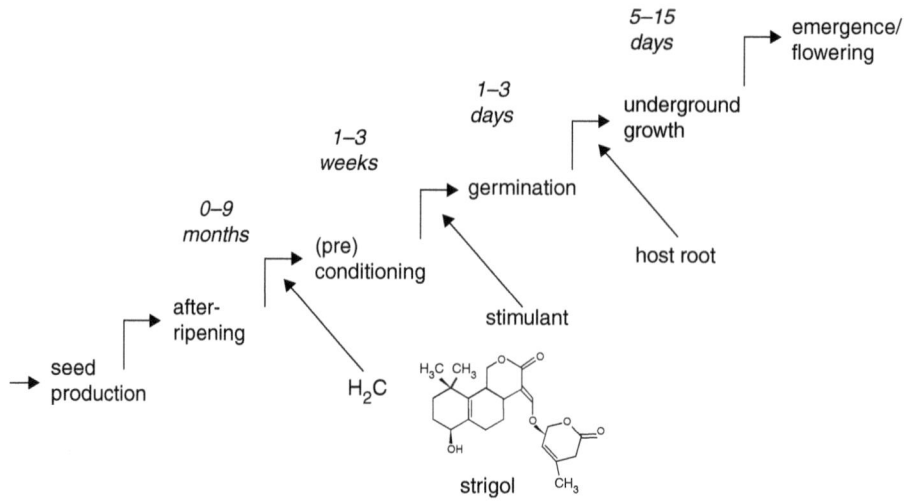

Fig 21.1 Schematic representation of the steps involved in germination and development of *Striga*.

phase begins with seed imbibition and germination. When the seed germinates, a radicle emerges out of the seed growing only to few millimeters. This process lasts a couple of days, during which the seedling utilizes the stored materials (e.g. lipids) in the seed, until a host root is found for attachment.

A chemical "cross-talk" between host and parasite triggers the transition from auto-trophic to heterotrophic growth by signaling haustorium development. The chemistry of haustorial induction is distinct from that of germination. The growing *Striga* seedling produces H_2O_2 (Keyes *et al.*, 2001), which is generally the limiting substrate in perox-idase oxidation of cell-wall-localized phenolics into benzoquinones in the host root (Kim *et al.*, 1998). The accumulating quinones diffuse back to the parasite seedling and initiate haustoriogenesis. Depending on the xenognosin concentration, several hours of exposure time are necessary for an irreversible commitment to haustorium development (Chang and Lynn, 1986). The heterotrophic phase, during which *Striga* becomes dependent on nutrients derived from the host, starts as soon as the haustorium invades the host root, forming a physiological bridge between the vascular system of the host and that of the parasite (Joel, 2000). Host-derived materials may then be transferred from the source (host plant) to the sink (parasite) through straw-like penetrations referred to as oscula into the host vascular system (Dörr, 1997). *Striga* then develops a shoot that grows under-ground for 4–7 weeks. In some species, lateral adventitious roots that emerge from a young parasite can also develop haustoria whenever a host root is available in the vicinity. A single *Striga* plant can bear several haustoria, connecting them to roots of one or more host plants. Flowering time of the parasite is dependent on species and environment. For example, *S. gesnerioides* flowers as soon as it emerges from soil (Emechebe *et al.*, 1991; Parker and Riches, 1993). *Striga asiatica* and *S. hermonthica* flower about four weeks after emergence and set seeds one month later.

Most *Striga* species are hemiparasites, because they photosynthesize upon emergence. However, growth and photosynthesis measurements of *S. hermonthica* on cereal hosts suggest that the parasite cannot sustain growth without carbon derived from the host plant (Graves *et al.*, 1990). Stable isotopic measurements show that *S. hermonthica* draws 100% of its carbon requirements from a maize host before emergence, and up to 59% after emergence (Aflakpui *et al.*, 2005).

21.2 Taxonomy

Striga is a taxon of obligate root parasitic plants. Until recently, *Striga* used to be classified as a member of Scrophulariaceae; however, latest molecular phylogenetic analyses place all parasitic members of Scrophulariaceae and Orobanchaceae in a single clade that is distinct from the nonparasitic taxa, indicating that the genetic pathway for haustorium development evolved once in the evolutionary history of these families (dePamphilis, 1995; Young *et al.*, 1999). Therefore all parasitic plants of both families are now included in Orobanchaceae.

Currently it is estimated that *Striga* includes about 40 species, although different authors cite species numbers between 25 and 60 (Kuiper, 1997). *Striga* is characterized by minute seeds containing a reduced embryo that consists of a short radicle without plumule and two cotyledons, by a herbaceous habitus, zygomorphic flowers with a corolla separated into a tube and spreading lobes, and by narrow leaves (Musselman, 1987). All species are parasitic.

In the following sections, we describe only the weedy species of *Striga*.

21.2.1 Striga asiatica *(L.) Kuntze (asiatic witchweed)*

An annual, erect, branched herb, up to 50 cm height. **Stem** quadrangular, slender, scabrid-pubescent with whitish hairs. **Leaves** nearly opposite or alternate, narrowly linear or lanceolate (4×0.5 cm); acute or obtuse, entire, sessile, scabrid-pubescent, with glandular pubescent buds. **Inflorescence** in terminal spikes, 10–15 cm long, with flowers 6–9 mm wide, varying from yellow, white or red, sessile, axillary, solitary. Identified usually by 10 (5–17) distinct ribs on the calyx which is 5–8 mm in length. Corolla 5–10 mm in diameter, two-lipped, corolla tube 6–12 mm long, slender, smooth or modestly pubescent, straight and cylindrical, but distinctly curved and inflated at the apex. Bracts linear, about 5 mm long, scabrid-pubescent. **Fruits** oblong or ellipsoid capsules, (4×2 mm) each with about 800 seeds. **Seeds** minute, golden brown, ellipsoid, 0.2–0.3 mm long. Seed surficial area features a distinct network structure. Self-pollinating; is widely distributed.

21.2.2 Striga forbesii Benth. *(giant mealie witchweed)*

Stems 30–40 cm high, leaves broader than in *S. hermonthica* and *S. asiatica*, **flowers** pale salmon pink (occasionally white), only 2–6 open at a time, self-pollinating.

21.2.3 Striga gesnerioides *(Willd.) Vatke (cowpea witchweed)*

Small plant, densely branched from the ground, succulent appearance, often pale green or purplish. **Stems** 15–30 cm tall. **Leaves** scale-like, 0.5–0.7 cm long, appressed to the stem. **Inflorescence** from just above ground level, varying from creamy white, light pink to dark purple, rarely yellow. Flowers showy with two or three flowers open at the peak of flowering. Corolla 3–5 mm in diameter, turning blue-black as it withers. **Fruits** ovoid-oblong capsules, (*c.* 5×3 mm). **Seeds** dust-like, 0.2–0.3 mm long, dark brown with a characteristic surface pattern of ridges. **Haustorium** up to 3cm in diameter.

21.2.4 Striga hermonthica *(Del.) Benth. (purple witchweed)*

Annual erect, slender herb with showy inflorescence, up to 100 cm height. **Stem** quadrangular, loosely rough-haired with whitish hairs. **Leaves** alternate, sessile, narrowly oblanceolate, acute to acuminate at the apex, tapering towards the base; hispid and scabrous especially along the midrib and margins. **Inflorescence** dense, in terminal spikes with many purple flowers, rarely white. Calyx distinctly five-ribbed. Corolla tube about 11–17 mm in length, glabrous, bending characteristically at an angle immediately over the tip of the calyx. Bracts below each flower fringed with hairs. **Fruits** blackish, conical capsules, containing about 700 seeds. **Seeds** minute, glabrous, marked with striate lines, dark brown to black, 0.2–0.4 mm long. Cross-pollinating.

21.3 Geographical distribution

All the known species of *Striga* are native to Old World tropics (Mohamed *et al.*, 2001). The epicenter of their distribution is located in the arid and semi-arid tropics, in areas receiving an average annual precipitation of 500–1000 mm. These regions have a predominant vegetation of natural and/or anthropogenic savanna. A majority of *Striga* species occur in Africa, with the greatest diversification in West and Central Africa (Raynal-Roques, 1994). Few species occur in Asia and Madagascar, and three are endemic to Australia (Raynal-Roques, 1994). Two species, *S. asiatica* and *S. gesnerioides*, were accidentally introduced to the southeastern USA. Maize fields in North Carolina were already reported to be infested by *S. asiatica* in the late 1950s, while *S. gesnerioides* was detected for the first time in Florida in the late 1970s. The principal distribution area of *S. hermonthica* is in Africa, between 20° N and 20° S (Fig. 21.2). It extends from Senegal in the west to Somalia in the east, from Sudan in the north to Angola and Mozambique in the south. It also occurs in the southwestern Arabian Peninsula (Yemen). Although *S. hermonthica* is considered a tropical weed, it occurs at altitudes up to 2500 m in Ethiopia, in a region climatically resembling the temperate zone, with occasional night frosts.

Striga forbesii occurs throughout West and East Africa, South Africa, and Madagascar. It prefers areas with relatively wet soil conditions. *Striga asiatica* is more widespread than *S. forbesii* and occurs in lands between 35° N and 35° S. The areas of distribution of

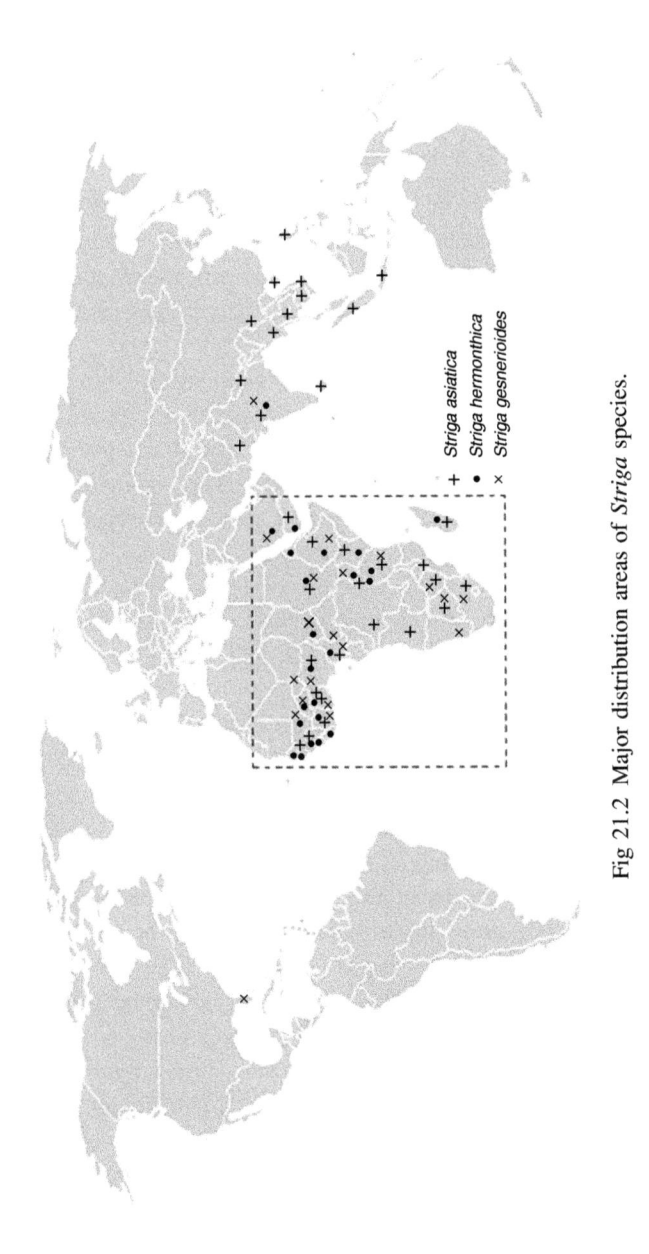

Fig 21.2 Major distribution areas of *Striga* species.

S. asiatica include much of Africa and parts of southwest Arabia. It occurs extensively in Asia, including India and China, and in Australia. *Striga gesnerioides* occurs in the Cape Verde Islands and is widespread throughout Africa. It also occurs in the Arabian Peninsula and appears in western and southern India, as well as in Sri Lanka. Populations of this species have established in Florida.

21.4 Economic consequences

Although most *Striga* species are of no agricultural importance, those that parasitize crop plants usually cause extreme damage. By reducing crop yields, weedy *Striga* species have a greater impact on humans worldwide than any other parasitic plant and impact the lives of people of the semi-arid tropics, where cereal and legume crops susceptible to *Striga* are major sources of energy and protein in diets.

Species of *Striga* exert the greatest damage prior to emergence, often before the farmer can recognize an infection. Weedy *Striga* species develop high sink capacity, which enables them to draw water, mineral nutrients, and assimilates from their host plants. Parasitized crops usually suffer slow growth and, depending on the severity of infection, biomass is also reduced. The loss of crop biomass induced because of *Striga* infection may even exceed the parasite's biomass, indicating the involvement of other than source/sink-based relations, such as the inhibition of photosynthesis in the host plant (Frost *et al.*, 1997). Crop growth is probably also affected by toxic substances released by the parasite shortly after attachment (Press *et al.*, 2001). Thus, infection by *Striga* impairs the ability of hosts to grow and yield.

Species of primary economic importance on cereals are *S. hermonthica* and *S. asiatica*. Of secondary importance in Africa are *S. aspera* (Willd.) Benth. and *S. forbesii* (Parker and Riches, 1993). The economic impact of *Striga* species is the greatest in grain grasses cultivated in Africa. Maize (*Zea mays* L.), sorghum (*Sorghum bicolor* (L.) Moench.), pearl millet (*Pennisetum glaucum* (L.) R. Br.), upland rice (*Oryza sativa* L., *O. glaberrima* Steud.), fonio (*Digitaria exilis* (Kippist) Stapf), and finger millet (*Eleusine coracana* (L.) Gaertn.) (Poaceae) are all parasitized by one or more species of *Striga*. In Ethiopia, teff (*Eragrostis tef* (Zuccagni) Trotter and barley (*Hordeum vulgare* L.) (Poaceae) are attacked by *S. hermonthica*. *Striga gesnerioides* has a host range different from those of the other species: it is specialized to attack dicotyledons, such as cowpea (*Vigna unguiculata* (L.) Walp.), bambara groundnut (*Vigna subterranea* (L.) Verdc.) (Fabaceae), sweet potato (*Ipomoea batatas* (L.) Poir.) (Convolvulaceae), and tobacco (*Nicotiana tabacum* L.) (Solanaceae). Among these, cowpea is an economically important plant in West Africa. With more cowpea production and an increasing proportion of monocultures, *S. gesnerioides* damage in cowpea has become more acute, particularly in areas with sandy, infertile soils and low rainfall (Singh and Emechebe, 1997). Under these circumstances, the damage caused by weedy *Striga* species can prove disastrous to resource-poor farmers, whose lives can be threatened through complete yield loss in both cereal and legume crops.

The extent of crop damage by *Striga* depends on crop variety, soil fertility, rainfall pattern, and level of infestation in the field. Mainly because of irregular amount and distribution of rainfall during crop season, poor soil biological activity, and monoculture, *Striga* infection is erratic across seasons. Based on data available from Benin, Cameroon, Gambia, Ghana, Nigeria, and Togo, 34–75% of the grain cultivation area was estimated to be infested by *Striga*, on an average of 48% (Sauerborn, 1991). Yield losses ranged from 6% to 26% of total grain production, but in areas of heavy infestation, complete yield loss is not uncommon. Twenty-one million hectares of arable land in sub-Saharan Africa (SSA) were estimated to be infested with *Striga*, resulting in an annual grain yield loss of 4.1 million tons (Sauerborn, 1991). Even 13 years later similar estimations of crop land infested by *Striga* spp. throughout SSA prevail (Gressel *et al.*, 2004), but grain production loss amounted to 8 million tons. These figures indicate that the severity of *Striga* infestation across fields has increased in the last years. Yield loss due to *Striga* infestation is now estimated at 40% on average. In West Africa, *Striga* is the most common weed and has infested 17 million ha of arable land. In eastern and central Africa, *Striga*-contaminated land covers 3 million ha, whereas in southern Africa, the infested area is estimated at 1.6 million ha (Gressel *et al.*, 2004). The increasing severity of *Striga* infestation in Africa – both in terms of area affected and intensity of infestation – results from ongoing drastic changes in farming practices followed by the intensification of cereal production, principally through monocropping in an attempt to produce adequate food for increasing human populations (Doggett, 1984; Sauerborn, 1991; Butler, 1995; Berner *et al.*, 1996). Levels of infestation are often so severe that continued grain production becomes impractical. Farmers often abandon these fields for less infested areas (Doggett, 1984; Lagoke *et al.*, 1991).

21.5 Management

Striga seeds are easily dispersed from one field to another by wind, water, soil movement, human activities, and by clinging to animals, vehicles and farm machines, tools, shoes, and clothing. Movement of seeds likewise occurs through *Striga*-contaminated crop seed lots (Berner *et al.*, 1994). Harvesting parasite shoots in forage crops may also assist in spreading the parasite with contaminated manure since the seeds are resistant against digestion by animals (Jacobsohn *et al.*, 1987). Prevention of *Striga* seed spread is, thus, the important first step in avoiding new infestations or reinfestations and in making sustainable control feasible. So far the effectiveness of conventional control methods has remained restricted due to factors such as lack of appropriate agricultural extension services, scant funds for agricultural inputs, and the complex nature of the host–parasite relationship. It is also the intimate relationship between host and parasite that hinders efficient control such as by herbicides. Although several potential control measures have been developed in past decades, those approaches applied as a single means often are only partially, and sometimes inconsistently, effective since they are altered by environmental conditions. Moreover, it needs to be considered that most damage to the crop is done before the parasite emerges above ground. Therefore, control methods should aim at reducing the soil seed bank and disrupt the parasite's early developmental stages. Because of the close alliance between the parasitic weed and its host, the use of biocontrol organisms (fungi, bacteria, arthropods), which are very host specific, can be considered a promising alternative since these organisms may operate where other weed control options have failed.

21.5.1 Conservation biological control

Biologically active soils naturally suppress the build-up of *Striga* populations (Berner *et al.*, 1996; Gbehounou *et al.*, 1996; Ransom, 2000). The causal factors involved in these soils are not yet known, but suppressiveness is associated with the microbial activity (Pieterse *et al.*, 1996; Sauerborn *et al.*, 2003; Ahonsi *et al.*, 2004). For example, rhizobacteria capable of destroying *Striga* seeds are particularly promising biological control agents (Berner *et al.*, 1995; Miché *et al.*, 2000). With arbuscular mycorrhizal (AM) fungal inoculation, a significant reduction in the number of *S. hermonthica* shoots was achieved (30% on maize and more than 50% in sorghum; Lendzemo *et al.*, 2005), along with increased crop growth. These studies indicate that microorganisms have the potential to reduce damage by *Striga* to crops. Hence, a simple means of maximizing suppressiveness is to modify soil conditions such that the growth of the microbial agents is encouraged. Improved soil fertility, addition of organic matter, and rotations were found to increase the natural *Striga* suppressiveness of soils (Ransom, 2000; Sauerborn *et al.*, 2000, 2003; Ahonsi *et al.*, 2004).

21.5.2 Mycoherbicides

In recent years, the isolation of microorganisms occurring on *Striga* and the evaluation of their pathogenicity have been carried out more systematically. However, although numerous microorganisms have been isolated and reported in the past, none has achieved continuous widespread use.

About 16 fungal genera from Africa, India, and the USA are reported to appear on *Striga* spp. (Nag Raj, 1966; Meister and Eplee, 1971; Zummo, 1977; Greathead, 1983; Abbasher and Sauerborn, 1992, 1995, Kirk, 1993; Ciotola *et al.*, 1995; Abbasher *et al.*, 1995; 1998; Kroschel *et al.*, 1996; Marley *et al.*, 1999; Hess *et al.*, 2002; Yonli *et al.*, 2006). Surveys for fungal pathogens of *Striga* spp. in West Africa indicated that *Fusarium* species were the most prevalent fungal species associated with diseased *Striga* plants. Of these, *F. oxysporum* was the predominant species (Abbasher *et al.*, 1998). Numerous *Fusarium* species (*F. acuminatum*, *F. equiseti*, *F. nygamai*, *F. oxysporum*, *F. semitectum* var. *majus*, *F. verticillioides*) have been categorized as potential biocontrol agents.

Fusarium *(Nectriaceae)*

Trials to evaluate *F. nygamai*, *F. oxysporum*, and *F. semitectum* var. *majus* pathogenic on *S. hermonthica* yielded good results under controlled and field conditions (Abbasher and Sauerborn, 1992; Ciotola *et al.*, 1995, 2000; Kroschel *et al.*, 1996; Marley *et al.*, 2004,

2005; Schaub *et al.*, 2006). All development stages of the parasite, from seeds to flowering shoots, are attacked. Thus, *Fusarium* spp. can be very effective in reducing the parasite seed bank by both direct destruction of the seeds in the soil and prevention of reproduction. Since *Striga* seeds in the soil can be directly infected by *Fusarium* spp., it might not be necessary to apply the inoculum during crop growth (Sauerborn *et al.*, 1996). The fungus could reduce the parasite seed bank every season, even if there is no host plant for *Striga* in the field, which could give the mycoherbicide strategy a special advantage.

It is well known that members of the genus *Fusarium* produce a range of chemically diverse phytotoxic compounds. These include, for example, enniatin, fumonisin, fusaric acid, moniliformin, and trichothecenes that exert a broad range of biological activities and metabolic effects. Some of these compounds have been considered and proposed for use as natural herbicides, as alternatives to or in addition to the use of weed pathogens (Strobel *et al.*, 1991; Duke and Lydon, 1993). Fusaric acid and 9,10-dehydrofusaric acid reduced *Striga* seed germination at low doses (Zonno *et al.*, 1996; Zonno and Vurro, 1999). *Fusarium nygamai* is known to produce fumonisin B_1 (Capasso *et al.*, 1996). Recently, it was shown that fumonisin B_1 impairs development and growth of both *S. hermonthica* and *S. asiatica* (Kroschel and Elzein, 2004). Otherwise, mycotoxin production might be an important constraint to the use of *Fusarium* species as mycoherbicides, since mycotoxins including some trichothecene derivatives, zearalenone and zearalenols, when contaminating food and feed, are often associated with chronic or acute mycotoxicoses in humans and farm animals. Fumonisin B_1, for example, showed cancer-promoting activity in rats after application at high doses over a long period of time (Gelderblom *et al.*, 1988). However, in tests on the possible production of mycotoxins by *F. oxysporum*, *F. nygamai*, and *F. semitectum* var. *majus* isolated from *Striga*, none of the most important *Fusarium* mycotoxins was detected (Abbasher, 1994, Ciotola *et al.*, 1995).

As promising as some results obtained from laboratory, greenhouse, and field trials may seem, prospects for success in controlling *Striga* species with the help of microorganisms are still in question. The main reason for this is the lack of reliable field efficacy under certain conditions. To find a remedy, careful attention should be given to how these organisms compete with indigenous soil microorganisms, and to finding an organism that will spread rapidly together with the rhizosphere of the host root system.

21.5.3 Herbivores

Surveys for natural antagonists of *Striga* in Africa and India have revealed the presence of numerous insects that can inflict damage to different species of *Striga*. However, due to the short life-span and abundant seed production of the parasitic weeds, and the great damage caused to the host crop by underground stages of *Striga* plants, the parasites probably cannot be considered ideal target organisms for biological control by insects.

Most of the insects which have been reported to occur on *Striga* species are polyphagous without any host specificity. For biological control, oligophagous and monophagous insects that are regularly found on *Striga* are required. Herbivores of great interest for biological control of *Striga* include weevils of the genus *Smicronyx* that form galls in *Striga* capsules, thus reducing seed production. Other potential species are *Junonia orithya* reported from different parts of the world, *Ophiomyia strigalis* Spencer (Diptera: Agromyzidae) in Africa, and *Eulocastra argentisparsa* and *E. undulata* in India.

The first report on insects weakening *Striga* originates from India: larvae of *Junonia orithya* L. (Lepidoptera: Nymphalidae) had been observed to feed on *S. asiatica* (Murthy and Rao, 1949). Stem and root galls on *S. asiatica* induced by *Smicronyx albovariegatus* Faust (Coleoptera: Curculionidae) have also been found (Khan and Murthy, 1955). *Striga* seed capsules were reported to be infested by *Smicronyx* in Nigeria (Williams and Caswell, 1959). Occurrence of the shoot miner *O. strigalis* and a gall-inducing wasp, *Eurytoma* sp. (Hymenoptera: Eurytomidae), have been observed in Kenya (Davidson, 1963). After the discovery of *S. asiatica* in the USA in 1956, extensive surveys of insects were conducted by the Commonwealth Institute of Biological Control (CIBC) stations in India (Sankaran and Rao, 1966) and in Tanzania (Greathead and Milner, 1971). Bashir and Musselman (1984) reported the occurrence of *Smicronyx umbrinus* Hustache, *J. orithya*, and larvae of the feather moth *Stenoptilodes taprobanes* Felder (Lepidoptera: Pterophoridae) attacking *S. hermonthica* in Sudan.

Smicronyx *(Coleoptera: Curculionidae)*

Weedy *Striga* species are known to be attacked by at least four *Smicronyx* species (*S. albovariegatus* Faus, *S. dorsomaculatus* Cox, *S. guineanus* Voss, *S. umbrinus* Hustache), which are thought to be highly specific. The adult weevil is black to reddish black and densely covered by grayish hairs. The size of the weevil ranges from 2.5 to 5.0 mm, depending on sex and species. One week after the female has been fertilized it lays its eggs into the flowers of *Striga*. Typically, one tiny egg is deposited in a single ovary. Four to eight days after egg deposition, the larva starts to hatch and develop within the formed gall. A full-grown larva reaches 3 to 4 mm in length. Finally the larva bites through the gall wall and emerges. It drops to the soil, where it buries itself up to 15 cm deep. By cementing soil particles around the body, the larva pupates. Under the conditions of West Africa, pupation occurs from October through November and lasts until July of the following year. *Smicronyx* species have only one generation per year.

Through the development of larvae inside the seed capsules of their target hosts *Striga*, these insects prevent seed production and thus contribute to reducing the parasite's reproductive capacity and spread. However, research has revealed that their efficacy in preventing seed set is limited. In a survey conducted in northern Ghana, *Smicronyx* galls were found in only 22.5% of *S. hermonthica* plants. In infected *Striga* plants 77.3% of the seed capsules were transformed into fruit galls. The impact of *Smicronyx* on *S. hermonthica* seed production based on this contamination was calculated to be a mere 17.4%, which will not be enough to significantly diminish the soil seed bank (Jost *et al.*, 1996). Smith and Webb (1996) estimated that approximately 70 to 80% of capsules would need to be abolished each year to impair *S. hermonthica* seed load in the soil. A factor that may further limit the effect of *Smicronyx* spp. is soil cultivation. Hibernating pupae can be destroyed and/or buried, which would prevent insect emergence. Another limiting factor is pesticide application against crop pests where it coincides with the flight periods of the beneficial insect. Moreover, *Smicronyx* can be attacked by a number of parasitic wasps: braconids, eupelmids, pteromalids, (Hymenoptera: Braconidae, Eupelmidae, Pteromalidae), which demonstrably reduce the population in the field (Greathead, 1983; Jost, 1997). Nevertheless, *Smicronyx* spp. routinely reduce the amount of new *Striga* seeds being produced in some locations and years (Kroschel *et al.*, 1995, 1999; Traore *et al.*, 1996) and can be important as a component of an integrated *Striga* control program.

References

Abbasher, A. A. (1994). Microorganisms associated with *Striga hermonthica* and possibilities of their utilization as biological control agents. Ph.D. thesis, University of Hohenheim, Germany.

Abbasher, A. A. and Sauerborn, J. (1992). *Fusarium nygamai*, a potential bioherbicide for *Striga hermonthica* control in sorghum. *Biological Control*, 2, 291–296.

Abbasher, A. A. and Sauerborn, J. (1995). Pathogens attacking *Striga hermonthica* and their potential as biological control agents. In *Proceedings of the Eighth International Symposium on Biological Control of Weeds*, ed. F. S. Delfosse and R. R. Scott, Melbourne, Australia: DSIR/CSIRO, pp. 527–533.

Abbasher, A. A., Hess, D. E. and Sauerborn, J. (1998). Fungal pathogens for biological control of *Striga hermonthica* on sorghum and pearl millet in West Africa. *African Crop Science Journal*, 6, 179–188.

Abbasher, A. A., Kroschel, J. and Sauerborn, J. (1995). Micro-organisms of *Striga hermonthica* in northern Ghana with potential as biocontrol agents. *Biocontrol Science and Technology*, 5, 157–161.

Aflakpui, G. K. S., Gregory, P. J. and Froud-Williams, R. J. (2005). Carbon (13C) and nitrogen (15N) translocation in a maize–*Striga hermonthica* association. *Experimental Agriculture*, 41, 321–333.

Ahonsi, M. O., Berner, D. K., Emechebe, A. M. and Lagoke, S. T. (2004). Effects of ALS-inhibitor herbicides, crop sequence, and fertilization on natural soil supressiveness to *Striga hermonthica*. *Agriculture, Ecosystems and Environment*, 104, 453–463.

Akiyama, K., Matsuzaki, K. and Hayashi, H. (2005). Plant sesquiterpenes induce hyphal branching in arbuscular mycorrhizal fungi. *Nature*, 435, 824–827.

Atsatt, P. R. (1977). The insect herbivore as a predictive model in parasitic seed plant biology. *The American Naturalist*, 111, 579–586.

Bashir, M. O. and Musselman, L. J. 1984. Some natural enemies of *Striga hermonthica* in the Sudan. *Tropical Pest Management*, 30, 211.

Bebawi, F. F., Eplee, R. E., Harris, C. E. and Norris, R. S. (1984). Longevity of witchweed (*Striga asiatica*) seed. *Weed Science*, 32, 494–497.

Berner, D. K., Cardwell, K. F., Faturoti, B. O., Ikie, F. O. and Williams, O. A. (1994). Relative roles of wind, crop seeds, and cattle in the dispersal of *Striga* species. *Plant Disease*, 78, 402–406.

Berner, D. K., Carsky, R. J. Dashiell, K. E. Kling, J. G. and Manyong, V. M. (1996). A land management based approach to integrated *Striga hermonthica* control in sub-Saharan Africa. *Outlook on Agriculture*, **25**, 157–164.

Berner, D. K, Kling, J. G. and Singh, B. B. (1995). *Striga* research and control: a perspective from Africa. *Plant Disease*, **79**, 652–660.

Butler, L. G. (1995). Chemical communication between the parasitic weed, *Striga* and its crop host: a new dimension in allelochemistry. In *Allelopathy: Organisms, Processes, and Applications*, ed. Inderjit, K. M. M. Dakshini and F. A. Einhellig. ACS Symposium Series 582. Washington, DC: American Chemical Society, pp.158–168.

Capasso, R., Eviedente, A., Cutignano, A., *et al.* (1996). Fusaric acid 9, 10–dehydrofusaric acids and their methyl esters from *Fusarium nygamai*. *Phytochemistry*, **41**, 1035–1039.

Chang, M. and Lynn, D. G. (1986). The haustorium and the chemistry of host recognition in parasitic angiosperms. *Journal of Chemical Ecology*, **12**, 561–579.

Ciotola, M., DiTommaso, A. and Watson, A. K. (2000). Chlamydospore production, inoculation methods and pathogenicity of *Fusarium oxysporum* M12-4A, a biocontrol for *Striga hermonthica*. *Biocontrol Science and Technology*, **10**, 129–145.

Ciotola, M., Watson, A. K. and Hallett, S. G. (1995). Discovery of an isolate of *Fusarium oxysporum* with potential to control *Striga hermonthica* in Africa. *Weed Research*, **35**, 303–309.

Davidson, A. (1963). Insects attacking *Striga* in Kenya. *Nature*, **197**, 923.

dePamphilis, C. W. (1995). Genes and genomes. In *Parasitic Plants*, ed. M. C. Press and J. D. Graves. London: Chapman and Hall, pp.177–205.

Doggett, H. (1984). *Striga* its biology and control an overview. In *Striga Biology and Control*, ed. E. S. Ayensu, H. Doggett, R. D. Keynes, *et al.* International Council of Scientific Union, Paris, France, and the International Research and Development Center, Ottawa, Canada. Paris: ICSU Press, pp. 27–36.

Dörr, I. (1997). How *Striga* parasitizes its host: a TEM and SEM study. *Annals of Botany*, **79**, 463–472.

Duke, S. O., and Lydon, J. (1993). Natural phytotoxins as herbicides. In *Pest Control with Enhanced Environmental Safety*, ed. S. O. Duke, J. J. Mann and J. R. Plimmer. ACS Symposium Series 524. Washington, DC: American Chemical Society, pp. 110–124.

Emechebe, A. M., Singh, B. B., Leleji, O. I., Atokple, I. D. K. and Adu, J. K. (1991). Cowpea *Striga* problems and research in Nigeria. In *Combating Striga in Africa*, ed. S. K. Kim. Ibadan, Nigeria: IITA, pp. 18–28.

Frost, D. L., Gurney, A. L. Press, M. C. and Scholes, J. D. (1997). *Striga hermonthica* reduces photosynthesis in sorghum: the importance of stomatal limitations and a potential role for ABA? *Plant Cell and Environment*, **20**, 483–492.

Gbehounou, G., Pieterse, A. H. and Verkleij, J. A. C. (1996). The decrease in seed germination of *Striga hermonthica* in Benin in the course of the rainy season is due to a dying off process. *Experientia*, **52**, 264–267.

Gelderblom, W. C. A., Jaskiewicz, K., Marasas, W. F. O., *et al.* (1988). Fumonisins novel mycotoxins with cancer-promoting activity produced by *Fusarium moniliforme*. *Applied Environmental Microbiology*, **54**, 1806–1811.

Graves, J. D., Wylde, A., Press, M. C. and Stewart, G. R. (1990). Growth and carbon allocation in *Pennisetum typhoides* infected with the parasitic angiosperm *Striga hermonthica*. *Plant, Cell and Environment*, **13**, 367–373.

Greathead, D. J. (1983). The natural enemies of *Striga* spp. and the prospects for their utilization as biological control agents. In *Striga Biology and Control*, ed. E. S. Ayensu, H. Doggett, R. D. Keynes, *et al.* Oxford: IRL Press Ltd, pp.133–160.

Greathead, D. J., and Milner, J. E. D. (1971). A survey of *Striga* spp. (Scrophulariaceae) and their natural enemies in east Africa with a discussion on the possibilities of biological control. *Tropical Agriculture* (Trinidad), **48**, 111–124.

Gressel, J., Hanafi, A., Head, G., *et al.* (2004). Major heretofore intractable biotic constraints to African food security that may be amenable to novel biotechnological solutions. *Crop Protection*, **23**, 661–689.

Hess, D. E., Kroschel, J., Traoré, D., *et al.* (2002). *Striga*: Biological control strategies for a new millennium. In *Sorghum and Millet Diseases 2000*, ed. J. F. Leslie. Ames, IA: Iowa State Press, pp. 165–170.

Jacobsohn, R., Ben-Ghedalia, D. and Marton, K. (1987). Effect of the animal's digestive system on the infectivity of *Orobanche* seeds. *Weed Research*, **27**, 87–90.

Joel, D. M. (2000). The long-term approach to parasitic weeds control: manipulation of specific developmental mechanisms of the parasite. *Crop Protection*, **19**, 753–758.

Jost, A. (1997). *Integrierter Getreideanbau in Nord-Ghana unter besonderer Berücksichtigung der* Striga *Problematik.* Plits 15(4). Stuttgart, Germany: W. & S. Koch.

Jost, A., Kroschel, J. and Sauerborn, J. (1996). Studies on *Smicronyx* spp. and *Junonia orithya* and their potential for biological control of *Striga hermonthica* in northern Ghana. In *Advances in Parasitic Plant Research: Proceedings of the VI International Parasitic Weed Symposium*, ed. M. T. Moreno, J. I. Cubero, D. Berner, *et al.* Cordoba, Spain: Junta de Andalacia, pp. 888–889.

Keyes, W. J., Taylor, J. V., Apkarian, R. P. and Lynn, D. G. (2001). Dancing together. Social controls in parasitic plant development. *Plant Physiology*, **127**, 1508–1512.

Khan, M. Q. and Murthy, D. V. (1955). *Smicronyx albovariegatus* Faust (Curculionidae: Coleoptera) on *Striga* spp. *Indian Journal of Entomology*, **17**, 362.

Kim, D., Kocz, R., Boone, L., Keyes, W. J. Lynn, D. G. (1998). On becoming a parasite: evaluating the role of wall oxidases in parasitic plant development. *Chemistry & Biology*, **5**, 103–117.

Kirk, A. A. (1993). A fungal pathogen with potential for control of *Striga hermonthica* (Scrophulariaceae). *Entomophaga*, **38**, 459–460.

Kroschel, J. and Elzein, A. (2004). Bioherbicidal effect of Fumonisin B1, a phytotoxic metabolite naturally produced by *Fusarium nygamai*, on parasitic weeds of the genus *Striga*. *Biocontrol Science and Technology*, **14**, 117–128.

Kroschel, J., Abbasher, A. A. and Sauerborn, J. (1995). Herbivores of *Striga hermonthica* in northern Ghana and approaches of their use as biological control agents. *Biocontrol Science and Technology*, **5**, 163–164.

Kroschel, J., Hundt, A., Abbasher, A. A. and Sauerborn, J. (1996). Pathogenicity of fungi collected in northern Ghana to *Striga hermonthica*. *Weed Research*, **36**, 515–520.

Kroschel, J., Jost, A. and Sauerborn, J. (1999). Insects for *Striga* control – possibilities and constraints. In *Advances in Parasitic Weed Control at On-farm Level. Vol I. Joint Action to Control Striga in Africa*, ed. J. Kroschel, H. Mercer-Quarshie and J. Sauerborn. Weikersheim, Germany: Margraf Verlag, pp. 117–132.

Kuiper, E. (1997). *Comparative Studies on the Parasitism of* Striga aspera *and* Striga hermonthica *on Tropical Grasses.* Delft, The Netherlands: Eburon P&L.

Lagoke, S. T., Parkinson, V., and Agunbiade, R. M. (1991). Parasitic weeds and control methods in Africa. In *Combating Striga in Africa*, ed. S. K. Kim. Proc. Int. Workshop organized by IITA, ICRISAT, and IDRC, 22–24 Aug 1988. Paris, France: ICSU Press, pp. 3–15.

Lendzemo, V. W., Kuyper, T. W., Kropff, M. J., and van Ast, A. (2005). Field inoculation with arbuscular mycorrhizal fungi reduces *Striga hermonthica* performance on cereal crops and has the potential to contribute to integrated *Striga* management. *Field Crops Research*, **91**, 51–61.

Lynn, D. G., Steffens, J. C., Karnat, V. S., *et al.* (1981). Isolation and characterization of the first host recognition substance for parasitic angiosperms. *Journal of the American Chemical Society*, **103**, 1868–1870.

Marley, P. S., Aba, D. A., Shebayan, J. A. Y., Musa, R. and Sanni, A. (2004). Integrated management of *Striga hermonthica* in sorghum using a mycoherbicide and host plant resistance in the Nigerian Sudano-Sahelian savanna. *Weed Research*, **44**, 157–162.

Marley, P. S., Ahmed, S. M., Shebayan, J. A. Y. and Lagoke, S. T. O. (1999). Isolation of *Fusarium oxysporum* with potential for biocontrol of the witchweed (*Striga hermonthica*) in the Nigerian savanna. *Biocontrol Science and Technology*, **9**, 159–163.

Marley, P. S., Kroschel, J. and Elzein, A. (2005). Host specificity of *Fusarium oxysporum* Schlecht (isolate PSM 197), a potential mycoherbicide for controlling *Striga* spp. in West Africa. *Weed Research*, **45**, 407–412.

Meister, C. W. and Eplee, R. E. (1971). Five new fungal pathogens of witchweed (*Striga lutea*). *Plant Disease Reporter*, **55**, 861–863.

Miché, L., Boillant, M.-L. Rohr, R. Sallé, G. and Bally, R. (2000). Physiological and cytological studies on the inhibition of *Striga* seed germination by the plant growth-promoting bacterium *Azospirillum brasilense*. *European Journal of Plant Pathology*, **106**, 347–351.

Mohamed, K. J., Musselman, L. J. and Riches, C. R. (2001). The genus *Striga* (Scrophulariaceae) in Africa. *Annals of the Missouri Botanical Garden*, **88**, 60–103.

Murthy, D. V., and Rao, A. S. (1949). *Precis orythia* Swinhoei L. (Fam. Nymphalidae) feeding on *Striga* spp. – the phanerogamic parasite of sugarcane and jowar. *Current Science*, **9**, 342.

Musselman, L. J. (1987). Taxonomy of witchweeds. In *Parasitic Weeds in Agriculture* ed. L. J. Musselman. Boca Raton, FL: CRC Press, pp. 3–12.

Nag Raj, T. R. (1966). Fungi occurring on witchweed in India. (Commonwealth Mycological Institute of Biological Control) *Technical Communications*, **7**, 75–80.

Parker, C. and Riches, C. R. (1993). *Parasitic Weeds of the World: Biology and Control.* Wallingford, UK: CAB International, 332 pp.

Pieterse, A., Verkleij, J. A. C., Den Hollender, N. G., Odhiambo, G. D. and Ransom, J. K. (1996). Germination and viability of *Striga hermonthica* seeds in Western Kenya in the course of the long rainy season. In *Advances in Parasitic Plant Research: Proceedings of the Sixth Parasitic Weed Symposium*, ed. M. T. Moreno, J. I. Cubero, D. Berner, *et al.* Cordoba, Spain: Junta de Andalucia, pp. 457–464.

Press, M. C., Scholes, J. D. and Riches, C. R. (2001). Current status and future prospects for management of parasitic weeds (*Striga* and *Orobanche*). In *The World's Worst Weeds*, ed. C. R. Riches. Proceedings BCPC/ Monograph Series 77. Bracknell, UK: British Crop Production Council, pp. 71–90.

Ransom, J. K. (2000). Long-term approaches for the control of *Striga* in cereals: field management options. *Crop Protection*, **19**, 759–763.

Raynal-Roques, A. (1994). Répartition géographique et spéciation dans le genre *Striga* (Scrophulariaceae parasites). *Memoirs of the Society of Biogeography*, **4**, 83–94.

Sand, P. F. (1990). Discovery of witchweed in the United States. In *Witchweed Research and Control in the United States*, ed. P. F. Sand, R. E. Eplee and R. G. Westerbrooks. Champaign, IL: Weed Science Society of America, pp. 1–6.

Sankaran, T. and Rao, V. P. (1966). Insects attacking witchweed (*Striga*) in India. *Commonwealth Institute of Biological Control Technical Bulletin*, **7**, 63–73.

Sauerborn, J. (1991). *Parasitic Flowering Plants: Ecology and Management.* Weikersheim, Germany: Verlag Josef Margraf, 127 pp.

Sauerborn, J., Dörr, I., Abbasher, A. A., Thomas, H. and Kroschel, J. (1996). Electron microscopic analysis of the penetration process of *Fusarium nygamai*, a hyperparasite of *Striga hermonthica. Biological Control*, **7**, 53–59.

Sauerborn, J., Kranz, B. and Mercer-Quarshie, H. (2003). Organic amendments mitigate heterotrophic weed infestation in savannah agriculture. *Applied Soil Ecology*, **23**, 181–186.

Sauerborn, J., Sprich, H. and Mercer-Quarshie, H. (2000). Crop rotation to improve agricultural production in Sub Saharan Africa. *Journal of Agronomy and Crop Science*, **184**, 67–72.

Schaub, B., Marley, P., Elzein, A. and Kroschel, J. (2006). Field evaluation of an integrated *Striga hermonthica* management in Sub-Saharan Africa: synergy between *Striga*-mycoherbicides (biocontrol) and sorghum and maize resistant varieties. *Journal of Plant Disease and Protection* (Special Issue), **20**, 691–699.

Singh, B. B. and Emechebe, A. M. (1997). Advances in research on cowpea *Striga* and *Alectra*. In *Advances in Cowpea Research*, ed. B. B. Singh, D. R. Mohan Raj, K. E. Dashiell and L. E. N. Jackai. Ibadan, Nigeria: IITA/JIRCAS, pp. 215–224.

Smith, M. C. and Webb, M. (1996). Estimation of the seedbank of *Striga* spp. (Scrophulariaceae) in Malian fields and the implications for a model of biocontrol of *Striga hermonthica. Weed Research*, **36**, 85–92.

Strobel, G. A., Kenfield, D., Bunkers, G., Sugawara, F. and Clardy, J. (1991). Phytotoxins as potential herbicides. *Experientia*, **47**, 819–826.

Traore, D., Vincent, C. and Stewart, R. K. (1996). Association and synchrony of *Smicronyx guineanus* Voss, *S. umbrinus* Hustache (Coleoptera: Curculionidae), and the parasitic weed *Striga hermonthica* (Del.) Benth. (Scrophulariaceae). *Biological Control*, **7**, 307–315.

Williams, C. N. and Caswell, G. H. (1959). An insect attacking *Striga. Nature*, **184**, 1668.

Yasuda, N., Sugimoto, Y., Kato, M., Inanaga, S. and Yoneyama, K. (2003). (+)−Strigol, a witchweed seed germination stimulant, from *Menispermum dauricum* root culture. *Phytochemistry*, **62**, 1115–1119.

Yonli, D., Traoré, H., Hess, D., Sankara, P. and Sérémé, P. (2006). Effect of growth media, *Striga* seed burial distance and depth on efficacy of *Fusarium* isolates to control *Striga hermonthica* in Burkina Faso. *Weed Research*, **46**, 73–81.

Young, N. D., Steiner, K. E. and de Pamphilis, C. W. (1999). The evolution of parasitism in Scrophulariaceae/Orobanchaceae: plastid gene sequences refute an evolutionary transition series. *Annals of the Missouri Botanical Garden*, **86**, 876–893.

Zonno, M. C. and Vurro, M. (1999). Effect of fungal toxins on germination of *Striga hermonthica* seeds. *Weed Research*, **39**, 15–20.

Zonno, M. C., Vurro, M. R., Capasso, R., *et al.* (1996). Phytotoxic metabolites produced by *Fusarium nygamai* from *Striga hermonthica*. In *Proceedings of the IX International Symposium on Biological Control of Weeds*, ed. V. C. Moran and J. H. Hoffman. Stellenbosch, South Africa: University of Cape Town, pp. 223–226.

Zummo, N. (1977). Diseases of giant witchweed, *Striga hermonthica* in West Africa. *Plant Disease Reporter*, **61**, 428–430.

22

Biological control of weeds in India

Jebomoni Rabindra and Basavaraj S. Bhumannavar

22.1 Introduction

Large-scale movement of vegetable, fruit, and ornamental plants between nations entails the danger of accidental introduction of insect pests, nematodes, plant pathogens, and weeds. The problems due to accidental introduction of weeds are manifold. A pest organism or weed, thus introduced, finds the new habitat conducive for breeding and establishment without any regulation by the natural enemies that would have kept the introduced species under check in their original ranges. Dominance of the invasive species in the new habitat causes immense damage to the native fauna and flora, thus upsetting the natural balance within the new habitat. Conventional methods of weed control are difficult for such invasive weeds and the use of chemical herbicides in uncultivated areas generally is uneconomical and can have ill effects on nontarget organisms.

An ideal way of managing invasive species, whether insects, mites or weeds, would be to introduce and establish effective natural enemies from their native home range. Biological weed control involves the deliberate use of natural enemies (e.g. plant-feeding and disease-causing organisms) to reduce the densities of weeds to economically or aesthetically tolerable limits, which need not necessarily lead to complete eradication. This chapter has been written to supplement the information of the review by Sankaran (1973) and Jayanth (2000).

22.2 History of biological control of weeds

The first outstanding success in biological control of weeds in India was achieved when *Opuntia monacantha* (Wildenow) Haworth (Cactaceae) was controlled in central and north India by the introduction of the mealy bug *Dactylopius ceylonicus* Green (Hemiptera: Dactylopidae) from Brazil for the commercial production of cochineal dye, confusing it for *Dactylopius coccus* Costa (Hemiptera: Dactylopiidae), in 1795. This effort was not a

Biological Control of Tropical Weeds using Arthropods, ed. R. Muniappan, G.V.P. Reddy, and A. Raman. Published by Cambridge University Press. © Cambridge University Press, 2009.

deliberate attempt of biological control of weeds employing insects. Nevertheless, the potential of classical biological control came to light for the first time by using an insect to control a weed (Pruthi, 1969). The weed-infested area became fit for cultivation in 5–6 years (Pruthi, 1969). The subsequent introduction of *D. ceylonicus* to Sri Lanka in about 1865 and the successful control of *O. monacantha* constituted the first international transfer of a natural enemy for biological control of weeds (Goeden, 1988). *Dactylopius ceylonicus* restricted to *Opuntia* failed when introduced in southern India to suppress *Opuntia stricta* Haworth (= *Opuntia dillenii* Haworth) (Cactaceae). The intentional introduction of *Dactylopius opuntiae* (Cockerell) (Hemiptera: Dactylopiidae) (a North American species) from Sri Lanka into India in 1926 resulted in spectacular suppression of *O. stricta* and related *Opuntia elatior* Miller (Kunhikannan, 1928; Ayyar, 1931). This was the first successful intentional use of an insect to control a weed in India and an area covered by *O. elatior* of more than 40 000 ha was thus cleared (Narayanan, 1954).

In the early 1900s, with the exception of the introduction of lantana seed fly, *Ophiomyia lantanae* Froggatt (Diptera: Agromyzidae), from Hawaii against lantana, most other introductions – such as: the fish, *Ctenopharyngodon idella* (Cuvier and Valenciennes) (Pisces: Cyprinidae) against *Hydrilla* from China in 1959; *Procecidochares utilis* Stone (Diptera: Tephritidae) against *Ageratina adenophora* (Sprengel) King and Robinson (Asteraceae) from New Zealand in 1963; and *Pareuchaetes pseudoinsulata* Rego Barros (Lepidoptera: Arctiidae) against *Chromolaena odorata* (L.) King and Robinson (Asteraceae) – were carried out in cooperation with the Commonwealth Institute of Biological Control (CIBC) (now CAB International) Indian Station (Rao *et al.*, 1971; Sankaran and Rao, 1972; Sankaran, 1973). With the initiation of the "All India Coordinated Research Project on Biological Control of Crop Pests and Weeds" (AICRP–BC), concerted efforts were made from 1982 for the biological control of *Eichhornia crassipes* (Martius) Solms-Laubach (Pontederiaceae), *Salvinia molesta* D. S. Mitchell (Salviniaceae), *C. odorata* and *Parthenium hysterophorus* Linnaeus (Asteraceae) (Jayanth, 2000; Singh, 2001).

22.3 Biological control of alien weeds

22.3.1 Ageratina adenophora

The Crofton weed, *Ageratina* (= *Eupatorium*) *adenophora*, a native of Central America and Mexico, was accidentally introduced into India in the early twentieth century and has become a serious weed in India, especially in the Nilgiris (a state of Tamil Nadu) and the hilly areas of the state of West Bengal. It has occupied vacant land in tea, teak, rubber, and other forest plantations including wastelands and roadsides. A tephritid stem gall fly, *P. utilis* (from Mexico), was introduced from New Zealand in 1963 and released in the Nilgiris, Darjeeling, and Kalimpong areas (West Bengal). Although the insect has established (Rao *et al.*, 1971) and spread into Nepal (Kapoor and Malla, 1978), it has failed to create any substantial impact in control of the weed due to heavy parasitism by indigenous natural enemies (Sankaran, 1973).

22.3.2 Chromolaena odorata

The Siam weed, *Chromolaena odorata*, a native of the neotropics, was introduced into India in the mid 1800s and spread into the state of Kerala after World War II (Bennett and Rao, 1968). Its infestation in 1933–1934 in plantations of Buxa and Jalpaiguri divisions of Assam resulted in suppression of *Acacia catechu* and *Dalbergia sissoo* Roxb. (Fabaceae) regenerations in forests (Sen Gupta, 1949). In India, *C. odorata* is well distributed in areas receiving an annual rainfall of 150 cm and above (Muniappan *et al.*, 1989) and extensively evident in northeastern and southern states. The weed has become a serious problem in coconut, rubber, oil palm, tea, teak, coffee, cardamom, citrus and other plantations, orchards and forests (Muniappan and Viraktamath, 1993), and turns out to be a serious fire risk in the forests during the dry season (Singh, 1998).

Classical biological control was attempted from 1970 onwards through introduction of natural enemies from the native range of *C. odorata*. A host-specific arctiid defoliator, *P. pseudoinsulata* was imported by the Commonwealth Institute of Biological Control (CIBC, Indian Station, Bangalore) from Trinidad in 1970. Host-specificity tests were conducted using 13 plant species by Giriraj and Bhatt (1970) and 95 plant species by Sankaran and Sugathan (1974). After confirming that the insect was safe to other plants, field releases commenced in Kodagu where eggs (6700), larvae (33 000), and adult moths (600) were released. The insect, however, failed to establish probably due to predatory ants (Sankaran and Sugathan, 1974). Renewed efforts were made under the AICRP-BC by releasing about 20 750 larvae of different developmental stages and 600 gravid females at Chettalli, Kushalnagar, and Gonikoppal between September 1978 and April 1979. The failure to establish was attributed to predation by ants in the field and also due to granulovirus infection (Singh, 1980). Attempts to establish strains of this moth from Venezuela and Trinidad also failed. A Sri Lankan strain, introduced in 1984, was established in the Kerala Agricultural University campus (Joy *et al.*, 1985a). Further releases in rubber plantations resulted in large-scale defoliation of this weed (AICRP, 1984). Following further releases, it established in the fields at Mallesara near Teerthahalli in Shimoga district and Sullia Dakshina Kannada district, Karnataka State (AICRP, 1987). In 2005 it was observed defoliating chromolaena in Walayar, Nilambur, and Kottappara in Kerala (Varma *et al.*, 2006).

Adults of *Apion brunneonigrum* Beguin Billecocq (Coleoptera: Apionidae), a seed-feeding weevil, were obtained from Trinidad by CIBC Indian Station and supplied to Kerala Agricultural University, Thrissur, and Central Horticultural Experimental Station, Chettalli, in 1982 and 1983. About 800 weevils were released in both these places. Periodic observations revealed feeding holes on leaves but grubs did not develop. Establishment in the field has not been reported (Singh, 1989). A laboratory culture of *Mescinia parvula* Zeller (Lepidoptera: Pyralidae) introduced in 1986 from Trinidad could not be established (Jayanth, 2000).

To supplement the effects of *P. pseudoinsulata*, a stem gall fly, *Cecidochares connexa* Macquart (Diptera: Tephritidae) was introduced from Indonesia into India in 2002. A culture of the tephritid was established on *C. odorata* in the laboratory in the Project Directorate of Biological Control at Bangalore (Bhumannavar *et al.*, 2004). Host-specificity tests on 76 host plants belonging to 29 families revealed that the gall fly was capable of feeding and reproducing only on *C. odorata* (Bhumannavar *et al.*, 2004). Field releases of the gall fly were made at the University of Agricultural Sciences, GKVK, Bangalore, in July 2005. The gall fly established and caused a significant reduction in plant height, number of branches per plant, panicles per plant, capitula per panicle, and seeds per head (Bhumannavar and Ramani, 2007).

22.3.3 Eichhornia crassipes

Water hyacinth, *Eichhornia crassipes*, a native of South America, was introduced into India in the 1900s in Bengal as an ornamental plant. It has spread throughout India and occupies several freshwater ponds, tanks, lakes, reservoirs, streams, rivers, and irrigation channels. In some areas in Kerala, *E. crassipes* has been known to encroach into paddy fields.

In 1982, two weevils, *Neochetina eichhorniae* Warner (Coleoptera: Curculionidae) and *Neochetina bruchi* Hustache (Coleoptera: Curculionidae) and a mite, *Orthogalumna terebrantis* Wallwork (Acarina: Galumnidae) of Argentinean origin, were imported from USA. Host-specificity tests were conducted with 76 plant species belonging to 42 families for the weevils and with 88 plant species belonging to 42 families for the mite (Nagarkatti and Jayanth, 1984; Jayanth and Nagarkatti, 1987a, 1988). Based on the results of the host-specificity tests permission to field release these organisms in India was granted.

The weevils established readily under field conditions and were effective in suppressing water hyacinth either individually or in combination around Bangalore. The suppression, however, was slow in water bodies that were sedimented, as the silt adhering to the roots prevented pupation (Jayanth, 1987a, 1988a, b). In the water bodies that dry up in the summer, *N. eichhorniae* and *N. bruchi* probably survived as adults by remaining either beneath plant debris or in cracks in the hard soil (Jayanth and Ganga Visalakshy, 1990). *Orthogalumna terebrantis* did not appear to be capable of suppressing water hyacinth. Dense populations of the mite, however, occurred in all the water bodies, when they were released along with the weevils, and the mites induced browning of the laminae (Jayanth and Ganga Visalakshy, 1989a).

Water hyacinth was successfully controlled in the 286-sq-km Loktak lake in Manipur, when three-quarters of the lake was covered by it, in three years after the release of about 18 500 adults of both weevils in 1987–1988 (Jayanth and Ganga Visalakshy, 1989b). These weevils were also released in 15 other states and the weed is under control in water bodies in Hyderabad (Gupta *et al.*, 1993) and Gorakhpur (Misra *et al.*, 1989). According to Jayanth (1988a) a single release of 2000 adults per water body, irrespective of the area of weed coverage, was needed to establish an effectively breeding weevil population.

22.3.4 Lantana camara

Lantana camara Linnaeus (Verbenaceae), a native to tropical America introduced into India in 1809 as an ornamental plant in Kolkatta, has become a serious weed in many parts of the country (Muniappan and Viraktamath, 1986). It is a troublesome weed on vacant lands as well as forests and plantations in Karnataka and Tamil Nadu (Tadulingam and Venkatanarayana, 1932). Muniappan and Viraktamath (1986) have reviewed the status of this weed in India. A survey of natural enemies of lantana conducted in 1918 yielded 148 species of insects of which *Lantanophaga pusillidactyla* (Walker) (Lepidopterra: Pterophoridae) was considered to be of some value (Rao, 1920). However, *L. pusillidactyla* had a number of natural enemies, which impaired its efficiency.

Sankaran (1973) and Singh (1994) have reviewed natural enemies against *L. camara* in India. Biological control against this weed was initiated in 1921 with the introduction of the seed fly *O. lantanae* from Hawaii, which established, but failed to offer any recognizable weed suppression. Reports indicate that *Epinotia lantana* (Busck) (Lepidoptera: Tortricidae) (accidentally introduced) and *O. lantanae* inflicted damage to 95% lantana berries around Bangalore (Muniappan and Viraktamath, 1986).

The lace-wing bug, *Teleonemia scrupulosa* Stal (Hemiptera: Tingidae), was introduced from Australia in 1941; however, due to apprehensions about its capability to attack teak in plantations, releases were not attempted (Roonwal, 1952). Cultures of *T. scrupulosa* were supposed to have been destroyed as the adults fed on teak flowers in the quarantine at Dehra Dun, but the insect escaped quarantine, and currently it is recorded on lantana in all parts of the country. Joshi (1969) reported that *T. scrupulosa* to suppress *L. camara* in Bhimtal, Nainital, Uttar Pradesh, and Bisht and Bhatnagar (1978) contended that despite defoliation caused by the bug, the plants were not killed. The effectiveness of *T. scrupulosa* is impaired by an egg parasitoid *Erythmelus teleonemiae* (Hymenoptera: Mymaridae), which parasitises up to 85% of the eggs (Jayanth and Ganga Visalakshy, 1992; Visalakshy, 1998).

Salbia (= *Syngamia*) *haemorrhoidalis* Guenée (Lepidoptera: Pyralidae), *Leptobyrsa decora* Drake (Hemiptera: Tingidae), and *Diastema tigris* Guenée (Lepidoptera: Noctuidae) (origin: Mexico) were introduced during 1969–1971 (Sankaran, 1973), but none established. Two leaf miners of lantana, *Octotoma scabripennis* Guerin-Meneville (Coleoptera: Chrysomelidae) and *Uroplata girardi* Pic (Coleoptera: Chrysomelidae) were imported from Australia in 1971–1972 by the Forest Research Institute, Dehra Dun, and released in Haldawani and Bhopal between 1972 and 1975 (Julien and Griffiths, 1998). They have since established in northern India, but have not spread to southern India (Sankaran, 1973; Muniappan and Viraktamath, 1986).

22.3.5 Mikania micrantha

Mikania micrantha Kunth (Asteraceae) (the mile-a-minute weed), a perennial South American weed introduced during World War II, first noticed in the western Ghats in 1940s, has spread to southwestern and northeastern states of India causing damage by smothering trees and subsistence crops (Choudhury, 1972; Muniappan and Viraktamath, 1993). *Mikania micrantha* is a fast-growing vine and the nodes root readily when they come into contact with soil. Biological control is the only option for suppression of *M. micrantha* in India as other methods of weed control such as mechanical removal are expensive and ineffective, and chemical control is not practical.

Recently, the Project Directorate of Biological Control, Bangalore, in association with CAB International has introduced the rust fungus *Puccinia spegazzinii* into India for suppression of this weed (Ellison *et al.*, 2007). The establishment and spread of the rust in Assam and Kerala are being monitored.

22.3.6 Opuntia *spp.*

Prickly pear cacti, *Opuntia* spp. (Cactaceae) (Origin: New World) are weeds on vacant wasteland and pasture. The details of biological control efforts against *O. monacantha* (central India), *O. stricta* and *O. elatior* (southern India) have already been presented in Section 22.2.

22.3.7 Orobanche *spp.*

Orobanche spp. (Orobanchaceae) are weeds parasitizing tobacco, eggplant, tomato, sunflower, and other economically important crops. They derive their nutrition from the roots of crops, thus affecting adversely the growth and yield of their hosts. *Phytomyza orobanchia* Kaltenbach (Diptera: Agromyzidae), a fly attacking *Orobanche*, has been recorded from Anand (Gujarat) (Manjunath and Nagarkatti, 1977), but it was heavily parasitized. The maggots feed mainly on the seeds and hence the species has potential to destroy substantial quantities of seed. In 1982, a culture of *P. orobanchia* was introduced from Yugoslavia but puparia received were in diapause, the adults that emerged were weak, and no culture was established.

22.3.8 Parthenium hysterophorus

The carrot weed or parthenium, *Parthenium hysterophorus* is an annual herbaceous plant originating in northeast Mexico (Haseler, 1976). In the past hundred years, it has spread to Australia, Africa, and Asia (Towers *et al.*, 1977). It was first brought to India as an ornamental plant in 1910 (Prashar, 1989), but failed to establish. However, in the 1950s it was introduced into India and Australia along with wheat imported from the United States of America (Rao, 1956). It was first reported from Pune in 1955 (Rao, 1956) and later spread to almost all parts of India (Krishnamurthy *et al.*, 1977).

Biocontrol efforts were initiated against this weed with the introduction of the leaf-feeding beetle *Zygogramma bicolorata* Pallister (Coleoptera: Chrysomelidae), a flower-feeding weevil *Smicronyx lutulentus* Dietz (Coleoptera: Curculionidae), and a stem gall-inducing moth *Epiblema strenuana* (Walker) (Lepidoptera: Tortricidae) from

Mexico in 1983–1985. A shipment of 307 adults of *Z. bicolorata* was received in August 1983. Cultures of *E. strenuana* and *Z. bicolorata* were established under quarantine conditions, while *S. lutulentus* died during the shipment. The culture of *E. strenuana* was terminated as it fed on niger *Guizotia abyssinica* (Jayanth, 1987b). Host-specificity tests carried out on 40 species of plants belonging to 22 families in India (Jayanth and Nagarkatti, 1987b) confirmed that *Z. bicolorata* was specific to parthenium. However, it completed its life cycle on another weed, *Xanthium strumarium* (Viraktamath *et al.*, 2004).

Zygogramma bicolorata established readily under field conditions around Bangalore after releases were made in 1984 (Jayanth, 1987c). It established in an area of 10 ha in August 1988 and increased to 400, 5000, 20 000, and 50 000 sq km by October 1989, 1990, 1991, and 1992, respectively (Jayanth and Ganga Visalakshy, 1994). Currently the beetle has dispersed to over more than 200 000 km^2 in Karnataka, Tamil Nadu, and Andhra Pradesh, causing large-scale defoliation. It has been established in Jammu, Punjab, Haryana, Himachal Pradesh, Maharashtra, and Madhya Pradesh in northern India (Sushilkumar, 2005).

Defoliation of parthenium due to feeding by larvae and adults of *Z. bicolorata* induced up to 98% reduction in flower production, even though the insect did not feed directly on flowers (Jayanth and Geetha Bali, 1994). The early-stage larvae congregate and feed on the terminal and axillary vegetative and floral buds, thus preventing the emergence of flowers. Extensive defoliation of parthenium in and around Bangalore city has caused an overall reduction in flower production by the weed, which in turn has reduced pollen density in the atmosphere (Jayanth, 1996). Consequently, 40 different species of plants displaced by *P. hysterophorus* have also recolonized in the area (Jayanth and Ganga Visalakshy, 1996).

Reports of *Z. bicolorata* feeding on sunflower in 1994–1995 caused panic and resulted in a setback for further releases. However, field studies from 1994–1996 showed that the beetles nibbled only on sunflower leaves when contaminated with pollen from adjoining parthenium stands. Laboratory studies confirmed that the beetles fed on sunflower leaves, smeared with either an aqueous suspension of pollen or an extract of leaves of parthenium. However, the beetle did not complete its life cycle on sunflower under field conditions (Bhumannavar and Balasubramanian, 1998; Bhumannavar *et al.*, 1998; TNAU, 1998; Jayanth *et al.*, 1993). Consequently, further release and distribution were resumed.

There was reduction in number of flowers per plant, plant height, biomass, and root length in plants fed on *Z. bicolorata* compared with non-fed plants (Table 22.1) at Acharya N. G. Ranga Agricultural University, Hyderabad (ANGRAU, 2004). Similar results were also obtained (Table 22.2) at Tamil Nadu Agricultural University (TNAU, 2004).

22.3.9 Salvinia molesta

The water fern, *Salvinia molesta*, is a free-floating weed from Brazil, first noticed in Veli Lake, Kerala, in 1955. By 1964, it assumed the status of a serious aquatic weed all over Kerala. In the Kuttanad area, it choked the rivers, canals, lagoons, and covered Kakki and

Table 22.1 *Effect of* Zygogramma bicolorata *feeding on growth of* Parthenium hysterophorus

Growth parameters	Control	Beetle fed
No. of flowers/plant	853.9	537.1
Plant height (cm)	106.7	97.3
No. of adults/2.25 m^2	1.5	3.7
No. of grubs/2.25 m^2	1.5	2.9
No. of eggs/2.25 m^2	9.5	25.9
Root length (cm)	16.2	15.3
Plant biomass (gm/2.25 m^2)	8.0	8.6
No. of seedlings germinated/2.25 m^2	61.3	47.0

Table 22.2 *Effect of* Zygogramma bicolorata *feeding on growth of* Parthenium hysterophorus

Growth parameters	Control	Beetle fed[a]
Percent flowering plants (%)	83.5	3.5
Plant height (cm)	48.0	28.2
Root length (cm)	18.3	13.2
Defoliation (%)	3.6	81.2
Seedling density/m^2	22.8	29.6
Mean plant biomass (g)/2.25 m^2	2512.0	972.0

[a] Differences between control and beetle fed significant ($P = 0.05$) by least significant difference.

Idukki reservoirs, hindering navigation, irrigation, fishing, and other operations. In some areas, cultivation of rice had to be abandoned because of salvinia infestation (Joy *et al.*, 1985b).

Paulinia acuminata (De Geer) (Orthoptera: Pauliniidae) was introduced at the beginning of the biological control program in 1976, but the grasshopper did not establish. *Cyrtobagous salviniae* (Calder & Sands) (Coleoptera: Curculionidae) was introduced into India from Australia in 1982. Quarantine screening was done at Bangalore and host-specificity tests were conducted with 75 plants belonging to 41 families. Under multiple-choice tests the insect could feed and breed only on *S. molesta* (Jayanth and Nagarkatti, 1987c). A pond near Bangalore infested with salvinia was cleared within 14 months of release of *C. salviniae* (Jayanth, 1987d).

In Kerala, 4202 weevils were released in October 1983, at the Trichur, Kottayam, and Alleppey districts. The release of weevils provided spectacular results in many parts of

Kerala, and in some areas up to 99% suppression was achieved in 12–16 months (Joy *et al.*, 1985b). Savings due to the weevils has been estimated to be Rs.6.8 million (= US $0.15 million) every year that was spent on labor alone for manual clearing of the weed (AICRP, 1987). The majority of the water bodies in Kerala, such as Kuttanad and Kole land paddy fields, Vembanad Lake, and numerous canals in the region have since remained relatively free of any serious salvinia accumulation.

22.4 Biological control efforts on indigenous weeds

Sankaran and Rao (1972) have provided a long list of insects feeding on 36 species of aquatic and terrestrial indigenous weeds in India.

22.4.1 Cyperus rotundus

The purple nut sedge, *Cyperus rotundus* Linnaeus (Cyperaceae), is a serious weed throughout the world (Holm *et al.*, 1977). *Bactra venosana* (Zeller) (Lepidoptera: Tortricidae) was identified as a potential natural enemy of this weed (Sankaran and Rao, 1972; Habib, 1976). Life table studies, however, revealed a high mortality due to egg parasitism by *Trichogrammatoidea bactrae* Nagaraj (Hymenoptera: Trichogrammatidae). This rendered *B. venosana* ineffective against the weed in India (Ganga Visalakshy and Jayanth, 1995). An artificial diet (Ganga Visalakshy, 2001) was developed for mass production and augmentative releases of larvae of *B. venosana* for the control of the weed (Ganga Visalakshy and Jayanth, 2002).

22.4.2 Ipomoea carnea

Ipomoea carnea Jacq. (Convolvulaceae) is a serious weed found along the riverbeds, canals, ponds, lakes, and agricultural ecosystems in India. The tortoise beetle, *Aspidomorpha miliaris* Fabricius (Coleoptera: Chrysomelidae*)*, feeds on this weed (Oudhia and Ganguli, 1999; Oudhia, 2000). Ramesh (1996) reported that *A. miliaris* did not feed on agricultural crops like soybean, rice, maize, mung bean, groundnut, sesamum, castor, sorghum, and cotton. Host-specificity studies conducted on 47 plant species revealed that *A. miliaris* fed on *I. carnea*, *I. aquatica* Forsk., *Convolulus arvensis* Linnaeus, *I. palmata* Forsk. and *I. reniformis* Choisy (Oudhia, 2000). Murugesan and Paulraj (2003) studied the feeding potential of *A. miliaris* on *I. carnea*. However, no substantial control capability has been demonstrated.

22.4.3 Portulaca oleracea

Portulaca oleracea Linnaeus (Portulacaceae), a plant of South American origin, considered as the world's ninth worst weed, affects about 85 crops in 45 countries (Holm *et al.*, 1977). It is a rainy-season weed of vegetable gardens, vineyards, and banana orchards. It acts as alternate host to varied pests and diseases, and has allelopathic effects on many crops (Waterhouse, 1993). *Ceutorhynchus portulacae* Marshall (Coleoptera: Curculionidae) was identified as a potential indigenous biocontrol agent that could be utilized for the biological suppression of the weed (Ganga Visalakshy and Jayanth, 1997). Adults remain inactive during the winter and were not capable of undergoing diapause (Ganga Visalakshy and Krishnan, 2001).

22.5 Future strategies

Although *Z. bicolorata* can actively defoliate and suppress *P. hysterophorus* in the rainy season, it is unable to suppress the weed in winter and summer. It is desirable to import additional host-specific natural enemies such as the seed-feeding weevil *Smicronyx lutulentus* Ditz. (Coleoptera: Curculionidae), stem gall-inducing weevil *Conotrachelus albocinereus* Fielder (Coleoptera: Curculionidae), root boring moth *Carmenta ithacae* Beutenmuller (Lepidoptera: Sesiidae), and the rust fungus *Puccinia abrupta* var. *partheniicola* for the biological control of *P. hysterophorus*. Similarly, there is an urgent need to introduce the host-specific insect, *Heteropsylla spinulosa* Muddiman (Homoptera: Psyllidae) for the biological control of *Mimosa diplotricha* Sauvalle (Mimosaceae), a problem weed in Kaziranga National Park, Assam, and rubber plantations in Kerala in India.

22.6 Conclusions

Classical biological control of weeds has made a significant contribution to the control of introduced weeds in many countries including India and none of the organisms introduced for weed control has become a pest of crops. The risks involved in biological weed control agents temporarily feeding on nonhosts are negligible, compared with those posed by chemicals; however, such instances become magnified due to high visibility. With an increase in the movement of plant material, each country is at risk of being invaded by new organisms and some of them may become global pests and weeds. Quarantine alone will not prevent the entry of invasive species. The impact on the environment and agricultural production in the first few years of invasion of a pest species is tremendous. Such impacts can be minimized with international cooperation through exchange of information on invasive weeds and their natural enemies. There is a need for coordinated interdisciplinary work among ecologists, agronomists, weed scientists, entomologists, and plant pathologists in identifying already invaded organisms and in assessing their ecological problems, environmental concerns in different ecosystems, economic damage, and methods of control.

References

AICRP (1984). *Annual Report*. Bangalore, India: All India Co-ordinated Research Project on Biological Control of Crop Pests and Weeds, Project Directorate of Biological Control, pp.16–19.

AICRP (1987). *Annual report*. Bangalore, India: All India Co-ordinated Research Project on Biological Control of Crop Pests and Weeds, Project Directorate of Biological Control, pp. 167–179.

ANGRAU (2004). Annual report of AICRP on Bio-control of Crop Pests and Weeds for the year 2003–04. Hyderabad, India: Acharya N. G. Ranga Agricultural University, pp. 188–189.

Ayyar, T. V. R. (1931). The coccidae of the prickly-pear in South India and their economic importance. *Agriculture and Livestock in India*, **1**, 229–237.

Bennett, F. D. and Rao, V. P. (1968). Distribution of an introduced weed, *Eupatorium odoratum*, in Asia and Africa and possibilities of its biological control. *PANS*, **14**, 277–281.

Bhumannavar, B. S. and Balasubramanian, C. (1998). Food consumption and utilization by the Mexican beetle *Zygogramma bicolorata* on *Parthenium hysterophorus*. *Journal of Biological Control*, **12**, 19–23.

Bhumannavar, B. S. and Ramani, S. (2007). Introduction of *Cecidochares connexa* into India for the biological control of *Chromolaena odorata*. In *Proceedings of the Seventh International Workshop in the Biological Control and Management of* Chromolaena odorata *and* Mikania micrantha, ed. P. Y. Lai, G. V. P. Reddy and R. Muniappan. Taiwan: National Pingtung University of Science and Technology, pp. 38–48.

Bhumannavar, B. S., Balasubramanian, C. and Ramani, S. (1998). Life table of the Mexican beetle *Zygogramma bicolorata* on parthenium and sunflower. *Journal of Biological Control*, **12**, 101–106.

Bhumannavar, B. S., Ramani, S., Rajeshwari, S. K. and Chaubey, B. K. (2004). Host-specificity and biology of *Cecidochares connexa* introduced into India for the biological suppression of *Chromolaena odorata*. *Journal of Biological Control*, **18**, 111–120.

Bisht, R. S. and Bhatnagar, S. P. (1978). Studies on the ecology and distribution of Lantana bug *Teleonemia scrupulosa* and of *Lantana camara* in Kumaon region. *Uttarkhand Bharti*, **3**, 10–13.

Choudhury, A. K. (1972). Controversial Mikania (climber): a threat to the forests and agriculture. *Indian Forester*, **98**, 178–186.

Ellison, C. A., Puzari, K. C., Sreerama Kumar, P., *et al.* (2007). Sustainable control of *Mikania micrantha*: implementing a classical biological control strategy in India using the rust fungus, *Puccinia spegazzinii*. In *Proceedings of the Seventh International Workshop in the Biological Control and Management of* Chromolaena odorata *and* Mikania micrantha, ed. P. Y. Lai, G. V. P. Reddy and R. Muniappan. Taiwan: National Pingtung University of Science and Technology, pp. 94–105.

Ganga Visalakshy, P. N. (2001). An artificial diet for mass multiplication of *Bactra venosana*, a potential biocontrol agent of *Cyperus rotundus*. *Pest Management in Horticultural Ecosystems*, **7**, 134–136.

Ganga Visalakshy, P. N. and Jayanth, K. P. (1995). Suppression of *Bactra venosana*, potential biocontrol agent of *Cyperus rotundus* by *Trichogrammatoidea bactrae*. *Phytoparasitica*, **23**, 355–356.

Ganga Visalakshy, P. N. and Jayanth, K. P. (1997). *Ceutorhynchus portulucae*, a potential biological control agent of *Portulaca oleracea*. *Entomon*, **22**, 150–151.

Ganga Visalakshy, P. N. and Jayanth, K. P. (2002). Suppressing *Cyperus rotundus* by augmentative releases of *Bactra venosana*. *Journal of Biological Control*, **16**, 85–86.

Ganga Visalakshy, P. N. and Krishnan, S. (2001). Quiescence behaviour in *Ceuthorhynchus portulucae*, a potential biocontrol agent of the purslane weed, *Portulaca oleracea. Journal of Biological Control*, **15**, 27–30.

Goeden, R. D. (1988). A capsule history of biological control of weeds, *Biocontrol News and Information*, **9**, 55–61.

Giriraj, C. N. and Bhatt, V. K. (1970). Supply of natural enemies of the "Siam weed" *Eupatorium odoratum* (for Nigeria and Malaysia). Annual Report, Commonwealth Institute of Biological Control. Slongh, UK: Commonwealth Agriculture Bureaux, 112 pp.

Gupta, M., Rao, P. and Pawar, A. D. (1993). Suppression of water hyacinth *Eichhornia crassipes* by *Neochetina* spp. in Hyderabad. *Indian Journal of Plant Protection*, **21**, 23–25.

Habib, R. (1976). *Bactra* spp. in Pakistan and their potential as biocontrol agents of *Cyperus rotundus. PANS*, **22**, 499–508.

Haseler, W. H. (1976). *Parthenium hysterophorus* in Australia. *PANS*, **22**, 515–517.

Holm, L. G., Plucknett, D. L., Pancho, J. V. and Herberger, J. P. (1977). *The World's Worst Weeds, Distribution and Biology*. Honolulu, HI: University Press of Hawaii, 609 pp.

Jayanth, K. P. (1987a). *Biological Control of Water Hyacinth in India*. Technical Bulletin 3. Bangalore, India: Indian Institute of Horticultural Research, 28 pp.

Jayanth, K. P. (1987b). Investigations on the host-specificity of *Epiblema strenuana* introduced for biological control trials against *Parthenium hysterophorus* in India. *Journal of Biological Control*, **1**, 133–137.

Jayanth, K. P. (1987c). Introduction and establishment of *Zygogramma bicolorata* on *Parthenium hysterophorus* in Bangalore, India. *Current Science*, **56**, 310–311.

Jayanth, K. P. (1987d). Biological control of water fern, *Salvinia molesta* infesting a lily pond in Bangalore by *Cyrtobagous salviniae. Entomophaga*, **32**, 163–165.

Jayanth, K. P. (1988a). Successful biological control of water hyacinth (*Eichhornia crassipes*) by *Neochetina eichhorniae* in Bangalore, India. *Tropical Pest Management*, **34**, 263–266.

Jayanth, K. P. (1988b). Biological control of water hyacinth in India by release of exotic weevil, *Neochetina bruchi. Current Science*, **57**, 968–970.

Jayanth, K. P. (1996). Status of biological control trials against *Parthenium hysterophorus* by *Zygogramma bicolorata* in India. *Madras Agriculture Journal*, **83**, 672–678.

Jayanth, K. P. (2000). Biological control of weeds in India. In *Bicontrol Potential and Its Exploitation in Sustainable Agriculture*, Vol. 1, ed. R. K. Upadhyay, K. G. Mukerji and B. P. Chamola. New York: Kluwer Academic/Plenum Publishers, pp. 207–221.

Jayanth, K. P. and Ganga Visalakshy, P. N. (1989a). Establishment of the exotic mite, *Orthogalumna terebrantis* on water hyacinth in Bangalore, India. *Journal of Biological Control*, **3**, 75–76.

Jayanth, K. P. and Ganga Visalakshy, P. N. (1989b). Introduction and establishment of *Neochetina eichhorniae* and *N. bruchi* on water hyacinth in Loktak lake, Manipur. *Science and Culture*, **55**, 505–506.

Jayanth, K. P. and Ganga Visalakshy, P. N. (1990). Studies on drought tolerance in the water hyacinth weevils *Neochetina eichhorniae* and *N. bruchi* (Coleoptera: Curculionidae). *Journal of Biological Control*, **4**, 116–119.

Jayanth, K. P. and Ganga Visalakshy, P. N. (1992). Suppression of the lantana bug, *Teleonemia scrupulosa* by *Erythmelus teleonemiae* in Bangalore, India. *FAO Plant Protection Bullettin*, **40**, 164.

Jayanth, K. P. and Ganga Visalakshy, P. N. (1994). Dispersal of the parthenium beetle, *Zygogramma bicolorata* in India. *Biocontrol Science and Technology*, **4**, 363–365.

Jayanth, K. P. and Ganga Visalakshy, P. N. (1996). Succession of vegetation after suppression of parthenium weed by *Zygogramma bicolorata* in Bangalore. *Biological Agriculture and Horticulture*, **12**, 303–309.

Jayanth, K. P. and Geetha Bali (1994). Biological control of *Parthenium hysterophorus* by the beetle, *Zygogramma bicolorata* in India. *FAO Plant Protection Bulletin*, **42**, 207–213.

Jayanth, K. P. and Nagarkatti, S. (1987a). Host-specificity of *Neochetina bruchi* introduced into India for the biological control of water hyacinth. *Entomon*, **12**, 385–390.

Jayanth, K. P. and Nagarkatti, S. (1987b). Investigations on the host-specificity and damage potential of *Zygogramma bicolorata* introduced into India for the biological control of *Parthenium hysterophorus*. *Entomon*, **12**, 141–145.

Jayanth, K. P. and Nagarkatti, S. (1987c). Host-specificity of *Cyrtobagous salivinae* introduced into India for the control of *Salvinia molesta*. *Entomon*, **12**, 1–6.

Jayanth, K. P. and Nagarkatti, S. (1988). Host-specificity of *Orthogalumna terebrantis* introduced for biological control of water hyacinth in India. *Journal of Biological Control*, **2**, 46–49.

Jayanth, K. P., Sukhada, M., Asokan, R. and Ganga Visalakshy, P. N. (1993). Parthenium pollen induced feeding by *Zygogramma bicolorata* on sunflower. *Bulletin of Entomological Research*, **83**, 595–598.

Joshi, D. P. (1969). Eradication of *Lantana camara* by lantana bug. *Indian Forester*, **95**, 152–154.

Joy, P. J., Sathesan, N. V. and Lyla, K. R. (1985a). Biological control of weeds in Kerala. In *Proceedings of National Seminar on Entomophagons Insects*, held in University of Calicut, India, pp. 247–251.

Joy, P. J., Sathesan, N. V., Lyla, K. R. and Joseph, D. (1985b). Successful biological control of the floating weed *Salvinia molesta*, using the weevil, *Cyrtobagous salviniae* in Kerala (India). In *Proceedings of the 10th Asian Pacific Weed Science Society Conference*, pp. 622–626.

Julien, M. H. and Griffiths, M.W. (1998). *Biological Control of Weeds: A World Catalogue of Agents and Their Target Weeds*. Wallingford, UK: CAB International, 223 pp.

Kapoor, V. C. and Malla, Y. K. (1978). The infestation of the gall fruit-fly, *Procecidochares utilus* on Crofton weed, *Eupatorium adenophorum* in Kathmandu. *Indian Journal of Entomology*, **40**, 337–339.

Krishnamurthy, K., Ramachandra Prasad, T. V., Muniyappa, T. V. and Venkata Rao, B. V. (1977). Parthenium, a new pernicious weed in India. Technical Series 17. Bangalore, India: University of Agricultural Sciences, 46 pp.

Kunhikannan, K. (1928). The introduction of a new insect into Mysore (India). *Agricultural Journal*, **8**, 141–148.

Manjunath, T. M. and Nagarkatti, S. (1977). Natural enemies of *Orobanche* in India and possibilities of its biological control. *CIBC Technical Bulletin*, **18**, 75–83.

Misra, M. P., Pawar, A. D. and Ram, N. (1989). Establishment of exotic phytophagous weevils *Neochetina eichhorniae* and *N. bruchi* for the biological control of water hyacinth in north India. *Journal of Advanced Zoology*, **10**, 136–138.

Muniappan, R. and Viraktamath, C. A. (1986). Status of biological control of the weed, *Lantana camara* in India. *Tropical Pest Management*, **32**, 40–42.

Muniappan, R. and Viraktamath, C. A. (1993). Invasive introduced weeds in the Western Ghats. *Current Science*, **64**, 555–557.

Muniappan, R. and Sundaramurthy, V. T. and Viraktamath, C. A. (1989). Distribution of *Chromolaena odorata* and biology, utilization of food and defoliation by *Pareuchaetes pseudoinsulata* in India. In *Proceedings of the VII International Symposium on Biological Control of Weeds*, ed. E. S. Del Fosse. Melbourne, Australia: CSIRO, pp. 401–410.

Murugesan, A. G. and Paulraj, M. G. (2003). Feeding potential of fool's gold beetle, *Aspidomorpha miliaris* on *Ipomoea carnea*. *Journal of Biological Control*, **17**, 171–174.

Nagarkatti, S. and Jayanth, K. P. (1984). Screening biological control agents of water hyacinth for their safety to economically important plants in India. 1. *Neochetina eichhorniae*. In *Water Hyacinth*, ed. G. Thyagarajan. Nairobi, Kenya: United Nations Environmental Programme, pp. 868–883.

Narayanan, E. S. (1954). (Discussion on biological control of weeds). *Report of Sixth Commonwealth Entomological Conference*, ed. W. J. Hall. London: Commonwealth Institute of Entomology, p. 100.

Oudhia, P. (2000). Studies on host specificity and preference of the metallic coloured tortoise beetle (*Aspidomorpha miliaris*). *Ecology and Environment Conservation*, **6**, 357–359.

Oudhia, P. and Ganguli, J. (1999). Outbreak of tortoise beetle, *Aspidomorpha miliaris* in Chhattisgarh plains. *Insect Environment*, **5**, 110–111.

Prashar, A. S. (1989). Congress grass: start chemical fight. *The Tribune*, **109**, 17.

Pruthi, H. S. (1969). *Text Book of Agricultural Entomology*. New Delhi, India: Indian Council of Agricultural Research, 977 pp.

Ramesh, P. (1996). Host specificity of the tortoise beetle, *Aspidomorpha miliaris* and its possible role in the management of *Ipomoea carnea* and *I. aquatica*. *Indian Journal of Entomology*, **58**, 140–142.

Rao, R. S. (1956). *Parthenium hysterophorus* a new record for India. *Journal of Bombay Natural History Society*, **54**, 218–220.

Rao, Y. R. (1920). Lantana insects in India. *Memoirs Department of Agriculture India, Entomological Series*, **5**, 239–314.

Rao, V. P., Ghani, A. M., Sankaran, T. and Mathur, K. C. (1971). *A Review of the Biological Control of Insects and Other Pests in South East Asia and the Pacific Region*. Technical Communication 6. Slongh, UK: The Commonwealth Institute of Biological Control, 149 pp.

Roonwal, M. L. (1952). The natural establishment and dispersal of an imported insect in India. The lantana bug, *Teleonemia scrupulosa* with a description of its egg, nymph and adult. *Journal of the Zoological Society of India*, **4**, 1–16.

Sankaran, T. (1973). Biological control of weeds in India. A review of introductions and current investigations of natural enemies. In *Proceedings of II International Symposium on Biological Control of Weeds*, ed. P. H. Dunn. Slongh, UK: Commonwealth Agricultural Bureau, pp. 82–88.

Sankaran, T. and Rao, V. P. (1972). *An Annotated List of Insects Attacking Some Terrestrial and Aquatic Weeds in India, With Records of Some Parasites of the Phytophagous Insects*. Technical Bulletin 15. Slongh, UK: Commonwealth Institute of Biological Control, pp. 131–157.

Sankaran, T. and Sugathan, G. (1974). Host specificity tests and field trials with *Ammalo insulata* in India. Report of the Commonwealth Institute of Biological Control, 11 pp.

Sen Gupta, J. N. (1949). The growing menace of *Assam lota* (*Eupatorium* spp.) and how to control it? *Indian Forester*, **75**, 351–353.

Singh, S. P. (1980). Experiments on propagation and field release of *Ammalo insulata* for the biological control of *Chromolaena odorata*. In *Proceedings of the third Workshop held at Punjab Agricultural University, Ludhiana*. All India Coordinated Research Project on Biological Control of Crop Pests and Weeds, pp. 173–175.

Singh, S. P. (1989). *BiologicalSuppression of Weeds*. Technical Bulletin 1. Bangalore, India: Biological Control Centre (National Center for Integrated Pest Management), 27 pp.

Singh, S. P. (1994). Biological suppression of *Lantana camara* in India. In *National Seminar on Weed Management in Hills*. Palampur, India: Himachal Pradesh Agricultural University, 8 pp.

Singh, S. P. (1998). A review of biological suppression of *Chromolaena odorata* in India. In *Proceedings of the Fourth International Workshop on Biological Control and Management of* Chromolaena odorata, ed. P. Ferrar, R. Muniappan, and K. P. Jayanth. Mangilau: Agricultural Experiment Station, University of Guam, pp. 86–92.

Singh, S. P. (2001). Biological control of invasive weeds in India. In *Alien Weeds in Moist Tropical Zones: Banes and Benefits*, ed. K. V. Sankaran, S. T. Murphy and H. C. Evans. Peechi, India: Kerala Forest Institute, pp. 1–11.

Sushilkumar (2005). *Biological Control of Parthenium Through* Zygogramma bicolorata. NRCWS Technical Bulletin 5. Jabalpur, India: National Research Center for Weed Science, 89 pp.

Tadulingam, C. and Venkatanarayana, G. (1932). *A Handbook of Some South India Weeds*. Madras, India: Government Press, 356 pp.

TNAU (1998). Final report of ICAR Ad-hoc Research Project on 'Studies on the pest potential of the Mexican beetle, *Zygogramma bicolorata* introduced for biocontrol of *Parthenium*'. Coimbatore, India: Tamil Nadu Agricultural University, 72 pp.

TNAU (2004). Annual report of AICRP on bio-control of crop pests and weeds, for the year 2003–04. Coimbatore, India: Tamil Nadu Agricultural University, 190 pp.

Towers, G. H. N., Mitchell, J. C., Rodriguez, E., Bennett, F. D. and Subba Rao, P. V. (1977). Biology and chemistry of *Parthenium hysterophorus*, a problematic weed in India. *Journal Scientific and Industrial Research*, **36**, 672–684.

Varma, R. V., Shetty, A., Swaran, P. R., Puduvil, R. and Shamsudeen, R. S. M. (2006). Establishment of *Pareuchaetes pseudoinsulata*, an exotic biocontrol agent of the weed, *Chromolaena odorata* in the forests of Kerala, India. *Entomon*, **31**, 49–52.

Viraktamath, C. A., Bhumannavar, B. S. and Patel, V. N. (2004). Biology and ecology of *Zygogramma bicolorata*. In *New Developments in the Biology of Chrysomelidae*, ed. P. Jolivet, J. A. Santiago-Blay and M. Schmitt. The Hague, Netherlands: Academic Publishing Bv., pp. 767–777.

Visalakshy, P. N. G. (1998). Factors influencing the effectiveness of the tinged bug, *Teleonemia scrupulosa* on lantana in Bangalore, India. *Pest Management in Horticultural Ecosystems*, **4**, 29–31.

Waterhouse, D. F. (1993). *Biological Control – Pacific Prospects*. Supplement 2. Canberra, Australia: Australian Centre for International Agricultural Research, 138 pp.

23

The role of International Institute of Tropical Agriculture in biological control of weeds

Fen Beed and Thomas Dubois

23.1 International Institute of Tropical Agriculture (IITA)

IITA is one of Africa's leading research organizations in finding solutions for the devastating social issues of hunger and poverty. IITA was established in 1967 with a mission to enhance food security and improve livelihoods for the people of Africa through research for development. Operating from a number of stations across sub-Saharan Africa, its scientists work towards the development of technologies that reduce risk for producers and consumers, increase local production and wealth generation. It is the largest among several agricultural research centers across the world, supported by the Consultative Group on International Agricultural Research (CGIAR). IITA recognizes the agricultural sector as a vital element to sub-Saharan Africa's economic development employing nearly two-thirds of its population. IITA also recognizes that agriculture is a complex network of skills and expertise which includes the conception of an idea for a specific agricultural product until it nourishes a satisfied customer. This process may be as different as a farmer knowing when to plant a cassava crop to be able to prepare a nutritious family meal following a bountiful harvest, to the investment in the infrastructure and organization needed for African cocoa to be marketed throughout the world for the benefit of consumers who are willing to pay a premium for luxury products.

Agriculture covers a multiplicity of stakeholders and systems, which lead from the "soil to supper." IITA works with partners within Africa and beyond, enhancing crop quality and productivity to create impact in the lives of poor people, both rural and urban, within Africa. In addition, IITA develops technologies for Africans who have the expertise, initiative, and resources to go beyond food security and produce enough to achieve a financial profit. If African agriculture is to serve as an engine of economic development, research is required to develop models that define how industries and enterprises succeed. Research for development at IITA is a process where scientific principles are used to identify specific needs and constraints in order to generate development solutions. IITA's technical expertise combined with its long-term experience with and knowledge of African food and agricultural systems allows it to work with investors

Biological Control of Tropical Weeds using Arthropods, ed. R. Muniappan, G. V. P. Reddy, and A. Raman. Published by Cambridge University Press. © Cambridge University Press, 2009.

and local, national, regional, and international partners including advanced research institutes, other CGIAR centers, national agricultural research organizations, universities, governments, policy makers, nongovernmental organizations, private sector, and networks such as ASARECA, COMESA, FARA, IAPSC, and NEPAD. The combined goal is to enhance the security and profitability of Africa's agricultural sector through synergy between partners with different roles, along the research for development continuum.

Research at IITA starts with identifying development needs, which are often incorporated in international, regional, and national priority-setting activities. This is followed by identifying the specific problems that can be addressed by IITA and partners, which then results in detailing the research approach and strategy. In collaboration with research and development partners, IITA subsequently develops technologies and shares information through scientific publications, and takes on an advocacy and monitoring role to facilitate their implementation and promotion as international public goods (Alene *et al.*, 2007).

Further to efforts to improve crop germplasm, IITA has made significant strides to develop sustainable methods for germplasm management in the field. Focus has been on the development of biologically based integrated pest management (IPM) options as they provide more environmentally benign and durable control than reliance on single compounds, such as active ingredients in synthetic pesticides. Moreover, IPM options are better adapted to the needs of African farmers due to prohibitive cost, and poor availability and quality of synthetic pesticides, together with lack of either appropriate equipment or relevant instruction for their application. A survey in Benin to determine perception of farmers to speargrass (*Imperata cylindrica* (L.) Beauv.) (Poaceae) and methods for its control demonstrated that despite it being considered the most troublesome weed leading to abandonment of agricultural land, none of the questioned 300 farmers used synthetic herbicides to combat this weed (Ayeni *et al.*, 2004). However, in sub-Saharan Africa, a barrier towards the acceptance, and consequently the implementation of plant protection strategies, exists (Yaninek and Schulthess, 1993), because small-scale farmers are reluctant to accept changes that increase their exposure to non-traditional methods, which are readily perceived as risky. Farmers initiate actions that they consider are either less risky or risk free, such as overstocking livestock and sowing the same crops continuously. Supporting systems to provide advice or credit are either limited or totally absent, and consequently a farmer usually selects an earlier maturing variety that is not only disease prone but also low yielding, so that some "profit" is ensured rapidly. Furthermore, labor available for demanding farm management tasks, such as weeding, is in acute shortage in many rural areas because of the emigration of youth to cities and the impact of diseases (e.g. malaria and AIDS). While farmers demand crop protection solutions, they are unaware of all options available. In such a complex context, organizations such as IITA function in developing, testing, publicizing, and implementing crop protection strategies for the farmers. The use of natural enemies as a crop protection option provides reasonable dividends to farmers as their use has no negative impact on the ecology, while natural biodiversity is preserved to the maximum.

However, the effective design and implementation of biological control technologies can be only mediated by an organization such as IITA, which bears a mandate across communities and countries. IITA has the capacity to engage the participation of relevant partners from within and outside Africa and to source support from a range of donors.

23.2 Classical biological control in Africa and role of IITA

IITA's mandate to develop biologically based solutions, supported by the Commonwealth Institute of Biological Control (CIBC) and others, led to some of the largest classical biological control programs ever established (Greathead, 2003). Examples include the establishment of biological control agents for control of mealybugs (*Phenacoccus manihoti* Matile-Ferrero (Hemiptera: Pseudococcidae) and *Rastrococcus invadens* Williams (Hemiptera: Pseudococcidae)) on cassava (*Manihot esculaenta* Crantz) (Euphorbiaceae) and mango (*Mangifera indica* L.) (Anacardiaceae). Nonetheless, this required biologically robust evidence: proof of the magnitude of the pest problem, selection of potential biological control agents from Central America and India, host-specificity tests and quarantine clearance from CIBC headquarters in the UK, tests to determine establishment, ecological studies, such as hyperparasitism, durability of control efficacy under African conditions and development of predictive weather-driven models. Moreover such efforts required diplomacy to convince governments to permit the import and release of exotic species, to encourage a series of donors to invest in the process (rather than the deployment of nonspecific pesticides), and to inform and train stakeholders. The knowledge accrued from such a thorough research program raised awareness of the need to study the implementation of a biological control technology as part of a farming systems approach (Herren and Neuenschwander, 1991; Neuenschwander, 1996, 2001; Schulthess *et al.*, 1997). In terms of economic sustainability (the ratio between "dollars gained" and "dollars invested") control of cassava mealybug was assessed at 200–500 to 1 (depending on the agricultural scenario adopted) (Zeddies *et al.*, 2001) and 145 to 1 for mango mealybug biological control (Bokonon-Ganta *et al.*, 2002). Similarly, tremendous economic impact of weed biological control mediated by IITA has been demonstrated: De Groote *et al.*, (2003) performed a survey in southern Benin in 1999 following the release of arthropod biological control agents for the control of water hyacinth (*Eichhornia crassipes* (Mart.) Solms) (Pontederiaceae) between 1991 and 1993 and reported a benefit-cost ratio of 124:1 and predicted increased income from fishing of US$260 over the following 20-year period. However, it must be recognized that such monetary valuations underestimate the "true" benefit of biological control of pests because the ecological and social benefits are hard to quantify. For example, many of the farmers that benefited from the establishment of control agents did not in fact invest in them but did benefit from an increased understanding of ecological interactions relevant to farming and there were significant impacts on capacity building for many partners and institutions involved (Neuenschwander and Markham, 2001).

23.3 Inundative biological control in Africa and role of IITA

Successes in classical biological increased the awareness and capacity of IITA and its partners, both within and outside Africa, and created the opportunity to move forward towards exploiting inundative biological control agents. Biopesticides using fungal pathogens have been shown to be more efficient at controlling target pests than their natural counterparts because they are applied in larger concentrations and because environmental constraints to infection are overcome through the use of appropriate chemical formulations. IITA, CABI, and a network of partners and donors developed a biopesticide for control of desert locusts and grasshoppers using an isolate of *Metarhizium anisopliae* var. *acridum* (Metchnikoff) Sorokin (Hypocreales: Clavipitaceae) (Lomer *et al.*, 2001). The research for a development continuum approach led to official registration of Green Muscle™ (Green Guard in Australia). In terms of oil formulation tests to increase spore infectivity through overcoming humidity requirements and carrying ultra-violet protectants, application methods and ecological impact under Sahelian conditions, Green Muscle™ is the most thoroughly tested biopesticide in the world. Furthermore, by comparison with synthetic chemical insecticides, Green Muscle™ has been shown to provide increased sustainable control efficacy of target pests without negative impacts on the environment or other natural enemies, including insect predators, protozoan parasites, and birds (Langewald *et al.*, 2003). Efforts to effect implementation have been enhanced by the recent decision of the government of Senegal to create a factory for the mass production of Green Muscle™ and other biopesticides. This model study has also stimulated the development of other biopesticides at IITA, such as the use of an isolate of *Beauveria bassiana* (Balsamo) Vuillemin (Hypocreales: Clavicipitaceae) for the control of dia-mondback moth *Plutella xylostella* L. (Lepidoptera: Plutellidae) which devastates global cabbage production. A further example of inundative biocontrol led by IITA has been developed in response to the damaging effects of mycotoxins, in particular aflatoxins, on human health (Gong *et al.*, 2003). Atoxigenic races of *Aspergillus flavus* Johann Heinrich Friedrich (Eurotiales: Trichocomaceae) have been selected based on safety, vegetative compatibility grouping, and efficacy in laboratory and field trials to competitively exclude toxigenic races. This biological control approach has reduced aflatoxin contamination in maize by up to 99.8% in field trials (Bandyopadhyay *et al.*, 2005).

23.4 A case study for the biological control of water hyacinth and role of IITA in achieving sustainability

Research activities for control of *Eichhornia crassipes* are described by Coetzee *et al.* this volume. The purpose of this section is to outline how IITA as an organization approached the management of this aquatic weed. Earliest releases by IITA (in collaboration with CABI, CSIRO, and PPRI) in Benin included *Neochetina eichhorniae* Warner (Coleoptera: Curculionidae) in 1991, *Neochetina bruchi* Hustache (Coleoptera: Curculionidae) in 1992, *Niphograpta (Sameodes) albiguttalis* Warren (Coleoptera: Curculionidae) in 1993

(Van Thielen *et al.*, 1994), and *Eccritotarsus catarinensis* (Carvahlo) (Hemiptera: Miridae) in 2000. Starter cultures and capacity building on mass production and monitoring of *N. eichhorniae* and *N. bruchi* were provided in 1993 to Ghana, Kenya, Nigeria, Uganda, and Zimbabwe, in 1995 to Tanzania, in 1997 to Côte d'Ivoire, in 1998 to Burkina Faso, and in 2001 to Togo.

Monitoring the impact of released biological control agents to ensure establishment and sustainable pest control was extremely difficult under the short-term mandates of donors. However, IITA invested core funds to determine the success of classical biological control of water hyacinth. One study that merits mention is the survey in Congo Basin by IITA and the Ministry of Forestry and Environment of the Democratic Republic of Congo. This study aimed to determine the establishment and spread of classical biocontrol agents released against water hyacinth and also water lettuce (*Pistia stratiotes* L.) (Araceae) and water fern (*Azolla filiculoides* Lam.) (Azollaceae) (Mbati and Neuenschwander, 2005). Recoveries of specific exotic weevils; *N. eichhorniae, N. bruchi* from water hyacinth, *Neohydronomus affinis* Hustache (Coleoptera: Curculionidae) from water lettuce and *Cyrtobagous salviniae* Calder & Sands (Coleoptera: Curculionidae) from water fern were made from all but one of the 24 release sites. Control of water fern and water lettuce was so successful that fishing and navigation could be resumed, while reductions of water hyacinth populations were less successful. Studies in Benin showed that levels of establishment for *N. eichhorniae* were superior to *N. bruchi* and that establishment of the other insect agents had failed, often due to either competition or hyperparasitism. Furthermore, differences in control efficacy mediated by *N. eichhorniae* were found to vary between sites, and long-term monitoring evaluations showed that site topography was a determining factor. Deep-lake sites with vertical banks and minimal water flow, such as Tèvèdji and Lihu, experienced 80% control of water hyacinth populations within nine years, but sites with rapid water flow and in particular those with shallow banks that dried out or where water was either minimal or turbid suffered reductions in weevil populations. Studies have demonstrated that *Neochetina* populations could not pupate in roots when water was absent or turbid (Visalakshy and Jayanth, 1996; Ajuonu *et al*, 2003).

IITA's link with governments across West Africa revealed similar discrepancies in levels of water hyacinth control by weevils. Control efficacy was noticeably reduced in coastal lagoons due to the weevil's greater sensitivity to saline conditions than water hyacinth plants (Nwankwo and Akinsoji, 1988). National Agricultural officers in Burkina Faso failed to establish species of *Neochetina* in artificial irrigation ponds. In contrast, IITA's collaboration with Water Research Institute in Ghana showed that in the deep and slow-moving lagoon system bordering Côte d'Ivoire control of water hyacinth had been successful. However, where water hyacinth was eradicated hippo grass (*Vossia cuspidata* (Roxb.) Griff) (Poaceae) had invaded which posed a further problem as no control option is available for this weed. In the same region control of water hyacinth was less successful in the neighboring and faster moving watercourses such as the Tano River. Monitoring studies also revealed the impact of humans; some populations of plants infested with weevils were inadvertently cleared and noninfested plants were spread to new locations by the fishing practice of *Akadja* (provision of plant surface cover held in place by sticks to protect fish from predatory birds) and through the use of water hyacinth plants to camouflage smuggled fuel drums upriver. Despite the undoubted impact of *Neochetina* spp. weevils on reducing vigor and number of water hyacinth, infestations have continued to spread. For example, in Benin, the most important river systems, the Ouemé and the Sô, are choked during the dry season. On Lake Kainji in Central Nigeria despite the release of weevils, the construction of a physical boom and the removal of 1 million tonnes/annum of water hyacinth by local communities, weed populations have been increasing and pose a threat to hydroelectric power generation (Olokor *et al.*, 1998; Farri and Boroffice, 1999). A survey in 2001 by the National Institute for Freshwater Fisheries Research (NIFFR) in Nigeria showed water hyacinth was the nation's most common aquatic weed, being recorded in 26 out of 36 states. The Institut d'Economie Rurale (Bamako, Mali) reported that water hyacinth limited annual production of rice (by 500 000 t), sugar (25 000 t) and market garden produce with a commercial value of US$66 850 000. La Société Energie, Mali, in 2003 spent US$28 500 to produce a boom and remove accumulated plants, and the Office du Perimètre Irrigué spent US$287 400 on the manual removal of plants each year.

Future efforts to provide increased efficacy of water hyacinth control are thus required to complement the action of the weevils. In addition to efforts to establish further insect agents, I ITA has also focused on the potential for synergism using fungal pathogens and *Neochetina* spp. based on their different modes of action. Studies over nine years in Benin showed that the correlation coefficient between the mean number of adult weevil feeding scars per leaf and the percentage of water hyacinth plants with fungal infection (mainly due to *Myrothecium roridum* Tode (Hypocreales: Incertae sedis)) was remarkably high (0.74) (Ajuonu *et al.*, 2003). The inclusion of a fungal pathogen in the biocontrol package against water hyacinth seemed realistic because of historical evidence of the relatively cheap mass production costs demonstrated for *M. anisopliae* compared with mass-rearing costs for insects. Furthermore, evidence accrued from field surveys in Africa demonstrated the presence of naturally occurring fungal pathogens causing disease for water hyacinth (Daddy *et al.*, 2003). Fungal pathogens exotic to Africa were not considered due to apprehensions from governments in sub-Saharan Africa of introducing microscopic organisms that would be difficult to detect. Further justification for not studying the potential of exotic pathogens was provided by field surveys in Brazil, a region close to the centre of origin of water hyacinth, which did not provide evidence of a single fungal pathogen with sufficient virulence to justify changing this approach. In fact, no single insect pest or fungal pathogen was responsible for control of water hyacinth in Brazil, but, rather, several fungal pathogens played a role following feeding by a range of arthropods. Moreover, in its native environment, growth conditions of water hyacinth were limited due to low nutrient levels in water, shade provided by several plants, and competition from other plants, in particular *Eichhornia azurea* (Swartz) Kunth (Pontederiaceae), suggesting that the optimal control approach was to develop a method based on combined biocontrol agents to reduce the rate of weed growth.

The International Mycoherbicide Program for *E. crassipes* control in Africa (IMPECCA) led by CABI, IITA, and PPRI, set out to contribute to this approach by developing a mycoherbicide. The most promising candidate was *Alternaria eichhorniae* Nag Raj & Ponnappa (Pleosporales: Pleosporaceae) on the basis that it was virulent and host specific (Shabana *et al.*, 1995). Because it was indigenous to Africa and widely distributed, the likelihood of regulatory authorities approving field tests and eventually its registration was high. The requirements for overcoming long dew periods for infection were achieved by mixing fungal spores or mycelial fragments in an oil emulsion prior to application. Infection rates were further enhanced when silica-based abrasive agents were added to formulations along with nutritive ingredients such as sugars, in addition to surfactants (Beed, 2003). However, despite the development of efficient methods for mass production of *A. eichhorniae* using cheap and local materials and the selection of the most virulent isolate control, efficacy of the mycoherbicide under field conditions was variable. Studies showed that a factor contributing to fluctuations in control was varied levels of nutrients in water. Disease severity, for both inoculated leaves and those produced after inoculation, and disease incidence for all leaves, was significantly reduced under high nutrient concentrations (Avocanh *et al.*, 2003). The implication of this finding is that the mycoherbicide is best suited to control water hyacinth populations in environments (sites and seasons) where nutrient levels are reduced or where weed growth is limited for other reasons. Commonly an environment where water hyacinth grows poorly is during dry seasons on the perimeter of waterways where water is absent or turbid. Such sites are ideal targets for mycoherbicide applications for two reasons: firstly, weevils cannot pupate under these conditions, and secondly these "reservoir" sites, following rain, will supply diseased water hyacinth plants across interconnected waterways to newly infested sites.

Studies combining both insect and fungal biocontrol agents have shown control to be optimal if *N. eichhorniae* is introduced a minimum of one week prior to the application of *A. eichhorniae* (or another fungal pathogen: *Rhizoctonia solani* Kühn (Polyporales: Corticiaceae)). Interestingly, synergism was due to the combined impact on reducing the growth of water hyacinth rather than an association between epidemiological factors such as adult weevil feeding scars on leaves facilitating fungal infection (as has been reported for *Acremonium zonatum* (Sawada) W. Gams (Hypocreales: Hypocreaceae) in Brazil). It is perhaps disappointing that isolates of *R. solani* (belonging to AG 1-IA based on characterization using pectic zymograms, REP and ERIC PCR and ITS sequencing; H. Schneider unpublished) are not host specific as this pathogen is extremely virulent against water hyacinth. While this pathogen produces no spores that can be easily disseminated by wind it does produce large chlamydospores which can be transferred to other plants by contact or even following immersion in water. However, it can be argued that the use of *R. solani* still holds potential if it can be prevented from spreading from the site of its application, by selecting mutants that are dependent on the exogenous supply of a nutrient specific to water hyacinth. Other mycoherbicdes have been developed without the ability for cycling in the environment as has been achieved in Holland, Canada, and South Africa, for example, for *Chondrostereum purpureum* Pouzar (Polyporales: Meruliaceae).

23.5 Conclusions, future perspectives and role of IITA

IITA and its network of diverse partners catalyze classical biological control strategies and recognize the need to address the factors identified in Table 23.1. Similarly, inundative biological control options can be developed and implemented if the following are satisfied;

Technical efficacy: Weed populations reduced with a resulting positive and proven impact on crop yield

Practical efficacy: Methods for mass production and application are simple, robust, repeatable and reliable and subject to quality controls

Commercial viability: Dependent on cost and availability of materials selected for mass production, registration and interest from private sector

Sustainability: Host specific and durable across sites and seasons, provide ecological balance with survival of other natural enemies

Public benefit: Value added market for crops and their products as synthetic pesticides not used; health benefit to producers and consumers.

For meaningful impacts, despite the appropriate steps being taken to develop a classical or inundative biological control option, a wholehearted commitment and support from different governments and donors in Africa is imperative in order to legislate for the implementation of a biological control strategy. Ideally the private sector should be stimulated to produce and distribute biological control options ensuring sustainability and avoiding donor-funded project dependency. For a biological control technology to be introduced in an environmentally and economically sustainable manner there must be regulation and technical support provided by national agricultural research organizations (NARS) or preferably by a regional or continental consortium of them. Currently, the technical competence of some NARS requires further support from specialist international scientists coordinated via the likes of IITA in order to provide capacity building. Thereafter NARS can support regional public and private partnerships for the development, refinement, and deployment of a biocontrol strategy. Furthermore, links need to be strengthened between all stakeholders–researchers, commercial biocontrol companies, farmers, and extension workers – to ensure that biocontrol technologies are implemented on a large scale and in an appropriate manner. While NARS must optimally provide the research base support for a commercial partner to facilitate the scaling up of mass production and distribution, a major constraint in sub-Saharan Africa is the country-specific and high cost of registration for biocontrol agents. Based on previous experiences this is not a major concern for arthropod classical biocontrol agents but does pose concern if the agents are microbial pathogens. Farmers also need to be more fully exposed to the benefits of biocontrol through farmers' participatory learning approaches, field days, farmer group visits, radio and television broadcasts, extension manuals, and other means of technology implementation and dissemination. True acceptance will result from practical demonstrations that show that through adopting biocontrol technologies farmers

Table 23.1 *Overview of constraints, solutions and expertise and contributions of International Institute of Tropical Agriculture for a classical biological control program of weeds*

Constraint	Solution	IITA's expertise and contribution
Alert systems and quarantine systems	Cross-nation alert system	• Regional mandate that can liaise between and group together national programs and governments • Experience in diagnosing new invasions
	Uniform national quarantine measures	• Regional mandate that can liaise between and group together national programs and governments • Experience in setting up sub-Saharan quarantine systems
	Comprehensive and up-to-date baseline study of exotic weeds	• Regional mandate that can liaise between and group together national programs and governments • Experience in developing baseline studies
Agent selection	Carefully selected expedition site	• Experience in mounting expeditions to areas of origin
	Correct taxonomic classification of the biological control agent	• Experience in using genetic tools for species and subspecies identification
	Correct taxonomic classification of the invasive weed	• Experience in using genetic tools for species and subspecies identification
	Access to advanced genetic tools for taxonomic classification	• Advanced genetic tools (including sequencing) available in both East and West Africa
Agent testing	Meticulous host range testing and using predictive modeling and observations of the native herbivore populations to link laboratory results with field situations	• Experience in conducting all stages of a classical biological control programs, from laboratory host range testing to estimating the effect in the field
	Predicting nontarget effects based on cross-nation lists of native fauna	• Regional mandate that can liaise between and group together national programs and governments • Experience in conducting nontarget studies of a classical biological control program
	Quarantine facilities prior to field release	• Expertise in constructing and managing quarantine facilities according to international standards • Availability of quarantine facilities at selected locations

Table 23.1 (cont.)

Constraint	Solution	IITA's expertise and contribution
Conflict of interest	Biological control should be conducted under the auspices of independent and cross-boundary stakeholders	• Regional mandate that can liaise between and group together national programs and governments
Bureaucracies, inconsistencies or lack of legal frameworks	Reduction in duplication of national requirements	• Regional mandate that can liaise between and group together national programs and governments
	Standardization of national requirements according to the FAO Code of Conduct	• Regional mandate that can liaise between and group together national programs and governments
Demonstrating impact	Combining incomplete national data into a cross-nation impact study	• Experience in conducting post-release impact studies of classical biological control programs
	Incorporation of GIS in impact studies	• Experience in using geographic information systems (GIS) in Africa
Interdisciplinary teams	Incorporation of interdisciplinary teams that work together on an international basis	• Regional mandate that can group together specialists from national programs and governments, and where lacking, complement with in-house specialists
Scope of a classical biological control program	Availability of money and time, and connections with stakeholders worldwide	• Strong and long-term connections with donors and national and international partner networks worldwide

will increase substantially their agricultural production in a sustainable manner, with consequent increase in income and enhancement of rural livelihoods. Clearly any bio-control technology needs to be carefully integrated with other control methods and farming systems and ecologies in order to understand the key interactions and to therefore recommend the optimal combination of practices across the biophysical range where the target weed exists. IITA's role is unique in that it can coordinate all of the aforementioned activities by engaging partnerships. The strategy of pursuing the research for development approach enables IITA to aim for an exit strategy whereby the transfer of responsibility is handed over to the private sector supported by NARS in order to achieve full-scale implementation of biocontrol technologies in sub-Saharan Africa.

Acknowledgment

We thank Eric Koper (IITA, Ibadan) for valuable editorial comments.

References

Ajuonu, O., Chade, V., Veltman, B., Sedjro, K. and Neuenschwander, P. (2003). Impact of the exotic weevils *Neochetina eichhorniae* and *N. bruchi* (Coleoptera: Curculionidae) on water hyacinth, *Eichhornia crassipes* (Liliales: Pontederiaceae) in Bénin, West Africa. *African Entomology*, **11**, 153–161.

Alene, A. D., Manyong, V. M., Tollens, E. F. and Abele, S. (2007). Targeting agricultural research based on potential impacts on poverty reduction: strategic program priorities by agro-ecological zone in Nigeria. *Food Policy*, **32**, 394–412.

Avocanh, A., Senouwa, V., Diogo, R. and Beed, F. D. (2003). Use of *Alternaria eichhorniae* to control the invasive aquatic weed water hyacinth in Africa. In *Proceedings of the 8th International Congress of Plant Pathology*, held in February 2003 in Christchurch, New Zealand, p. 52.

Ayeni, S., Avocanh, A., and Beed, F. D. (2004). Perception of farmers in Bénin, West Africa to *Imperata cylindrica*. In *Proceedings of the 4th International Weed Science Congress*, held 20–24th June, Durban, South Africa, p. 45.

Bandyopadhyay, R., Kiewnick, S., Atehnkeng, J., *et al.* (2005). Biological control of aflatoxin contamination in maize in Africa. Invited presentation at the Deutscher Tropentag, The Global Food & Product Chain – Dynamics, Innovations, Conflicts, Strategies, October 11–13, Hohenheim (http://www.tropentag.de/2005/abstracts/full/398.pdf).

Beed, F. (2003). A mycoherbicide for water hyacinth. Biocontrol of weeds with pathogens. In *Proceedings of the 14th Biennial Australasian Plant Pathology Conference*, held 1–2 February, Lincoln University, New Zealand, pp. 8–10.

Bokonon-Ganta, A. H., de Groote, H. and Neuenschwander, P. (2002). Socio-economic impact of biological control of mango mealybug in Bénin. *Agriculture, Ecosystems and Environment*, **93**, 367–378.

Daddy, F., Ladu, B. M. B., Beed, F. D., Birnin-Yauri, Y. A. and Owotunse, S. (2003). Surveillance of potential pathogenic fungi associated with water hyacinth in lake Kainji, Nigeria. *Journal of Aquatic Sciences* **18**, 125–130.

De Groote, H., Ajuonu, O., Attignon, S., Djessou, R. and Neuenschwander, P. (2003). Economic impact of biological control of water hyacinth in Southern Bénin. *Ecological Economics*, **45**, 105–117.

Farri, T. A. and Boroffice, R. (1999). An overview on the status and control of water hyacinth in Nigeria. In *Proceedings of the First IOBC Water Hyacinth Working Group*, held in Harare, Zimbabwe in November 1998, pp. 18–24.

Gong, Y. Y., Egal, S., Hounsa, A., *et al.* (2003). Determinants of aflatoxin exposure in young children from Bénin and Togo, West Africa: the critical role of weaning. *International Journal of Epidemiology*, **32**, 556–562.

Greathead, D. J. (2003). Historical overview of biological control in Africa. In *Biological Control in IPM Systems in Africa*, ed. P. Neuenschwander, C. Borgemeister and J. Langewald. Wallingford, UK: CABI Publishing, pp. 1–26.

Herren, H. R. and Neuenschwander, P. (1991). Biological control of cassava pests in Africa. *Annual Review of Entomology*, **36**, 257–283.

Langewald, J., Stolz, I., Everts, J. and Peveling, R. (2003). Towards the registration of microbial insecticides in Africa: non-target arthropod testing on Green Muscle™, a grasshopper and locust control product based on the fungus *Metarhizium anisopliae* var. *acridum*. In *Biological Control in IPM Systems in Africa*, ed.

P. Neuenschwander, C. Borgemeister and J. Langewald. Wallingford, UK: CABI Publishing, pp. 207–225.

Lomer, C. J., Bateman, R. P., Johnson, D. L., Langewald, J. and Thomas, M. (2001). Biological control of locusts and grasshoppers. *Annual Review of Entomology*, **46**, 667–702.

Mbati, G. and Neuenschwander, P. (2005). Biological control of three floating water weeds, *Eichhornia crassipes*, *Pistia stratiotes*, and *Salvinia molesta* in the Republic of Congo. *BioControl*, **50**, 635–645.

Neuenschwander, P. (1996). Evaluating the efficacy of biological control of three exotic Homopteran pests in tropical Africa. *Entomophaga*, **41**, 405–424.

Neuenschwander, P. (2001). Biological control of cassava mealybug in Africa: a review. *Biological Control*, **21**, 214–229.

Neuenschwander, P. and Markham, R. (2001). Biological control in Africa and its possible effects on biodiversity. In *Evaluating Indirect Effects of Biological Control*, ed. E. Wajnberg, J. K. Scott and P.C. Quimby. Wallingford, UK: CAB International, pp. 127–146.

Nwankwo, D. I, and Akinsoji, A. (1988). Tolerance to salinity and survivorship of *Eichhorniae crassipes* (Mart) solms. growing in a creek around Lagos. In *International Workshop on Water Hyacinth: Menace and Resource*, ed. O. L. Oke, A. M. A. Imevbore, and T.A. Farri. Lagos, Nigeria: Nigerian Federal Ministry of Science and Technology, pp. 85–87.

Olokor, J. O., Daddy, F. and Adesina, G. O. (1998). Mapping the extent of spatial spread of water hyacinth (*Eichhornia crassipes*) across Kainji Lake. In *Annual Report of NIFFR: 1998*. pp. 15–16.

Schulthess, F., Neuenschwander, P. and Gounou, S. (1997). Multi-trophic interactions in cassava, *Manihot esculenta*, cropping systems in the subhumid tropics of West Africa. *Agriculture, Ecosystems and Environment*, **66**, 211–222.

Shabana, Y. M., Charudattan, R. and Elwakil, M. A. (1995). Identification, pathogenicity and safety of *Alternaria eichhorniae* from Egypt as a bioherbicide agent for water hyacinth. *Biological Control*, **5**, 123–135.

Van Thielen, R., Ajuonu, O., Schade, V., *et al.* (1994). Importation, release, and establishment of *Neochetina* spp. for the biological control of water hyacinth, *Eichhornia crassipes*, in Bénin, West Africa. *Entomophaga*, **39**, 179–188.

Visalakshy, P. N. G. and Jayanth, K. P. (1996). Effect of silt coverage of water hyacinth roots on pupation of *Neochetina eichhorniae* Warner and *Neochetina bruchi* Hustache (Coleoptera: Curculionidae). *Biocontrol Science and Technology*, **6**, 11–13.

Yaninek, J. S. and Schulthess, F. (1993). Developing an environmentally sound plant protection for cassava in Africa. *Agriculture, Ecosystems and Environment*, **46**, 305–324.

Zeddies, J., Schaab, R. P., Neuenschwander, P. and Herren, H. (2001). Economics of biological control of cassava mealybug in Africa. *Agricultural Economics*, **24**, 209–219.

24

The role of Secretariat of the Pacific Community in the biological control of weeds in the Pacific Islands region – past, present, and future activities

Warea Orapa

24.1 Introduction

The Pacific Community comprises small islands spread over 30 million square kilometers of ocean, with land making up a fraction of this area. The region (Fig. 24.1), which stretches from Palau and Papua New Guinea (PNG) in the west, the Commonwealth of the Northern Mariana Islands in the northwest, to New Caledonia in the south and French Polynesia and the Pitcairn Island in the east, offers a suitable environment for the proliferation of many tropical, alien invasive weeds. Increasing volumes of trade and movement of people to and from the Pacific render the threat of new plant invaders an ever-increasing problem. Many weeds have been introduced into the region in the last 100 years following European contact. As a result, the region is already inundated by several invasive weeds that cause significant environmental, economic, and social problems for the local people (Meyer, 2001; Orapa, 2001; Dovey *et al.*, 2004). Many of these weeds are difficult to control. On the other hand, the geographic nature of islands also offers unique opportunities for successful management of some weeds using relatively cheap and sustainable pest management methods such as biological control.

Weed biological control capacity in the Pacific region is not as well developed as in developed countries, which have a long history of biological control. There have been difficulties in managing weed problems using biological control due to low levels of national capacity. Necessary resources such as finance, trained personnel, and infrastructure for undertaking weed biological control are lacking in Pacific Island countries and territories (PICTs). In addition, some PICTs lack national policies on the management of either invasive pests or weeds. However, all is not depressing as biological control is well regarded regionally and in some island countries is considered an important method in pest management. The use of biological control to address agricultural pest and weed problems began over a century ago in the Pacific with varying degrees of success.

Biological Control of Tropical Weeds using Arthropods, ed. R. Muniappan, G. V. P. Reddy, and A. Raman. Published by Cambridge University Press. © Cambridge University Press, 2009.

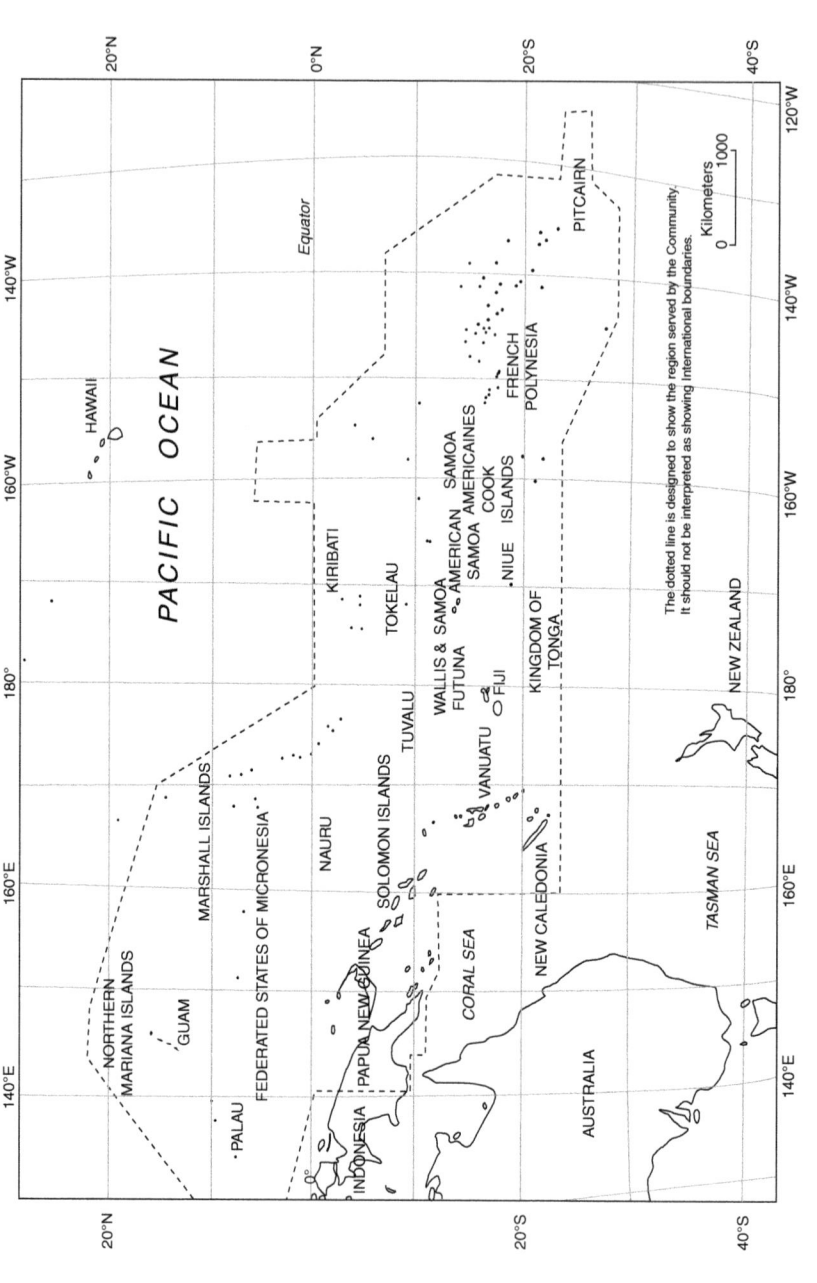

Fig. 24.1 Pacific Islands region covered by SPC (SPC's headquarters are in New Caledonia). Weed biological control efforts are coordinated from SPC's subregional offices in Fiji and the Federated States of Micronesia.

Collectively, the Pacific region was among the early users of biological control for weed management. The region also has a regional technical agency that can undertake such work on behalf of PICTs, either individually or in more than one country at a time. The Secretariat of the Pacific Community (SPC) is a regional, intergovernmental, technical agency serving all 22 PICTs[1] (Fig. 24.1). In the last 15 years, SPC has been involved in the promotion and implementation of biological control programs targeting over a dozen weeds. In this chapter, we will look at the role SPC plays in promoting and implementing biological control programs in the Pacific. We will also discuss the status of some past and present weed biological control programs. At the end of the chapter, we will look at some current invasive weed problems that require the use of biological control methods for long-term, ecosystem-wide solutions.

24.2 The Secretariat of the Pacific Community and its roles

Established by six colonial powers (and founding members)[2] in 1947, the SPC is a regional organization mandated to serve the developing PICTs. It provides technical and scientific support, research, and training for the sustainable conservation and utilization of all land, marine, and social resources. SPC's Plant Protection Service began in 1952 and in 1997, Dr. Robert Dun, then Director-General, wrote in his foreword to the SPC Guidelines for Biological Control Projects in the Pacific, that ". . . from the outset, plant quarantine and biological control of pests were of paramount importance". The strategic roles of the Secretariat in relation to invasive species including weed management have been discussed by Orapa (2003). Biological control is promoted by SPC's Plant Protection Service as an appropriate method for pest management in the islands.

The main roles of SPC are: (1) to provide advice to PICTs in implementing biological control programs and ensure that the PICTs follow acceptable protocols; and (2) to work with PICTs to develop and implement biological control programs, whereever a national capacity is lacking.

24.3 Coordination and advisory roles

SPC's regional responsibility is endorsed by PICTs under the aegis of the Pacific Plant Protection Organisation (PPPO). SPC acts as the secretariat of the PPPO, which is one of nine regional plant protection organizations under the International Plant Protection Convention (IPPC) of the Food and Agricultural Organization of the United Nations (FAO). SPC has coordinated most of the Pacific region's efforts in biological control through its extensive Plant Protection Service and now under two thematic groups that deal with Biosecurity and Plant Health. The Secretariat and the PPPO work together in contributing to the development and implementation of International Standards for Phytosanitary Measures (ISPM) including the guidelines on biological control. The important ISPM relevant to biological control is ISPM No. 3 (1996, revised 2005), which is the *Code of conduct for the import and release of exotic biological control agents*. SPC and the FAO's South Asia Pacific program jointly provide technical advice to member PICTs on the implementation of the relevant ISPM standards including the offer of guidelines for the movement of biological control agents within the region. Other relevant FAO guidelines that relate to biological control include ISPM Nos. 2, 11, and 21 (FAO, 2006).

The first Pacific guidelines for biological control were those developed for SPC by the late D. F. Waterhouse following a regional workshop in October 1985 (Waterhouse, 1991). Dossiers on biological control of important pests and weeds were also prepared (Waterhouse and Norris, 1987; Waterhouse, 1997). The Pacific guidelines endorsed by SPC and revised in 1997 also conform to the FAO guidelines referred to earlier in this article.

SPC has increased its advisory and coordination roles in invasive weed management for PICTs as a result of increased funding from the European Union since 2002. Biological control is one of many activities that the Secretariat undertakes and this will continue for as long as it is needed by regional governments. Much of the SPC's effort will continue to revolve around providing advice to regional governments on suitable biological control agents available for weeds of concern to the region, import risk analysis, and basic training in weed biological control.

24.4 National and regional weed biological control projects

24.4.1 Past projects

While weed biological control programs can be described as having been ad hoc or intermittently carried out in most of the small island PICTs, the Pacific region has experienced exemplary successes in weed biological control, compared with anywhere in the tropics, particularly in the last three decades. PICTs have generally benefited from having strong historical linkages to technical research and development agencies in the United Kingdom, United States of America (Hawaii and Guam), Australia, France, and New Zealand. The first reported example of weed biological control in the Pacific was against *Lantana camara* L. (Verbenaceae) in Fiji (1911), and later in Samoa, New Caledonia, Tonga, and the Micronesian Islands in the 1920s, 1930s, and 1940s (Kamath, 1979; Day *et al.*, 2003), well before the establishment of SPC. Several attempts at introducing different agents for the control of *L. camara* were also made in these islands and other island countries and territories in the 1960s, 1970s, and even as recently as 1994 in Niue by SPC (Waterhouse and Norris, 1987; Julien and Griffiths, 1998; Day *et al.*, 2003).

As most biological control programs have involved an elementary "technology transfer" of known classical biological control agents used elsewhere, successes have been many. Useful information on past weed biological control in Papua New Guinea (PNG), Fiji, and the Pacific in general, is available (Kamath, 1979; Young, 1982; Waterhouse and Norris, 1987; Julien and Griffiths, 1998; Julien *et al.*, 2007). Notable past examples of successful biological control resulting from "technology transfer" initiatives undertaken by PICTs include management of *Clidemia hirta* (L.) D. Don. (Melastomataceae) (S. N. Lal, SPC, personal communication), *Sida acuta* Burm.f. (Malvaceae) and *Sida rhombifolia* L. (Malvaceae) (Kuniata and Korowi, 2004; *L. camara* (Day *et al.*, 2003), *Mimosa diplotricha* C. Wright ex Sauvalle (Fabaceae) (Kunitata and Korowi, 2004), *Salvinia molesta* D.S. Mitchell (Salviniaceae) (Room and Thomas, 1985), *Pistia stratiotes* L. (Araceae) (Laup, 1986), *Tribulus cistoides* L. (Zygophyllaceae) (Young, 1982), and *Eichhornia crassipes* (Mart.) Solms (Pontederiaceae) (Julien and Orapa, 1999, 2001). Much of this work has been done in Fiji and PNG, and to some extent in the Solomon Islands, Cook Islands, Samoa, Guam, and Vanuatu.

Three large national and regional projects were implemented in the 1990s. A general biological control project funded by the German Agency for Technical Cooperation (GTZ) in the 1990s included distribution of some of the agents available for *L. camara* and *M. diplotricha* in the Solomon Islands, Fiji, Niue, Samoa, and Cook Islands. Biological control efforts against *Chromolaena odorata* (L.) King and Robinson (Asteraceae) in PNG supported by the Australian Centre for International Agricultural Research (ACIAR) (Orapa *et al.*, 2004; Bofeng *et al.*, 2004), and in Micronesia (Palau, Guam, the Federated States of Micronesia (FSM) and the Commonwealth of the Northern Mariana Islands) (Esguerra, 2001; Muniappan *et al.*, 2004) resulted in varying levels of control following releases of *Pareuchaetes pseudoinsulata* Rego Barros (Lepidoptera: Arctiidae) and *Cecidochares connexa* (Macquart) (Diptera: Tephritidae). A six-year (1993–1998) biological control project undertaken by the PNG Department of Agriculture and Livestock and the Commonwealth Scientific and Industrial Research Organisation (CSIRO) of Australia, using three host-specific biological control agents and AU$1.5 million, resulted in the successful reduction of huge water hyacinth infestations not only in the vast Sepik River wetlands but in nearly 100 separate infestations in that country (Julien and Orapa, 1999, 2001).

24.4.2 Current SPC projects

Current biological control activities supported by SPC are targeted against *C. odorata* in Palau and FSM, and against *E. crassipes*, three species of *Sida* and *P. stratiotes* in Vanuatu. SPC established a minimum security quarantine facility in Vanuatu to hold imported agents for a generation before release, following the FAO guidelines. The intention was to increase activity in biological control of weeds and arthropod pests in Vanuatu. *Calligrapha pantherina* Stål, (Coleoptera: Chrysomelidae) for species of *Sida*, *Neochetina eichhorniae* Warner (Coleoptera: Curculionidae) for *E. crassipes*, and *Neohydronomus affinis* Hustache (Coleoptera: Curculionidae) for *P. stratiotes* were imported into Vanuatu, reared, and released from this facility between August 2004 and December 2006.

A research project funded by ACIAR and aimed at managing *Mikania micrantha* Kunth (Asteraceae) in Fiji and PNG is currently being implemented by SPC with national collaborating agencies in the two countries. It is hoped that the rust fungus *Puccinia spegazzinii*, and the two butterflies *Actinote anteas* (Doubleday) (Lepidoptera:

Nymphalidae) and *A. thalia pyrrha* L. (Lepidoptera: Nymphalidae) will be released to control *M. micrantha* following limited host-specificity tests in Fiji and PNG. Following the initial work in the two countries, SPC will be able to transfer the agents to other PICTs with *M. micrantha* problems (Table 24.1).

24.5 Challenges and opportunities for weed biological control

24.5.1 National capacity

Weed biological control programs in PICTs are seriously hindered by capacity and financial constraints. The capacity of many island countries to implement weed biological control projects is either limited or nonexistent. Many PICTs have only a handful of suitable people who can skillfully wear many "hats," so their time and effort are often spread thinly over many areas such as agronomy and extension. Specialists in pest management or weed management are few and far between. This is true of American Samoa, Kiribati, Marshall Islands, Nauru, Niue, Pitcairn Islands, Solomon Islands, Tokelau, Wallis and Futuna, and Vanuatu. Capacity is better in PNG and Fiji, and high in Guam. Only a few PICTs undertake biological control programs and have some capacity to undertake biological control against weed problems.

Very little original biological control research or development of weed control programs have been attempted in PICTs because of the lack of capacity, large initial cost, lack of national agencies backed by governments, inexperience in international collaboration, and length of time that may be involved. The only attempt at initiating biological control research for the region was a preliminary exploration for natural enemies of *Clerodendron chinense* (Osbeck) mabb. (Verbenaceae) in Vietnam and southern China, but this did not proceed to the next steps (Julien, 1993). Weed biological control programs in the Pacific have therefore been "technology-transfer" projects because it is easier to import proven agents used in other countries or regions. Recently, only Guam (which in fact benefits from being a United States territory) has been able to undertake "full-scale" biological control research commencing with exploratory research on a weed's native range. The adoption of biological control agents from well-researched or proven programs from other regions has worked well for other PICTs (Julien *et al.*, 2007) and this aspect is likely to continue, limiting any new region or country-specific weed biocontrol ventures.

24.5.2 Regional capacity

SPC faces two main limitations when implementing biological control activities for PICTs. SPC was established as a service delivery agency and not a research agency. Therefore, the infrastructure and long-term technical expertise required to provide continuity of biological control programs for PICTs are unreliably available regionally. Secondly, all plant protection programs, including weed biological control, rely heavily on aid projects funded by a very narrow group of international donors such as GTZ,

Table 24.1 *Weed biological control targets in the Pacific islands in which SPC was involved*

Target weed	Biological control agent species introduced or released	Year	Countries where BCAs have been introduced	Status	Agencies involved[a]
Chromolaena odorata (Asteraceae)	Pareuchaetes pseudoinsulata (Lepidoptera: Arctiidae)		Palau, Federated States of Micronesia (FSM)	Good control in FSM. Results promising on several islands	SPC, UOG
	Cecidochares connexa (Diptera: Tephritidae)		Palau, FSM	Good control in most areas	UOG, SPC
Eichhornia crassipes (Pontederiaceae)	Neochetina eichhorniae (Coleoptera: Curculionidae)	Mid 1990s	Solomon Is., Vanuatu	Good to excellent control of the weed	PNG DAL, CSIRO
		2004			SPC, VQIS
Lantana camara (Verbenaceae)	Uroplata girardi (Col.: Chrysomelidae)	1993	Niue	Established and spread. Providing good control	SPC, GTZ, DAFF
	Octotoma scabripennis (Col.: Chrysomelidae)	1993, 1994	Niue, Fiji, Solomon Is., Samoa	Establishment not confirmed	SPC, GTZ
	Teleonemia scrupulosa (Hemiptera: Tingidae)	1993	Solomon Is.	Established	SPC, DAL, GTZ
	Calycomyza lantanae (Diptera: Agromyzidae)	1994	Niue	Established and effective	SPC, GTZ
		1996	Fiji	Established and giving good control	MOA, SPC
	Charidotis pygmaea (Col.: Chrysomelidae)	1995	Fiji	Not established	MOA, SPC
Mikania micrantha (Asteraceae)	Actinote anteas (Lepidoptera Nymphalidae: Acraenae)	2006	Fiji (introduced ex. Indonesia, July 2006)	Under evaluation for host specificity in quarantine	SPC, MOA
	Actinote thalia pyrrha (Lepidoptera: Nymphalidae Acraenae)	2006	Fiji	Under evaluation for host specificity in quarantine	SPC, MOA

Table 24.1 (cont.)

Target weed	Biological control agent species introduced or released	Year	Countries where BCAs have been introduced	Status	Agencies involved[a]
	Puccinia spegazzinii (Uridinales: Basidiomycetes)	2007	Fiji and PNG in 2007–2008	Planned introduction in November 2008	SPC, MOA, CABI
Mimosa diplotricha (Fabaceae)	*Heteropsylla spinulosa* (Hemiptera: Psyllidae)	1992–1996	Fiji, Niue, Solomon Is., Samoa, Vanuatu, Cook Is.	Excellent to good control	SPC, GTZ, MOA
Pistia stratiotes (Araceae)	*Neohydronomus affinis* (Col.: Curculionidae)	2006	Vanuatu	Under evaluation	SPC, VQIS
Sida spp. (Malvaceae)	*Calligrapha pantherina* (Col.: Chrysomelidae)	2002, 2004	Fiji, Vanuatu	Good control, particularly on *S. acuta* and *S. rhombifolia*	SPC, MOA, RSL, (PNG), VQIS

[a] CABI, formerly known as Commonwealth Agricultural Bureaux International; CSIRO, Commonwealth Scientific and Industrial Research Organisation; DAFF, Australian Department of Agriculture, Fisheries and Forestry; DAL, Department of Agriculture and Livestock (PNG); GTZ, Deutsche Gesellschaft für Technische Zusammenarbeit; MOA, Ministry of Agriculture (Fiji); PNG, Papua New Guinea; RSL, Ramu Sugar Ltd (PNG); SPC, Secretariat of the Pacific Community; UOG, University of Guam; VQIS, Vanuatu Quarantine Inspection Services.

ACIAR, and the United States Department of Agriculture, which has been focusing on some Micronesian islands. Recently (2002–2006), support from the European Union for SPC's regional plant protection program in 17 PICTs has rejuvenated some biological control activities, but overall, weed biological control has been attempted only intermittently in the region.

24.6 Managing negative perceptions

Weed biological control efforts in the region also face other constraints in the form of bad publicity, lack of awareness among decision makers, and the low priority given to invasive weed management compared with other agricultural pest problems. Negative perceptions of biological control promoted by self-interest groups with limited knowledge of the science and principles of the subject can have damaging consequences for current and future weed biological control efforts. The most familiar fears about biological control in the Pacific stem from cases where unexpected impacts on nontargets, particularly insects, vertebrates, and mollusks have occurred. The release of *Bessa remota* (Ald.) (Diptera: Tachinidae) in Fiji in 1925 is thought to have led to the extinction of *Levuana iridescens* Betheune-Baker (Lepidoptera: Zygaenidae). This could be the singular instance in the world where a biological control agent may have exterminated its host (Kuris, 2003). The predatory snail *Euglandina rosea* (Ferussac) (Stylommatophora: Spiraxidae) and two flatworm species of *Platydemus* have been implicated in reducing populations of nontarget native partulid land snails in some Pacific islands (Cowie and Cook, 2001). This has, to some extent, raised negative perceptions of biological control in general, although no examples of adverse impacts of weed biological control in the Pacific exist, perhaps because all of such biological control efforts to date have involved only basic "technology transfer." Lack of awareness of the benefits of well-researched weed biological control among policy makers and some environmental groups will continue to affect its implementation in some PICTs. For example, the Samoan government authorities were not prepared to undertake biological control programs against two species of *Sida* that are weeds in Samoa because of fears of host shifts in 2006 (Billy Fuifatu, Samoa Ministry of Agriculture, Forestry and Fisheries, personal communication), although the leaf beetle *C. pantherina* is known to be highly host specific to the two species (Forno *et al.*, 1992).

Despite these obstacles, SPC, along with a few PICTs (Guam, Fiji, and PNG), continues to stress that weed management centered on the use of biological control is appropriate for Pacific islands and where necessary implements projects. SPC will continue to work with PICTs to explore opportunities where weed management based on biological control is feasible. This will, however, ultimately depend on the availability of resources, particularly of skilled workers keen to explore the use of biological control, money, infrastructure, and the general desire of institutions within countries to utilize biological control. New regional collaborations on invasive species management in the Pacific between environmental, health, and economic sectors are beginning to break down some of the barriers raised by negative perceptions of biological control. The initiative by SPC to use biocontrol against mikania, for instance, has received support from the regional environmental agency, the Secretariat of the Pacific Regional Environment Programme based in Samoa. In fact, biological control is now seen as the key to solving many of the region's existing invasive plant problems (Dovey *et al.*, 2004).

SPC is increasingly mindful that improving levels of public awareness by involving policy makers and anti-biological-control groups in PICTs very early in any planned project is the key to reducing negative perceptions about biological control. Interestingly most antagonistic attitudes towards biological control in the Pacific islands nearly always stem from poor understanding of the benefits of biological control backed by good research. SPC now incorporates public awareness programs on plant pests, diseases, and weeds as part of biological control projects to inform both the public and policy makers.

24.7 Opportunities for weed biological control

Prior to the development of a weed biological control program, gathering of accurate information on the weed and its performace in PICTs is important. SPC has made preliminary attempts to develop prioritized lists of agricultural weeds for the region, starting with the work by Waterhouse (1997) and regional technical meetings on plant protection in 2002 and 2004 (SPC, unpublished data). During the two meetings held in Fiji, the region's 45–50 most important weeds were identified and ranked according to the importance placed on them by PICTs. For the majority of the weeds listed (69%), no known suitable biological control agents are available, 18% have had at least one natural enemy released in or outside the Pacific region with no follow-up work or evaluation, whereas 13% have suitable biological control agents already available in some PICTs or outside the region that could be used in affected PICTs (Table 24.2).

Weeds that are still problematic in one or more PICTs and that are also known to have one or more suitable biological control agents either released in one of the PICTs or in other tropical regions of the world are also listed in Table 24.2. Examples include *L. camara*, *S. rhombifolia*, *Parthenium hysterophorus*, *Xanthium strumarium*, *E. crassipes*, *M. diplotricha*, *M. micrantha*, and *C. hirta*.

Little information on the potential natural enemies of some of the most serious invasive weeds that affect economic and environmental sectors in the region exists. These weeds include *Spathodea campanulata* P. Beauv. (Bignoniaceae), *Cyperus rotundus* L. (Cyperaceae), *M. micranatha* and *Merremia peltata* (L.) Merr. Convolvulaceae), *Sphagneticola trilobata* (L. C. Rich.) Pruski (Asteraceae), and *Sorghum halepense* (L.) Pers. (Poaceae). These and many others impact on agriculture and the fragile environment and biological diversity associated with the Pacific islands (e.g. Auld and Nagatalevu-Seniloli, 2001; Meyer, 2001; Orapa, 2001). These weed species present unique opportunities for SPC and PICTs to undertake management based on the use of biological control.

Table 24.2 *Some weeds of interest to PICTs and their biological control status. Weeds with no known biological control agents are stated as "biological control agents required"*

Weed species	PICT where biological control has been attempted	Status of biological control in PICT where released	Status of biological control in Pacific countries
Acacia farnesiana Fabaceae/Mimosaceae			Biological control required (Fiji, New Caledonia, PNG)
Amaranthus spinosus Amaranthaceae			Biological control required (New Caledonia, PNG, Vanuatu)
Bidens pilosa Asteraceae			Biological control required (most PICTs)
Chromolaena odorata Asteraceae	Guam	Good control	Releases of two agents underway (PNG, CNMI, Palau, FSM. Introduction of a third agent into PNG failed
Clerodendrum chinense Verbenaceae			Biological control required (Fiji, French Polynesia, Samoa, Niue, Vanuatu). Potential agents likely to be found in Indochina
Clidemia hirta Melastomataceae	Fiji	Good control	The thrips *Liothrips urichi* needs to be introduced into Wallis and Futuna and Samoa
Commelina benghalensis Commelinaceae			Biological control required, especially in Tonga
Cyperus rotundus Cyperaceae	Fiji	No control	Biological control required. Early attempt using *Bactra* sp. not successful. Remains a major weed in 10 PICTs
Dissotis rotundifolia Melastomataceae			Biological control required (Fiji, Fr. Polynesia, Samoa, PNG, Vanuatu)
Eichhornia crassipes Pontederiaceae	PNG, Fiji, Solomon Is., Vanuatu	Good control	Four agents released in PNG; *Neochetina eichhorniae* providing control in Solomon Is. and Fiji; released in Vanuatu in 2004
Hyptis spp. Lamiaceae			Biological control required (many PICTs)
Kyllingia polyphylla Cyperaceae			Biological control required (Fiji, Fr. Polynesia, Samoa, Vanuatu)

Table 24.2 (cont.)

Weed species	PICT where biological control has been attempted	Status of biological control in PICT where released	Status of biological control in Pacific countries
Lantana camara Verbenaceae	Most Pacific countries	Good to moderate control	Some agents are widespread. Redistribution of one agent in Cook Is. in 2005. Additional agents required
Melinis minutiflora Poaceae			Biological control required (highland areas of Fr. Polynesia and PNG)
Merremia peltata Convolvulaceae			Biological control required (major plant invader in 10 PICTs)
Mikania micrantha Asteraceae			Biological control agent required. Three potential agents are planned to be used in Fiji and PNG under a 3-year project beginning 2006. Serious weed in 12 PICTS
Mimosa diplotricha Fabaceae/Mimosaceae	Fiji, Niue, PNG, Samoa, Solomon Is.	Partial to good control	*H. spinulosa* needed in PICTs where mimosa is still a problem.
Mimosa pudica Fabaceae/ Mimosaceae			Biological control required (Seven PICTs reported mimosa as serious but present in most PICTs)
Mormodica charantia Cucurbitaceae			Biological control required (Fiji, PNG, Solomon Is., Tonga)
Piper aduncum Piperaceae			Biological control required (Fiji, PNG, Solomon Is., Vanuatu)
Psidium cattleianum Myrtaceae			Biological control required (Cook Is., Fr. Polynesia)
Salvinia molesta Salviniaceae	PNG, Fiji	Excellent to good control	Excellent control in PNG and Australia. Present in Fr. Polynesia, Cook Is. and New Caledonia
Sida acuta Malvaceae	PNG, Fiji, Vanuatu	Excellent control	*Calligrapha pantherina* was introduced to Vanuatu in August 2004
Sida rhombifolia Malvaceae	PNG, Fiji, Vanuatu	Good control	Transfer of agent (*C. pantherina*) required in Samoa, Solomon Is., Tonga, Fr. Polynesia

Table 24.2 (cont.)

Weed species	PICT where biological control has been attempted	Status of biological control in PICT where released	Status of biological control in Pacific countries
Solanum torvum Solanaceae			Biological control required
Sorghum halepense Poaceae			Biological control required
Spathodea campanulata Bignoniaceae			Biological control required (Fiji, PNG)
Sphagneticola triloba Asteraceae			Biological control required. Serious weed in at least five PICTs
Stachytarpheta urticifolia S. jamaicensis Verbenaceae S. cayennensis			Biological control required (several PICTs)
Xanthium strumarium Asteraceae	PNG		Biological control required (Fiji, New Caledonia, PNG). Moth *Epiblema strenuana* failed during lab rearing in PNG. Rust fungus released in Australia needs to be considered.

Adapted from Dovey *et al.* (2004) and Julien *et al.* (2007)

24.8 Conclusion

SPC and its member PICTs face ongoing challenges in addressing serious weed problems using biological control methods, given their limited capacity and resources. Previous successful biological control activities have been the result of implementing technology-transfer type projects using natural enemies of weeds that have been proven elsewhere. Funding for new weed research is difficult to obtain and if available, comes only from ad-hoc donor aid. Most of the major weeds known today were never considered troublesome weeds in the past because many of them (e.g. African tulip tree) entered the Pacific islands as seemingly harmless ornamental plants. Such weeds present opportunities for new biological control research in the Pacific islands, but firstly the question of national and regional capacity needs to be addressed. On the other hand, biological control agents have already been released in PICTs or outside the region for some weeds, so these could be targeted at relatively little cost in other PICTs. SPC will continue to facilitate the movement of these biological control agents in the region using established regional and international guidelines for biological control.

Acknowledgments

I thank my colleagues Richard Davis, Angela Templeton, and Sidney Suma for reviewing the manuscript, and my employer, the Secretariat of the Pacific Community, for approving this chapter for publication.

Notes

1. PICTs: American Samoa, Commonwealth of the Northern Mariana Islands, Cook Islands, Federated States of Micronesia, Fiji, French Polynesia, Guam, Kiribati, Nauru, New Caledonia, Niue, Palau, Papua New Guinea, Pitcairn, Republic of the Marshall Islands, Samoa, Solomon Islands, Tonga, Tokelau, Tuvalu, Wallis and Futuna, and Vanuatu.
2. Other SPC members: Australia, France, New Zealand, United States of America, and until 2004 the United Kingdom.

References

Auld, B. A. and Nagatalevu-Seniloli, M. (2001). African tulip tree in the Fijian Islands. http://www.fao.org/docrep/006.

Bofeng, I., Donnelly, G., Orapa, W. and Day, M. (2004). Biological control of *Chromolaena odorata* in Papua New Guinea. In *Chromolaena in the Asia-Pacific Region: Proceedings of the 6th International Workshop on the Biological Control and Management of Chromolaena*, held in Cairns, Australia, May 2003, ed. M. D. Day and R. E. McFadyen. ACIAR Technical Reports 55. Canberra, Australia: ACIAR, 14–16.

Cowie, R. H. and Cook, R. P. (2001). Extinction or survival: partulid tree snails of American Samoa. *Biodiversity Conservation*, **10**, 143–159.

Day, M. D., Wiley, C. J., Playford, J. and Zaluchi, M. P. (2003). *Lantana: Current Management Status and Future Prospects*. Monograph 102. Canberra, Australia: ACIAR, 128 pp.

Dovey, L., Orapa, W. and Randall, S. (2004). The need to build biological control capacity in the Pacific. In *Proceedings of the XI International Symposium on Biological Control of Weeds*, held in Canberra in May 2003, eds. J. M. Cullen, D. T. Briese, D. J. Kriticos, *et al.* 2003, Canberra, Australia: CSIRO Entomology, pp. 36–41.

Esguerra, N. (2001). Introduction and establishment of the tephritid gall fly *Cecidochares connexa* on Siam weed *Chromolaena odorata* in the Republic of Palau. In *Proceedings of the 5th International Workshop on Biological Control and Management of Chromolaena odorata*, held in Durban, South Africa, 23–25 October 2000, ed. C. Zacchariades, R. Muniappan, and L. W. Strathie. Pietermaritzburg, South Africa: Teeanem, pp.148–151.

FAO. (2006). *International Standards for Phytosanitary Measures (1 to 27)*. 2006 edn. Rome: FAO, http://www.fao.org/ippc/ispms.

Forno, I. W., Kassulke R. C. and Harley, K. L. S. (1992). Host specificity and aspects of the biology of *Calligrapha pantherina* (Coleoptera: Chrysomelidae), a biological control agent of *Sida acuta* (Malvaceae) and *Sida rhombifolia* in Australia. *Biocontrol*, **37**, 409–417.

Julien, M. H. (1993). Surveys for the native range of *Clerodendrum chinense* and its natural enemies. In *Proceedings of the 10th Australian Weeds Conference and 14th Asian Pacific Weed Science Society Conference*, held in Brisbane in September 1993, ed. J. T. Swarbrick, C. W. L. Henderson, R. J. Jettner, L. Streit and S. R. Walker. Brisbane, Australia: Weed Society of Queensland, Vol. II, pp. 39–43.

Julien, M. H. and Griffiths, M. W. (1998). *Biological Control of Weeds: A World Catalogue of Agents and their Target Weeds*. Forth edn. Wallingford, UK: CAB International, 223 pp.

Julien, M. H. and Orapa, W. (1999). Successful biological control of water hyacinth in Papua New Guinea (abstract). In *Proceedings of the X International Symposium on Biological Control of Weeds*, held in Bozeman, Montana, in July 1999, ed. N. R. Spencer. Sidney, MT: USDA-ARS, pp. 138–139.

Julien, M. H. and Orapa, W. (2001). Insects used for the control of the aquatic weed water hyacinth in Papua New Guinea. *Papua New Guinea Journal of Agriculture, Forestry and Fisheries*, **44**, 49–60.

Julien, M. H. Scott, J. K., Orapa, W. and Paynter, Q. (2007). History, opportunities and challenges for biological control in Australia, New Zealand and the Pacific Islands. *Crop Protection*, **26**, 255–265.

Kamath, M. K. (1979). A review of biological control of insect pests and noxious weeds in Fiji 1969–1978. *Agricultural Journal of Fiji*, **41**, 55–72.

Kuniata, L. S. and Korowi, K. T. (2004). Bugs offer sustainable control of *Mimosa invisa* and *Sida* spp. in the Markham Valley, Papua New Guinea. In *Proceedings of the XI International Symposium on Biological Control of Weeds*, held in Canberra, April–May 2003, ed. J. M. Cullen, D. T. Briese, D. J. Kriticos, *et al.* Canberra, Australia: CSIRO Entomology, pp. 567–573.

Kuris, A. M. (2003). Did biological control cause extinction of the coconut moth, *Levuana iridenscens*, in Fiji? *Biological Invasions*, **5**, 133–141.

Laup, S. (1986). Biological control of water lettuce: early observations. *Harvest*, **12**, 41–43.

Meyer, J.-Y. (2001). Preliminary review of the invasive plants in the Pacific Islands (SPREP Member Countries). In *Invasive Species in the Pacific: A Technical Review and Regional Strategy*, ed. G. Sherley. Samoa: Apia, SPREP.

Muniapan, R., Englberger, K., Bamba, J. and Reddy, G. V. P. (2004). Biological control of chromolaena in Micronesia. In *Chromolaena in the Asia-Pacific Region: Proceedings of the 6th International Workshop on the Biological Control and Management of Chromolaena*, held in Cairns, Australia in May 2003, ed. M. D. Day and R. E. McFadyen. ACIAR Technical Reports 55, Canberra, Australia: ACIAR, pp. 11–12.

Orapa, W. (2001). Impediments to food security: the case of exotic weeds in Papua New Guinea. In *Food Security for Papua New Guinea: Proceedings of the Papua New Guinea National Food Security Conference*, held in Lae, PNG, in June 2000, ed. R. M. Bourke, M. G. Allen, and J. G. Salisbury. ACIAR Proceedings 99. Canberra, Australia: ACIAR, pp. 308–315.

Orapa, W. (2003). The role of the Secretariat of the Pacific Community in addressing invasive alien species. In *Prevention and Management of Invasive Alien Species: Proceedings of the Workshop on Forging Cooperation throughout the Austral-Pacific*, held in Honolulu, 15–17 Oct, 2002, ed. C. Shine, J. K. Reaser and A. T. Gutierrez. Honolulu, HI: Bishop Museum Press, pp. 85–88 (http://www.gisp.org/publications/workshops).

Orapa, W., Englberger, K. and Lal, S. N. (2004). Biological control of *Chromolaena odorata* in Papua New Guinea. In *Chromolaena in the Asia-Pacific Region*: *Proceedings of the 6th International Workshop on Biological Control and Management of Chromolaena*, ed. M. D. Day and R. E. McFadyen. ACIAR Technical Reports 55. Conberra, Australia: ACIAR, 13 pp.

Room, P. M. and Thomas, P. A. (1985). Nitrogen and establishment of a beetle for biological control of the floating weed *Salvinia* in Papua New Guinea. *Journal of Applied Ecology*, **22**, 139–156.

Waterhouse, D. F. and K. R. Norris. (1987). *Biological Control: Pacific Prospects*. Melbourne, Australia: Inkata Press, 454 pp.

Waterhouse, D. F. (1991). *Guidelines for Biological Control Projects in the Pacific*. Information Document 57. Noumea, New Caledonia: South Pacific Commission, and Conberra, Australia: ACIAR, 30 pp.

Waterhouse, D. F. (1997). *The Major Invertebrate Pests and Weeds of Agriculture and Plantation Forestry in the Southern and Western Pacific*. Canberra, Australia: Australian Centre for International Agricultural Research, 93 pp.

Young, G. R. (1982). Recent work on biological control in Papua New Guinea and some suggestions for the future. *Tropical Pest Management*, **28**, 107–114.

Index

For EU product safety concerns, contact us at Calle de José Abascal, 56–1°, 28003 Madrid, Spain or eugpsr@cambridge.org.

www.ingramcontent.com/pod-product-compliance
Ingram Content Group UK Ltd.
Pitfield, Milton Keynes, MK11 3LW, UK
UKHW051007240426
470322UK00018B/552